AF247497

Electronics

Second Edition

Electronics

Second Edition

P. Arun

Alpha Science International Ltd.
Oxford, U.K.

Electronics
Second Edition
484 pgs. | 372 figs. | 8 tbls.

P. Arun
Department of Electronics
SGTB Khalsa College
University of Delhi
Delhi, India

Copyright © 2006, 2011
    Reprint 2008
Second Edition 2011

ALPHA SCIENCE INTERNATIONAL LTD.
7200 The Quorum, Oxford Business Park North
Garsington Road, Oxford OX4 2JZ, U.K.

www.alphasci.com

Printed from the camera-ready copy provided by the Author.

ISBN 978-1-84265-703-4

Printed in India

Preface to the Second Edition

It is a pleasing feeling when ones' contribution finds acceptance. Since the book was first published, I have received words of appreciation and encouragement from collegues and peers. I gratefully acknowledge their kindness. Some of my collegues from different colleges have used this book as a text and have put it through what one might call "field test". In due course they were able to point out some errors in the Ist edition, topics to be added and areas that required improvement. In this edition earnest efforts have been made to address these issues. Changes made in terms of topics added are

- Inclusion of coding and markings made on various type of capacitors in Chapter 1

- Chapter 2 has a section on Anderson Bridge. While this experiment is essentially done in Physics, the topic is included here to show how the various theorems of Network Analysis can be applied to improve the sensitivity of the experiment.

- Chapter 5 now introduces Light Emitting Diodes (LED).

- The percentage of regulation (POR) of zener diodes (Chapter 7) has been modified in such a manner that students can experimentally verify and establish POR of zeners in a lab.

- Chapter 8 now includes the Ebers-Moll model of the transistor.

- An interesting behaviour of the JFET by which it can be used as a voltage controlled resistance has been added in Chapter 14.

- The Chapter 18 on SCRs now discusses application circuits of SCR.

- The Chapter 20 of the first edition is now is a part of the Chapter on JFET.

With these additions I hope the book addresses needs of a student being introduced to Electronics. Finally, I would like to acknowledge the efforts made by my publisher "Narosa", in bringing this book to you.

P. Arun
arunp92@gmail.com

Preface to the First Edition

There is a common complaint made by the first-year under-graduates whom I have been teaching this course for some time now, that books available do not give them the confidence to design circuits on their own. Obviously, some authors have tried to address this problem as is evident from a number of practical books available. However, they seem to dissociate themselves with the concerned theory. Also, they are usually addressed to engineering students where while teaching how to design circuits, the authors use thumb rules. These thumb rules are used as black boxes without any justification, thus mystifying it. This present book tries to give both theoretical and practical aspects of the circuit design, interlaced with examples which should give the reader confidence to design their own circuit. Also, an attempt has been made to show every topic as a logical extension of the ideas developed in the previous topics. The book does not intend to substitute the teacher, classroom or the laboratory. All of them are important in their own way. No book has every taught a person how to ride a bicycle or how to swim. One always requires an instructor. Along with an instructor, one actually has *to jump into the water*, i.e. gain hands on experience. Thus, this book in no way is to undermine the importance of the teacher, classroom or the laboratory. The main objective of this book is to inspire confidence in its readers to test things they learn. To quote Galileo Galilee *"You can not teach a man anything; you can only help him find it within himself."* If the book develops and forwards the interest of electronics in the reader, I would consider my objective fulfilled.

The book has grown in terms of pages more than I had initially thought off and yet so much has to be said and done in this subject. It in a way puts to perspective the task ahead for the students. A logical way to learn any subject is similar to the process of building a wall, with one brick coming on top of the last brick. That is you learn and develop on what you already know. One can call this the vertical method of learning. However, many concepts of electronics call for a horizontal method of learning. Even before you have learnt of a concept you might have to use it to understand some other concept. Thus, the usage of a black box technique is not totally avoidable and you might have to go along. However, once you have attained a degree of comfort in the subject, an overlook would give a great satisfaction in learning. This is true for all subjects but more true in electronics.

The structure and tone adopted is quite different from the usual textbooks available on the subject. While the first two chapters build on the simplest of law (Ohms law) required to analyze a circuit and introduces various laws which enable one to solve more complex and involved circuits. The circuits can easily be assembled in any college lab and the laws verified with nothing more then a multimeter. The detailed working of the multimeter has not been discussed in this book. However, a short over view of the oscilloscope has been given in chapter 3.

Though the scope is now just another measuring device like the multimeter, it has been included since in all probability the students who have entered college would be using an oscilloscope for the first time (in developing countries). The next four chapters carry the reader in a step-by-step like method through the internal circuitry of a DC power supply. One hopes after going through these chapters a student would be confident to make his own power supply with a little help from his or her lab instructor.

From Chapter 8 onwards the book has been arranged so that each successive chapter travels through the history of semiconductor device physics, its uses, time and logical reason of research and development. With the invention of the transistor in 1948, the idea of miniaturization caught on. This would then enable the making of digital watches, walkmans etc. However, for portability you required good batteries. Research started in two directions, for better batteries or devices which used less current. This lead to the development of JFETs and other devices as listed in this book. Semiconductor purification and manufacturing processes improved with the years, allowing for the fabrication of devices which could control power transmission etc. Of those, the thyristor is the most important and has been discussed in chapter 18. The race for miniaturization really never ended and was restricted by technological limitations. These limitations when overcome allowed for the development of integrated circuits (ICs). Chapter 19 introduces the process of making IC. After which we discuss devices which were developed to suit the IC industry.

The exponential growth of electronic consumer goods and in turn the changes bought about by these are of far greater impact then the industrial revolution. All these developments are a result of the invention of the transistor and subsequent inventions of electronics. The changes are quite rapid. Documentation of the developments and changes in the industry, along with a vast amount of other related information are available on the internet. I was tempted to list some web sites which would be of interest to the audience in general. However, since most of the sites are by electronics hobby enthusiast, they are rarely maintained and the links close ever so rapidly. Another interesting use of the computer is the availability of softwares like PSpice and Electronic Workbench. These softwares allow you to design circuits and then test them with virtual oscilloscopes and multimeters. Serious students are encouraged to use these new methods of learning.

I would like to record my appreciation for my students at S.G.T.B. Khalsa College and Dr. Naveen Gaur (Department of Physics, Dyal Singh College, Delhi) whose insistence forced me to put my experiences in shape of this book. I would like to thank Ms Arti Dwivedi, Mr Sushil Kumar Singh, Mr Ashok Kumar Das and Dr. Adarsh Singh for going through some sections of the manuscript and the suggestions they offered. Also, I would like to record my gratitude to my teachers; Mr Chakraburty (NGFS, Saket), Mr. K. R. Lal and Dr. A. G. Vedeshwar. The book has been a cause of major distraction from my social obligations at home and hence, an apology is due to my mother and wife. It was a super-human effort on my mother's side to put up with my tantrums from the onset of this project. Finally, I would like to acknowledge the contributions made by my parents. It is a matter of personal sorrow that my father is no more to share the joy and pleasure associated with the writing of this book. I am sure the number of errors would have been bare minimum had he gone through the manuscript.

P. Arun

Contents

CONTENTS

xiii

Chapter 1

Kirchhoff's Theorems

Electronics is a branch of science which has effected the way modern civilization exists today. In fact the changes bought by the electronic revolution has far greater impact than the industrial revolution of the nineteenth century. It hence, becomes important to learn and appreciate electronics. To an extent learning electronics is as easy and as enjoyable as learning algebra.

One of the first law of electronics that students learn in school is Ohms law.[1] The law in itself can be classified as a law of electricity. This is essentially because of it's linear nature. If an AC signal, like a sine wave, is given as an input to a circuit made of devices described by Ohms law, the output received would be sine. The magnitude might vary, however, the shape of the output signal would be the same as the input signal. In other words, signal modification does not take place. Hence, electronics is different from the study of electricity. Electronics, essentially deals with the study of devices and circuits which are capable of signal modifications. Circuits are also called networks. Electronic circuits are made by a combination of many passive and active elements like resistances, capacitors, power supplies, IC's etc in series or parallel combination. From the physical inter-connectivity comes the name *networks.* *Active elements* are those that require energy sources like voltage supply, batteries and current sources for it's operation and has an output that is a function of the present and past input signal, example the transistor. *Passive elements* like resistances, capacitors etc do not require energy source for it's functioning. *Network Analysis* hence, is a subject of analyzing the circuit, predicting its performance by finding the currents, voltages and power consumption at various parts of the circuit. Thus, when the subject is about signal modification, analyzes and prediction then one can easily expect an over-dose of mathematics. However, this book has restricted the mathematics to algebra and trivial calculus that a student picks up as he graduates from school.

Though the Ohms law equation looks trivial, it is capable of analyzing many simple circuits.

---

[1] Georg Simon Ohm was born on March 16, 1789 (1789-1854) in the city of Erlangen in Bavaria, which is now in Germany. Ohm went on to be a mathematics teacher in a school. In 1826, Ohm gave a mathematical description of conduction in circuits modeled of Fourier's study of heat conduction. Ohm published a book in 1827 which contained what is now know as the 'Ohm Laws' ($V = I \times R$). In 1849 Ohm took up a post in Munich and began to lecture at the University of Munich. In 1852, two years before his death, he was appointed to the chair of physics at the University of Munich.

For example, if you are trying to analyze the circuit shown in fig(1.1), Ohms law is sufficient to find the voltage across the resistance R_2. First find the current in the circuit. Using Ohms law,

$$V = I(R_1 + R_2)$$
$$I = \frac{V}{(R_1 + R_2)}$$

this current flows through R_2 and gives rise to a potential drop (again using Ohms law), say V_o

$$V_o = IR_2$$
$$V_o = \frac{R_2 V}{(R_1 + R_2)} \tag{1.1}$$

This is the potential divider equation. Similarly a trivial and obvious result one can get using Ohms law is

$$V = I(R_1 + R_2)$$
$$V = IR_{net}$$
$$R_{net} = R_1 + R_2$$

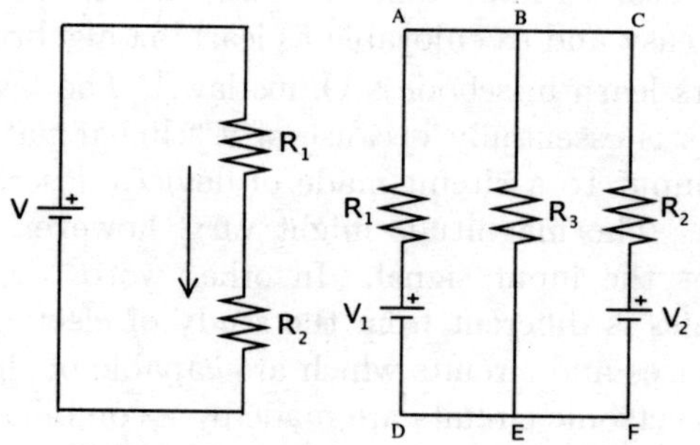

Figure 1.1: *Ohm's law can be used to find the potential drop across any of the two resistances. The second circuit has the nodes identified.*

This is the effective resistance in the circuit when resistive elements are kept in series. Hence for analyzing simple circuits Ohms law is good enough. However, with the increasing complexity of the network, one will find it near impossible to do any calculation based on Ohms law. Hence various new laws will be introduced which would make circuit analysis quite simple. Before introducing these laws, like learning any new language, one has to develop the subject's vocabulary.

Consider the second circuit of fig(1.1), the circuital elements between the terminals 'AD', 'BE' and 'CF' form a *branch*. A constant current flows in all the circuital elements of a branch. Hence, branch can be defined as a path of a circuit where the current flows without increase or decrease in its magnitude. Various branches meet at a common point called the *node*. The point where the three resistances in fig(1.1) meet is a node. Points 'A', 'B' and 'C' are the same and is a single node. Path 'AB' and 'BC' can not be considered branches as no circuital elements exists here, they are only connecting wires. By their very nature, the resistance of connecting wire is zero and hence the potential drop across 'AB' and 'BC' is zero. The point 'A', 'B' and 'C' hence, are treated as a terminal or a single nodal point. Defining point of inter-connection of circuital elements as a node, the point of connection between resistance R_1 and voltage source V_1 would also be a node. However, as far as circuit analysis is concerned since the current entering and leaving this node is the same, this point of inter-connection is not an unique node. In general it is at the nodes that the potential drop are experimentally determined. The current flowing in a branch is called *branch current* and its magnitude depends on the circuital elements of the branch and the potential difference between the nodes which enclose the branch. Also, notice in the circuit under discussion, certain set of branches form a closed path. For example the circuital elements enclosed in 'ABED' and 'BCFE' form a closed path for current flow and is called a *Loop or Mesh*. Removal of any circuital element would break the path of current and hence destroy the mesh.

Example 1.1: Determine the current and resistance R_1 of circuit shown in fig(1.2).

Ohms law gives the current as

$$\frac{30}{R_1 + 2} = I$$
$$30 = R_1 I + 2I$$

Since the voltage drop across R_1 is known ($R_1 I = 20v$), I is 5mA. Hence,

$$5 \times 10^{-3} R_1 = 20$$
$$R_1 = 4 \times 10^3$$

i.e. $R_1 = 4K\Omega$.

There is an alternative method, using the potential divider equation (1.1), just use it and see how simpler life is

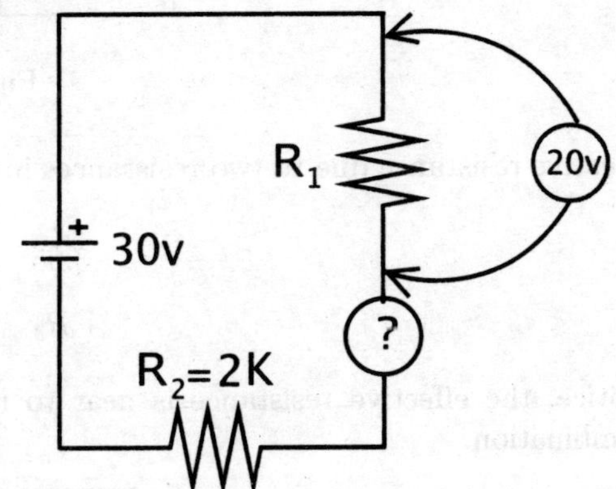

Figure 1.2:

$$20 = \left(\frac{R_1}{R_1 + 2}\right) 30$$
$$R_1 = \frac{40}{10} = 4k\Omega$$

Ohms law gives I $= 20/R_1 = 5mA$.

Answer I= 5mA and $R_1 = 4K\Omega$.

If you are still not convinced that the potential divider equation makes your life easier let us see the next example.

Example 1.2: Determine the resistance R_3 of circuit shown in fig(1.3).

Assume the parallel combination of the $18k\Omega$ resistance and of R_3 is R, then using the potential divider equation, we have

$$22.5 = \left(\frac{R}{R + 1.3 + 1.8}\right) 38$$
$$R = 3.05k\Omega$$

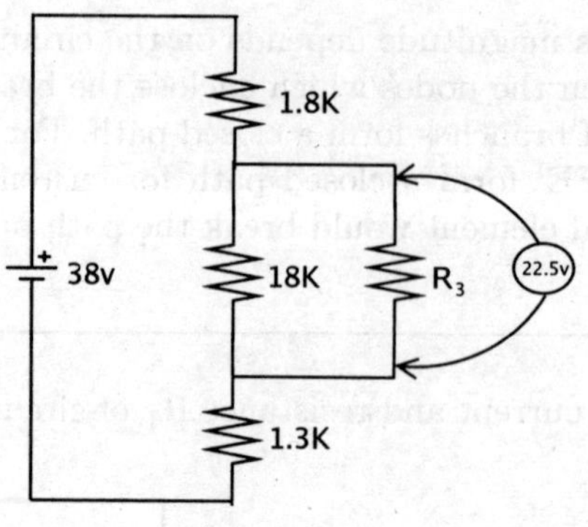

Figure 1.3:

effective resistance due to two resistances in parallel is given as

$$3.05 = \frac{18R_3}{18 + R_3}$$
$$R_3 = 3.67k\Omega$$

Notice, the effective resistance is near to the value of the smaller resistance in the parallel combination.

$$\boxed{\text{Answer } R_3 = 3.67K\Omega.}$$

$\boxed{\textbf{Example 1.3:}}$ Determine the potential V_{BD} in fig(1.4).

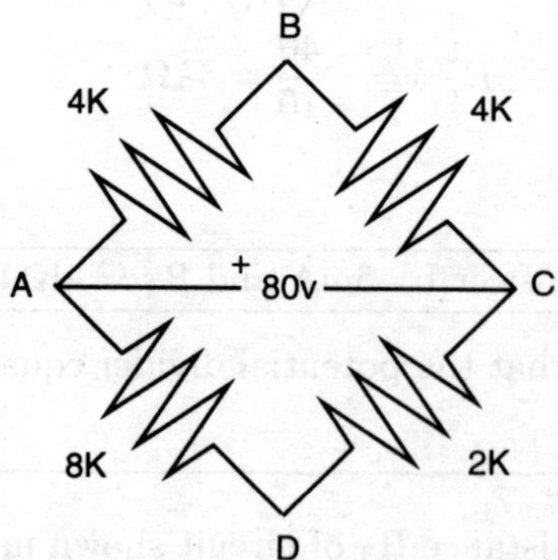

Figure 1.4:

Using the potential divider equation the potential drop across the 2kΩ resistance is

$$V_{DC} = \left(\frac{2}{2+8}\right)80 = 16v$$

Similarly, across the 4kΩ resistance

$$V_{BC} = \left(\frac{4}{4+4}\right) 80 = 40v$$

The potential V_{BC} is the potential difference between points B and C, i.e

$$V_{BC} = V_B - V_C$$

Similarly, $V_{DC} = V_D - V_C$. If we subtract the two equations, we have

$$V_{BC} - V_{DC} = (V_B - V_C) - (V_D - V_C) = V_B - V_D = V_{BD}$$

Hence, the potential difference V_{BD} is 40-16=24v.

$$\boxed{\text{Answer } V_{BD} = 24 \text{volts}}$$

With the introduction of various terms of the subject we now proceed to learn the most important laws of network analysis; the Kirchhoff's law.

1.1 Kirchhoff's Current Law (KCL)

Kirchhoff's Current Law or KCL states that the "algebraic sum" of the currents in all the branches which converge at a common node is equal to zero. The point of importance in this statement is the "algebraic sum". At a node, which is a point of converging of many branches, some currents may flow into the node while some away from the node. By convention currents flowing into the node is given a positive sign while currents flowing out of node is given a negative sign. Hence,

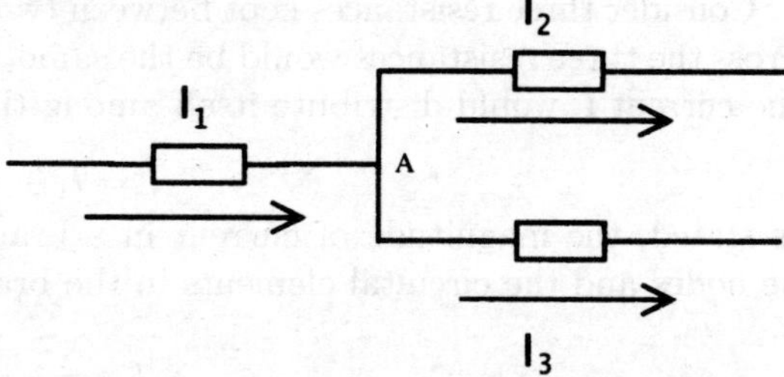

Figure 1.5: *The current I_1 coming into node 'A' breaks into two currents I_2 and I_3 which flow away from the node.*

$$\sum_i I = 0$$

$$\sum I_{in} - \sum I_{out} = 0$$

$$\sum I_{in} = \sum I_{out} \qquad (1.2)$$

The above calculation (eqn 1.2) shows that KCL may be redefined as the sum of currents flowing into a node is equal to the sum of currents flowing out of the node. Consider the case of a water work in which one pipe comes from the tank and bifurcates into two taps. If the two taps are open, the amount of water coming out of the tank must be equal to the amount of water coming from the two taps. Circuit shown in fig 1.5 can easily be seen as the schematics of the discussed water work. Hence, Kirchhoff's Current Law also states an obvious conservation law, the law of charge conservation. Equation 1.2 can be written in terms of rate of changes entering and leaving a node as

$$\sum \left(\frac{dq}{dt}\right)_{in} = \sum \left(\frac{dq}{dt}\right)_{out}$$

In unit time the number of charges entering and leaving the node would be the same. The positive and negative sign associated with current entering and leaving the node is purely convention. A note of caution to those who might what to adopt their own convention, whatever it may be, once started- stick to it.

Example 1.4: Determine the current I_2 of fig(1.5) if I_1 and I_3 is 8A and 3A respectively.

Using KCL (eqn 1.2),

$$I_1 = I_2 + I_3$$
$$8 = I_2 + 3$$

Answer $I_2 = 5A$

Consider three resistances kept between two nodes (fig 1.6a) fed with a current I. The voltage across the three resistances would be the same, since they are all between the same nodes ('AB'). The current I, would distribute itself among the three resistances. By KCL

$$I = I_1 + I_2 + I_3$$

As stated, the magnitude of current in a branch depends on the potential difference between the nodes and the circuital elements in the branch. The branch currents by Ohms law is

$$I = \frac{V}{R_1} + \frac{V}{R_2} + \frac{V}{R_3}$$
$$\frac{V}{R_{net}} = \frac{V}{R_1} + \frac{V}{R_2} + \frac{V}{R_3}$$
$$\frac{1}{R_{net}} = \frac{1}{R_1} + \frac{1}{R_2} + \frac{1}{R_3}$$
$$\frac{1}{R_{net}} = \sum_i \frac{1}{R_i}$$

The trivial result of parallel combinations of resistances come from the Kirchhoff's Current law.

Another useful result that one obtains using KCL is the current division equation. Consider a current source[2] supplying a current I to two parallel resistances R_1 and R_2 (fig 1.6b).

---

[2] An energy source that supplies a constant current to the load irrespective of the load's magnitude and potential drop across it. So if a current source was supplying current I to the three resistances of fig(1.6a), even on removing R_3 from the circuit, the source provides a current I to the load. It would be interesting for the reader to compute the potential V_{AB} in the two cases.

Again, the voltage drop across the two resistances would be the same. Along with Kirchhoff's Current Law, we can compute the currents in each resistance from the set simultaneous equation:

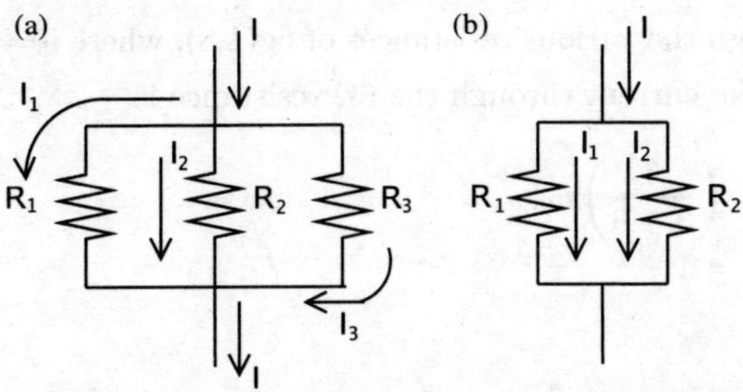

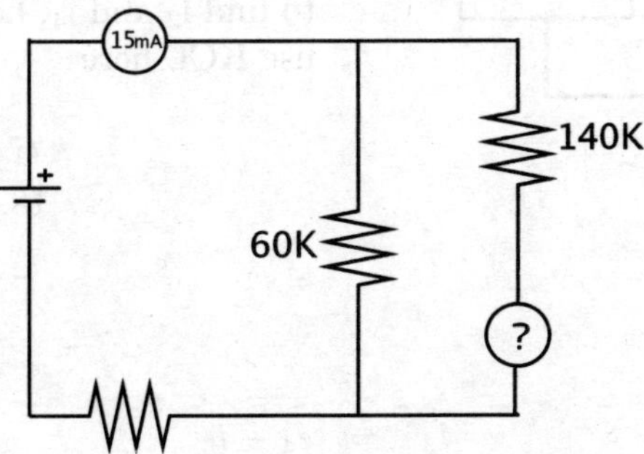

$$V_{R_1} = V_{R_2}$$
$$I_1 R_1 = I_2 R_2$$
$$(R_1)I_1 - (R_2)I_2 = 0 \qquad (1.3)$$

and

$$I_1 + I_2 = I$$

Solving the simultaneous equation we have the two currents as

$$I_1 = \frac{R_2 I}{R_1 + R_2}$$
$$I_2 = \frac{R_1 I}{R_1 + R_2} \qquad (1.4)$$

Figure 1.6: *In (a) the current* I *coming into the upper node breaks into various currents,* I_1, I_2 *and* I_3 *which flow away from the node into respective branches. At the lower node, they all merge together to give back the original current. In (b) the same is seen when one considers only two resistances.*

The above equations are called the current divider equation. Note the current in a resistance is proportional to the size of the parallel neighbor. A student of electronics should be well versed with this equation along with the voltage divider equation eq(1.1).

Example 1.5: Determine the current through the 140kΩ resistance of circuit shown in fig(1.7).

Figure 1.7:

Using the current divider equation(1.4) the current through the 140kΩ resistance is

$$I = \left(\frac{60}{60 + 140}\right) 15$$
$$= 4.5mA$$

Note, more current (15-4.5=11.5mA) flows through the lower resistive path, the path having the smaller resistance (60kΩ)

$$\boxed{\text{Answer } I = 4.5\text{mA}}$$

$\boxed{\text{Example 1.6:}}$ Determine the currents through the various resistances of fig(1.8), where $i_5=4A$.

Using the current divider equation(1.4) the current through the 6Ω resistance is

$$
\begin{aligned}
i_5 &= \left(\frac{3}{3+6}\right) i_3 \\
&= 4A
\end{aligned}
$$

Solving the above we have

$$i_3 = 12A$$

Similarly,

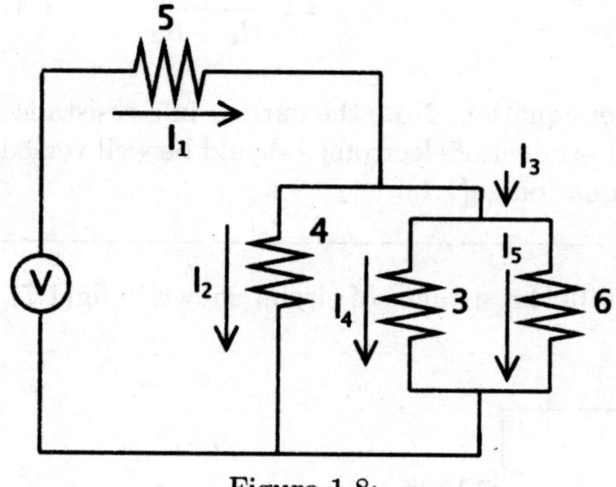

Figure 1.8:

$$
\begin{aligned}
i_3 &= \left[\frac{4}{4+(3\|6)}\right] i_1 \\
&= 12A
\end{aligned}
$$

Therefore,

$$i_1 = 18A$$

You can use the current divider equation again to find i_2 and i_4, however, it is more prudent to use KCL here

$$
\begin{aligned}
i_1 &= i_2 + i_3 \\
18 &= i_2 + 12 \\
i_2 &= 6A
\end{aligned}
$$

Also,

$$
\begin{aligned}
i_3 &= i_4 - i_5 \\
12 &= i_4 + 4 \\
i_4 &= 8A
\end{aligned}
$$

The currents flowing in the various resistances are

$$\boxed{\text{Answer } i_1 = 18A,\ i_2 = 6A,\ i_3 = 12A,\ i_4 = 8A \text{ and } i_5 = 4A}$$

1.2 Kirchhoff's Voltage Law (KVL)

Kirchhoff's Voltage Law states that the algebraic sum of the voltages between successive nodes in a closed path in the network is equal to zero.

Consider the circuit shown in fig(1.9). Assume a current flows in the circuit from the battery's (V_1) positive terminal (higher potential) to its negative terminal (lower potential), i.e. a clockwise current flows in the circuit. As the current flows through the resistances it generates a potential drop across them. The end of the resistance where the current is entering is at a higher potential while the end from which the current leaves the resistance is at a lower potential. One can for convenience imagine an equivalent circuit as shown in fig(1.9). Kirchhoff's voltage law states

$$V_1 - V_{R1} - V_{R2} - V_2 = 0 \qquad (1.5)$$

Again the point to be appreciated in the law's statement is *"algebraic sum"*. During any circuit analysis, according to the assigned current a potential rise is assigned a positive sign while a potential drop is assigned a negative sign. To appreciate this better, one could assign a counter-clockwise current in the above loop. The loop equation then would be written differently, since the potential rise and drop would be seen different by this newly assigned current. The loop equation would be

$$-V_1 - V_{R1} - V_{R2} + V_2 = 0 \qquad (1.6)$$

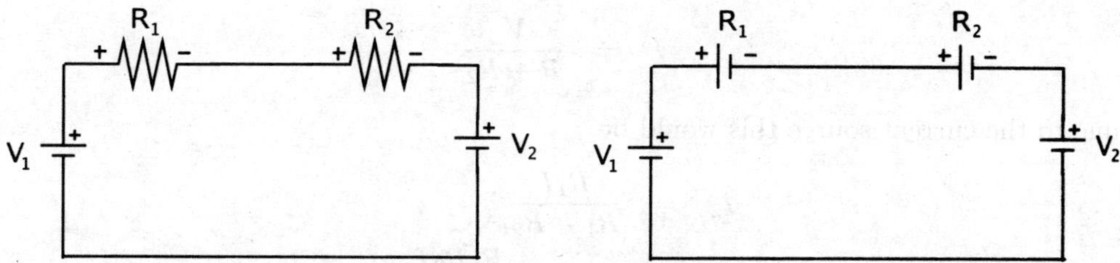

Figure 1.9: *The resistances of the left hand side circuit have been replaced by voltage sources whose magnitude is equal to the potential drops across the respective resistances. The polarity is also in accordance to the direction of current flow. Equation (1.5) is the loop equation using Kirchhoff's Voltage Law.*

Usually, our thought process, while understanding a circuit is that an applied voltage causes a current to flow through the resistance. In Ohms law

$$V = IR$$

we imagine the voltage as the cause and the current an effect of the cause. However, if the above equation is re-written

$$I = \frac{V}{R}$$

one may interpret that a current flowing through a resistance gives rise to a potential drop across it. Both explanations are valid. It is just our familiarity with batteries that make us prefer voltage sources. A reorientation is required to appreciate both arguments as the same. The ability to explain and analyze the same circuit either with current source or voltage source is called the **Principle of Duality**.

In short we are just inter-changing the dependent and independent variables of the Ohms law equation. Thus, one must be comfortable in converting a voltage source in a circuit to an equivalent current source and visa verse. For replacing a voltage source with an equivalent current source, we demand irrespective of what source maybe in use the current through and voltage across a test resistance is the same. Assume R_2 of fig(1.6) and fig(1.10) to be our test resistance. Let R be an unknown resistance in the potential divider that we can adjust to attain equivalence.

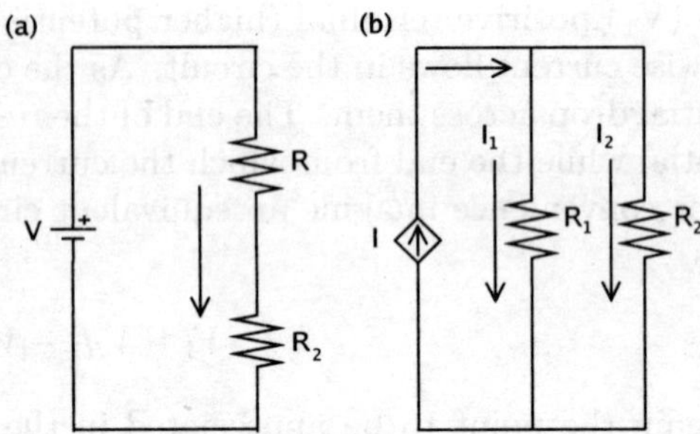

Figure 1.10: *By principle of Duality, a voltage source can be converted to a current source and visa verse.*

The voltage across R_2 and the current through it, in our potential divider circuit is given by

$$V_2 = \frac{R_2 V}{R + R_2}$$

$$I = \frac{V}{R + R_2}$$

while due to the current source this would be

$$I_2 = \frac{R_1 I}{R_1 + R_2}$$

$$V_2 = R_2 I_2 = \frac{R_1 R_2 I}{R_1 + R_2}$$

For the two energy sources to be equivalent, their impact/ effect on R_2 has to be the same, hence the above two set of equations have to be equal. The two equations give

$$V = R_1 I \ (or \ I = V/R_1)$$

$$R = R_1$$

Thus, for replacing a voltage source with an equivalent current source, we require a current source which gives current V/R_1 with a parallel resistance R across it (see fig 1.10b). For replacing a current source (I) with an equivalent voltage source, a resistance R_1 is put in series with a voltage source IR_1. The resistance R_1 is the internal resistance of the practical voltage source we had discussed.

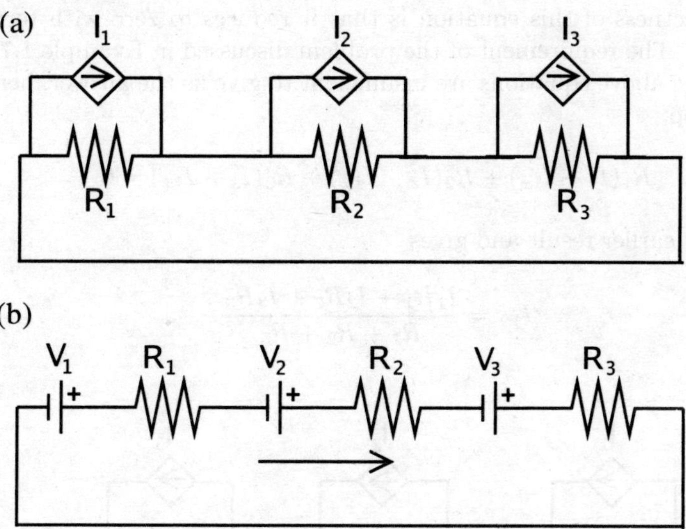

Figure 1.11: *Circuit for Example 1.7.*

Example 1.7: Consider the circuit shown in fig(1.11a). Find the current in the circuit.

The simplest method is to convert all the current sources into their respective voltage sources and redraw the circuit. The equivalent circuit is three voltage sources in series with three resistances fig(1.11b). The magnitude of the three voltage sources are I_1R_1, I_2R_2 and I_3R_3 respectively, hence, the net current is

$$\boxed{\text{Answer } I = (I_1R_1 + I_2R_2 + I_3R_3)/(R_1 + R_2 + R_3)}$$

We now proceed to develop methods to analyze more complex circuits. These methods are classified on the basis of whether KCL or KVL is being used. Sometimes, while analyzing both KCL and KVL can be used. The choice remains on the user based on convenience. Consider solving Example 1.7 again, however this time, without converting the current sources into voltage sources. Examine fig(1.12) and notice the distribution of the currents at various nodes. Nodes A, B and C are junctions to four branches. Following are the equations obtained on applying KCL on the nodes A, B and C

$$I_1 = I_2 + (I_1 - I_{1x}) - (I_2 - I_{1x})$$
$$I_2 = I_3 + (I_2 - I_{1x}) - (I_3 - I_{1x})$$
$$I_3 = I_1 + (I_3 - I_{1x}) - (I_1 - I_{1x})$$

The negative currents on the right hand side of the three equations represents the current flowing into the node while the positive currents represents the currents flowing away from the node.

A test of the correctness of this equation is that it reduces to zero with terms canceling each other (see eqn 1.2). The requirement of the problem discussed in Example 1.7 was to determine the current I_{1x}. The above equations are insufficient to give us the answer, hence we apply KVL on the circuital loop:

$$R_1(I_1 - I_{1x}) + R_2(I_2 - I_{1x}) + R_3(I_3 - I_{1x}) = 0$$

which confirms our earlier result and gives

$$I_{1x} = \frac{I_1 R_1 + I_2 R_2 + I_3 R_3}{R_1 + R_2 + R_3}$$

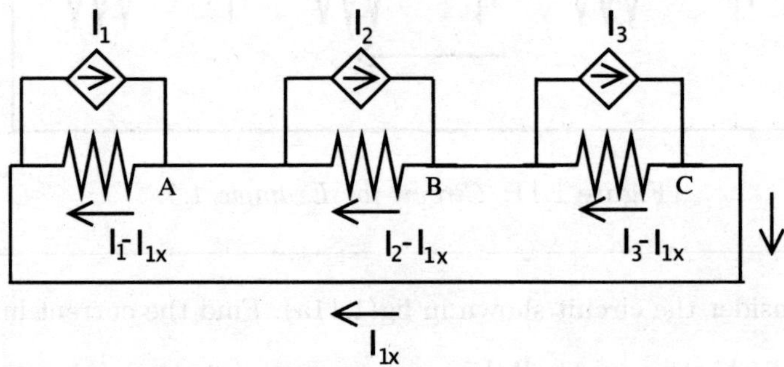

Figure 1.12: *Circuit of Example 1.7 indicating how current distributes itself at the nodes.*

The above example highlights that each problem need not require an unique method for solving, however, each example may have one method more simpler and easier compared to the other methods. We now examine three major methods of circuital analysis.

1.3   Branch Current Method

This method involves:

1. Assign a distinct current of arbitrary direction to each branch of the network.

2. Indicate polarities for each resistor/ circuital element by the assumed current direction.[3]

3. Apply Kirchhoff's Voltage Law around each loop and Kirchhoff's current law at the minimum number of nodes that will include all of the branch currents of the network.

4. Solve the set up simultaneous linear equation for the various branch currents.

---

[3]One does not assign polarity to voltage sources based on choice of current direction. Their polarity remain as it is.

Let us apply the Branch current method to the circuit shown in fig 1.13. The circuit has three branches between the two node 'A' and 'B'. The first branch has two circuital elements R_1 and V_1, while the second has only the resistance R_2. The third branch again consist two circuital elements R_3 and V_2. As per the first step, three arbitrary current I_1, I_2 and I_3 are assigned to the three branches. For the next step, we assign polarities for each resistance as per the assumed current direction. The current is assumed to flow from higher potential to the lower potential. Hence for resistance R_1 where the current flows from the left hand side to the right hand side, the left hand side is assigned a positive sign while the right hand side is given a negative sign. Similarly assign polarities for the remaining resistors. It should be noted that the polarity of the voltage sources remain unaffected irrespective of the current one assigns to the voltage source's branch.

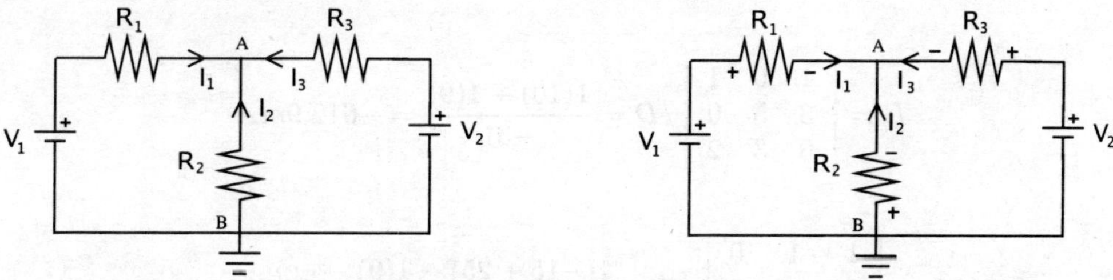

Figure 1.13: *Based on the assigned current direction, polarity's are denoted to show the current flow from higher to lower potential.*

The next step is to apply Kirchhoff's voltage law around all the loops. The circuit has two loops of importance, giving equations

$$V_1 - I_1 R_1 + I_2 R_2 = 0$$
$$V_2 - I_3 R_3 + I_2 R_2 = 0 \tag{1.7}$$

Now Kirchhoff's current law has to be applied at the minimum number of nodes that includes all of the branch currents of the network. Either node 'A' or 'B' would satisfy our requirement as all three currents in question exist on both these nodes. The Kirchhoff's current law at node 'A' gives

$$I_1 + I_2 + I_3 = 0 \tag{1.8}$$

The three equations form a set of linear equations which can be solved simultaneously.

Example 1.8: Consider the value of the two voltage sources V_1 and V_2 of fig(1.13) as 5v and 3v respectively. Find the value of various branch currents using the Branch Current Method for $R_1 = 3K\Omega$, $R_2 = 5K\Omega$ and $R_3 = 2K\Omega$. Substituting the values of the resistances and voltage sources in the equations (1.7 and 1.8) the problem reduces to that of solving the simultaneous equation.

$$I_1 + I_2 + I_3 = 0$$
$$3I_1 - 5I_2 + 0 = 5$$
$$0 - 5I_2 + 2I_3 = 3$$

One can solve the simultaneous equation using Crammer's rule (Determinants). Defining the common denominator determinant

$$D = \begin{vmatrix} 1 & 1 & 1 \\ 3 & -5 & 0 \\ 0 & -5 & 2 \end{vmatrix} = 1(-10) - 1(6) + 1(-15) = -31$$

The three currents are given as

$$I_1 = \begin{vmatrix} 0 & 1 & 1 \\ 5 & -5 & 0 \\ 3 & -5 & 2 \end{vmatrix} / D = \frac{-1(10) + 1(-25 + 15)}{-31} = 645.16mA$$

$$I_2 = \begin{vmatrix} 1 & 0 & 1 \\ 3 & 5 & 0 \\ 0 & 3 & 2 \end{vmatrix} / D = \frac{1(10) + 1(9)}{-31} = -612.9mA$$

$$I_3 = \begin{vmatrix} 1 & 1 & 0 \\ 3 & -5 & 5 \\ 0 & -5 & 3 \end{vmatrix} / D = \frac{1(-15 + 25) - 1(9)}{-31} = 32.25mA$$

The currents in various branches work out to be

$$\boxed{\text{Answer } I_1 = 645.16\text{mA}, \ I_2 = -612.9\text{mA and } I_3 = 32.25\text{mA}}$$

The negative signs in the worked out answers (Example 1.8) suggests that the directions assigned to the two currents I_2 and I_3 are in opposite to the actual direction which these currents are flowing. The magnitude is however, not effected by the choice made.

Crammer's method is easy for a 2×2 or 3×3 determinant. One can readily understand the difficulty involved in analyzing a circuit with more than three branches The difficulty involved in computing currents in various branches has been reduced with the introduction of computers. However, computer programs are more elegant and faster using matrices. The simultaneous equations (1.7 and 1.8) can be written in matrix form as

$$\begin{pmatrix} V_1 \\ V_2 \\ 0 \end{pmatrix} = \begin{pmatrix} R_1 & -R_2 & 0 \\ 0 & -R_2 & R_3 \\ 1 & 1 & 1 \end{pmatrix} \begin{pmatrix} I_1 \\ I_2 \\ I_3 \end{pmatrix}$$

or

$$V_i = R_{ij} I_j$$

To find the various branch currents involves doing a matrix inversion, i.e.

$$I_j = (R_{ij})^{-1} V_i$$

Let us setup the required simultaneous equations and matrix to analyze the circuit of fig(1.14).

$$\begin{aligned} V_1 &= R_1 I_1 - R_2 I_2 \\ 0 &= -R_2 I_2 + R_3 I_3 + R_4 I_4 \\ V_2 &= R_5 I_5 - R_4 I_4 \end{aligned}$$

The KCL applied on node 'A' and 'B' gives

$$\begin{aligned} I_1 + I_2 + I_3 &= 0 \\ I_4 + I_5 - I_3 &= 0 \end{aligned}$$

To solve by matrix inversion, we have to setup the matrix representation of the above set of simultaneous equation. We have

$$\begin{pmatrix} V_1 \\ V_2 \\ 0 \\ 0 \\ 0 \end{pmatrix} = \begin{pmatrix} R_1 & -R_2 & 0 & 0 & 0 \\ 0 & 0 & 0 & -R_2 & R_3 \\ 0 & -R_2 & R_3 & R_4 & 0 \\ 1 & 1 & 1 & 0 & 0 \\ 0 & 0 & -1 & 1 & 1 \end{pmatrix} \begin{pmatrix} I_1 \\ I_2 \\ I_3 \\ I_4 \\ I_5 \end{pmatrix}$$

| Example 1.9: | Repeat the last example (Example 1.8) and find the various branch currents using matrix inversion.

The matrix representation is given as

$$\begin{pmatrix} 0 \\ 5 \\ 3 \end{pmatrix} = \begin{pmatrix} 1 & 1 & 1 \\ 3 & -5 & 0 \\ 0 & -5 & 2 \end{pmatrix} \begin{pmatrix} I_1 \\ I_2 \\ I_3 \end{pmatrix}$$

The matrix inversion gives

$$\begin{pmatrix} I_1 \\ I_2 \\ I_3 \end{pmatrix} = -\frac{1}{31} \begin{pmatrix} -10 & -7 & 5 \\ -6 & 2 & 3 \\ -15 & 5 & -8 \end{pmatrix} \begin{pmatrix} 0 \\ 5 \\ 3 \end{pmatrix}$$

Solving this the values of current is same as that calculated in Example 1.8, i.e.

$$\boxed{\text{Answer } I_1 = 645.16\text{mA}, \ I_2 = -612.9\text{mA and } I_3 = 32.25\text{mA}}$$

1.4 Nodal Analysis

1. Convert all voltage sources to respective current sources.

2. Determine the number of nodes within the network.

3. Pick a reference node and assign each remaining node with a voltage, example V_1, V_2, ... etc.

4. Assign currents to various branches and write Kirchhoff's Current Law for each node except the reference node in terms of node voltages.

5. Solve the set up simultaneous linear equation for the various nodal voltages.

We apply Nodal analysis on the circuit shown in fig(1.14). The voltage sources when converted to current sources would result in the circuit shown in fig(1.15). Basically, the circuit has only three nodes (V_A, V_B and V_C) of which we pick V_C as the reference node (i.e. ground). Assign currents to various branches. As per Ohms law and current direction convention, the current I_2 is given as

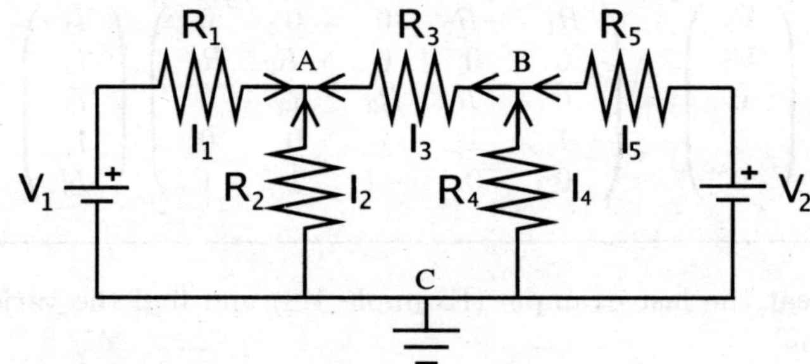

Figure 1.14: *The currents I_1, I_2 and I_3 flow into the node while currents I_4 and I_5 add at node 'B' to give I_3.*

$$I_2 = \frac{V_B - V_A}{R_3}$$

KCL applied on the two prominent nodes gives

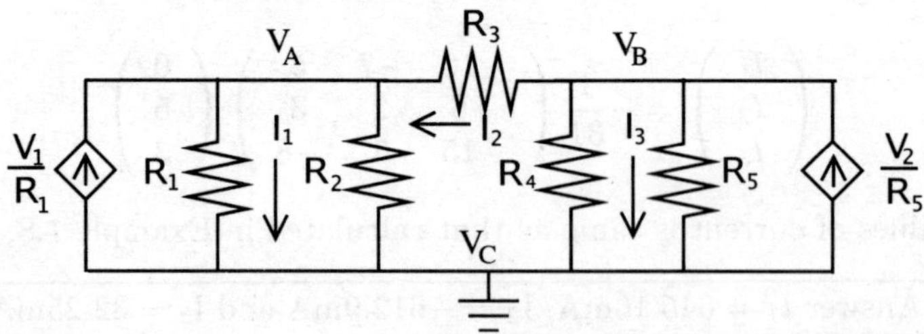

Figure 1.15: *The voltage sources of fig 1.14 have been converted to current sources. Note the current labellings are not related to those of fig 1.14.*

$$\frac{V_1}{R_1} + I_2 = \frac{V_A}{R_1 \| R_2}$$

$$\frac{V_2}{R_5} = I_2 + \frac{V_B}{R_4 \| R_5}$$

or

$$\left[\frac{1}{R_1\|R_2} + \frac{1}{R_3}\right] V_A - \left(\frac{1}{R_3}\right) V_B = \frac{V_1}{R_1}$$

$$-\left(\frac{1}{R_3}\right) V_A + \left[\frac{1}{R_3} + \frac{1}{R_4\|R_5}\right] V_B = \frac{V_2}{R_5}$$

Solving the above simultaneous equation for V_A and V_B, we analyze for currents based on Ohms law and current division equation. Below, we test our knowledge of Nodal method with an example.

| Example 1.10: | Use nodal analysis for finding I_o of the circuit shown in fig(1.16).

Ideally, one should convert all voltage sources to current source in nodal analysis. However,

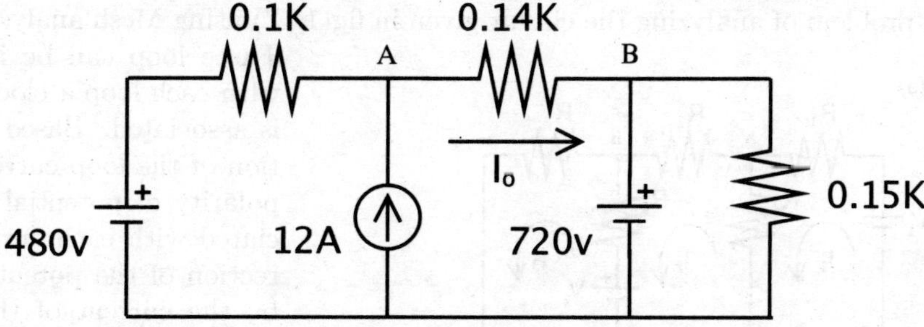

Figure 1.16:

as can be identified, the 720v source can not be converted to a current source. Hence, one can solve and analyze this circuit without converting the voltage sources to current sources. The following steps highlights this.

The voltage at node 'B' would be 720v. The current in the 100Ω (I_1) and 140Ω (i.e. I_o) have to add up as 12A. Hence, the currents

$$I_1 = \frac{480 - V_A}{140}$$

$$I_o = \frac{V_A - 720}{140}$$

have to satisfy

$$I_1 + 12 = I_o \quad or \quad I_o - I_1 = 12$$
$$\frac{V_A - 720}{140} - \frac{480 - V_A}{140} = 12$$

gives $V_A = 1280v$. Hence, the current through 140Ω resistance is

$$I_o = \frac{1280 - 720}{140} = 4A$$

1.5   Mesh Analysis

In "Mesh Analysis" the electric circuit analysis is done purely using Kirchhoff's voltage law, and hence the name Mesh Analysis. To analyze a circuit using this method, the following sequence of steps are followed:

1. Assign a distinct current in the clockwise direction to each independent closed loop.[4]

2. Indicate polarities within each loop for each resistor as determined by the assumed direction of mesh current for *that particular mesh*.

3. Apply KVL in clockwise direction. A resistor/ branch which is common in two mesh has one current in clockwise and second in counterwise direction. Also the polarity of the voltage is unaffected.

4. Solve the set up simultaneous equation.

Consider the problem of analyzing the circuit given in fig(1.17) using Mesh analysis method.

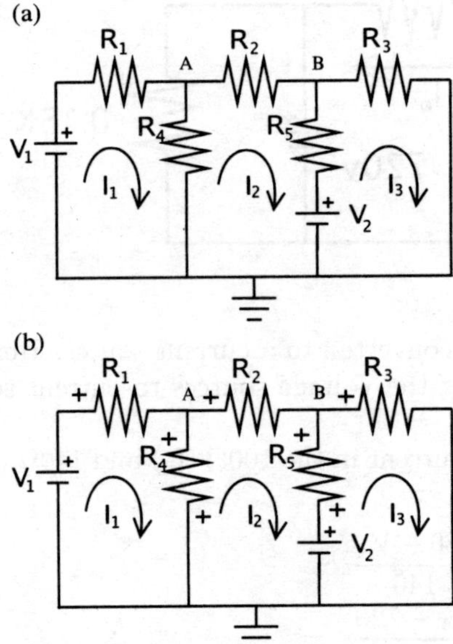

Three loop can be identified and with each loop a clockwise current is associated. Based on the direction of the loop currents assumed, polarity of potential drop is associated with each current. The direction of the potential is decided by the current of that particular loop. Note the polarities shown in fig(1.17b). Note the polarity associated with resistance R_4 due to current I_1 and I_2 in loop 1 and 2 respectively. Even though we are talking of the same resistance the polarity are different as seen in the two different loops. This is due to the fact in loop 1, the current is flowing from top to bottom while in loop 2 it is in the opposite direction, from bottom to top. Now we apply KVL in each loop. The convention remain the same as discussed above.

Figure 1.17: *Clockwise currents are attributed to each loop of the circuit in the Mesh method.*

Potential drops are given a negative sign while a potential rise is given a positive sign. Thus,

$$V_1 - R_1 I_1 - R_4(I_1 - I_2) = 0$$
$$-R_4(I_2 - I_1) - R_2 I_2 - R_5(I_2 - I_3) - V_2 = 0$$
$$V_2 - R_5(I_3 - I_2) - R_3 I_3 = 0$$

---

[4]There is a method called Loop Analysis (as the name "Loop" suggests KVL is used in this method) which is very similar to Mesh Analysis except clockwise, counterclockwise, or mixed currents can be defined.

Setting the simultaneous equation which when solved would give the three currents, we have

$$(R_1 + R_4)I_1 - R_4 I_2 + 0.I_3 = V_1$$
$$R_4 I_1 - (R_4 + R_2 + R_5)I_2 + R_5 I_3 = V_2$$
$$0.I_1 - R_5 I_2 + (R_3 + R_5)I_3 = V_2$$

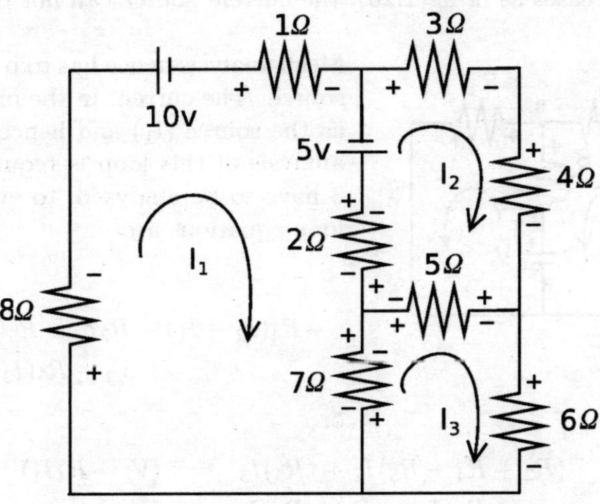

Figure 1.18:

Example 1.11: Use Mesh analysis for solving the circuit shown in fig(1.18).

Clock-wise direction currents are already assigned for the various loops in fig(1.18). Now assign polarities to various elements based on the direction of the currents. The loop equations would be given as

$$-7(I_3 - I_1) - 5(I_3 - I_2) - 6I_3 = 0$$
$$-5 - 3I_2 - 4I_2 - 5(I_2 - I_3) - 2(I_2 - I_1) = 0$$
$$-10 - I_3 + 5 - 2(I_1 - I_2) - 7(I_1 - I_3) = 0$$

Solving the simultaneous equation we have

Answer $I_3 = 0.0875$mA, $I_2 = -0.735$mA and $I_1 = 0.75$mA

One has to be judicious in identifying the number of loops in a circuit. Take the case of the circuit in fig(1.19). The loop with current 'I_1' is redundant from the point of view of mesh analysis, since all the contribution of voltage source (V_1) appears across R_1. Hence, the analysis of the circuit in hand is a two loop problem. The two loop equations being,

$$V_1 = R_2 I_2 + R_4(I_2 - I_3) + V_2$$
$$V_2 = R_4(I_3 - I_2) + R_3 I_3$$

To complete the analysis, solve for I_2 and I_3. I_1 is of course given as V/R_1. Ideally, while using Mesh method, all current sources should be converted to voltage sources.

However, in certain cases as in fig(1.20), the current source can not be converted to voltage source.

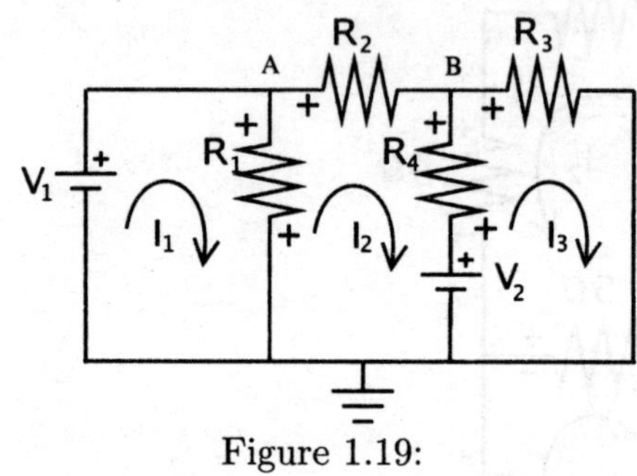

Figure 1.19:

Mesh analysis hence has to be done with the current source. The current in the first loop would be same as the source (I_1) and hence is known. No further analysis of this loop is required. Only loop 2 and 3 have to be analyzed, to evaluate I_2 and I_3. The loop equations are

$$-R_4(I_2 - I_1) - R_2 I_2 - R_5(I_2 - I_3) - V_2 = 0$$
$$V_2 - R_5(I_3 - I_2) - R_3 I_3 = 0$$

or

$$-(R_2 + R_4 + R_5)I_2 + (R_5)I_3 = (V_2 - R_4 I_1)$$
$$(-R_5)I_2 + (R_3 + R_5)I_3 = V_2$$

The task remaining of course is to solve the simultaneous equation for I_2 and I_3.

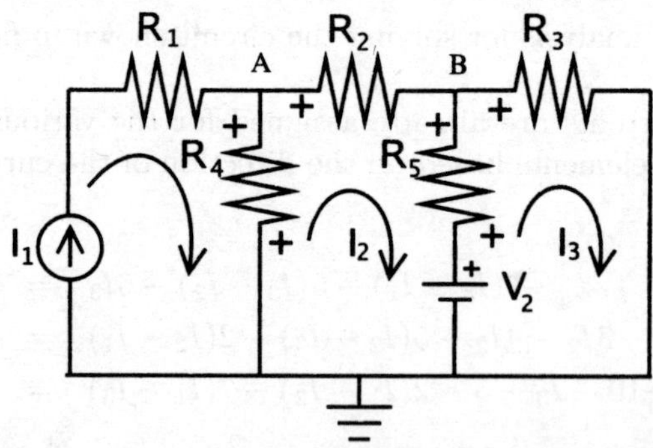

Figure 1.20:

1.6   Response to Sinusoidal Voltage

Until now we have been considering the source driving our circuits to be DC (direct current). Thus, we restricted ourselves to loads[5] which were resistive in nature. However, a practical

---

[5]Load is where the output is collected or where power is dissipated

circuit is made up of resistances, capacitors and inductors. While the capacitors and inductors only respond to DC while in transient, these element are usually used in AC (alternating current) circuits. We now investigate how these elements respond to AC signals.

1.6.1 Resistance

Assume an AC current i(t), given as

$$i(t) = I_m sin\omega t \tag{1.9}$$

is fed to a resistance. The potential drop across the resistance is given by Ohms law and is represented as

$$v(t) = i(t)R$$

$$v(t) = RI_m sin\omega t$$

RI_m has the dimensions of voltage and represents the maximum voltage drop appearing across the resistance when $\omega t = \pi/2$. Thus, the above equation may be written as

$$v(t) = V_m sin\omega t \tag{1.10}$$

What is interesting to note is the current and voltage equations (eqn 1.9 and 1.10) have similar sine terms. This might not look of a major interest right now, but read on to see what happens in case of a capacitor and an inductor.

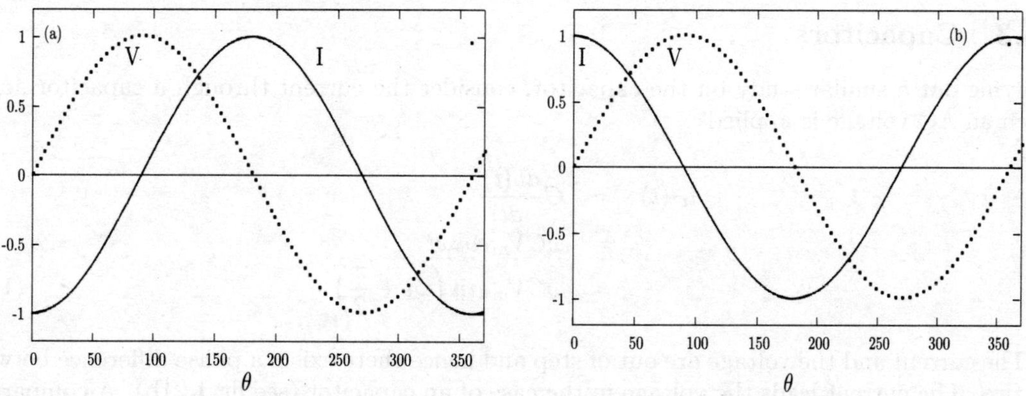

Figure 1.21: *The phase difference introduced between the current and voltage for (a) an inductor and (b) a capacitor.*

1.6.2 Inductors

One can find the current flowing through an inductor, if the voltage applied across it is given, following the steps below:

$$v_L(t) = L\frac{di(t)}{dt}$$

$$\frac{di(t)}{dt} = \frac{V_m}{L} sin\omega t$$

$$
\begin{aligned}
i_L(t) &= \frac{V_m}{L} \int sin\omega t \, dt = -\frac{V_m}{\omega L} cos\omega t \\
&= \frac{V_m}{\omega L} sin\left(\omega t - \frac{\pi}{2}\right)
\end{aligned}
\tag{1.11}
$$

The augment gives the maximum level of current and shows that it is controlled by the impedance offered by the inductor. Comparing with Ohms law, the impedance would be given as

$$X_L = \omega L$$

Note the difference in the sine term of the voltage and the current associated with the inductor. The current lags the voltage by $\pi/2$. That is, there is a phase difference of 90° between the current and voltage associated with an inductor.

This has been explicitly shown in fig(1.21a). Also note, the impedance offered by the inductor is frequency dependent. The graphical representation has been given in fig(1.22).

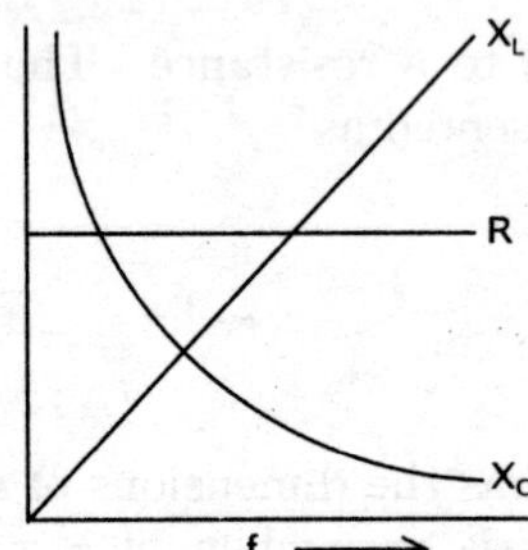

Figure 1.22: *The frequency dependence of impedance offered by a resistance, inductor and capacitor.*

1.6.3   Capacitors

Carrying out a similar study on the capacitor, consider the current through a capacitor across which an AC voltage is applied

$$
\begin{aligned}
i_C(t) &= C\frac{dv(t)}{dt} \\
&= \omega C V_m cos\omega t \\
&= \omega C V_m sin\left(\omega t + \frac{\pi}{2}\right)
\end{aligned}
\tag{1.12}
$$

The current and the voltage are out of step and hence there exists a phase difference between the two. The current leads the voltage in the case of an capacitor (see fig 1.21b). A comparison with Ohms law shows the impedance offered by the capacitor is also frequency dependent (fig 1.22).

$$X_C = \frac{1}{\omega C}$$

The important aspect we have introduced here is that the impedance offered by the capacitor and inductor are frequency dependent. Thus any circuit having these elements would respond differently to different frequency and the circuit is said to have a frequency response.

1.7 Complex Numbers

The above section also established the fact that inductors and capacitors introduced a phase difference between the voltage and current. Thus, to have a complete knowledge of the circuital performance both the magnitude of the impedance offered and the phase difference introduced between the voltage and current should be evaluated for a circuit working with AC supply.

With the problem of representing two entities, i.e. the magnitude and phase, we require a mathematical tool that would also allow for easy calculation. The solution to our representation problem lies in adopting complex number representation while dealing with ac circuits. Consider a circuit with an inductor and resistance, the loop equation would be given as

$$v = Ri + L\frac{di}{dt} \tag{1.13}$$

Approaching the problem of finding the magnitude of the net impedance offered and the phase difference introduced by this circuit as in the previous section, we have

$$
\begin{aligned}
v &= RI_m sin\omega t + \omega L I_m cos\omega t \\
&= V_m sin\left(\omega t + \phi\right)
\end{aligned}
$$

where

$$
\begin{aligned}
V_m &= aI_m \\
a &= \sqrt{R^2 + \omega^2 L^2} \\
\phi &= tan^{-1}\left(\frac{\omega L}{R}\right)
\end{aligned}
$$

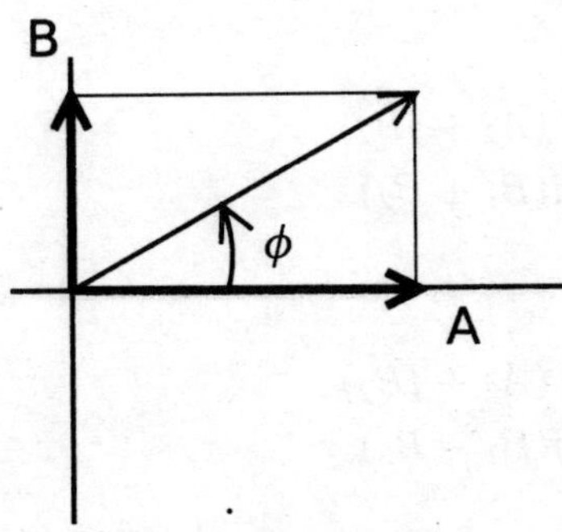

Figure 1.23: *The rectangular representation of effective impedance offered by an RL combination.*

We see from eqn(1.13), analyzing a R and L load circuit involves using differential equations. However, if we are considering the inductor as impedance in circuits (ωL), we would like to use the usual algebraic addition for circuit analysis, example

$$R_{net} = R + X_L$$

However, the phase information which came via differentiation, is lost in the above equation. However, if the phase information is carried along with X_L, then algebraic calculations may be done, i.e. the equation has to be written as

$$v = (R + X_L \angle 90)i$$

where $\angle 90$ implies the inductor introduces a phase difference of 90° between voltage and current. Instead of writing $\angle 90^0$, let us just represent it by 'j', i.e.

$$v = (R + jX_L)i$$

where $j = \sqrt{-1}$. That is the impedance are represented as complex numbers. Choice of 'j' in electronics is to distinguish this from the standard representation used for AC current (i.e. 'i'). The above shown method of representing complex numbers is called rectangular representation (rectangular representation is A + jB). The convention used to graphically display a complex number is to place the real number, i.e. A, on the horizontal x-axis, with positive numbers to the right and negative ones to the left. The imaginary numbers, i.e., numbers of the form 0 + jB, are placed on the vertical y-axis. The net impedance, (A+jB), is shown graphically in fig(1.23).

The effective impedance offered by the combination would be the length of the hypertenuse. Simple Pythagoras theorem gives this as

$$a = \sqrt{R^2 + (\omega L)^2} \tag{1.14}$$

and the angle suspended with the x axis is given as

$$\phi = tan^{-1}\left(\frac{\omega L}{R}\right) \tag{1.15}$$

The net impedance is then fully defined by the length of the hyperthunis ('a') and the angle (ϕ) suspended. Hence, the net impedance can be represented as A+jB (the rectangular representation) and a$\angle\phi$ (the polár representation) where $a = \sqrt{A^2 + B^2}$ and $\phi = tan^{-1}(B/A)$.

1.8   Arithmetic Operations

The operations of addition and subtraction are easily understood. To add or subtract two complex numbers, just add or subtract the corresponding real and imaginary parts. For instance, the sum of C_1 and C_2 is given as

$$
\begin{aligned}
C_1 + C_2 &= (A_1 + jB_1) + (A_2 + jB_2) \\
&= (A_1 + A_2) + j(B_1 + B_2)
\end{aligned}
$$

and the difference of C_1 and C_2 is

$$
\begin{aligned}
C_1 - C_2 &= (A_1 + jB_1) - (A_2 + jB_2) \\
&= (A_1 - A_2) + j(B_1 - B_2)
\end{aligned}
$$

The multiplication and division of two complex numbers is more involved. First consider the exercise of multiplying two complex numbers C_1 and C_2;

$$C_1.C_2 = (A_1 + jB_1).(A_2 + jB_2)$$

Multiplication follows the associative law, i.e.

$$
\begin{aligned}
C_1.C_2 &= A_1 A_2 + j A_1 B_2 + j B_1 A_2 + j^2 B_1 B_2 \\
&= (A_1 A_2 - B_1 B_2) + j(A_1 B_2 + B_1 A_2) \tag{1.16}
\end{aligned}
$$

Similarly, for division

$$\frac{C_1}{C_2} = \frac{A_1 + jB_1}{A_2 + jB_2}$$

To write the resulting term in the standard rectangular representation $(\alpha + j\beta)$, we multiply and divide the R.H.S. with such a term that makes the denominator a real number

$$\frac{C_1}{C_2} = \frac{A_1 + jB_1}{A_2 + jB_2} \times \frac{A_1 - jB_1}{A_2 - jB_2}$$
$$= \frac{(A_1 A_2 + B_1 B_2) + j(B_1 A_2 - A_1 B_2)}{A_2^2 + B_2^2}$$

Both multiplication and division in rectangular representation presents a non-trivial task. Let us see, if we can simplify things. Take eq(1.16) and find the magnitude of the products

$$|C_1.C_2| = \sqrt{(A_1 A_2 - B_1 B_2)^2 + (A_1 B_2 + B_1 A_2)^2}$$
$$= \sqrt{A_1^2 A_2^2 + B_1^2 B_2^2 - 2A_1 A_2 B_1 B_2 + A_1^2 B_2^2 + B_1^2 A_2^2 + 2A_1 A_2 B_1 B_2}$$
$$= \sqrt{A_1^2 A_2^2 + B_1^2 B_2^2 + A_1^2 B_2^2 + B_1^2 A_2^2}$$

The above equation can be re-written as

$$\sqrt{(A_1^2 + B_1^2).(A_2^2 + B_2^2)} = \sqrt{(A_1^2 + B_1^2)}.\sqrt{(A_2^2 + B_2^2)} \tag{1.17}$$

We find that the magnitude of the products of two complex numbers is same as the product of the magnitude of the individual complex numbers. That is

$$|C_1.C_2| = |C_1|.|C_2|$$

Now let us consider the angle suspended w.r.t. the x-axis by the product two complex numbers (eq 1.16)

$$(\phi) = tan^{-1}\left(\frac{B_1 A_2 + A_1 B_2}{A_1 A_2 - B_1 B_2}\right)$$

or

$$tan(\phi) = \left(\frac{B_1 A_2 + A_1 B_2}{A_1 A_2 - B_1 B_2}\right)$$
$$= \left(\frac{\frac{B_1}{A_1} + \frac{B_2}{A_2}}{1 - \frac{B_1}{A_1}\frac{B_2}{A_2}}\right) \tag{1.18}$$

substituting

$$tan(\phi_1) = \frac{B_1}{A_1}$$
$$tan(\phi_2) = \frac{B_2}{A_2}$$

we find

$$\phi = \phi_1 + \phi_2 \tag{1.19}$$

This, along with the result of eqn(1.17), allows us to write the product as

$$
\begin{aligned}
C_1.C_2 &= (|C_1|\angle\phi_1).(|C_2|\angle\phi_2) \\
&= |C_1|.|C_2|\angle(\phi_1 + \phi_2)
\end{aligned}
$$

Thus, multiplication is trivial when polar representation of the complex numbers are used. Division is similarly trivial in polar representation

$$
\begin{aligned}
\frac{C_1}{C_2} &= \frac{|C_1|\angle\phi_1}{|C_2|\angle\phi_2} \\
&= \frac{|C_1|}{|C_2|}\angle(\phi_1 - \phi_2)
\end{aligned}
$$

1.8.1 Equivalent Circuits

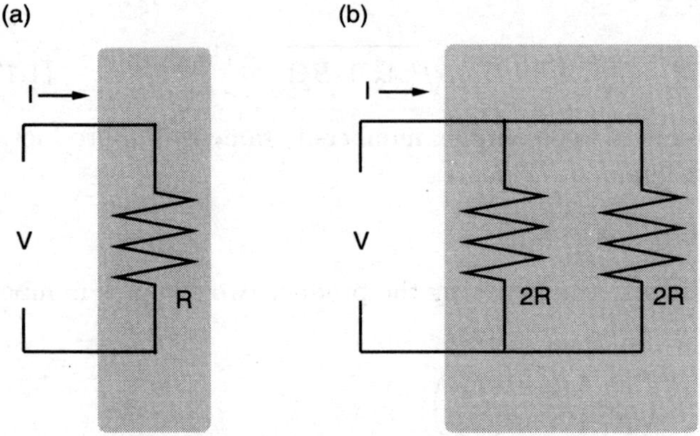

Consider the situation, where you would like to use a 10KΩ resistance with a 10v supply so that you draw a current of 1mA from it, as shown in fig(1.24a). However, as luck would have it, there are no 10KΩ resistances available. The only resistances available are of 20KΩ. By keeping two of these parallel, you obviously can get the required 10KΩ. Thus, the shaded regions of fig(1.24) a and b have the same effect on the supply, i.e. both demand the same current from the supply. The two circuits are thus said to be equivalent. The shaded circuits are trivial and are examples of one-port networks.

Figure 1.24: *The two circuits of (a) and (b) are equivalent as they draw the same amount of current from the power supply.*

Circuits we have dealt with until now were fairly simple and the topography of the network easy to identify which impedance is in series and which is in parallel. However, this may not be the case all the time. With increasing complexity of the circuit, it becomes impossible to resolve the which impedance are in series or in parallel. Figure(1.25) should be enough to prove our point. Hence, to analyze this circuit, the best alternative would be to consider a trivial but equivalent circuit. By equivalent circuit we imply the substitution does not change the voltages, currents or even the phases associated with these quantities at the input and output terminals (or more popularly the two "ports") of the circuits.

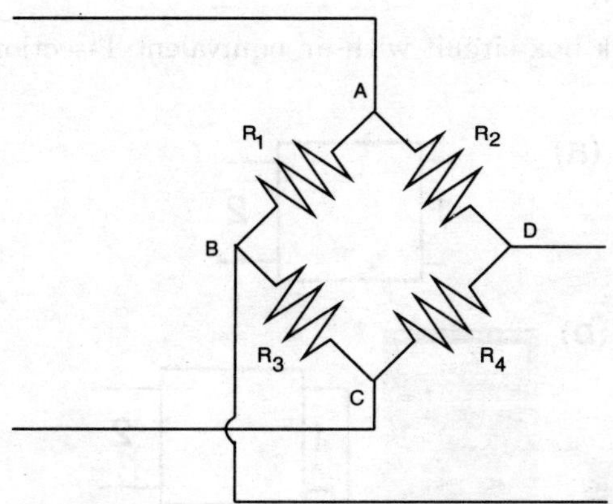

Figure 1.25: *Four circuital elements arranged in a bridge like circuit is also called a lattice network.*

We go to the extreme of saying, we do not want to know what circuit is under consideration or how complex is the given circuit. Just put it in a black box with four wires coming out, two acting as the input and two as the output wires. Consider, the black box we are discussing is shown in figure(1.26a). The input and output ports are labeled 1 and 2 respectively. One can give a probe voltage v_1 on the input side and look for the resulting input current (i_1), output voltage (v_2), output current (i_2) and their respective phase differences w.r.t. the input voltage. That is, seven (phases are w.r.t. input voltage whose phase we set zero) parameters can be identified. However, it should be noted that the output voltage, output current and their phases are essentially determined by the the load impedance that is applied and not by the circuital elements of the black box. Hence, only three variables are related to the elements inside the black box, namely, v_1, i_1 and the phase difference input current has with the input voltage. Thus, a minimum of three impedance are required to form the equivalent circuit. Three impedance can be arranged in two possible configurations which we shall see in the following passages. Measuring impedance would have information of both the input current and phase difference for a given input voltage. Hence, in actual practise one measures the input impedance when the output side is left open (z_{1oc}) as shown in figure(1.26b), and the input impedance when the output side is shorted (z_{1sc}) as seen in figure(1.26c). Similarly, the impedance can be measured at the output side by leaving the input side open or shorted, giving z_{2oc} and z_{2sc}.

T-section

We talked of two possible configurations in which three impedance can be arranged to fabricate equivalent circuits. The first of the two possibilities is called the T-section equivalent circuit, shown in figure(1.27). In the following section we develop the required mathematics from which we can calculate the three required impedance of our equivalent T-section from experimentally determined z_{1oc}, z_{1sc}, z_{2oc} and z_{2sc}.

From the arrangement of impedance in the T-section circuit, the expression of these four impedance can be compared to those of the black box circuit which were experimentally determined. Using simple arithmetic steps the required z_1, z_2 and z_3 can be computed by which we can replace the complex black box circuit with an equivalent T-section circuit.

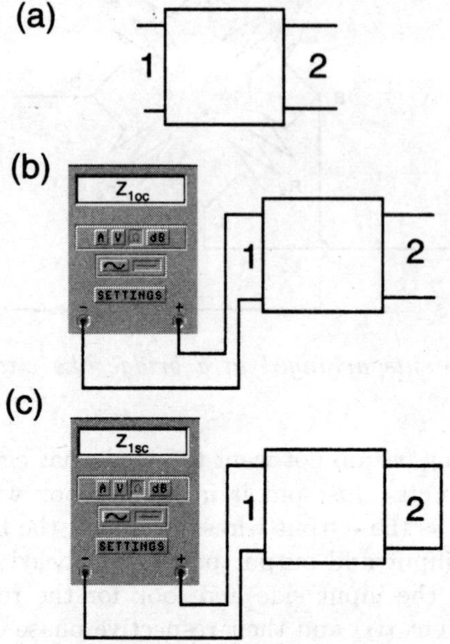

Figure 1.26: *Any complex circuit can be considered as a (a) black box where one can probe the circuit by measuring the impedance at the input and output side. From the input side one can measure the impedance at the input side with (b) the output kept open or (c) the output is shorted.*

$$z_{1oc} = z_1 + z_3 \tag{1.20}$$

$$z_{1sc} = z_1 + \frac{z_2 z_3}{z_2 + z_3} \tag{1.21}$$

$$z_{2oc} = z_2 + z_3 \tag{1.22}$$

$$z_{2sc} = z_2 + \frac{z_1 z_3}{z_1 + z_3} \tag{1.23}$$

Subtracting eqn(1.21) from eqn(1.20) we have

$$
\begin{aligned}
z_{1oc} - z_{1sc} &= z_1 + z_3 - z_1 - \frac{z_2 z_3}{z_2 + z_3} \\
&= z_3 - \frac{z_2 z_3}{z_2 + z_3} \\
&= \frac{z_3^2}{z_2 + z_3} = \frac{z_3^2}{z_{2oc}} \tag{1.24}
\end{aligned}
$$

$$z_3 = \pm\sqrt{z_{2oc}(z_{1oc} - z_{1sc})} \tag{1.25}$$

If the circuit in question has only resistances in it, then only the positive solution has any physical relevance. However, the negative sign has significance in AC circuits, where the negative sign implies an introduction of an $180°$ of phase difference.

$$\begin{aligned} z_1 &= z_{1oc} - z_3 \\ z_1 &= z_{1oc} \mp \sqrt{z_{2oc}(z_{1oc} - z_{1sc})} \end{aligned} \tag{1.26}$$

$$z_2 = z_{2oc} \mp \sqrt{z_{2oc}(z_{1oc} - z_{1sc})} \tag{1.27}$$

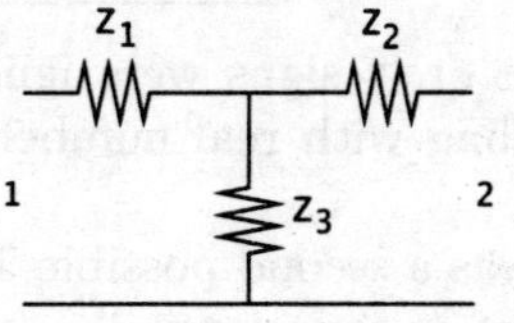

The steps of the above derivation could have proceeded by finding the difference of z_{2oc} and z_{2sc} which would have lead to the results

Figure 1.27: *A 'T'-section circuit.*

$$z_3 = \pm\sqrt{z_{1oc}(z_{2oc} - z_{2sc})}$$

$$z_1 = z_{1oc} + \sqrt{z_{1oc}(z_{2oc} - z_{2sc})}$$

$$z_2 = z_{2oc} \mp \sqrt{z_{1oc}(z_{2oc} - z_{2sc})}$$

This implies there are two possible equivalent circuits and the equations do not provide for a unique solution.

| Example 1.12: | The R_{1oc}, R_{1sc}, R_{2oc} and R_{2sc} resistances of a complex two port circuit are experimentally determined as 40/3KΩ, 20/3KΩ, 15KΩ and 7.5KΩ respectively. Find the equivalent T-section circuit.

Using eqn(1.25), eqn(1.26) and eqn(1.27) the three resistances can be calculated

$$\begin{aligned} R_3 &= \sqrt{R_{2oc}(R_{1oc} - R_{1sc})} \\ &= \sqrt{15\left(\frac{40}{3} - \frac{20}{3}\right)} \\ &= 10K\Omega \end{aligned}$$

Similarly,

$$\begin{aligned} R_1 &= R_{1oc} - R_3 \\ &= \frac{40}{3} - 10 \\ &= \frac{10}{3}K\Omega \end{aligned}$$

and

$$R_2 = R_{2oc} - R_3$$
$$= 5K\Omega$$

Answer $R_1 = 3.33K\Omega$, $R_2 = 5K\Omega$ and $R_3 = 10K\Omega$.

Note the $\pm$ or $\mp$ signs were ignored, since we are dealing with resistances or in other words we are dealing with real numbers. Physically, in a circuit, there is no such thing as negative resistance.

There was a second possible T-section as discussed above. Let us see the resulting T-section from this set of equations.

$$R_3 = \sqrt{R_{1oc}(R_{2oc} - R_{2sc})}$$
$$= \sqrt{\frac{40}{3}(15 - 7.5)}$$
$$= 10K\Omega$$

$$R_1 = R_{1oc} - R_3$$
$$= \frac{40}{3} - 10$$
$$= \frac{10}{3}K\Omega$$

$$R_2 = R_{2oc} - R_3$$
$$= 15 - 10 = 5K\Omega$$

That is one ends up with the same T-section equivalent circuit irrespective of which set of equations one uses.

The following example searches for an equivalent circuit of a seemingly complex circuits which has resistances and an inductor. Hence, the negative sign of eqn(1.25-1.27) can not be ignored. Before proceeding, it would be worthwhile to understand in figure(1.28) the inductor is marked to have an impedance of 4Ω. The inductors impedance is frequency dependent and hence this 4Ω is the inductors impedance at a particular frequency. The frequency is not mentioned as it is irrelevant for this example, however, whatever T-section equivalent circuit would be evaluated would be equivalent for that particular frequency. If one only knows the inductance of the inductor (in Henry), one while doing circuit analysis would have to compute ωL (in Ω), where ω is the frequency at which the analysis is to be done.

Example 1.13: Convert the shown circuit in figure(1.28) to an equivalent T-section equivalent circuit. The first step would be to calculate z_{1oc}, z_{1sc}, z_{2oc} and z_{2sc} impedance of the given circuit.

$$
\begin{aligned}
z_{1oc} &= \frac{4(2+4j)}{6+4j} \\
&= \frac{\sqrt{(8)^2 + (16)^2}\angle tan^{-1}(16/8)}{\sqrt{(6)^2 + (4)^2}\angle tan^{-1}(4/6)} \\
&= 2.48\angle 29.74^o \quad (polar) \\
&= 2.48cos(29.74) + j2.48sin(29.74) \\
&= 2.153 + 1.231j \quad (rectangular)
\end{aligned}
$$

Similarly,

$$
\begin{aligned}
z_{1sc} &= 2 + 2j \\
&- 2.828\angle 45^o
\end{aligned}
$$

$$
\begin{aligned}
z_{2oc} &= 1.538 + 0.3076j \\
&= 1.567\angle 11.31^o
\end{aligned}
$$

$$
\begin{aligned}
z_{2sc} &= 1.6 + 0.8j \\
&= 1.788\angle 26.56^o
\end{aligned}
$$

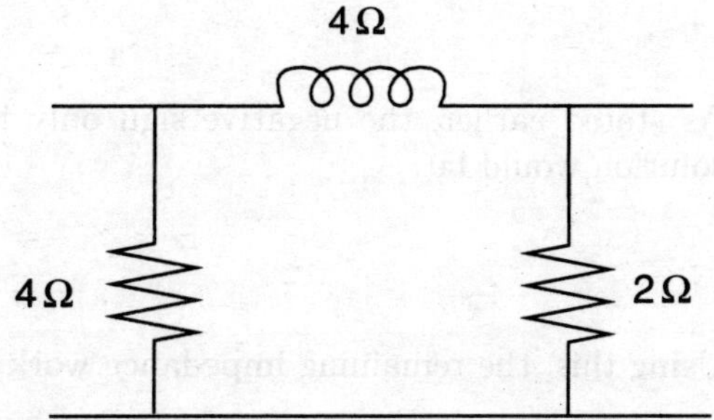

Figure 1.28: *Replace the shown circuit by a T-section circuit.*

Using these values and eqn(1.25)-(1.27) the three impedance can be calculated

$$
\begin{aligned}
z_3 &= \pm\sqrt{1.569\angle 11.31^o(2.153 + 1.231j - 2 - 2j)} \\
&= \pm\sqrt{1.569\angle 11.31^o(0.153 - 0.77j)} \\
&= \pm\sqrt{1.569\angle 11.31^o(0.153 - 0.77j)} \\
&= \pm\sqrt{1.569\angle 11.31^o(0.785\angle -78.76^o)} \\
&= \pm\sqrt{1.231\angle -67.38^o}
\end{aligned}
$$

Finding the square root of the above might look difficult but remember the complex number in its polar representation is a short form of Euler's expression

$$R\angle\theta = Re^{j\theta}$$

Hence, the under-root would be

$$
\begin{aligned}
\sqrt{R\angle\theta} &= \sqrt{R}e^{j\theta/2} \\
&= \sqrt{R}\angle\frac{\theta}{2}
\end{aligned}
$$

Hence, z_3 is

$$z_3 = \pm 1.109 \angle -33.69^o$$

The $\pm$ presents two possible solutions, firstly

$$z_3 = 1.109 \angle -33.69^o$$

or in rectangular representation

$$z_3 = 0.921 - 0.614j$$

giving z_1 and z_2 as

$$z_1 = 1.232 + 1.845j$$

$$z_2 = 0.616 + 0.922j$$

The second possibility would be

$$z_3 = -1.109 \angle -33.69^o$$

As stated earlier, the negative sign only implies a phase reversal of π, and hence the actual solution would be

$$z_3 = 1.109 \angle 33.69^o$$
$$= 0.921 + 0.614j$$

Using this, the remaining impedance work out to be

$$z_1 = 1.232 + 0.616j$$

$$z_2 = 0.616 - 0.307j$$

The first possibility gives an equivalent T-section with

> Answer $z_1 = (1.23 + 1.84j)\Omega$, $z_2 = (0.61 + 0.92j)\Omega$ & $z_3 = (0.92 - 0.61j)\Omega$.

The impedance z_1 results from a resistance of 1.232Ω being in series with an inductor of impedance 1.846Ω. Similarly, z_3 is a series combination of a 0.921Ω resistance and a capacitor of impedance 0.614Ω.

The second possibility found was

> Answer $z_1 = (1.23 + 0.61j)\Omega$, $z_2 = (0.61 - 0.30j)\Omega$ & $z_3 = (0.92 + 0.61j)\Omega$.

The second set of equations would similarly give two more possible T-section circuits.

The lack of an unique equivalent T-section circuit should not be considered a problem. For on the onset itself, our quest was to replace a complex circuit with a trivial and less complex one. Any of the T-section circuits would fulfill our requirement. We now proceed to study the second possible equivalent circuit which can be constructed with three impedance.

π-section

The steps of the last section is repeated again here after the required minimum three impedance are arranged in a different topography. The three impedance of fig(1.29) are arranged to look like the Latin alphabet π and is appropriately called the π-section equivalent circuit. The four impedance z_{1oc}, z_{1sc}, z_{2oc} and z_{2sc} of the complex black box circuit is experimentally determined and equation are set representing the same entities of the π-section circuit.

$$z_{1oc} = \frac{z_A(z_B + z_C)}{z_A + z_B + z_C} \tag{1.28}$$

$$z_{1sc} = \frac{z_A z_B}{z_A + z_B} \tag{1.29}$$

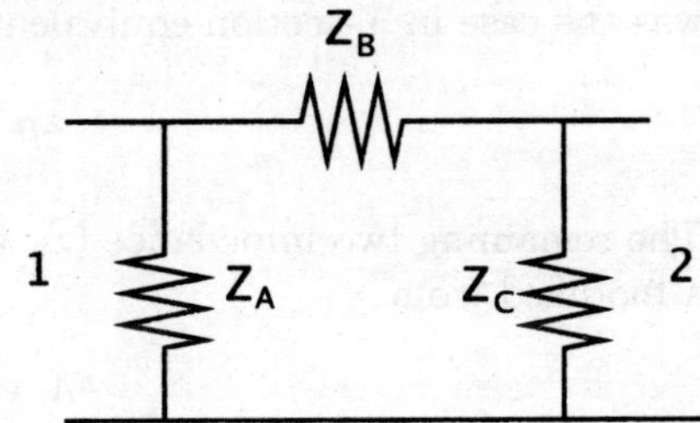

Figure 1.29: A 'π'-section circuit.

$$z_{2oc} = \frac{z_C(z_A + z_B)}{z_A + z_B + z_C} \tag{1.30}$$

$$z_{2sc} = \frac{z_B z_C}{z_B + z_C} \tag{1.31}$$

Similar arithmetic steps of the last section gives us the various impedance which go on to form the π section circuit.

$$
\begin{aligned}
z_{1oc} - z_{1sc} &= \frac{z_A(z_B + z_C)}{z_A + z_B + z_C} - \frac{z_A z_B}{z_A + z_B} \\
&= \frac{z_A^2 z_C}{(z_A + z_B + z_C)(z_A + z_B)}
\end{aligned}
$$

Multiplying the above equation with eq(1.30) would give an elegantly reduced expression, from which we can proceed to find the individual elements of our equivalent π circuit.

$$
\begin{aligned}
z_{2oc}(z_{1oc} - z_{1sc}) &= \frac{z_A^2 z_C}{(z_A + z_B + z_C)(z_A + z_B)} \times \frac{z_C(z_A + z_B)}{z_A + z_B + z_C} \\
&= \frac{z_A^2 z_C^2}{(z_A + z_B + z_C)^2}
\end{aligned}
$$

or

$$\pm\sqrt{z_{2oc}(z_{1oc} - z_{1sc})} = \frac{z_A z_C}{(z_A + z_B + z_C)} \tag{1.32}$$

To find the elements required for creating an equivalent π circuit we require terms like

$$\frac{z_A z_B z_C}{(z_A + z_B + z_C)}$$

This can be obtained from

$$z_{1sc}z_{2oc} = \frac{z_A z_B}{z_A + z_B} \times \frac{z_C(z_A + z_B)}{z_A + z_B + z_C}$$

$$= \frac{z_A z_B z_C}{z_A + z_B + z_C} = z_{1oc}z_{2sc} \qquad (1.33)$$

The second multiple shows a second set of equations can exist for an equivalent circuit, just as was the case in T-section equivalent circuits. Dividing eq(1.33) by eq(1.32) we get z_B

$$z_B = \pm \frac{z_{2oc}z_{1sc}}{\sqrt{z_{2oc}(z_{1oc} - z_{1sc})}} \qquad .(1.34)$$

The remaining two impedance (z_A and z_C) can be found using eqn(1.29) and eqn(1.31), both in a modified from

$$z_A = \frac{z_B z_{1sc}}{z_B - z_{1sc}}$$

$$= \frac{z_{2oc}z_{1sc}}{z_{2oc} \mp \sqrt{z_{2oc}(z_{1oc} - z_{1sc})}} \qquad (1.35)$$

Similarly

$$z_C = \frac{z_B z_{2sc}}{z_B - z_{2sc}}$$

$$= \frac{z_{1oc}z_{2sc}}{z_{1oc} \mp \sqrt{z_{2oc}(z_{1oc} - z_{1sc})}} \qquad (1.36)$$

$\boxed{\text{Example 1.14:}}$ Convert a T-section circuit having $R_1 = 60\Omega$, $R_2 = 60\Omega$ and $R_3 = 60\Omega$ into an equivalent π circuit.

First determine R_{1oc}, R_{1sc}, R_{2oc} and R_{2sc} of the given T-section circuit.

$$R_{1oc} = R_1 + R_3$$
$$= 120\Omega$$

$$R_{1sc} = R_1 + \frac{R_2 R_3}{R_2 + R_3}$$
$$= 120\Omega$$

$$R_{2oc} = R_2 + R_3$$
$$= 120\Omega$$

$$R_{2sc} = R_2 + \frac{R_1 R_3}{R_1 + R_3}$$
$$= 120\Omega$$

From these determined values we can find the equivalent pi-section circuit using eqn(1.34), eqn(1.35) and eqn(1.36)

$$\boxed{\text{Answer } R_1 = 3.33\text{K}\Omega, \ R_2 = 5\text{K}\Omega \text{ and } R_3 = 10\text{K}\Omega.}$$

Note the $\pm$ or $\mp$ signs were ignored, since we are dealing with resistances or in other words we are dealing with real numbers. Physically, in a circuit, there is no such thing as negative resistance.

The zeroth law of thermodynamics states that if A and B are in thermal equilibrium with C, then A is in thermal equilibrium with B. Similarly, if a circuit can be replaced by an equivalent T-circuit or equivalent π-circuit, then an T-circuit can be replaced by an equivalent π-circuit or visa-verse.

$$z_{1oc} = z_1 + z_3 = \frac{z_A(z_B + z_C)}{z_A + z_B + z_C} \tag{1.37}$$

$$z_{1sc} = z_1 + \frac{z_2 z_3}{z_2 + z_3} = \frac{z_A z_B}{z_A + z_B} \tag{1.38}$$

$$z_{2oc} = z_2 + z_3 = \frac{z_C(z_A + z_B)}{z_A + z_B + z_C} \tag{1.39}$$

$$z_{2sc} = z_2 + \frac{z_1 z_3}{z_1 + z_3} = \frac{z_B z_C}{z_B + z_C} \tag{1.40}$$

Following the same mathematical steps of the previous two sections we get

$$
\begin{aligned}
z_{1oc} - z_{1sc} &= z_1 + z_3 - z_1 - \frac{z_2 z_3}{z_2 + z_3} = \frac{z_A(z_B + z_C)}{z_A + z_B + z_C} - \frac{z_A z_B}{z_A + z_B} \\
&= \frac{z_3^2}{z_2 + z_3} = \frac{z_A^2 z_C}{(z_A + z_B + z_C)(z_A + z_B)}
\end{aligned}
\tag{1.41}
$$

Using eqn(1.39), we get

$$
\begin{aligned}
z_3^2 &= \frac{z_A^2 z_C}{(z_A + z_B + z_C)(z_A + z_B)} \times (z_2 + z_3) \\
&= \frac{z_A^2 z_C}{(z_A + z_B + z_C)(z_A + z_B)} \times \frac{z_C(z_A + z_B)}{z_A + z_B + z_C}
\end{aligned}
$$

giving

$$z_3 = \frac{z_A z_C}{(z_A + z_B + z_C)} \tag{1.42}$$

Using eqn(1.37) and eqn(1.42) we have

$$
\begin{aligned}
z_1 + z_3 &= \frac{z_A(z_B + z_C)}{z_A + z_B + z_C} \\
z_1 &= \frac{z_A(z_B + z_C)}{z_A + z_B + z_C} - \frac{z_A z_C}{(z_A + z_B + z_C)} \\
z_1 &= \frac{z_A z_B}{(z_A + z_B + z_C)}
\end{aligned}
\tag{1.43}
$$

Similarly, using eqn(1.39) and eqn(1.42) we get

$$z_2 = \frac{z_B z_C}{(z_A + z_B + z_C)} \tag{1.44}$$

Using above three equations (1.42-1.44), one can easily convert a π section to a T-section equivalent circuit. Similarly, it is sometimes useful in simplifying a circuit by converting a T-section circuit to an π-section equivalent circuit. This can be done eqn(1.38)

$$z_1 + \frac{z_2 z_3}{z_2 + z_3} = \frac{z_A z_B}{z_A + z_B}$$

$$\frac{z_1 z_2 + z_1 z_3 + z_2 z_3}{z_2 + z_3} = \frac{z_A z_B}{z_A + z_B}$$

$$z_1 z_2 + z_1 z_3 + z_2 z_3 = \frac{z_A z_B}{z_A + z_B} \times (z_2 + z_3)$$

Substituting the value of $z_2 + z_3$ in terms of the π circuital elements, using eqn(1.39)

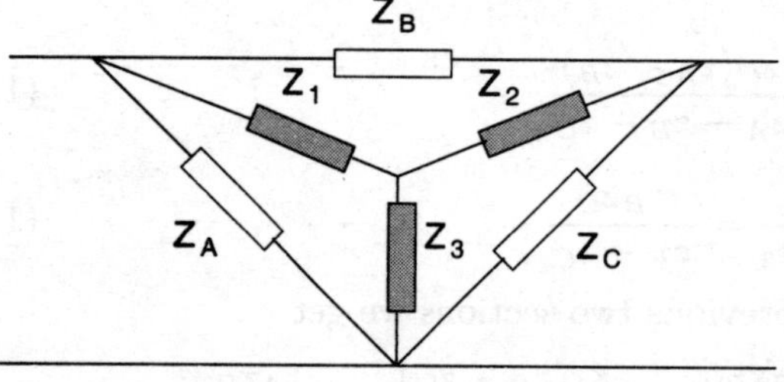

Figure 1.30: *Mnemonic figure to remember expression required to convert a T-section to π-circuit and visa verse.*

$$z_1 z_2 + z_1 z_3 + z_2 z_3 = \frac{z_A z_B z_c}{z_A + z_B + z_c}$$

Using eqn(1.42-1.44) one by one we get

$$z_A = \frac{z_1 z_2 + z_1 z_3 + z_2 z_3}{z_2} \tag{1.45}$$

$$z_B = \frac{z_1 z_2 + z_1 z_3 + z_2 z_3}{z_3} \tag{1.46}$$

$$z_C = \frac{z_1 z_2 + z_1 z_3 + z_2 z_3}{z_1} \tag{1.47}$$

Remembering all six equations (1.42-1.47) for converting a T-circuit to a π-circuit and visa verse would at first sight appear difficult. However, there are mnemonics to remember these equations. One feature to notice in both sets of equations is, in conversion of π to T equations (eqn 1.42-1.44), the denominator is the same. Similarly, for the equations (eqn 1.45-1.47) that is used to convert T-circuits to π, the numerator remains the same. If the two cases are remembered, it then becomes a case of remembering the varying part of the expression. For this, we look up fig(1.30).

When converting the π-circuit to T-section, the numerator has the product of the two neighboring impedance of the T-section impedance in question. That is, if one is calculating z_1 of the T-circuit, the two neighbors in fig(1.30) are z_A and z_B. Hence, the numerator has the product of z_A and z_B. Similar is the case of z_2 and z_3. For converting a T-circuit to its counter-part, the denominator of the expression will have the impedance of the T-circuit which is furthest from the π-circuits impedance in question. Thus, if one is calculating z_A, the furthest T-circuit impedance from z_A is z_2 (see fig 1.30). Hence, in the denominator of expression eqn(1.45) you have z_2. The remaining impedance can be found in this manner.

$\boxed{\text{Example 1.15:}}$ Convert the π-circuit shown in fig(1.28) into an equivalent T-section circuit.

We can now proceed to do so without following the methodology used in example and directly use eqn(1.42-1.44) for calculating the various impedance.

$$z_1 = \frac{z_A z_B}{(z_A + z_B + z_C)}$$
$$= \frac{4 \times 4j}{6 + 4j} = \frac{16 \angle 90^o}{7.21 \angle 33.7^o}$$
$$= 2.22 \angle 56.31^o = 1.23 + 1.84j$$

Similarly,

$$z_2 = \frac{z_B z_C}{(z_A + z_B + z_C)}$$
$$= \frac{8j}{6 + 4j}$$
$$= 0.61 + 0.92j$$

and

$$z_3 = \frac{z_A z_C}{(z_A + z_B + z_C)}$$
$$= \frac{8}{6 + 4j}$$
$$= 0.92 - 0.61j$$

$\boxed{\text{Answer } z_1 = (1.23 + 1.84j)\Omega, \ z_2 = (0.61 + 0.92j)\Omega \text{ and } z_3 = (0.92 - 0.61j)\Omega.}$

$\boxed{\text{Example 1.16:}}$ Convert the lattice network shown in fig(1.25) into an equivalent T-section circuit. The values of R_1, R_2, R_3 and R_4 are 4KΩ, 6KΩ, 5KΩ and 5KΩ respectively.

We proceed with the methodology used in characterizing a black box and find the various impedance related with the lattice network. To begin with we calculated the input impedance when the output is open. In that case R_1 comes in series with R_3 and this combination is parallel to the R_2 and R_4 combination, i.e.

$$z_{1oc} = \frac{(R_1 + R_3)(R_2 + R_4)}{R_1 + R_3 + R_2 + R_4}$$
$$= 4.95 K\Omega$$

On shorting the output pins (B and D are connected and are the same node). Then R_1 comes in parallel with R_2 and this combination is in series with the parallel combination of R_3 and R_4.

Hence,

$$z_{1sc} = \frac{R_1 R_2}{R_1 + R_2} + \frac{R_3 R_4}{R_3 + R_4}$$
$$= 4.9K\Omega$$

Similarly, the open and short circuited impedance on the output side may be calculated as

$$z_{2oc} = \frac{(R_1 + R_2)(R_3 + R_4)}{R_1 + R_3 + R_2 + R_4}$$
$$= 5K\Omega$$

$$z_{2sc} = \frac{R_1 R_3}{R_1 + R_3} + \frac{R_2 R_4}{R_2 + R_4}$$
$$\sim 4.95K\Omega$$

From these four impedance the required z_1, z_2 and z_3 for the T-section can be calculated using eqn(1.25-1.27). Eqn(1.25) allows a negative impedance ($\pm$), however, since the example deals only with resistances we can discard the negative possibility.

$$\boxed{\text{Answer } z_1 = 4.45K\Omega, \ z_2 = 4.5K\Omega \text{ and } z_3 = 0.5K\Omega.}$$

1.9   Resistance Coding

The magnitude or value of the resistance offered by the carbon film is decided by the first three color bands painted on it (see fig 1.31). The information is decoded as follows

$$R = (First\ Color\ Band)(Second\ Color\ Band) \times 10^{(Third\ Color\ Band)}$$

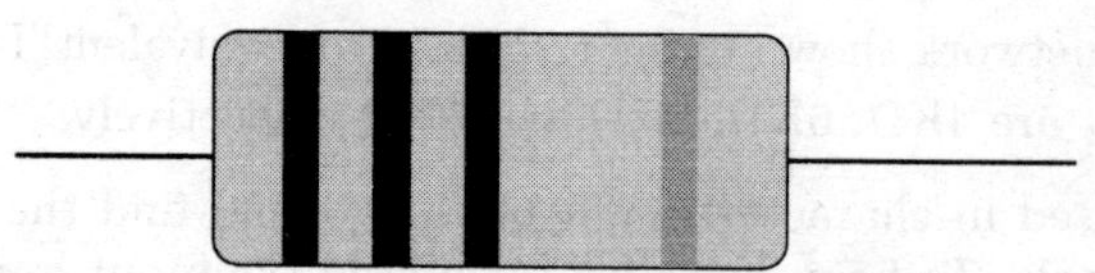

Figure 1.31: *Color coding is used in carbon resistances to iden-tity it's value.*

where each color is associated with a number. The color and its associated numbers are listed in Table(1.1). Hence if the colors on the resistance are Red, Red, Orange, the value would be (use the table for looking up now, however, it is best to memorize it)

$$R = 22 \times 10^3 = 22K\Omega$$

As another example consider the coloring on the resistance to be orange, white and red, then the value would be

$$R = 39 \times 10^2 = 3.9K\Omega$$

There is a forth band called the tolerance band which is represented by four colors; brown (1%), red (2%), gold (5%) and silver (10%). In case there is no forth band, the tolerance is (20%).

Table 1.1

Color	Value
Black	0
Brown	1
Red	2
Orange	3
Yellow	4
Green	5
Blue	6
Violet	7
Grey	8
White	9

A regular circuit designer should remember the values associated with these colors. There are short sentences which help one remember the sequencing, for example:

- **B**ad **B**eer **R**ips **O**ur **Y**oung **G**uts **B**ut **V**odka **G**ives **W**ell, and

- **B. B. ROY** of **G**reat **B**riton has a **V**ery **G**ood **W**ife.

1.10 Capacitor and Their Coding

A capacitor is a device that stores electrical charges or energy. They are made by placing two metal plates very close together. The charge or amount of electricity stored on the plates is proportional to the applied voltage and the capacitor's capacitance, i.e.

$$Q = C \times V$$

where 'C' is the capacitance (in Farads). The capacitance of a capacitor is given as

$$C = \epsilon \frac{A}{d}$$

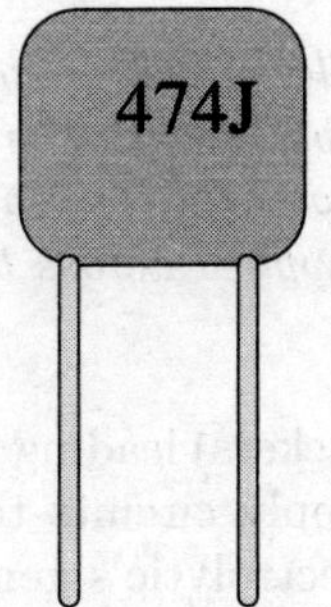

Figure 1.32: *Ceramic capacitor with coding written on it.*

The equation implies that the larger the plate area (A) and smaller the distance between the plates (d), larger is the capacitance. This would mean for storing large amount of charges, we would require very large plates and in turn very large capacitors. However, the capacitance can be increased by inserting an insulating material or dielectric material between the two plates. The insulator's dielectric constant (ϵ) would be more than that of air ($\epsilon_o = 85 \times 10^{12}$F/m or 85pF/m). As can be seen this number is quite small, hence material's dielectric constant is usually reported as multiples of that of air, i.e. $\epsilon_r\epsilon_o$. Table 1.2 gives ϵ_r of some commonly used dielectrics.

Table 1.2

Color	Value
Teflon	2.10
Paper	3.00
Bakelite	4.90
Porcelain	5.5
Glass	7.75

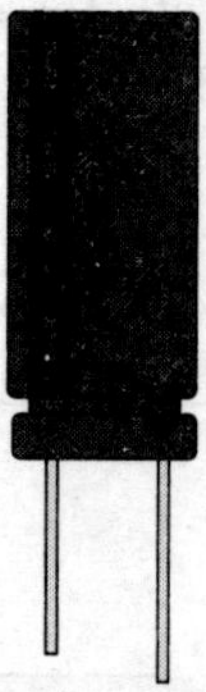

Based on the type of dielectric used, the capacitors come in a variety of models, shapes and sizes, for example electrolytic, ceramic and polyester capacitors. The name suggests the type of dielectric used. Electrolytic capacitors have a semi-liquid electrolyte solution in the form of a jelly or paste which serves as the second electrode (usually the cathode). The dielectric is a very thin layer of oxide which is grown electrochemically in production with the thickness of the film being less than ten microns. The majority of electrolytic types of capacitors are polarized, hence the voltage applied to the capacitor terminals must be of the correct polarity. An incorrect polarization will break down the insulating oxide layer resulting in excesive heating (sometimes they

Figure 1.33: *Electrolytic capacitor with coding written on it. Note the '-ive' sign shows pin to be kept at lowest potential and also maximum voltage that can be applied across the capacitor is also indicated.*

burst like fire crackers) leading to permanent damage. Electrolytic Capacitors are generally used in DC power supply circuits to help reduce the ripple voltage or for coupling and decoupling applications. Electrolytic's generally come in two basic forms, the Aluminum Electrolytic and Tantalum Electrolytic capacitors.

Table 1.3

3rd Digit	Multiplier	Letter	Tolerance (%)
0	1	D	0.5
1	10	F	1
2	10^2	G	2
3	10^3	H	3
4	10^4	J	5
5	10^5	K	10
		M	20

While the value of the capacitance is directly written on an electrolytic capacitor, in the small capacitors it becomes difficult to write the value. Hence codes are used. Usually in ceramic capcitors a three digit number is used. If a capacitor is marked 105, it would mean 10+5 zeros or 10,00000pF. That default the unit is in pF and 105 would mean 10^6pF of 1μF. Sometimes a letter is added to represent the value of its tolerance (see Table 1.3). So 474J would mean a 0.47μF capacitor of 5% tolerance.

Exercise

Q1. Find the 'potential at' and 'current through' node A (V_A) of circuit (fig 1.34).

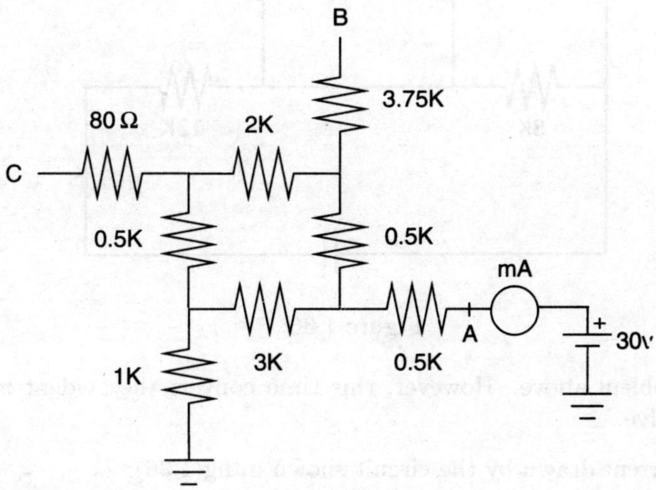

Figure 1.34:

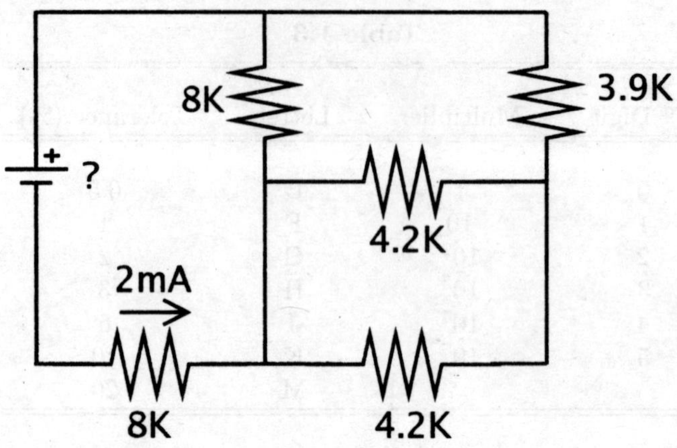

Figure 1.35:

Q2. State and prove the principle of Duality.

Q3. What voltage source should be applied in the circuit of fig(1.35) so that a net current of 2mA is drawn by the circuit.

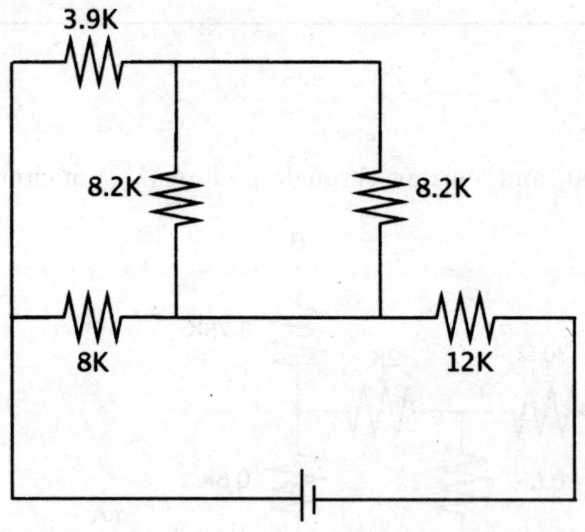

Figure 1.36:

Q4. Repeat the problem above. However, this time convert the evident π network to a 'T' network and solve.

Q5. What is the current drawn by the circuit shown in fig(1.36).

Q6. Compute the value of resistance R_3 and the applied voltage source of the circuit shown in fig(1.37).

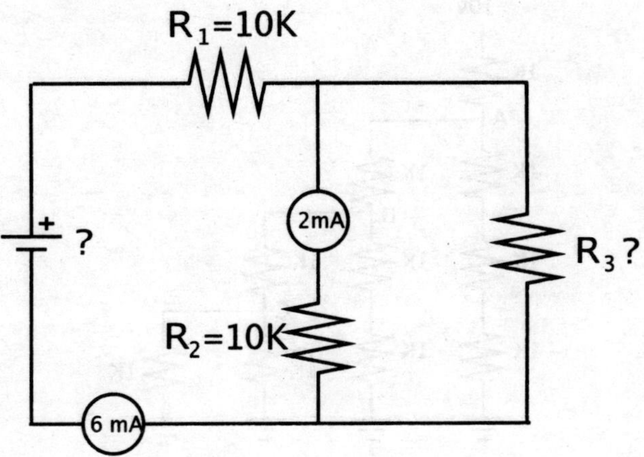

Figure 1.37:

Q7. Compute the value of load current in the circuit shown in fig(1.38).

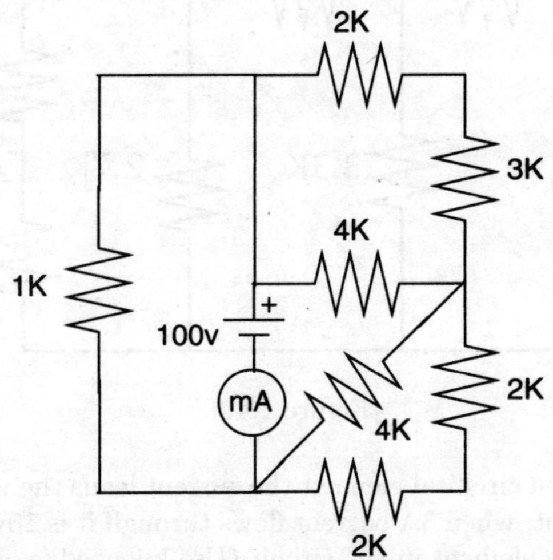

Figure 1.38:

Q8. Calculate the potential drops, V_A and V_B for the given circuit (fig 1.39).

Q9. Find the potential at node A (V_A) of circuit (fig 1.40).

Q10. Find the potential source used in the given circuit (fig 1.41).

Q11. What is the DC output voltage of the DAC (fig 1.42) for the given digital input 0100?

Q12. What is meant by an ideal voltage and ideal current source.

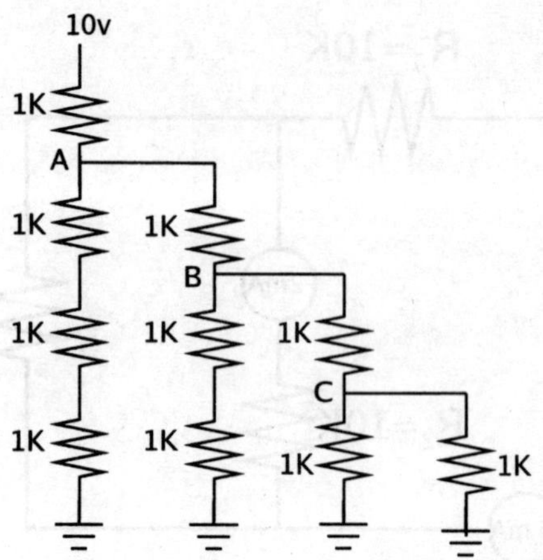

Figure 1.39:

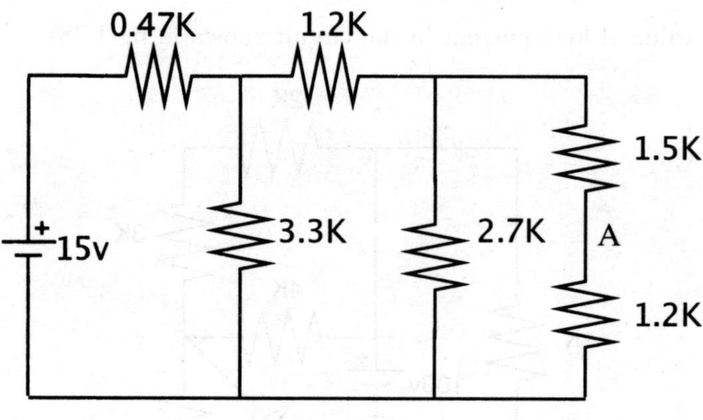

Figure 1.40:

Q13. Due to the presence of a circuital element the current leads the voltage by 90°. The volatge drop across the element, when 5A current flows through it is 10v. Identify the element and report the value of the element in the circuit (Use knowledge of complex numbers learnt).

Q14. Write the physical interpretation of the Kirchhoff's current law, i.e. at a given node $\sum_i I_i = 0$.

Q15. If two circuits are equivalent, is the same power dissipated in both? Prove.

Q16. List the difference between active and passive devices.

Q17. Convert the rectangular represented complex number 4+j3 into polar representation.

Q18. If the input voltage is $25\sin(\omega t + 30°)$, represent this as a complex number in polar and rectangular representation.

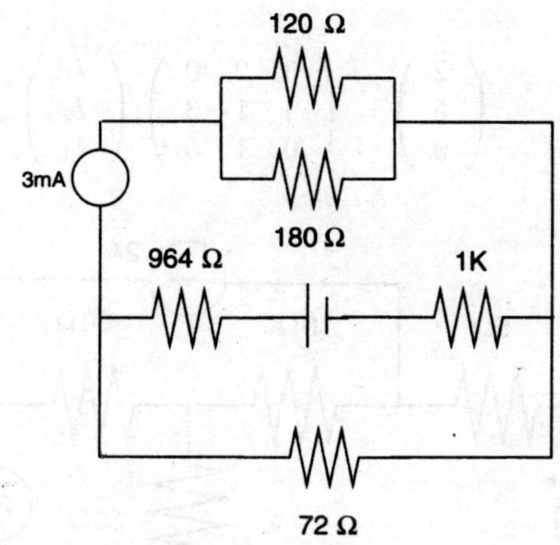

Figure 1.41:

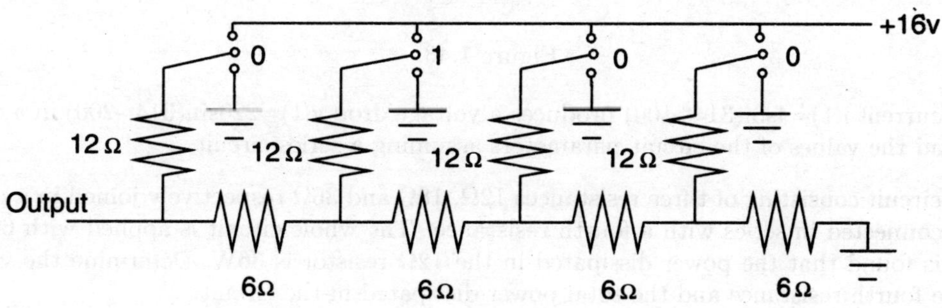

Figure 1.42:

Q19. On applying the voltage given in Q18 to an element, the current through the element is $2\cos(\omega t + 45°)$. What is the impedance offered by the element?

Q20. If the 'T' section of fig(1.20) is converted into a π section, the current source can be converted into a voltage source. Solve the new circuit using Mesh Analysis. Is the result of I_1, I_2 and I_3 same as calculated earlier.

Q21. Explain any difference in currents calculated in Q20.

Q22. Draw a circuit corresponding to the following Mesh matrix equation

$$\begin{pmatrix} 6 \\ 3 \end{pmatrix} = \begin{pmatrix} 3 & 2 \\ 2 & 3 \end{pmatrix} \begin{pmatrix} I_1 \\ I_2 \end{pmatrix}$$

Also, calculate the two currents.

Q23. Draw a circuit corresponding to the following Mesh matrix equation

$$\begin{pmatrix} 2 \\ 5 \\ 0 \end{pmatrix} = \begin{pmatrix} 3 & 2 & 0 \\ 4 & 1 & 3 \\ 0 & 1 & 5 \end{pmatrix} \begin{pmatrix} I_1 \\ I_2 \\ I_3 \end{pmatrix}$$

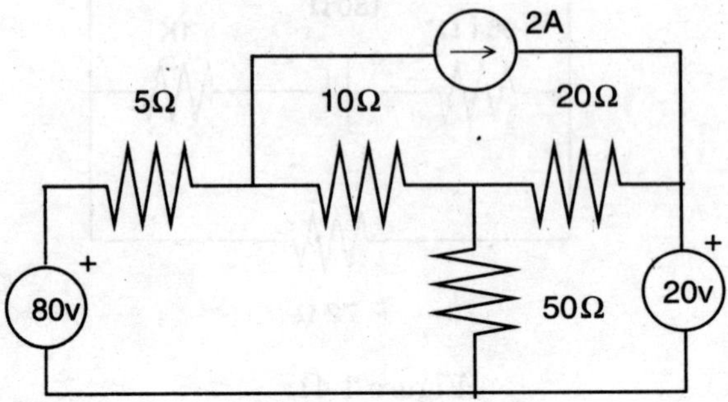

Figure 1.43:

Q24. A current i(t)=4sin(314t-100) produces a voltage drop v(t)=220sin(314t-200) in a circuit. Find the values of the circuit parameters assuming a series circuit.

Q25. A circuit consisting of three resistances 12Ω, 18Ω and 36Ω respectively joined in a parallel is connected in series with a fourth resistance. The whole circuit is applied with 60v and it is found that the power dissipated in the 12Ω resistor is 36W. Determine the value of the fourth resistance and the total power dissipated in the circuit.

Q26. Two resistors $R_1 = 2.5K\Omega$ and $R_2 = 4K\Omega$ are joined in series and connected to a 100v supply. The voltage drop across R_1 and R_2 are measured successively by a voltmeter having a resistance of 50KΩ. Find the sum of the readings.

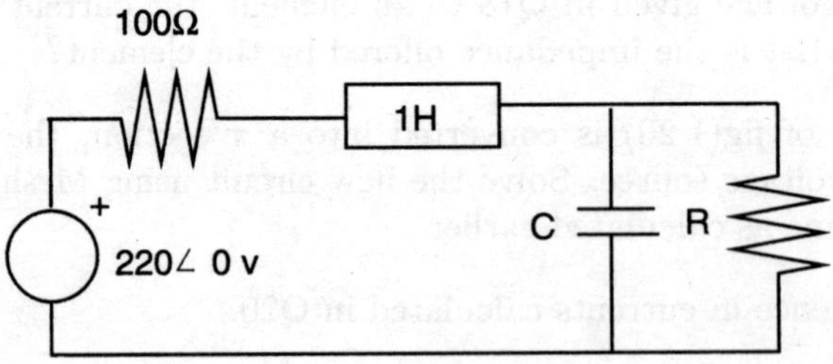

Figure 1.44:

Q27. Define

    (a) Frequency

    (b) Phase difference

Q28. Derive expressions for current and power in a capacitor supplied with an alternating sinusodial voltage.

Q29. Find the potential across the resistances 10Ω and 20Ω of circuit shown in fig(1.43). Show polarity also.

Q30. In the circuit shown in fig(1.44), determine the value of load impedance (combination of R & C) so that maximum power is transferred to load. State the power absorbed by the load.

Chapter 2

Theorems

Any circuit in principle can be analyzed using the Kirchhoff's laws. The two rules provide the basic framework to solve and understand the circuit. However, this usually involve solving simultaneous equation whose complexity (in terms of order) increases with increasing number of loops. Various additional theorems are now introduced which aid to solve and analyze circuits faster and elegantly. The following theorems will help in trivializing the calculations, and also in two port networks would allow to replace complex circuits with equivalent voltage and current sources. Thus, large circuits maybe reduced to circuits with single energy sources and impedance, along with the load impedance, whose load current and output voltage would be of interest.

2.1 Superposition Theorem

In the various methods we studied for solving various circuits, we ended up solving simultaneous equations. Depending on ones choice one ended up using 2×2 or 3×3 determinant or matrix for 2 or 3 loop circuits respectively. In circuits with more loops, the degree of difficulty increased with increasing size of matrix or determinant. Superposition Theorem presents a convenient method to determine circuital currents or voltages without requiring one to solve a simultaneous equation.

The current in or the voltage across any linear circuital device is equal to the algebraic sum of the currents or voltages produced independently by each source.

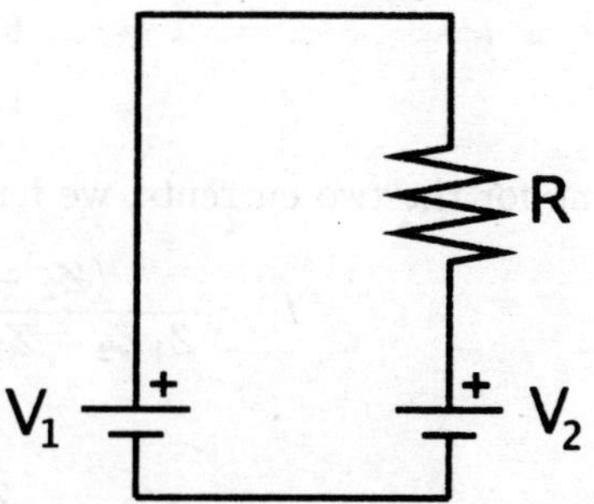

Figure 2.1: *The current flowing through the resistance* R $= 5\Omega$ *can be considered to be the algebraic sum of currents flowing due to the voltage sources* $V_1 = 15v$ *and* $V_2 = 10v$ *acting independently.*

Consider the simple circuit shown in figure(2.1). Without any theorem or elaborate calculation, one can see that the two voltage source are in anti-series to each other and hence the net voltage in the circuit is 15-10=5 volts, giving rise to a 1A current through the resistance.

However, let us apply superposition theorem to this trivial circuit.

We have to establish the current through the resistance resulting from the voltage supplies acting independently. Hence we have to remove one voltage source and calculate the current due the voltage source left in the circuit. For removing an energy source from a circuit, in case of voltage source you short circuit the source (see fig 2.2a). However, in case of a current source for removal you have to leave connections to source open (see fig 2.2b). Thus, the 10v source is short circuited (fig 2.2c). The current in the 5Ω resistance due to the lone 15v source is 3A in the direction shown. Next introduce the 10v supply back into the circuit and remove the 15v source. The current through the resistance is 2A in the direction shown in fig 2.2d. The net current is the circuit is the algebraic sum of these two currents, 3-2=1A. This might appear as an elaborate theorem to solve something as trivial as the above circuit, however fur-

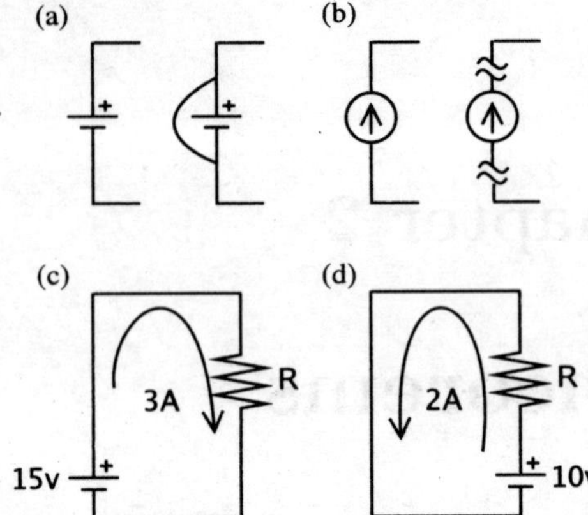

Figure 2.2: *The (a) voltage sources when are to be removed, are shorted while (b) the current sources are left open circuited when they have to be removed. On applying superposition theorem on circuit shown in fig(2.1), (c) 10v source is shorted and then (d) 15v source is removed and respective currents are computed.*

ther examples will show the use of this theorem. Before that, we proceed to prove this theorem.

Consider the circuit shown in figure(2.3). We can find the current through Z_3 using any of the methods introduced in Chapter 1. Here we analyze the circuit using loop method.

$$V_1 = (Z_1 + Z_3)I_1' + Z_3 I_2'$$
$$V_2 = Z_3 I_1' + (Z_2 + Z_3)I_2'$$

Solving for the two currents, we have

$$I_1 = \frac{(Z_2 + Z_3)V_1}{Z_1Z_2 + Z_1Z_3 + Z_3Z_2} - \frac{Z_3V_2}{Z_1Z_2 + Z_1Z_3 + Z_3Z_2}$$

and

$$I_2 = -\frac{Z_3V_1}{Z_1Z_2 + Z_1Z_3 + Z_3Z_2} + \frac{(Z_1 + Z_3)V_2}{Z_1Z_2 + Z_1Z_3 + Z_3Z_2}$$

Thus, the net current through Z_3 is

$$I = I_1 + I_2 = \frac{Z_2V_1}{Z_1Z_2 + Z_1Z_3 + Z_3Z_2} + \frac{Z_1V_2}{Z_1Z_2 + Z_1Z_3 + Z_3Z_2} \tag{2.1}$$

Now we investigate whether the methodology adopted by Superposition Theorem also comes to the same conclusion. First we remove V_2 from the circuit and find the current through impedance Z_3.

$$V_1 = (Z_1 + Z_3)I_1' + Z_3 I_2'$$
$$0 = Z_3 I_1' + (Z_2 + Z_3)I_2'$$

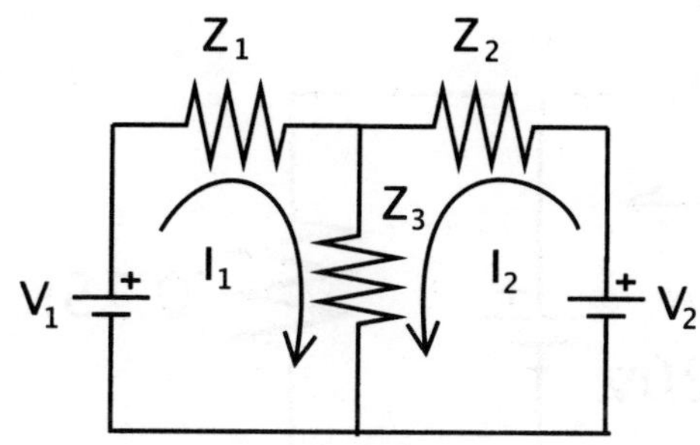

Figure 2.3: *Since any complex circuit can be reduced to a T section equivalent circuit, we consider such a 'T' circuit to verify the superposition theorem. If it is true for this circuit, then it holds for any circuit.*

Solving the above equation, we get the current through Z_3 due to V_1 as $I_{V_1} = I_1' + I_2'$

$$= \left[\frac{(Z_2 + Z_3)V_1}{Z_1 Z_2 + Z_1 Z_3 + Z_3 Z_2} \right] + \left[-\frac{Z_3 V_1}{Z_1 Z_2 + Z_1 Z_3 + Z_3 Z_2} \right]$$

Repeating the above exercise, we remove V_1 from the circuit to find current through Z_3 due to source V_2.

$$0 = (Z_1 + Z_3)I_1'' + Z_3 I_2''$$
$$V_2 = Z_3 I_1'' + (Z_2 + Z_3)I_2''$$

current due to V_2 is

$$I_{V_2} = I_1'' + I_2''$$

$$= \left[-\frac{Z_3 V_2}{Z_1 Z_2 + Z_1 Z_3 + Z_3 Z_2} \right] + \left[\frac{(Z_1 + Z_3)V_2}{Z_1 Z_2 + Z_1 Z_3 + Z_3 Z_2} \right]$$

The net current in the impedance due to both sources is $I_{V_1} + I_{V_2}$. This works out to be the same as that obtained in eqn(2.1) (remember that was obtained by Mesh analysis on which we have complete faith). Hence, superposition theorem is proved.

Consider just for the sake of argument, a new Ohms law exists of the form

$$V = RI^2$$

if we now use the same mathematical steps as above to prove Superposition theorem, we would get

$$(I_1 + I_2)^2 = I_1^2 + I_2^2$$

which is not true. Hence, Superposition theorem only holds because of the linearity of Ohms law. This explains why the statement emphasis that the law is applicable only to linear circuits.

In general, there might be 'N' energy sources in the circuit. To use superposition theorem, remove '(N-1)' sources and calculate effect due to the left source. Repeat steps one at a time for all 'N' sources. Let us develop a better understanding of this theorem by looking into various examples.

Example 2.1: | Use superposition theorem to determine current I_o in fig(2.4).

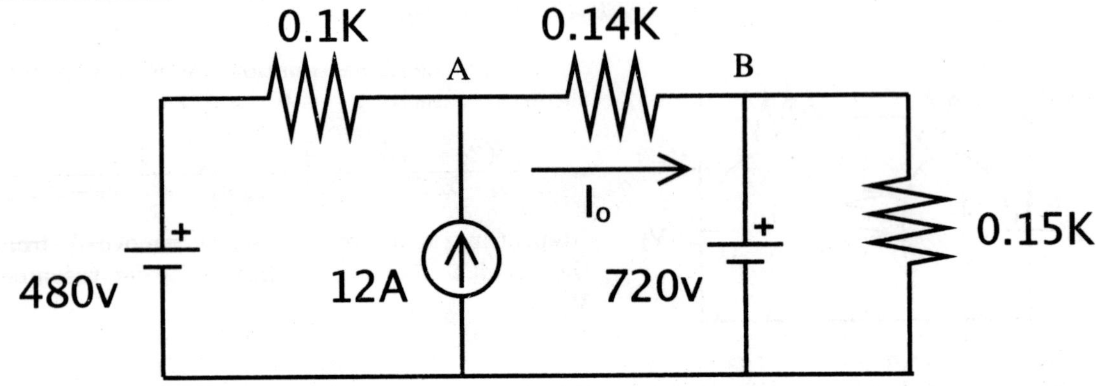

Figure 2.4: *Circuit for Example 2.1*

There are three energy sources in the given circuit. Thus, to use superposition theorem, two of the energy sources have to be removed for any given calculation. Let us start by removing the two voltage sources.

> ***Removing voltage sources, involves shorting them.***

That is, you will have to visualize it as if a connecting wire is kept across it. The 150Ω resistance comes in parallel with one of the shorts and hence does not contribute. The 12A current breaks into two currents, one flowing in 100Ω and the second in the 140Ω resistance. Using the current division equation, the current in 140Ω due to the current from the current source is given as

$$I_1 = \frac{100}{100 + 140} \times 12 = 5A$$

This current flows from node 'A' to 'B'.

Now assume the 720v and 12A supply is to be removed. The voltage source of course is shorted, but for removing the current source, it is removed by making it's path open.

> ***Removing current sources, involves opening them.***

The shorting of 720v leaves the 150Ω resistance still in parallel with the short and hence does not contribute. The circuit resulting circuit is two resistances (100Ω and 140Ω) in series with the 480v source. Hence the current due to this source is

$$I_2 = \frac{480}{100 + 140} = 2A$$

This current also flows from node 'A' to 'B'.

Finally, 480v and 12A supply is to be removed. Immediately, it is obvious that the 720v source pins the potential across the 150Ω resistance at 720v. However, the 720v is divided across 100Ω and 140Ω. Hence, using the potential divider equation, the voltage across the 140Ω resistance is

$$V = \frac{140}{100 + 140} \times 720 = 420v$$

Hence, the current through it is 3A ($420v/140\Omega$), however, this current flows from node 'B' to 'A'. The net current I_o (based on the worked directions) is

$$\begin{aligned} I_o &= I_1 + I_2 - I_3 \\ &= 5 + 2 - 3 = 4A \end{aligned}$$

$$\boxed{\text{Answer } I_o = 4A}$$

2.2   Thevenin's Theorem

By Thevenin's Theorem a complex circuit is reduced to an equivalent series circuit consisting of a single voltage source, V_{TH}, a series impedance, Z_{TH}, and a load impedance, Z_L. After creating the Thevenin equivalent circuit, you may then easily determine the load voltage V_L or the load current I_L.

Any voltage network which may be viewed from two terminals can be replaced by a voltage-source equivalent circuit comprising a single voltage source V_{TH} and a single series impedance Z_{TH}. The voltage V_{TH} is the open-circuit voltage between the two terminals and the impedance Z_{TH} is the impedance of the network viewed from the terminals with all energy sources removed from circuit.

Thus, what we require is a complex test circuit. What would be the most complex complex circuit one can think of? Figure(1.27) shows a possible complex circuit suitable to understand the Thevenin's Theorem. One might feel that the T-section is not complex enough to be worthy of ones effort. But remember we have already seen in the last chapter, any complex circuit can be replaced with an equivalent circuit. So, we have just done that, we considered a very complex circuit and found it's T-equivalent circuit and now proceed to understand Thevenin's Theorem with this circuit.

Now let us examine the Thevenin's Theorem statement. It states the steps required to find an equivalent circuit. As per the theorem, first remove all energy sources. In our circuit (figure 2.5), we short the voltage source and in place of the load impedance place a virtual multimeter (figure 2.5a), i.e. we ask what would be impedance of the circuit as seen by the load. This is the Thevenin's equivalent impedance Z_{TH}. As can be seen from fig(2.5a), the Thevenin's impedance works out as

$$Z_{TH} = Z_2 + \frac{Z_1 Z_3}{Z_1 + Z_3}$$

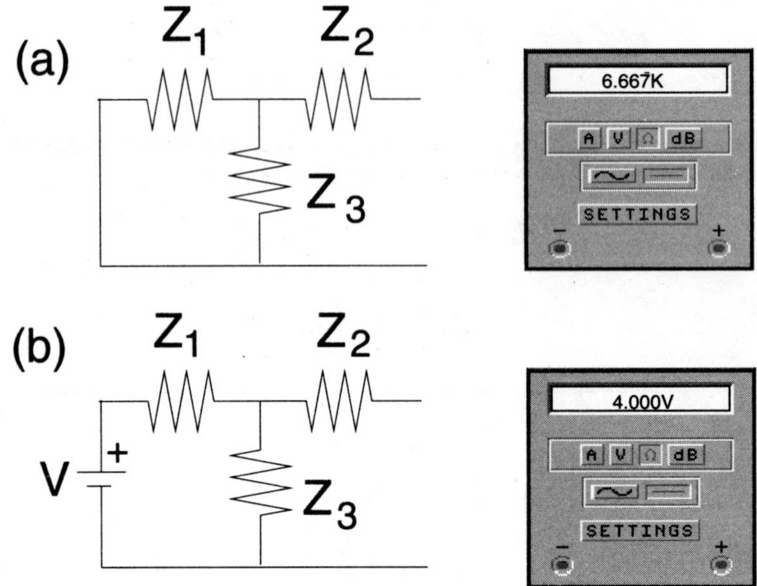

Figure 2.5: *The Thevenin's equivalent circuit is found by (a) removing all energy sources and finding the impedance as seen by the output side. As an example consider Z_1, Z_2 and Z_3 as 8KΩ, 4KΩ, and 4KΩ respectively. The second step (b) involves finding the open circuited output voltage, assume v=12volts.*

For the given values for impedance (given in caption of fig 2.5), the multimeter reads 6.667KΩ. The second step is, bring back the energy sources, and ask what would be the voltage as seen by the load. This would be the Thevenin's equivalent voltage V_{TH}. Going back to the statement of the theorem, it refers to this voltage as the open-circuit voltage. This implies what would be the voltage as seen by an infinite load drawing zero current.[1] So from figure 2.5(b), the open-voltage would be the voltage drop across Z_3 giving the Thevenin's equivalent voltage V_{TH} as

$$V_{TH} = \frac{Z_3 V}{Z_1 + Z_3}$$

For the 12v supply of figure 2.5, the multimeter reads 4.0volts. Thevenin's Theorem states that instead of considering the complex circuit shown, use a simple circuit of 4volts with a impedance of 6.67KΩ in series to it fig(2.6). The load would not be any wiser, since both circuits would give the same voltage drop across and same current through it. The current through a load Z_L would be

$$I_R \quad = \quad \frac{V_{TH}}{Z_L + Z_{TH}} \tag{2.2}$$

---

[1] Imagine you are measuring the voltage using a multimeter by placing it where the load impedance was. Also remember ideally multimeters would have infinite input impedance and hence draw no current

We now have to prove Thevenin's Theorem. Since we are totally confident of the Kirchhoff' laws and the methods which use Kirchhoff's rules to solve circuits, we use Mesh analysis to find current in Z_L of fig(2.6a)

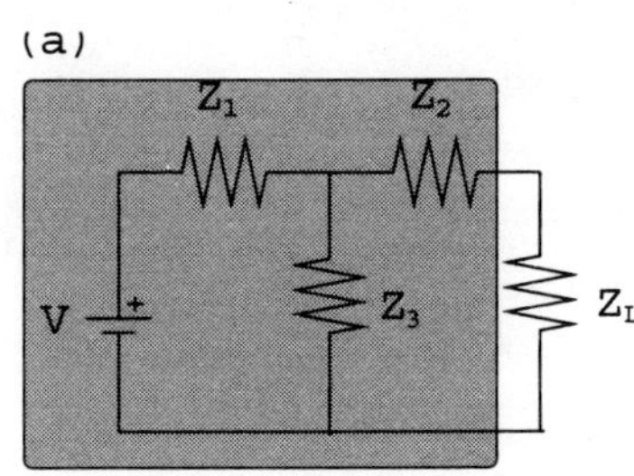

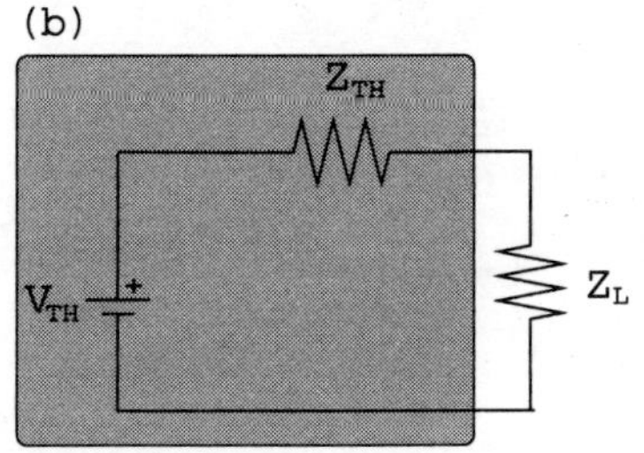

Figure 2.6: *The complex voltage network shown by the shaded region of (a) can be replaced by an equivalent Thevenin's circuit (shaded region of b) which would deliver the same current and voltage as the starting network.*

$$V = Z_1 I_1 + Z_3(I_1 - I_R)$$
$$0 = (Z_2 + Z_L)I_R + Z_3(I_R - I_1)$$

or

$$V = (Z_1 + Z_3)I_1 - Z_3 I_R$$
$$0 = -Z_3 I_1 + (Z_2 + +Z_3 + Z_L)I_R$$

giving

$$I_R = \frac{VZ_3}{(Z_1 + Z_3)(Z_2 + Z_3 + Z_L) - Z_3^2}$$
$$= \frac{VZ_3}{Z_2(Z_1 + Z_3) + Z_L(Z_1 + Z_3) + Z_3(Z_1 + Z_3) - Z_3^2}$$
$$= \frac{VZ_3}{Z_2(Z_1 + Z_3) + Z_L(Z_1 + Z_3) + Z_1 Z_3}$$

Dividing the above equation by $(Z_1 + Z_3)$, we have

$$I_R = \frac{V\left(\frac{Z_3}{Z_1+Z_3}\right)}{Z_2 + Z_L + \left(\frac{Z_1 Z_3}{Z_1+Z_3}\right)} \tag{2.3}$$

Substituting,

$$Z_{TH} = Z_2 + \frac{Z_1 Z_3}{Z_1 + Z_3}$$

$$V_{TH} = \frac{Z_3 V}{Z_1 + Z_3}$$

Equation(2.3) can be re-written as

$$I_R = \frac{V_{TH}}{Z_L + Z_{TH}} \tag{2.4}$$

The circuit corresponding to eqn(2.4) is shown in fig(2.6b). As far as the load impedance, Z_L is concerned the shaded region of fig(2.6b) is identical to the circuit preceding it in fig(2.6a). These expressions obtained from Mesh analysis are identical to those obtained from Thevenin's Theorem, eqn(2.2), which in turn gave us an equivalent circuit. ***Thus, the Thevenin's Theorem directly gives us a voltage source equivalent circuit for any complex voltage network.***

Example 2.2: Using Thevenin's theorem, reduce the circuit preceding R_L (fig 2.7) to a single voltage source and resistance.

The resistance R_L is removed and then the voltage source is shorted. The resistance as measured by a multimeter with infinite input impedance at terminal 'AB' is given as

$$R_{TH} = \frac{R_1 R_3}{R_1 + R_3} + \frac{R_2 R_4}{R_2 + R_4}$$

This is the Thevenin equivalent resistance. Fig(2.8), shows the equivalent circuit on shorting the voltage source, from which R_{TH} was calculated.

The Thevenin equivalent voltage is the open loop voltage, i.e. again measured when the resistance R_L between node 'A' and 'B' is removed. The potential at node 'A' with respect to ground is given by the potential divider equation

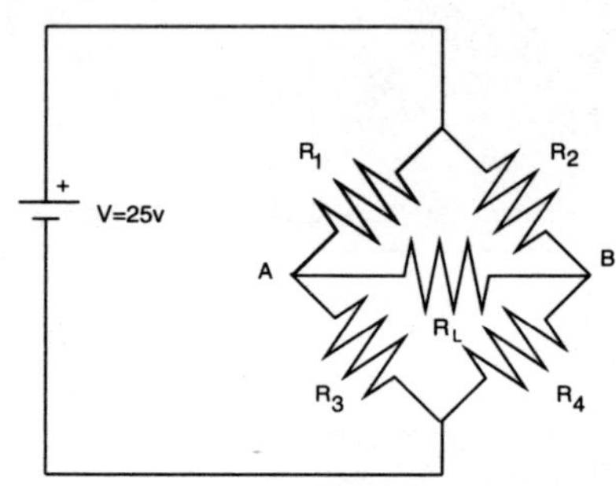

Figure 2.7: *Circuit for Example 2.2*

$$V_A = \left(\frac{R_3}{R_1 + R_3} \right) 25v$$

Similarly, the potential at 'B' is

$$V_B = \left(\frac{R_4}{R_2 + R_4} \right) 25v$$

The loop equation from which we get the required Thevenin voltage (in this case V_{AB}) is

$$V_A = V_{AB} + V_B$$

giving

$$
\begin{aligned}
V_{AB} &= V_A - V_B \\
&= \left[\frac{R_3}{R_1 + R_3} - \frac{R_4}{R_2 + R_4} \right] 25v \\
V_{TH} &= \left[\frac{R_2 R_3 - R_1 R_4}{(R_1 + R_3)(R_2 + R_4)} \right] 25v
\end{aligned}
$$

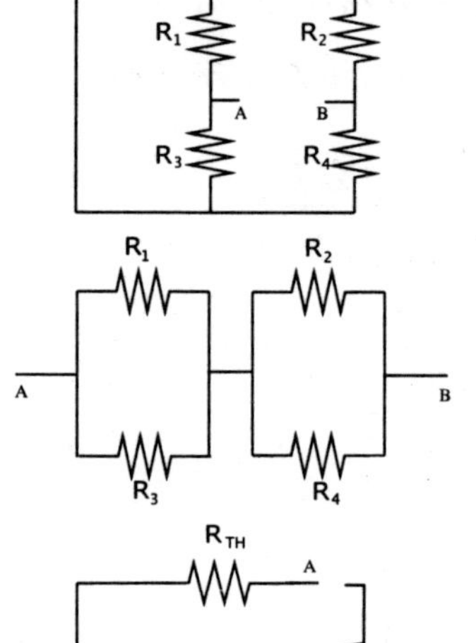

Figure 2.8:

$$\boxed{\text{Answer } V_{TH} = \left[\frac{R_2 R_3 - R_1 R_4}{(R_1 + R_3)(R_2 + R_4)} \right] 25v \text{ and } R_{TH} = \frac{R_1 R_3}{R_1 + R_3} + \frac{R_2 R_4}{R_2 + R_4}}$$

The Thevenin equivalent circuit can be drawn using these values and represented as shown in fig(2.8).

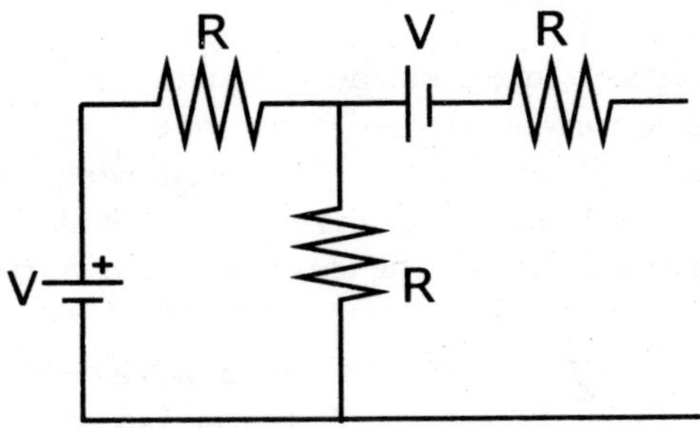

Figure 2.9: *Circuit for Example 2.3*

Example 2.3: Find the Thevenin equivalent for the circuit shown in fig(2.9), given V=25*volts* and R = 10Ω.

On removal of energy sources, the resistances are so arranged that, two of them come in parallel with the third resistance in series to this parallel combination. Hence R_{TH} is

$$R_{TH} = 10 + \frac{10 \times 10}{10 + 10} = 15\Omega$$

The circuit can be viewed as two circuits, with the first loop having two resistances in series. It would act as a potential divider, giving voltage across each resistance as 12.5*volts*. Infact, the loop can be thought as being replaced with a Thevenin equivalent circuit with circuit elements V/2 and R/2. This voltage source adds in series with 'V'v source of the second section of the original circuit. Hence, the Thevenin equivalent voltage and resistance would be

Answer V_{TH}=-12.5volts and $R_{TH} = 15\Omega$

2.3   Norton's Theorem

Using the *Principle of Duality*, the practical voltage source of fig(2.6b) can be replaced by a current source which would also deliver the same current into the load and develop the same voltage across it as the voltage source. Converting the voltage source shown in figure (2.6b) into a current source, we have

$$I = \frac{V_{TH}}{Z_{TH}}$$

with Z_{TH} coming parallel to the current source. Thus, using the above equation and results from the Thevenin's Theorem, the complex source of fig(2.6a) can be replaced by a current source:

$$I = \frac{\frac{Z_3 V}{Z_1 + Z_3}}{Z_2 + \frac{Z_1 Z_3}{Z_1 + Z_3}} \tag{2.5}$$

Thus, depending on the designers choice or the demand of the situation, one can analyze a circuit using voltage sources or current sources. Here we introduce another theorem by which any complex energy source network can be directly replaced by an equivalent current source with a parallel impudence. This saves the circuit analyzer few steps as compared to using Thevenin's Theorem followed by Principle of Duality. The theorem called Norton's Theorem states:

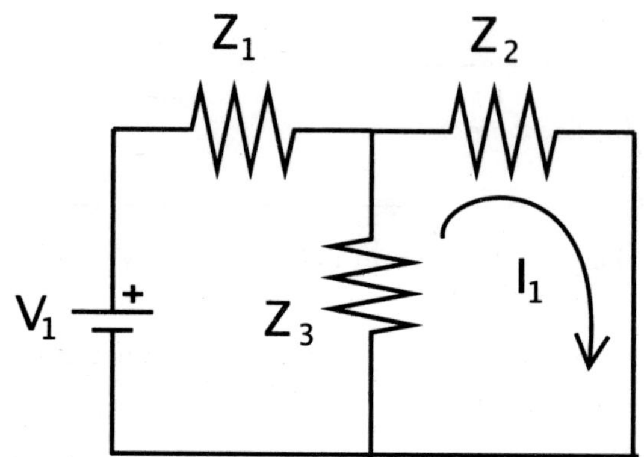

Any network containing energy sources which may be viewed from two terminals can be replaced by a current-source equivalent circuit comprising a single current source I_N and a single shunt conductance G $(1/Z_N)$. The current I_N is the short-circuit current between the two terminals and the conductance G is the conductance of the network viewed from the terminals with all energy sources removed.

Figure 2.10: *Since any complex circuit can be reduced to a T section equivalent circuit, we consider one such 'T' circuit to verify the Norton's theorem. If it is true for this circuit, then it would hold for any circuit.*

Consider the circuit shown in fig(2.6a). Following the statement of Nortons Theorem, remove the load. In place of it, place the multimeter and after removing all energy sources (shorting the voltage sources, and in case there are current sources replace them with an open), look in through the terminals and find R_{TH}, the Thevenin equivalent resistance. The reciprocal would give you the conductance of the network. Now remove the multimeter and short the output.

We now have to calculate the current in this arm of the circuit, after re-introducing the energy sources. The current in the short is the same as the current in impedance Z_2 (fig 2.10). The current is given as

$$I = \frac{Z_3}{Z_2 + Z_3}\left(\frac{V}{Z_1 + \frac{Z_2 Z_3}{Z_2 + Z_3}}\right)$$

$$= \frac{Z_3 V}{Z_1 Z_2 + Z_1 Z_3 + Z_2 Z_3} \tag{2.6}$$

Comparing eqn(2.5) with this (eqn 2.6), we see again the Mesh analysis results[2] are the same as those of the Norton's Theorem statement. Hence, we have proved the Norton's Theorem.

Example 2.4: Find the Norton equivalent circuit for the circuit shown in fig(2.9), given V=25*volts* and R = 10Ω.

A straight conversion of the Thevenin equivalent circuit's voltage source to a current source gives the Norton's equivalent current source and admittance as $I_N = \frac{12.5}{15} = \frac{5}{6}$A and $G = \frac{1}{15}$mho.

---

[2]Remember, converting voltage source to current source and visa verse was done with the help of Kirchhoff's law in the first chapter

However, a direct conversion of the circuit, using the statement of Nortons Theorem, without resorting to using Thevenin Theorem would call for the following steps. Energy source is removed and R_{TH} is found. The inverse of this gives the required admittance.

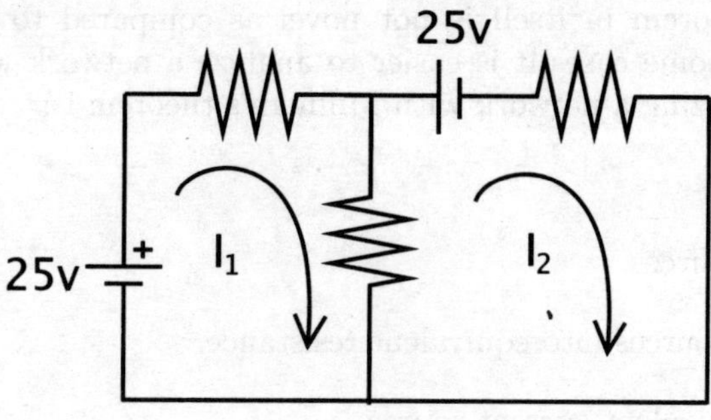

Figure 2.11:

The Norton equivalent current is the short-circuit current between the terminals. This requires us to analyze a two loop circuit and I_2, the current of the second loop (fig 2.11) would be the Norton's equivalent current source. Mesh analysis gives the simultaneous equation as

$$25 = 20I_1 - 10I_2$$
$$25 = 10I_1 - 20I_2$$

Solving the above set of equations, the value of current I_2 gives us the "Norton" current.

$$\boxed{\text{Answer } I_{Nor}=-5/6\text{A and } R_{TH} = 15\Omega}$$

$\boxed{\text{Example 2.5:}}$ Find the Norton equivalent circuit as seen by R_L for the circuit shown in fig(2.12). For analysis, remove R_L, i.e. open circuit the load. On shorting the voltage source, the resistance 100Ω is also shorted and hence does not play a role in the equivalent resistance. The equivalent resistance is a parallel combination of 120Ω and the 180Ω resistance, i.e. $R_{eq} = 72\Omega$.

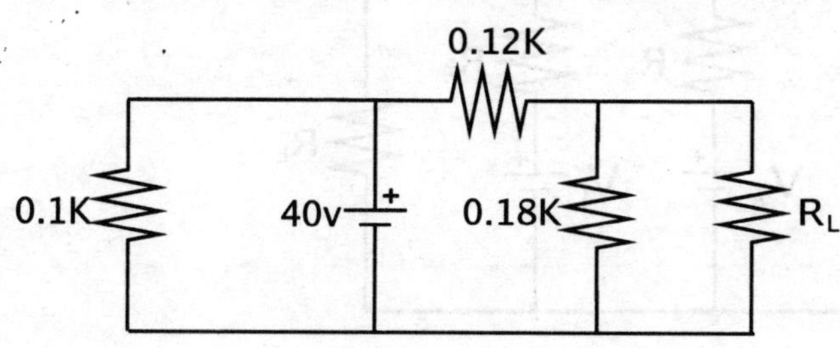

Figure 2.12:

For finding the Norton's equivalent current source, the energy source is kept back and a short is kept where R_L existed. The current in the short is obtained after calculating the net current in the circuit.

$$I = \frac{40(120 + 100)}{100 \times 120}$$

This current is split into the 100Ω and 120Ω resistance. Thus, using the current division equation, current in the 120Ω resistance (this will be the current in the short) is

$$
\begin{aligned}
I_N &= \frac{100}{100 + 120} I \\
&= \left(\frac{100}{100 + 120}\right) \times \left[\frac{40(120 + 100)}{100 \times 120}\right] \\
&= \frac{1}{3} A
\end{aligned}
$$

$$\boxed{\text{Answer } I_N=0.33\text{A and } R_{TH} = 72\Omega}$$

2.4 Millman's Theorem

Millman's theorem helps in reducing any number of parallel voltage sources into an equivalent circuit containing only one source. The theorem in itself is not novel as compared to the Thevenien or Norton's theorem, however, in some cases it is easier to analyze a network with this theorem. The procedure followed in analyzing a network with Millman's theorem is:

  i. Identify the load terminals.

 ii. Convert all voltage sources to current sources.

iii. Combine the resistances of the current sources into equivalent resistance.

 iv. Combine the current sources into an equivalent current source.

  v. Draw an equivalent circuit with the equivalent current source, the equivalent resistance and the load connected in parallel.

 vi. Solve the equivalent circuit for the load current, the load voltage and the load dissipation if required.

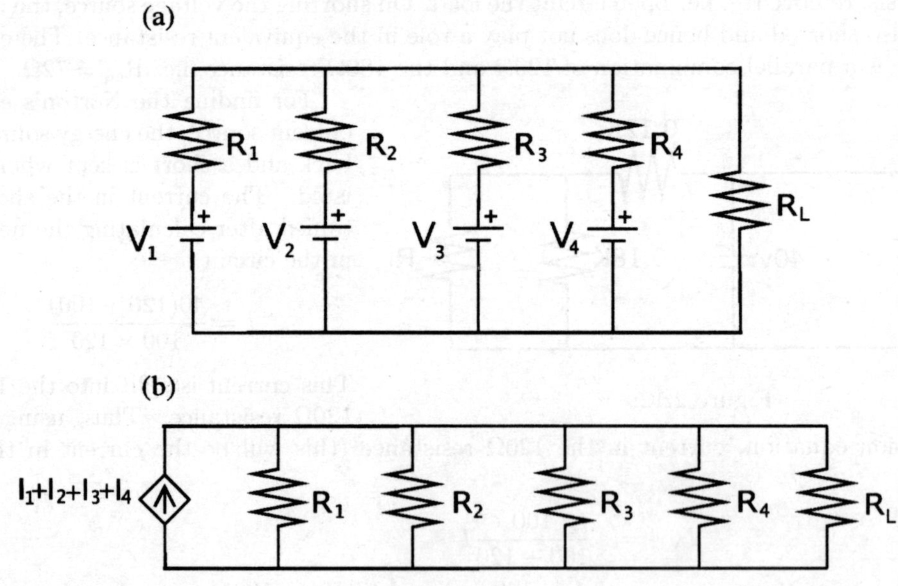

Figure 2.13:

Consider the circuit shown in fig(2.13a). If we have to calculate the current flowing in the load resistance R_L, we can go about doing so using Mesh analysis. However, that would involve writing four loop equations and solving a 4×4 matrix. The problem is simplified by using Millman's theorem, where the voltage sources are converted to current sources. The circuit reduces to that shown in fig(2.13b). The net current in the circuit (I_{eq}) is given as

$$I_{eq} = \frac{V_1}{R_1} + \frac{V_2}{R_2} + \frac{V_3}{R_3} + \frac{V_4}{R_4}$$

or in case there are 'n' current sources

$$I_{eq} = \sum_{i=0}^{n} \frac{V_i}{R_i}$$

The numerous resistances associated with the current sources are in parallel to each other and can be considered to be a single equivalent resistance R_{eq}, where for 'n' constituent resistances

$$\frac{1}{R_{eq}} = \sum_{i=0}^{n} \frac{1}{R_i}$$

The current through and in turn the voltage across the load resistance would be

$$I_{R_L} = \frac{R_{eq}}{R_{eq} + R_L} I_{eq}$$

and

$$V_{R_L} = \frac{R_{eq} R_L}{R_{eq} + R_L} I_{eq}$$

The equivalent resistance is a parallel combination of many resistances and would satisfy $R_{eq} \ll R_L$ (you could achieve this by demanding the load to be very large). Hence, the load voltage would be

$$\begin{aligned}
V_{R_L} &= R_{eq} I_{eq} \\
&= \frac{\sum_{i=0}^{n} \frac{V_i}{R_i}}{\sum_{i=0}^{n} \frac{1}{R_i}}
\end{aligned} \tag{2.7}$$

The ease involved in solving circuits with Millman's Theorem is best highlighted while dealing with weighted Digital to Analog Converter (DAC). These are circuits of digital electronics, used to convert digital signals to analog signals (fig 2.14).

Example 2.6: Figure(2.14) shows a DAC. The switch of resistance R/2 is connected to 5v while the other resistances are connected to ground. This corresponds to the digital input 0010. Using Millman's theorem find the analog output. Using eqn(2.7) the output voltage would be

$$\begin{aligned}
V_o &= \frac{\frac{0}{R/8} + \frac{0}{R/4} + \frac{5}{R/2} + \frac{0}{R}}{\frac{8}{R} + \frac{4}{R} + \frac{2}{R} + \frac{1}{R}} \\
&= \frac{10}{8 + 4 + 2 + 1} = 0.667 volts
\end{aligned}$$

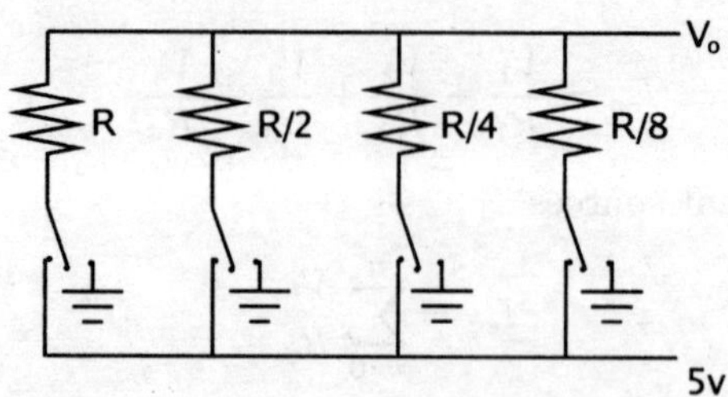

Figure 2.14: *A DAC circuit.*

Since, the calculation is very trivial, to understand the DAC's working, we might as well calculate
the analog output of 1000 and 1111.

$$V_{1000} = \frac{\frac{5}{R/8} + \frac{0}{R/4} + \frac{0}{R/2} + \frac{0}{R}}{\frac{8}{R} + \frac{4}{R} + \frac{2}{R} + \frac{1}{R}}$$

$$= \frac{40}{8+4+2+1} = 2.667 volts$$

$$V_{1111} = \frac{\frac{5}{R/8} + \frac{5}{R/4} + \frac{5}{R/2} + \frac{5}{R}}{\frac{8}{R} + \frac{4}{R} + \frac{2}{R} + \frac{1}{R}} = \frac{75}{8+4+2+1} = 5 volts$$

One is welcome to try and obtain the above results using Mesh Analysis.

2.5   Maximum Power Transfer Theorem

*The Maximum Power Transfer Theorem states that maximum power would be
transferred from a network to an load impedance* Z_R, *if the load impedance is the
complex conjugate of the voltage network's Thevenin equivalent impedance* Z_{TH} *of
the given circuit.*

Again to prove this theorem one would have to consider a complex voltage network circuit.
However, complex it might be, it can be replaced by a Thevenin's equivalent circuit as shown
in figure(2.15). To make the theorem general, we assume the load and Thevenin's equivalent
impedance of the form

$$Z_{TH} = R_{TH} + jX_{TH}$$
$$Z_R = R_R + jX_R$$

The current in the circuit is given as

$$I = \frac{E_{TH}}{Z_{TH} + Z_R}$$

$$= \frac{E_{TH}}{(R_{TH} + R_R) + j(X_{TH} + X_R)}$$

The current and voltage in a capacitor or inductor are out of phase by $\pi/2$. Hence, the power drawn by such impedance is zero.[3]

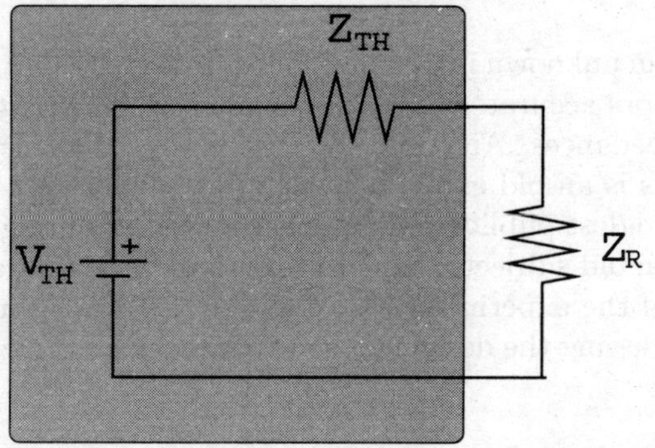

Figure 2.15: *A complex circuit (shaded region shows complex network replaced by Thevenin equivalent circuit) delivering power to load* Z_R.

$$
\begin{aligned}
P &= I^2 R_R \\
&= II^* R_R
\end{aligned}
$$

where for complex numbers, the square of a number is found from the product of the number with it's complex conjugate $(A \times A^*)$. Therefore,

$$
P = \frac{E_{TH}^2 R_R}{(R_{TH} + R_R)^2 + (X_{TH} + X_R)^2}
$$

To find the extrema, we differentiate the expression of power with respect to X_r and R_r.

$$
\frac{dP}{dX_R} = \frac{E_{TH}^2 R_R \times 2(X_{TH} + X_R)}{[(R_{TH} + R_R)^2 + (X_{TH} + X_R)^2]^2} = 0
$$

giving,

$$
X_R = -X_{TH} \tag{2.8}
$$

Also,

$$
\left(\frac{dP}{dR_R} \right)_{X_R = -X_{TH}} = \frac{E_{TH}^2 \left[(R_{TH} + R_R)^2 - 2R_R(R_{TH} + R_R) \right]}{(R_{TH} + R_R)^4} = 0
$$

gives

$$
\begin{aligned}
(R_{TH} + R_R) - 2R_R &= 0 \\
R_R &= R_{TH} \tag{2.9}
\end{aligned}
$$

Thus, maximum power transfer results for

$$
\begin{aligned}
Z_R &= R_R + jX_R \\
&= R_{TH} - jX_{TH} \\
&= (R_{TH} + jX_{TH})^* \\
&= Z_{TH}^*
\end{aligned}
$$

One might be tempted to ask "why maximize the power transfer to the load?" For larger power dissipated to the load would result in larger electricity bills. A very unpleasant thought. So one might say its more important to find a "**Power Minimizing Theorem**". This theorem finds its use in communication electronics like telephone industry. Since the length of the telephone copper cables are large, one wants to minimize the loss along the lines and impart maximum power to the ear-piece (telephone instrument) at the receiver side. Here, the cost of the power is not the concern but the difficulty in generating the signal sustainable to transverse the large distance.

---

[3]The power dissipated across a circuital element is given by $VI\cos\theta$, where θ is the phase difference between the current and the voltage. Since in capacitors and inductors, the phase difference is $\pi/2$, the cosine falls zero.

2.6   Application of Network Theorems

Case Study of The Anderson Bridge

AC bridges are often used to measure the value of an unknown impedance for example self/mutual inductance of inductors or capacitance of capacitors accurately. A large number of AC bridges are available for the accurate measurement of impedances. An Anderson Bridge is used to measure the self inductance of a coil (Fig. 2.16). This is an old experiment and has been a part of the graduation curriculum since ages. In fact the oldest publication on sensitivity of AC bridges is Rayleigh's paper.[4] As it usually happens in such old subjects, modern textbooks have diluted the attention paid to the details and intricacies of the experiment. Most of the textbooks tend to merely state the balance condition without discussing the design of the experiment for greater sensitivity.

Over the years of instructing young graduate students in the lab, I and my collegues[5] have observed unsatisfactory results reported by the students in terms of accuracy. One reason reported by the students was an inability to get a mute on balance condition in the head-phone. That is, the human perception of point of minima rendered results inaccurate. The fact that human ears perceive in decibels makes the situation worse. A question that invariably arose was whether "urbanisation and sound pollution" was effectively contributing to the inaccuracies of this experiment (physiological constraints) or rather it was ignorance of the relevant physics contributing.

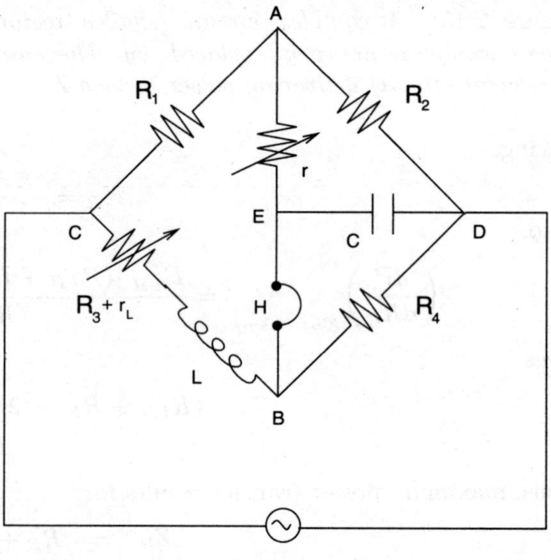

Below the analysis of Anderson Bridge is done using the basic theorems taught in this Chapter highlighting the utility of the various Network Theorem in a simple circuit. Before dwelling on the mathematics and circuit design considerations, for completeness and rendering the article self sufficient for the reader, we explain in short here, the Wheatstone Bridge and standard steps followed for DC and AC balancing of the bridges for the determination of the unknown impedances.

Figure 2.16: *The circuit schematics for Anderson Bridge. The resistance* r_L *is the resistance of the coil whose self-inductance is being determined.*

Most of the AC bridges are based on a generalised Wheatstone Bridge circuit. As shown in Fig. 2.17, the four arms of the D.C. Wheatstone Bridge are replaced by impedances (Z_A, Z_B, Z_C and Z_D), the battery by an A.C. source and the D.C. galvanometer by an A.C. null detector (usually a pair of headphones). Using Kirchhoff's Laws, it can be easily shown that the balance or null condition (i.e. when no current flows through the detector or the potential at the point P becomes equal to that at point R) is given by

---

[4]Lord Rayleigh, Proc. Roy. Soc., Vol 49, p203 (1891).

[5]P.Arun, Kuldeep Kapil and Mamta, Resonance Vol 15, p 244 (2010).

$$\frac{Z_A}{Z_B} = \frac{Z_C}{Z_D} \qquad (2.10)$$

Eqn (2.10) is a complex equation i.e. it represents two real equations obtained by separately equating the real and imaginary parts of the two sides. It follows from the fact that both amplitude and the phase must be balanced. This implies that to reach the balance condition two different adjustments must be made. That is, the DC and AC balance conditions have to be obtained one by one. The DC balance is obtained using a DC source and moving coil galvanometer by adjusting one of the resistances. After which the battery and the galvanometer are replaced with an AC source and a headphone respectively, without changing the resistances set earlier, the other variable impedance is varied to obtain minimum sound in the headphone.

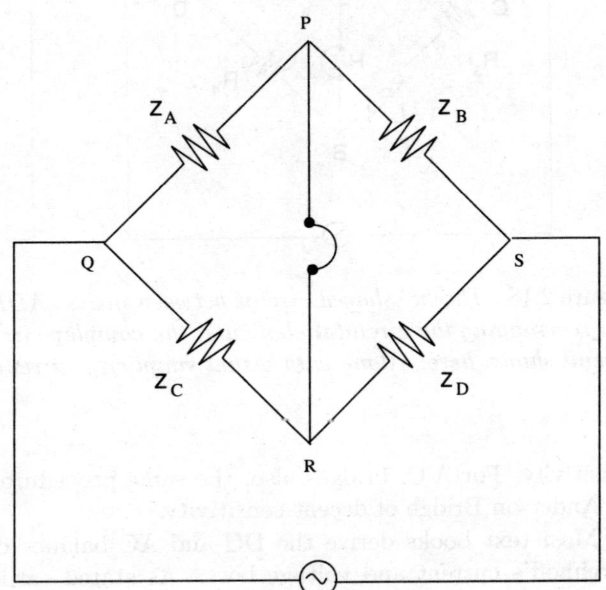

Figure 2.17: *A simple Wheatstone Bridge Arrangement.*

Circuit Designing

Various combinations of the impedances can satisfy the balance conditions of a bridge. For example, in a DC Wheatstone Bridge using either $R_A = R_B = R_C = R_D = 10\,\Omega$ or $R_A = R_B = 10\,\Omega$ and $R_C = R_D = 1000\,\Omega$ would balance the bridge as indicated by eqn. (2.10). However, the bridge is not equally sensitive in both cases. Also interchanging the positions of the source and detecting instrument do not alter the balance conditions (Reciprocity Theorem) but again the bridge may not be equally sensitive in the two cases. The bridge will be more 'sensitive' if, in a balanced bridge, changing the impedance in any one of the arms of the bridge from the value Z required by the balance conditions to value $Z + \delta Z$ (making the bridge off balance) results in a greater current through the detecting instrument. As explained by Yarwood,[6] a DC bridge is more sensitive if, *whichever has the greater resistance, galvanometer or battery, is put across the junction of the two lower resistances to the junction of the two higher resistances. Also the unknown resistance should be connected in the Wheatstone bridge between a small ratio arm resistance and a large known variable resistance.* However, it is common practice to make all the four arms of Wheatstone Bridge equal or make them of the same order for the optimum

---

[6] J.H. Fewkes and John Yarwood, *Electricity and Magnetism and Atomic Physics*, Vol 1 (2nd Ed), Oxford University Press 1965.

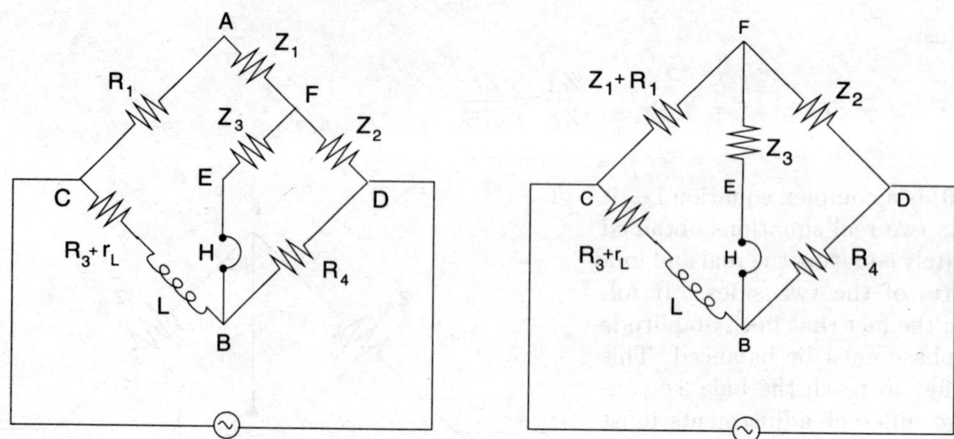

Figure 2.18: *The 'π' shaped circuit between nodes 'ADE' is replaced by an equivalent 'T' circuit. On rearranging the circuital elements, the complex circuit of fig 1 reduces to a apparently simple circuit shown here. Along with visual simplicity, it renders the computation of potentials trivial.*

sensitivity. For A.C. bridges also, the same procedure is followed. Below we discuss designing an Anderson Bridge of decent sensitivity.

Most text books derive the DC and AC balance conditions of the Anderson Bridge using Kirchhoff's current and voltage laws. As stated earlier, we plan to use various theorems introduced in this Chapter to the same. Hence, to analyse the bridge, we convert the π network consisting of r, R_2, and X_c and replace it with an equivalent 'T' network, whose three impedances would be given as

$$Z_1 \;=\; \frac{rR_2}{r + R_2 + X_c} \tag{2.11}$$

$$Z_2 \;=\; \frac{R_2 X_c}{r + R_2 + X_c} \tag{2.12}$$

$$Z_3 \;=\; \frac{rX_c}{r + R_2 + X_c} \tag{2.13}$$

The circuit shown in Fig. 2.16 can now be viewed as shown in Fig. 2.18. The circuit now looks more like the fundamental Wheatstone bridge, which purely due to its topology looks trivial and overcomes a mental barrier a student might have due to a complex looking circuit.

Balancing Conditions

For the bridge to be balanced, the current in arm 'EB' should be zero. This demands the potential V_{FB} to be equal to zero. This potential can be worked out as

$$V_{FB} = V_F - V_B$$

where, using potential divider expression, we can compute V_F and V_B. They are expressed as

$$V_F = \left(\frac{Z_2}{R_1 + Z_1 + Z_2} \right) V_{ac}$$

$$V_B = \left(\frac{R_4}{R_3 + r_L + X_L + R_4} \right) V_{ac}$$

giving the potential V_{FB} as

$$V_{FB} = \left(\frac{Z_2}{R_1 + Z_1 + Z_2} - \frac{R_4}{R_3 + r_L + X_L + R_4} \right) V_{ac} \tag{2.14}$$

On attaining balance of the bridge

$$\frac{Z_2}{R_1 + Z_1 + Z_2} = \frac{R_4}{R_3 + r_L + X_L + R_4} \tag{2.15}$$

which is same as the condition given by the eqn(2.10) with the four impedances replaced by the corresponding impedances in the four arms of Fig. 2.18.

Substituting the expressions of eqn(10.27), eqn(10.28) and eqn(10.29) in eqn(2.15), we have

$$\frac{R_2 X_c}{R_1(r + R_2 + X_c) + rR_2 + R_2 X_c} = \frac{R_4}{R_3 + r_L + X_L + R_4} \tag{2.16}$$

Collecting and simplifying the real terms of the above equation, we get

$$\frac{R_2}{R_4} = \frac{R_1}{R_3 + r_L} \tag{2.17}$$

Expression given by eqn(2.17) is the DC balance condition of the Anderson Bridge. Now collecting the imaginary terms of eqn(2.16), we have

$$\frac{L}{C} = \frac{R_4}{R_2}[R_1 R_2 + r(R_1 + R_2)] \tag{2.18}$$

Eqn (2.18) is the AC balance condition of the bridge.

The unknowns in the two eqns (2.17) and (2.18) are r_L and L respectively. To obtain the D.C. balance of the bridge, the circuit is made with 'r' shorted, capacitance C open, headphone replaced by a galvanometer and the A.C. source by a battery or D.C. power supply. Now R_3 is varied until galvanometer shows zero deflection. Eqn. (2.17) then gives the value of r_L. Now using the complete circuit given in Fig. 2.16 and leaving the resistances set R_1, R_2 and R_3 undisturbed, AC balance is obtained by varying the resistance 'r'. Eqn. (2.18) can now be used to compute the value of self inductance L applied in the circuit. Thus the unknown impedance $\sqrt{(L\omega)^2 + r_L^2}$ is computed. For the simplicity of circuit designing we suggest taking $R_2 = R_4$, which results in $R_1 = r_L + R_3$ from eqn. (2.17).

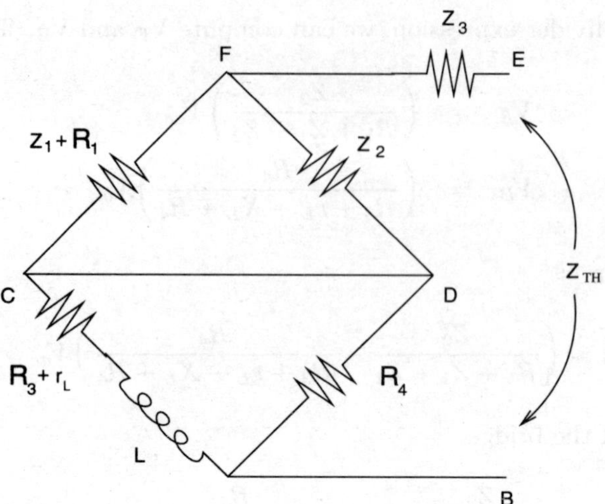

Figure 2.19: *The Thevenin impedance is calculated by shorting nodes 'C' and 'D' and evaluating the impedance as seen from nodes 'EB'.*

Applying Maximum Power Transfer

For a good audio signal at the head phone, we require that the circuit transfers a substantial amount of power to the head phone. Since the ear senses the intensity of sound, the head phone as a measuring device is being used as a power meter. Good sensitivity hence would be attained as large power is being transfered to the head phone for both balanced and unbalanced (AC) condition. Unbalanced AC condition would be represented by replacing 'r' (of eqn. 2.18) with '$r + dr$'.

The "Maximum Power Transfer Theorem" demands that the load impedance applied to a circuit should be the complex conjugate of the circuit's Thevenin impedance. Since the headphone's impedance is inductive in nature, i.e. it can be represented as '$a + jb$', the theorem demands the Thevenin impedance should be capacitive in nature, i.e. of the form of '$a - jb$'. The Thevenin impedance of the Anderson Bridge circuit (Z_{TH}) as seen by the head phone is theoretically determined by shorting the AC source (i.e. we neglect the impedance of the source), giving (see Fig. 2.19)

$$Z_{TH} = [\{(R_1 + Z_1)\|Z_2\} + \{(R_3 + r_L + X_L)\|R_4\} + Z_3]$$
$$= \left[\frac{(R_1 + Z_1)Z_2}{R_1 + Z_1 + Z_2} + \frac{R_4(R_3 + r_L + X_L)}{R_3 + r_L + X_L + R_4} + Z_3\right]$$

Using the D.C. balance condition given by eqn. (2.17) and imposing $R_2 = R_4$, we have

$$Z_{TH} = \left[\frac{(R_1 + Z_1)Z_2}{R_1 + Z_1 + Z_2} + \frac{R_2(R_1 + X_L)}{R_1 + X_L + R_2} + Z_3\right]$$

The best way to simplify the above expression is to select $X_c \sim 0$. This means the AC audio frequency (f) should be kept large and the capacitor used should have a very large value of 'C'.

From eqn (2.18), it naturally flows that X_L has to be considerably large. Selecting a high audio frequency automatically achieves this. Further, $X_c \sim 0$ might be difficult to attain given the fact that frequency is to be in audible range and very high capacitances are usually electrolytic, thus introducing problems of it's own. Hence it would be sufficient to select 'C' and 'f' such that $|X_c| << |X_L|$ and $(rR_1 + R_1R_2 + rR_2) >> R_2|X_c|$. Further, let $R_1 >> R_2$. The justification and advantage of this assumption would be evident later. With these conditions imposed, the Thevenin equivalent impedance reduces to

$$Z_{TH} = \left[R_2 + \frac{(R_2 + r)X_c}{(r + R_2 + X_c)} \right] \tag{2.19}$$

Thus, the basic constraints imposed on the circuit designing has made the Bridge's Thevenin equivalent impedance capacitive in nature, (eqn. 2.19 is of the form a-jb).

$$Z_{TH} = \left[R_2 + \frac{(r + R_2)|X_c|^2}{(r + R_2)^2 + |X_c|^2} \right] + \left[\frac{(r + R_2)^2}{(r + R_2)^2 + |X_c|^2} \right] X_c \tag{2.20}$$

As stated earlier, for the maximum power to be transferred, the Bridge's Thevenin equivalent impedance should be capacitive in nature since the head-phone's impedance is inductive in nature. Also, head phones are usually low impedance devices and hence X_c and R_2 should be kept as small as possible. While selection of small X_c was already imposed, an additional requirement is demanded, i.e. R_2 should also be of small value.

Sensitivity

To appreciate the approach of the AC balance point, the sensitivity of the circuit should be good. Here we define the bridge's sensitivity as (dP/dr), i.e. the power developed across the head-phone should show large change for a small change in 'r'. Upon the application of the *Maximum Power Transfer* theorem, the expression of current can be written as

$$i = \frac{V_{TH}}{2Re(Z_{TH})}$$

where V_{TH} is the Thevenin equivalent voltage. If the Thevenin impedance doesn't match the impedance of the headphone exactly, the current expression for a circuit designed along the lines discussed above reduces to (the Thevenin voltage would be of the same form as that given by eqn. 2.14. Notice V_{DC} has been replaced by V_{ac}, the applied A.C. voltage)

$$i = \left[\frac{1}{\left(R_H + \frac{r|X_c|^2}{r^2 + |X_c|^2} \right) + j \left(Z_H - \frac{r^2|X_c|}{r^2 + |X_c|^2} \right)} \right] V'_{ac} \tag{2.21}$$

where R_H and Z_H are the real and imaginary part of the impedance associated with the head phone being used. The term V'_{ac} is given as

$$V'_{ac} = \left(\frac{Z_2}{R_1 + Z_1 + Z_2} - \frac{R_4}{R_3 + r_L + X_L + R_4} \right) V_{ac} \tag{2.22}$$

Using the constraints imposed and eqn (2.20), this expression reduces to

$$V'_{ac} = R_2 \left[\frac{X_c}{R_1(r + R_2 + X_c)} - \frac{1}{R_1 + X_L} \right] V_{ac}$$

Hence, the power dissipated across the head phone is given as

$$P = ii^* R_H$$

$$P \approx \left[\frac{R_H}{\left(R_H + \frac{r|X_c|^2}{r^2 + |X_c|^2}\right)^2 + \left(Z_H - \frac{r^2|X_c|}{r^2 + |X_c|^2}\right)^2} \right] V'^2_{ac}$$

Differentiating w.r.t. 'r' (for simplifying the expression we can assume R_H and $Z_H \sim 0$ which is generally true since head phones are low impedance devices. Further approximation can be made that V'_{ac} is very small and hence is a shallow function of 'r' due to the small values of R_2 and X_c in the numerator and large R_1 and X_L in the denominator.),

$$\frac{1}{R_H V'^2_{ac}} \left| \left(\frac{dP}{dr} \right) \right| \approx \frac{2}{r^3} \tag{2.23}$$

The above expression demands 'r' to be of small value for good sensitivity of the bridge. However, till now we have been silent on the nature of 'r'. For this we revisit eq (2.18), to understand the effects of our considerations on the nature of 'r'. For a bridge designed along our suggestions, the AC balance condition (eq 2.18) gives

$$r \approx \frac{L}{R_1 C} \approx X_c \left(\frac{X_L}{R_1} \right) \tag{2.24}$$

The selection of high frequency, large capacitance with large R_1 makes sure that 'r', the resistance used to AC balance the bridge, is proportional to the impedance offered by the capacitor. As the case is, in our design AC balance should be achieved with small 'r' which would result in good sensitivity. The interpretations do not vary even if the approximations listed above eqn 2.23 are not made.

Another design consideration necessary is to make R_1 large. We summarize here the circuitial conditions one must maintain to get good sensitivity for the Anderson Bridge

1. $(R_1 = R_3) >> (R_2 = R_4)$ with R_2 being very small

2. Select $X_L >> X_c$.

3. Audio frequency to be large

Based on various theorems of *Network Analysis*, the above derivation shows how to design an Anderson Bridge with good sensitivity. Varous considerations show that for a better sensitivity, one should work at high frequencies and use a low R_2 and high capacitance (making $X_L >> X_c$).

Exercise

Q1. A manufacturer supplies resistors with various power dissipation ratings. Resistors with higher power dissipation ratings are larger in size and more expensive. If you are required to connect a 100Ω resistor between two nodes whose potential difference is 10V, which type of resistor would you chose to minimize cost?

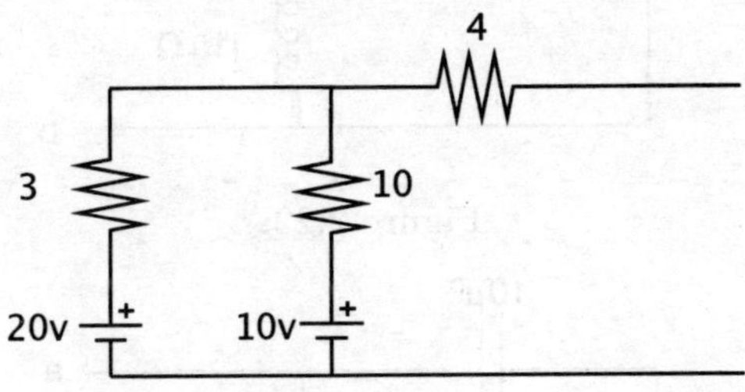

Figure 2.20:

Q2. State and prove Thevenin Theorem.

Q3. Find the Thevenin equivalent circuit for the circuit shown in fig(2.20). The values of resistances are given in ohm.

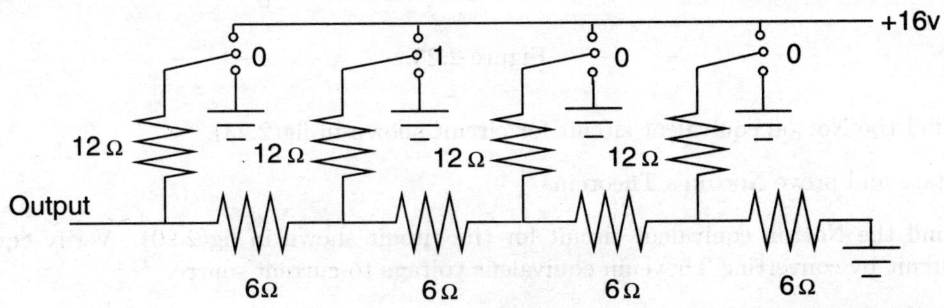

Figure 2.21:

Q4. Use Thevenin theorem to find the DC output voltage of the DAC (fig 2.21), this is called a binary ladder and is easier to design then the weighted ladder discussed in Millman's Theorem) for the given digital input 0100?

Q5. Find the Thevenin equivalent circuit for circuit shown in fig(2.22).

Q6. Find the Thevenin equivalent circuit for circuit shown in fig(2.23). Notice the basic difference between the information given in fig 2.22 and fig 2.23.

Q7. State and prove Superposition Theorem.

Q8. Find the Norton equivalent circuit for circuit shown in fig(2.22).

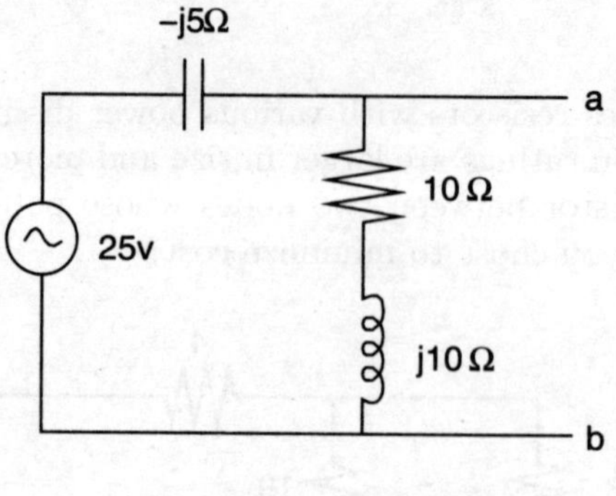

Figure 2.22:

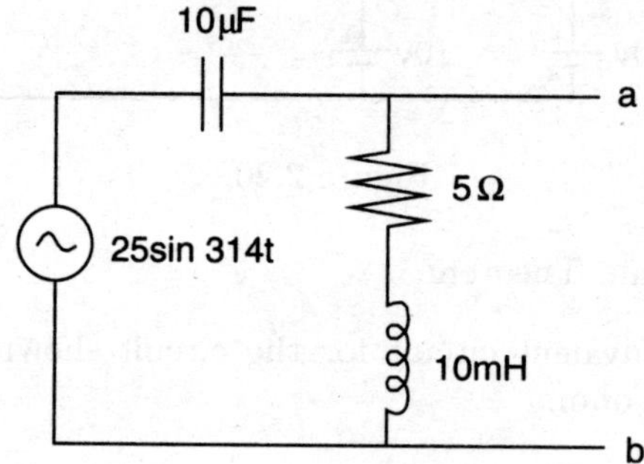

Figure 2.23:

Q9. Find the Norton equivalent circuit for circuit shown in fig(2.23).

Q10. State and prove Norton's Theorem.

Q11. Find the Norton equivalent circuit for the circuit shown in fig(2.20). Verify equivalent circuit by converting Thevenin equivalent voltage to current source.

Q12. Is the same power dissipated in a circuit and it's (i) Thevenin equivalent and (ii) Norton's equivalent circuit?

Q13. Find the Thevenin equivalent circuit for the circuit shown in fig(2.12) and verify the answer of Example(2.5).

Q14. Find the load to be applied at terminals 'ab' of circuit shown in fig(2.22) such that maximum power is transferred to it.

Chapter 3

The Oscilloscope

In the previous chapters we tested and analyzed circuits using a multimeter. Instead of a multimeter one could have used ammeters and voltmeters. However going by the axiom "seeing is believing" one would love to view the various waveforms at the input or output of the circuit. In fact it is easier to debug or correct a faulty circuit by looking for error in the waveform. But how does one look up the signal waveform? One uses an oscilloscope.

An oscilloscope is easily the most powerful instrument available for testing circuits and usually the most costly instrument in a college laboratory. In this chapter a short description of the oscilloscope is given without going into great detail about the internal circuitry of the oscilloscope. The basic idea behind doing so is to give an overview of how to use the oscilloscope in the electronic laboratory. On first usage, students typically either twiddle every knob or press every button on the oscilloscopes panel[1] (and there are quite a few of them, see fig 3.1), or adopt a dazed expression. Neither approach is right, especially the physical rough usage, considering the cost of the instrument. Hopefully, after this chapter a more systematic approach would be adopted and also the reader would appreciate topics explained in the following chapters. Even though there are many manufacturers across the world that make and sell oscilloscopes, different models essentially have the same functionality. Thus, a hands on experience on one model would given sufficient confidence to use any other model available in the market.

3.1   Overview of the Internal Circuits

Basically the oscilloscope's output, displayed on a **C**athode **R**ay **T**ube (CRT), is a plot of voltage against time, the voltage on the vertical or Y-axis, and time on the horizontal or X-axis. This is achieved by a smart bit of circuitry whose block diagram is shown in fig(3.2). The electrons

---

[1]If someone else has been using the oscilloscope before you and may have been rotating all the knobs and pressing all the buttons, before you start using the oscilloscope, check that all the controls are in their 'normal' positions. Usually, this means that all the push button switches are in the OUT position, all slide switches are in the UP position. Set both volts/div knob to 1 V/div and the time/div knob at 1ms/div. Once the oscilloscope is ON bring the spot or line at the center of the vertical axis using the Y-position control knob. The Y-pos control allows you to move the spot up and down the screen. Use the INTENSITY and FOCUS controls to make the spot of line visible and sharp.

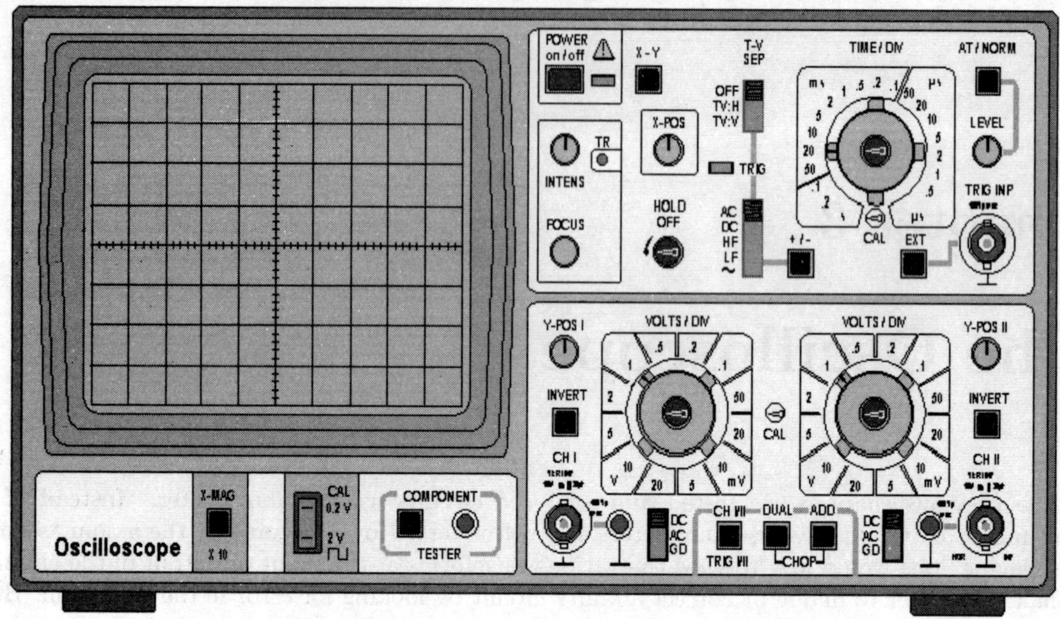

Figure 3.1: *The front panel of an average lab oscilloscope.*

required for display are obtained from the tube's cathode by heating it. This phenomenon of getting electrons by heating a metal is called **Thermionic emission**.

The hot electrons (electrons with high kinetic energy) are further accelerated towards the screen by hollow tube anodes. The anodes are kept at several thousands of volts. When the electrons hit the phosphor coated screen, the electron's kinetic energy is released as light. The image intensity seen by the user depends on

- the number of electrons hitting the spot,

- the spot size,

- speed with which the spot moves,

- the energy with which it hits the spot and

- the phosphor material's character itself (i.e. ability to convert kinetic energy to light) etc.

Of these, the first three are controllable by the oscilloscopes user namely by the controls **INTENSITY, FOCUS** (see fig 3.3) and **SWEEP RATE**.

The intensity is varied by varying the number of electrons reaching the phosphor coated screen. This is done by a control grid placed between the cathode and the anode. The grid is a metal which has a net like appearance. It is given a negative potential and hence repels electrons. However, all electrons are not repelled as most electrons pass through it unhindered

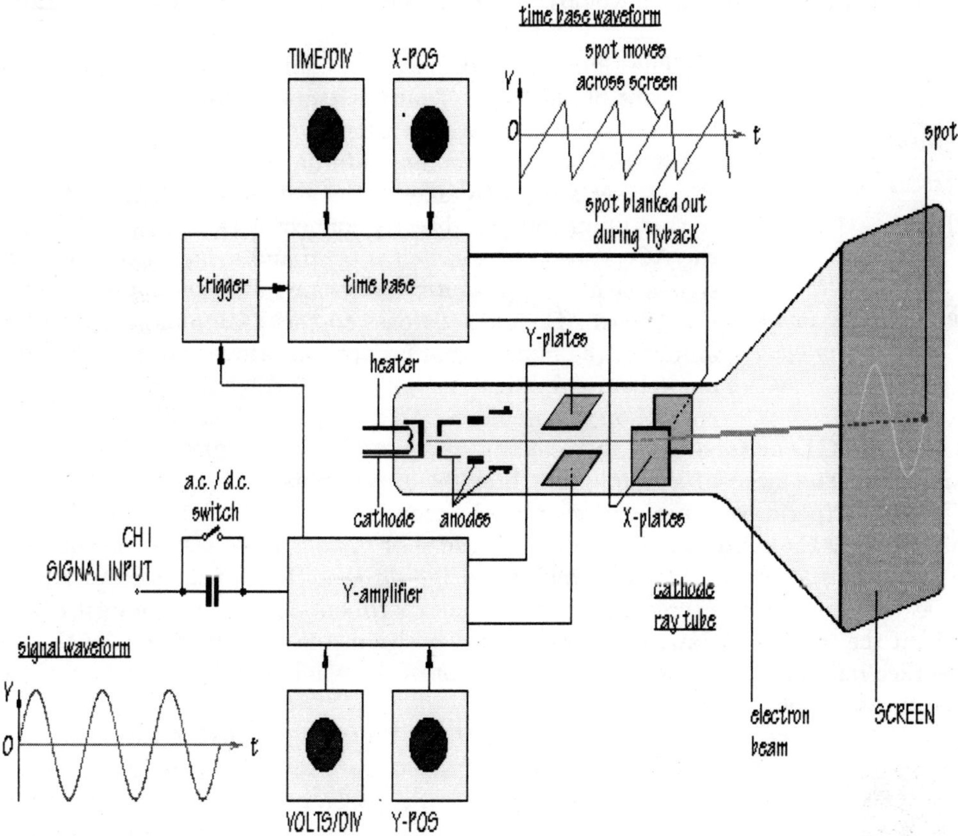

Figure 3.2: *The important blocks of circuit that are present in the oscilloscope.*

due to it's porous structure. The larger the negative potential given to this grid, more electrons are repelled and hence the intensity of the image is brought down.

The anode hollow tube, mentioned above, is not a monolithic piece but is made up of three different hollow tubes. As the electrons move to the screen they first encounter the pre-accelerator anode followed by the focusing anode and finally the accelerating anode. The pre-accelerating and accelerating anodes are kept at a higher positive potential then the focusing anode. Although the intermediate tube is called the focusing anode, all three anodes contribute in focusing the electrons from the cathode to form a dot on the screen. Deviating and thus focusing electrons by making them travel through varying electric field is called electric focusing. The focusing anode is aptly named as by varying it's potential the users can change focusing as per his/ her requirement. When both the intensity and focusing are correctly set, the spot would be reasonably bright but not glaring, and as sharply focused as possible. In actual practice (under normal setting), one gets a line on the screen. However, we still have to explain how the electrons coming from a point source cathode to form a spot on the CRT, becomes a line in the oscilloscope. Till the full picture develops, we shall continue assuming a spot formation on the CRT's screen.

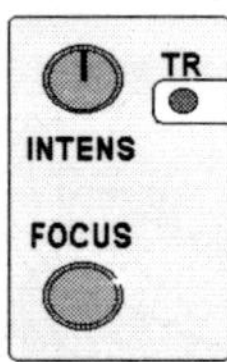

Figure 3.3: *The control for focusing spot/ image and to vary the image's intensity.*

The signal to be displayed is connected to the input. If the user is aware of the type of output signal the circuit is going to give, the user selects AC or DC using the slide switch on the front panel (fig 3.4). The corresponding internal circuit can be seen in fig(3.2). When user selects DC mode, the switch shown in fig(3.2) is closed so that input signal is directly connected to the Y-amplifier, while in the AC position, the switch is open and a capacitor is placed in the signal path. The capacitor blocks any DC signal present in the input but allows AC signals to pass to the Y-amplifier. The Y-amplifier amplifies the input signal for appropriate viewing of the signal. The amount of amplification can be varied by the user by the volts/div control on the oscilloscope's front panel. The third possibility shown in fig(3.4) is the **GD** or **GrounD**. On selecting this option, input signal is disconnected from the Y-amplifier. The input of Y-amplifier is brought to zero potential (ground). This is usually used before any experimental input is given to the scope to set the reference vertical position of the beam (usually the center of the screen). If the spot is not at the center, using the Y-POS one can move the spot up and down, bringing the spot to the center of the screen. After setting the spot at the center of the screen, now when the experimental input is given with the control set at DC, if the stop moves away from the initial position, the magnitude of the DC potential can be worked out. Fig(3.5) shows the displacement of the whole waveform when one goes from DC mode to AC mode.

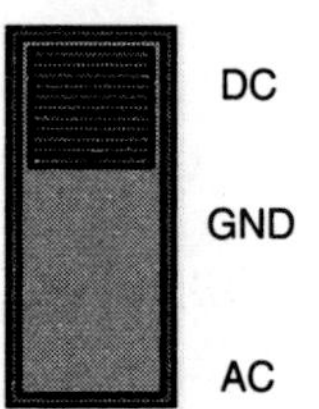

Figure 3.4: *The output of a circuit can be viewed in DC or AC mode.*

The output of the Y-amplifier is fed to two deflection plates called "Y-plates" since any potential appearing on these plates deflects or bends the electron beam's path in the vertical direction (i.e. along the screen's y-axis). Thus, when a sine wave is given in the scopes input, the Y-amplifier amplifies the signal and the Y-plates deflect the electron beam such that at $\theta = 0, \pi$ etc, the spot is not deflected and is at the screen's center while for $\theta = \pi/2, 5\pi/2$ etc the spot is displaced the maximum in the upward direction. However, since the input is a smoothly varying sine wave, with time, the potential across the Y-plate would smoothly move the spot up and down in accordance to the input. For a fast varying sine input, the user would see (since we still have not introduced the horizontal deflection system) a vertical line. It is just like one is writing on the same point (better calling it scribbling rather than writing) without one's arm moving left to right along the paper.

The user must be careful in inferring as to why the waveform has been displaced. The displacement of the waveform may be due to a DC signal being present, which would be understood on viewing in DC mode. However, a obvious displacement can be seen even in AC mode, as seen in fig(3.6). This displacement is brought about by rotating the Y-pos control. This control allows the user to displace the waveform so as to allow him or her to estimate the peak-to-peak voltage. Ideally, before using the oscilloscope to make measurements, the user should view in GD mode and using the Y-pos control move the spot or line to the center of the screen.

Followed which making any observation in DC mode would tell the user the displacement of the signal with respect to zero or ground potential. This aids the user to make more precise measurements.

The screen's display of vertical line for the user obviously does not reflect the input wave. To get a sine wave display, the oscilloscope requires the horizontal deflection plates, or X-plates to produce side to side movement. As you can see (fig 3.2), the X-plates are linked to a system block called the time base. The time base produces a sawtooth waveform. During the rising phase of the sawtooth, the

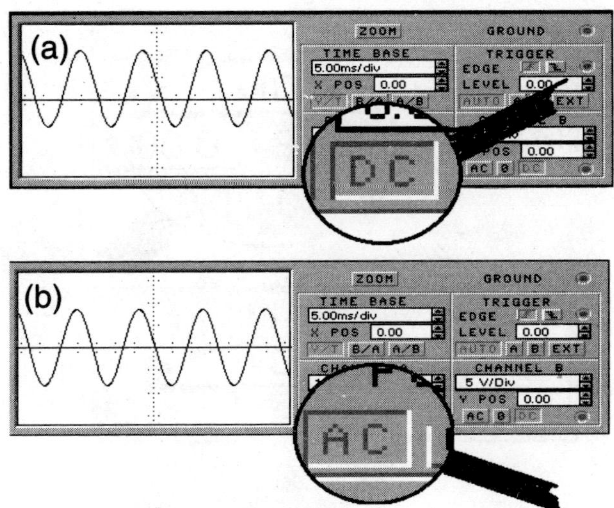

Figure 3.5: *The waveform viewed in (a) DC and (b) AC mode.*

spot is driven at a uniform rate from left to right across the front of the screen. During the falling phase, the electron beam returns rapidly from right to left, but the spot is 'blanked out' so that nothing appears on the screen. Associated with the time base block is another control, the **"TIME/DIV"** control. This varies the rise time of the sawtooth waveform and it's slope, thus allowing the user to vary the horizontal scale of the graph which appears on the oscilloscope screen.

To understand how the saw-tooth's rise time and slope varies the horizontal scale, consider that the when no horizontal voltage exists the electrons are near the left hand horizontal plate (i.e. x=0) and as the potential of the plate increases, the electrons are deflected towards the right hand plate. In other words, the horizontal deflection produced on the electrons are directly proportional to the potential applied

$$\begin{aligned} x \quad &\propto \quad V_x \\ &= \quad KV_x \end{aligned} \tag{3.1}$$

where 'K' is a proportionality constant. The horizontal deflection is produced by the time varying saw-tooth waveform which can be represented by the equation

$$V_x(t) = mt \tag{3.2}$$

where 'm' is the slope of the saw-tooth waveform. Consider that on the completion of one cycle (called the trace time, T_t), the maximum horizontal deflection potential is V_m, the slope then would be given as

$$m = \frac{V_m}{T_t} \tag{3.3}$$

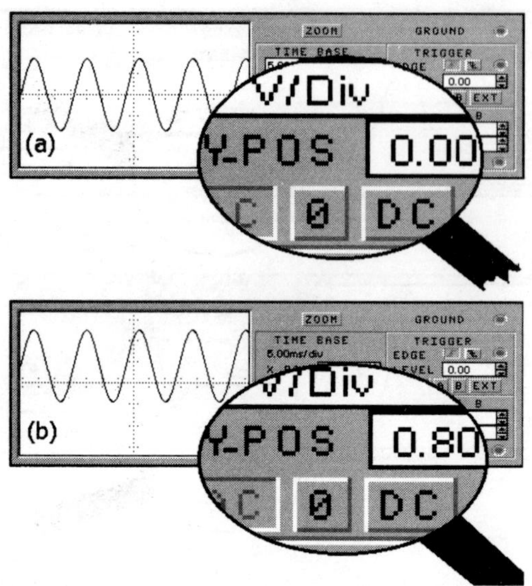

Figure 3.6: *The waveform can be moved up or down using the Y-pos switch, note this displacement is when you are still in AC mode.*

Combining the three equation, we can find the displacement of the electron beam from the left hand horizontal plate (x=0) with time, which is given as

$$x = \frac{KV_m}{T_t}t \qquad (3.4)$$

On applying the maximum horizontal deflection potential, the electron beam deflection is complete and it is deflected towards the right hand plate. Thus, from eqn(3.1) it follows, that kV_m is the width of the CRT screen. Usually, for the convenience of measurement by the user the CRT screen has 10 divisions made on it along the 'x' direction (see fig 3.1). Hence eqn(3.4) can be written as

$$x = \frac{10div}{T_t}t \qquad (3.5)$$

The above equation only represents the question (simple math done in primary school), if the electron beams moved 10 divisions in time T_t, then how much time does it take to move through a distance 'x'? That is

$$10div \quad \longrightarrow \quad T_r$$
$$x \quad \longrightarrow \quad ?$$

The answer to this question is

$$t = \left(\frac{T_t}{10div}\right)x \qquad (3.6)$$

The terms in the bracket is the time/div given as a control for the oscilloscope user. Note, the time per division depends on both the rise time and the slope (through V_m) of the saw-tooth waveform. It would be interesting to use some numbers in the equations above. With 10 squares across the screen and the spot moving at 0.2 s/DIV, how long does it take for the spot to cross the screen? The answer is $0.2 \times 10 = 2$ s (use eqn 3.6). Similar to the Y-POS control a X-POS control is provided, which allows the whole voltage versus time graph to be moved from side to side (horizontal displacement) on the oscilloscope screen. This is useful when you want to use the horizontal and vertical grid markings on the screen to make measurements, for example, to measure the period of a waveform that we shall discuss in the next section.

The only block of fig(3.2) left to discuss is that of the trigger. Sometimes, while displaying a wave that is varying with time, the display is not stable, it would appear as the wave is either moving left to right or right to left.

Sometimes the beam can actually appear to be a jumble of multiple traces jumping around on the screen. What is going on is that as the beam sweeps from left to right it does not always start at the same point on the input voltage. To explain this statement consider a sine wave given as an input to the scope while the time/div is set as shown in fig(3.7). At instant t=0, the electron beam is at it's initial position x=0, i.e. near the left hand horizontal plate. In the figure, at this instant the vertical deflection is such that the input sine wave's zero coincides with the oscilloscopes center axis marking (as mentioned above, the user has set the oscilloscope by pressing GD, then moving the line up/down with the Y-pos followed by viewing the waveform in AC mode).

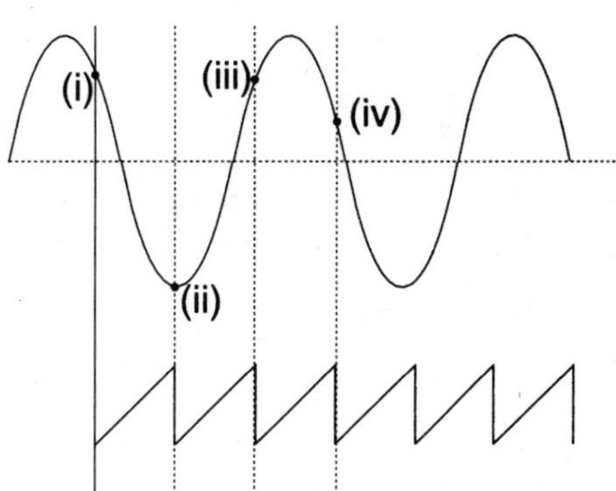

Figure 3.7: *If triggering is not done properly, when the deflected electron comes back to the left hand 'X' plate (instant when does so is the maxima of sawtooth), the value of the sine input (and hence 'Y' axis deflection) would be different with each cycle.*

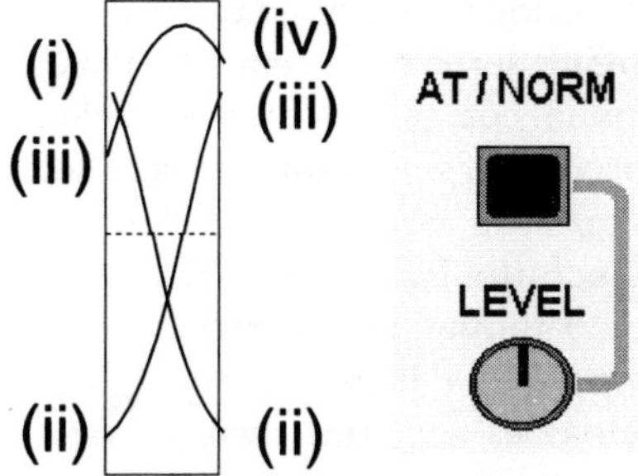

Figure 3.8: *For the cases depicted in fig(3.7), one would see a non-stationary sine wave as shown. The front switch of Automatic/ Normal with level is used to make the wave stationary.*

Now as the saw-tooth voltage rises, the electron beam is being deflected to the right and the input sine wave is resulting in a proportional deflection along the vertical axis. Both deflections acting simultaneously gives the sine wave display of the CRT screen. The second cycle of the saw-tooth repeats this displaying action. However, at the instant of start of the second saw-tooth pulse, the sine wave input is at point (ii) (i.e. a non-zero voltage). Point (ii) in the second cycle of the ramp signal becomes vertical position of the electron beam near the left hand horizontal plate x=0. Note with each new cycle of the saw-tooth pulse, the sine wave's amplitude at x=0 is different. The display would hence appear as shown in fig(3.8). If the display is not jumbled as shown in fig(3.8), then the display seen would be a free running sine wave (like a traveling wave) either moving right or left. This makes it impossible to make any measurement of time period or frequency. For making these measurements the display should not be jumbled and should be stationary. In order to produce a stationary waveform image of the input signal the input should be synchronized with the saw-tooth voltage of the sweep generator (also called ramp generator). Figure(3.9) shows two ramp signals of rise time T_{t1} and T_{t2}. Notice every time the ramp signal reaches maxima, the input signal wave instantaneous voltage is same as that at t=0. Hence, when the new rising ramp voltage arrives the input wave starts with the same voltage at the horizontal deflection's left hand plate, i.e. v(x=0, t)=v(x=0,t=0)=V$_r$. Note from the example in fig(3.9)

$$T_{t1} = T$$
$$T_{t2} = 2T$$

or in general to achieve a stationary display, we require the sweep signals time period T$_t$ (assuming the fall time to be zero) to be an integral multiple of the input signal's time period, i.e.

$$T_t = nT \qquad (3.7)$$

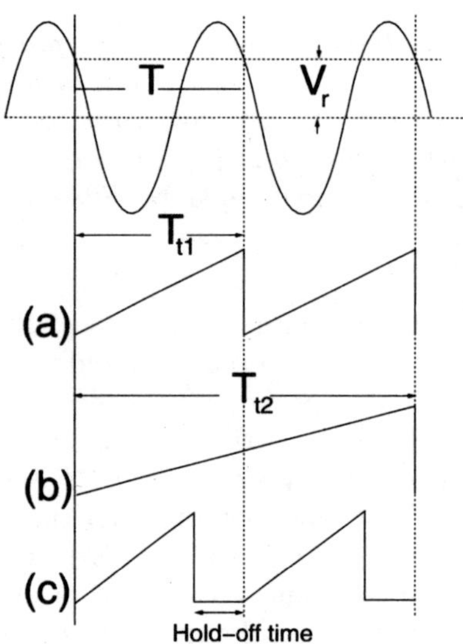

Figure 3.9: *If the sawtooth time period (a & b) is an integral multiple of the waveform under study, then a stationary waveform is obtained on the screen. The third waveform shows how the "trigger" introduces a Hold-off time by which stationary wave is obtained.*

This condition is similar to the condition of attaining stationary waves on a plucked string of length 'L' (L = nλ). Since the number of possible time/div that an oscilloscope user can select from from the scope's front panel is limited, with no such restrictions made on the time period of the wave that one would be interested in studying, the chances of the condition represented by eqn(3.7) always being satisfied is remote. Thus, an additional control has been provided for the process of synchronization called the **Triggering** control. In this the sweep generator is so designed that it is not free running, i.e. on the completion of one linear increasing cycle the next cycle does not start own it's own (as is the case in a sine wave function generator where after the positive cycle, the negative cycle comes about followed by a positive cycle and so on till the function generator is switched off).

The modified circuit is such that the next ramp cycle comes about only when a pre-decided voltage level is given at the input. When the Y-amplifier's output attains the pre-decided voltage level, a signal goes to the ramp generator (see fig 3.2) and triggers it to generate a ramp signal. Figure(3.9c) shows that the ramp signal only appears after the pre-decided voltage V$_r$ is attained. Since every time ramp starts, signal is at the same potential at x=0, a stationary display is obtained. This pre-decided voltage level is set by the manufacturer and hence triggering done is said to be in **AUTO** mode. In **NORMAL** mode, the triggering level (earlier V$_r$) can be altered by the user using the level knob.

In short the Trigger Controller is actually a set of controls that determines when each sweep of the beam begins. In either case, adjusting the Trigger Controller is how to make the display stable. The trigger circuit is used to delay the time base waveform so that the same section of the input signal is displayed on the screen each time the spot moves across.

When the AT/NORM button is in the OUT position, triggering is automatic. This works for most signals. If you change the AT/NORM button to its IN position, the most likely result is that the signal will disappear and the oscilloscope screen will be blank. However, if you now adjust the **LEVEL** control, the display will be reinstated. As you adjust the LEVEL control, the display starts from a different point on the signal waveform. This makes it possible for you to look in detail at any particular part of the waveform. The **EXT** button should normally be in its OUT position. When it is pushed IN, triggering occurs from a signal connected to the trigger input, TRIG INP, socket. Similarly, the **LINE** switch allows from synchronization with the line frequency (50Hz as in India).

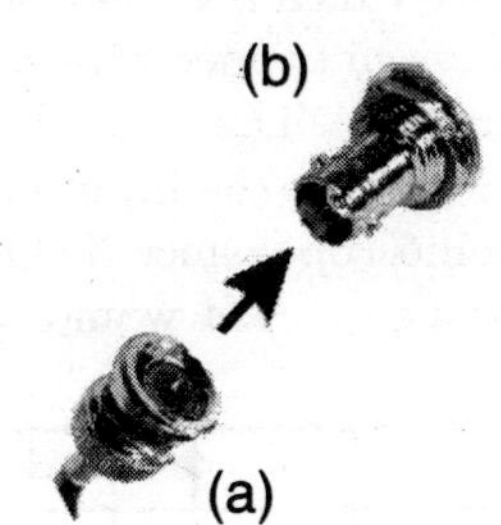

Figure 3.10: *The (a) BNC connector and (b) the socket on the scopes front panel where the BNC is clasped too (bayonetted). The center pin is the positive wire while the body of the bnc and the socket is connected to the ground of the circuit.*

3.2 Using the Oscilloscope

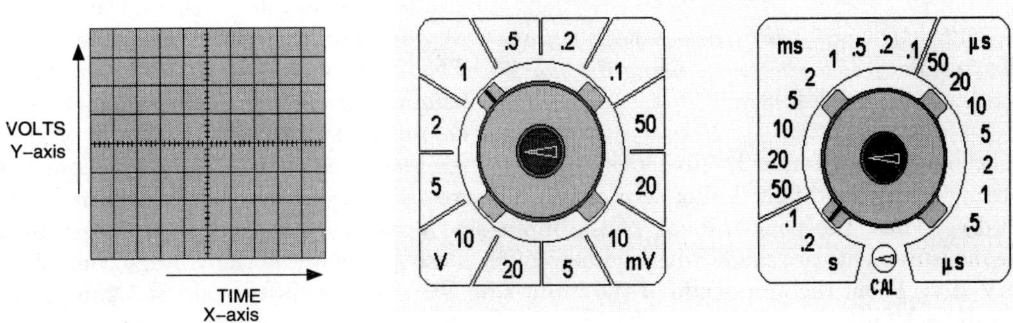

Figure 3.11: *The marking made on the CRT which enables an oscilloscope user to make measurements. Also, shown are the volts/div and time/div knobs of the oscilloscope. If not mentioned explicitly on the panel, they are identified by the units written, V or mV for the volts/div knob and ms or μs for time/div knob.*

First thing first, if one wants to examine the input or output of a circuit, one has to feed the signal to the oscilloscope. Hence, it would be useful to examine how input signals are fed to oscilloscopes. Usually the input is given by BNC's. **BNC** are essentially co-axial cables and stands for **B**ayone **N**eill **C**oncelman and are named after the engineers who developed it.

However, different books give different full-forms of BNC. One usage is British Naval Connector, while the popular American usage is "Baby N" Connector. Fig(3.10) shows the (a) BNC connector and (b) the socket on the scopes front panel where the BNC is clasped to. Coaxial cables are used to connect input signals to the oscilloscope. Essentially coaxial cable is a strand of copper wire which is covered by a plastic insulator followed by the ground wire consisting of a sheath or mesh of copper wire wound on the plastic insulation. The whole thing is then encased by a rubber layer. The coaxial cable by it's peculiar geometry eliminates any stray signal that might interfere with the input signal. The BNC was invented for proper coupling of the coaxial with the oscilloscope, since if the BNC is not used the point of contact of the coaxial cable with the oscilloscope's socket would act as the site of stray signals interference with the input.

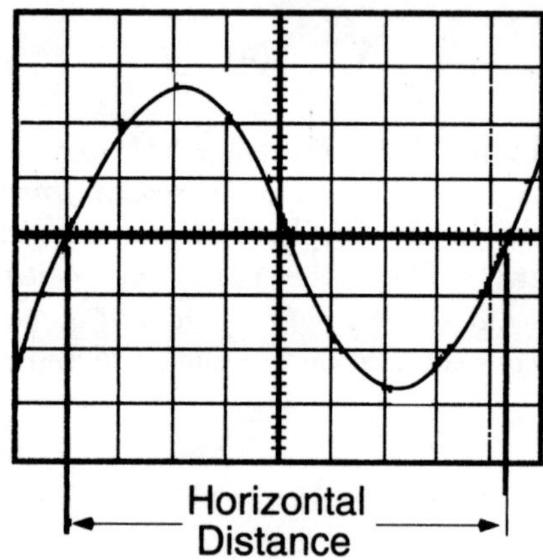

Figure 3.12: *Measuring the peak-to-peak voltage and the frequency of a sine wave using the markings made on the CRT screen.*

For making measurements of the waveform, the CRT screen of the oscilloscope has squares or divisions on the two axes, as shown in fig(3.11). Usually, these squares are 1 cm in each direction, with five smaller divisions within each square. Using the divisions on the vertical scale and information as to the setting of the volts/division (volts/div) knob (see fig 3.11) on the panel, one can find the amplitude of the input waveform. Similarly, using the divisions on the horizontal scale and the information of the set time/division (tim/div) knob, one can find the input waves time period or in turn it's frequency.

As an example consider the sine wave as viewed on the oscilloscope (see fig 3.12). The sine wave is symmetrical about the horizontal marked line (scale), which a user by default uses as the zero potential or ground line. The sine wave extends 2.6 divisions above this reference line as well as 2.6 divisions below the reference or ground line. Using sign convention, we say the sine wave spreads from +2.6div to -2.6 div. Thus, the peak-to-peak of the input sine wave is 5.2 divisions. It illustrate how voltage measurements are made with the help of oscilloscopes, assume the voltage control is set at 10mV/div. Then, the amplitude of the input sine wave (it's peak-to-peak) is 5.2 divisions $\times$ 10 mV/div = 52 mV.

The importance in proper selection of volts/division can be understood from making the peak-to-peak measurement of the same input signal with a higher volts/div. Higher volts/div makes the spread of the sine wave on the screen smaller. Let us say the same input signal of the above example is used measured with the volts/division set at 20mV/div. The sine wave spread would be from between +1.2 and +1.4 division to -1.2 and -1.4 division. Since the wave's peak is in between two divisions, for the purpose of measurement, one takes the wave's spread from +1.2 division to -1.2 division (The lowest division that has been crossed). The measured peak-to-peak would be 2.4 divisions $\times$ 20mV/div= 48mV. From this example, it is clear that the accuracy of measurement is better with lower volts/div settings. This is obviously due to

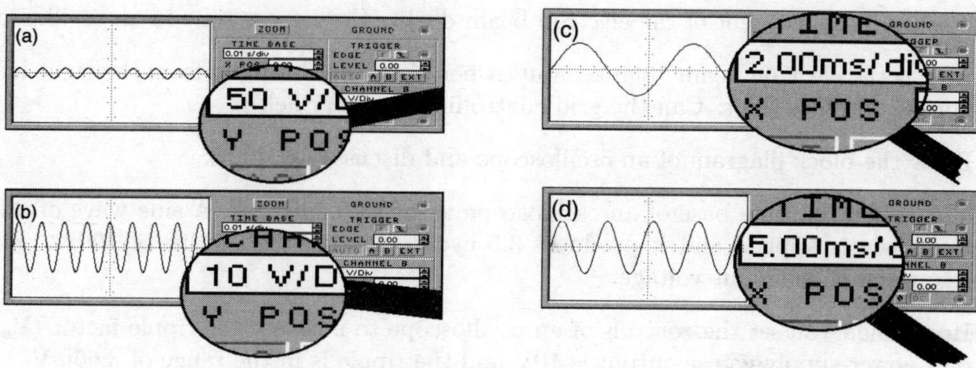

Figure 3.13: *The same waveform viewed with (a) 50v/div and (b) 10v/div. Lower volts/div makes the waveform appear larger on the screen. Similarly, a (c) lower tim/div (compared to that shown in d) allows for more accurate estimation of the input signal's frequency.*

the increased spread of the voltage waveform over a wider range of the screen for lower values of volts/div (size of the wave in the vertical direction increases) as can be seen from fig(3.13). The only trivial precaution to be made is that the volts/div is not set so small as to make the spread of the wave so large that it is beyond the range of the screen.

The frequency of the sine wave shown in fig(3.12) is also measured in a similar fashion using the horizontal scale and the tim/div setting. Assume that the scope's tim/div is set to $20\mu s$/div, from the figure (fig 3.12) one can measure the horizontal distance between the $\theta = 0$ and $\theta = \pi$ position. Using the markings on the screen we find it to be 8.2 divisions. Using eqn(3.6) we have the wave's time period as $20\mu s$/div $\times$ 8.2div $= 164\mu s$ or 0.164ms. The frequency of the viewed signal is thus 6.1KHz.

Note the front panel shown in fig(3.1) has two volt/div knobs, labeled I and II. These are to control the amplitude of the waveform or signal given in two different channels (CH1 and CH2, two inputs). These oscilloscopes are of particular use and are called 'Dual trace' oscilloscopes, which display two waveforms (Volts versus time) simultaneously. By this one can compare the input and output signals of any circuit.

Exercise

Q1. Two sinusoidal waves of voltage $Y_1(t)$ and $Y_2(t)$ are visible on the oscilloscope screen. In which mode is the oscilloscope operating? How can frequencies and time periods T of the signals be determined? How can the amplitudes of the voltage signals be determined?

Q2. The time needed for the electron beam to cover a distance of unit length in the horizontal direction on the oscilloscope screen is determined by the time/div switch. How is the time base (time/div) related with the rise time (T_t) of the saw-tooth wave?

Q3. Specify the movement of the electron beam during the retrace time in an oscilloscope?

Q4. How is the electron beam "blanked out" when they travel rapidly from the right-left hand horizontal plate (hint: Can the grid controlling intensity help)?

Q5. Draw the block diagram of an oscilloscope and discuss each block.

Q6. The calibrated time base of an oscilloscope is set at 0.2ms/div. A sine wave of unknown frequency is applied and it produces 3.5 cycles over a sweep width of 10div. Find the frequency of the input voltage.

Q7. How would you set the controls of an oscilloscope to measure the ripple factor (V_{ac}/V_{dc}) of a power supply whose output is 10V and the ripple is in the range of ±60mV.

Q8. What parameters of a saw-tooth waveform is controlled by the tim/div control of an oscilloscope?

Q9. Discuss how the electrons in the oscilloscope are (a) emitted and (b) then accelerated.

Q10. With the same signal applied to the oscilloscope (fig 3.12), the time-base setting is altered to 20ms/div. Discuss the effect this has on the trace on the CRT screen and measurement.

Chapter 4

Transformers

In this chapter we briefly discuss the invention of Lucien Gaulard and John Gibbs. In 1881, they demonstrated the first working transformer in London city, a device capable of either stepping up or down the input AC voltage level. The working of a transformer is explained using the various laws of electricity and magnetism one learns in physics. Before introducing the role of transformers in electronic circuits, it would be a good idea to go through the physics of this device.

When a current flows in a conductor a magnetic field (whose strength is pictorially represented by the number of flux lines) is generated around the conductor. This is nothing new and it's intricacies are explained by *Amperes Law*. Magnetic flux is analogous to current through a conductor, in the sense that there must be some force to drive it. In electric circuits, the motivating force is voltage (or the electro-motive force, EMF). In magnetic "circuits," this motivating force is magneto-motive force, or MMF. Magneto-motive force generated by a coil is equal to the amount of current through that coil (in amps) multiplied by the number of turns (n) of the coil (in a straight wire n=1). The magnetic flux (ϕ) generated is directly proportional to the MMF, i.e.

$$MMF \propto nI$$
$$\phi \propto MMF \tag{4.1}$$

for a coil with large number of turns and strong current in it, the MMF and in turn the magnetic flux generated is large.

Another law from the subject of electricity and magnetism is *Faraday's Law* which states that an induced emf is generated in a conductor kept in the presence of a varying magnetic field. Both, Ampere's and Faraday's laws come into play in the working of a transformer. Place two conducting coils near each other. Apply a time varying current, i.e. an AC in one of the coils. The varying current results in a varying magnetic field around the first coil.[1] The second coil if kept very near to the first would sense the varying magnetic field and an induced emf would be generated in the second coil as per Faraday's law. Thus, a time varying current is required in the first coil to generate an induced emf in the second coil (or more precisely secondary coil, fig 4.1

---

[1] A constant current, as in the case of DC, would give a constant magnetic field.

shows the circuital diagram of a transformer). For an induced emf to appear in the secondary coil, it should see the flux (ϕ) cutting it to be changing with time. If the primary coil has a *DC* current, ϕ is constant with respect to time. Thus, in this case, the only way the secondary would see a varying ϕ is when the secondary coil is moving with respect to the primary. That obviously is not a practical solution, hence, the transformer is essentially only useful for *AC* circuits.

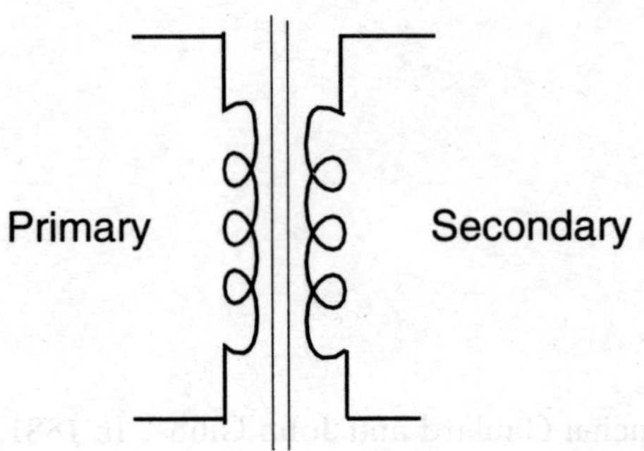

Figure 4.1: *Figure shows the circuital representation of a transformer. By convention, the input is fed into the primary coil, while the output is taken across the secondary coil.*

The induction of a voltage in one coil in response to a change in current in other coil is called mutual inductance. Mutual inductance is symbolized by "**M**" and share the same unit of Henry with normal (or self) inductance (L). Mutual inductance between two coils is given as

$$M = k\sqrt{L_1 L_2}$$

where L_1 and L_2 are the inductance of the two coils being used. The proportionality constant varies between 0 and 1 depending on the how well the two coils are coupled. Obviously, if the two coils are far apart, k=0. The induced emf in the secondary coil is given as

$$e = M\frac{di_{pri}}{dt}$$

To understand and develop equation of use to an average circuit designer we have to know the voltages that would appear on the primary and the secondary side of the transformer. This can be done by treating the two coils independently as inductor coils. One would be well aware of the voltage drop across an inductor, which is given as

$$v = L\frac{di}{dt}$$

Using the equation (eqn 4.1), the rate of change of current can be written as

$$\frac{di}{dt} \propto \frac{1}{n}\frac{d\phi}{dt}$$

giving the voltage across the inductor as

$$v \propto L\frac{1}{n}\frac{d\phi}{dt}$$

The inductance of the coil is related to the number of turns by

$$L = \frac{n^2 \mu A}{l} \tag{4.2}$$

where μ is the permeability of the core material, 'A' the area of cross section of the coil (in m^2) and 'l' the length of the coil in meters. The permeability is the ability of a material to allow the magnetic flux to pass through it. If no core is present (i.e. the coil is trapping air within it) $\mu = 1$.

Thus, the instantaneous voltage (voltage drop at any instant in time) across a coil is proportional to the number of turns of that coil around the core multiplied by the instantaneous rate-of-change in magnetic flux (dϕ/dt) linked with the coil. This was easily deduced from the relation between magnet flux, MMF and the current in the coil. Thus, the voltage drop across the primary and secondary coil would be given as

$$v_{pri} \quad \propto \quad \left(n \frac{d\phi}{dt} \right)_{pri}$$

$$v_{sec} \quad \propto \quad M \frac{di_{pri}}{dt}$$

$$\propto \quad M \frac{1}{n_{pri}} \frac{d\phi}{dt}$$

$$\propto \quad k \sqrt{L_1 L_2} \frac{1}{n_{pri}} \frac{d\phi}{dt} \propto k' n_{pri} n_{sec} \frac{1}{n_{pri}} \frac{d\phi}{dt} \tag{4.3}$$

$$\propto \quad \left(n \frac{d\phi}{dt} \right)_{sec} \tag{4.4}$$

If k=1, all the flux generated from the primary cuts the secondary and it only requires elementary skills in calculus to understand that if the input voltage (v_{pri}) is sinusoidal, the output would also be sinusoidal. Also, since we are dealing with ideal coils, no resistance is associated with the wire making up the coils, i.e. we are dealing with inductors. Hence, the primary current would be sinusoidal but lagging the applied voltage by 90°.

As stated, the coupling coefficient (K) depends on the geometry of the placement of the two coils. Better the coupling, better is the transformer functioning. However, a good coupling of the two coils is difficult to achieve. We now discuss two methods used to overcome the non-prefect coupling of the primary and secondary coils. Again, if k $\neq$ 1, then the primary and secondary inductance aren't perfectly linked, which would result in some loss of flux. That is, some of the magnetic field isn't linking with the secondary coil, and thus cannot couple energy to it. This lost flux is pictured as contributing an induced emf to a "stray" or "leakage" inductance present in series, both in the primary and secondary circuits (see fig 4.2). The self-inductance of the "leakage" inductance results in a reduced load voltage resulting from the voltage drop across it. Ideally we would like no voltage drop across the leakage inductance. A compromise solution is to design both primary and secondary coils with less inductance, the logic being that less inductance leads to less "leakage" inductance ($L_{leakage}$), for any given degree of magnetic coupling inefficiency. Thus, the voltage loss across this inductance would be small and in turn flux loss would be low. This results in a load voltage that is closer to the ideal. This method is not without flaw and would be discussed in the section "Losses in Transformer".

One way to increase the number of magnetic flux lines of the primary coil that cuts the secondary coil is to use a common core. The flux lines of the primary coil is trapped and amplified by the core which is carried over to the secondary coil. Thus, a concentration of

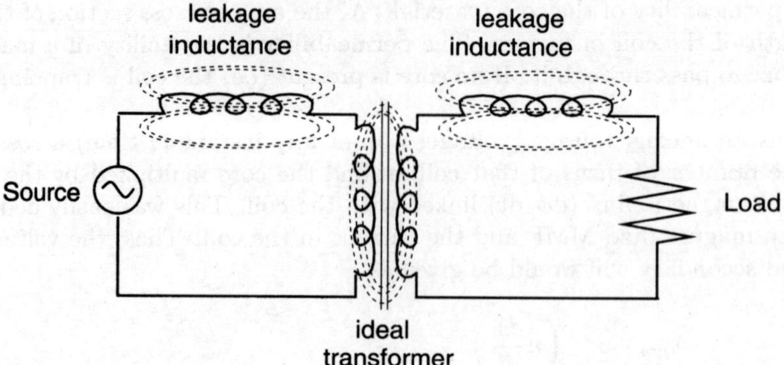

Figure 4.2: *Circuital representation of transformer which has not achieved prefect coupling between the primary and secondary coil (k ≠ 1). The loss of magnetic flux is represented by the leakage inductance.*

the flux lines can be achieved by winding both the primary and the secondary on a magnetic material like iron. Figure (4.3) gives an idea of how primary coil is wound on a magnetic *core*. Because of the core, coupling between the two coils can be as close as 100%. It is important to note transformer action would be possible between two coils with or without a ferrite core.

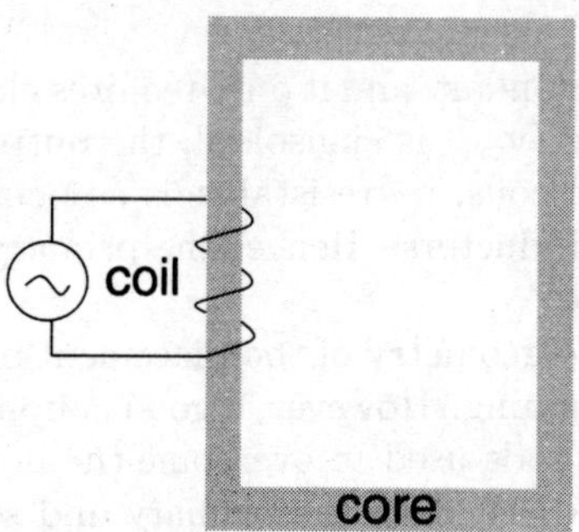

Figure 4.3: *To improve coupling between the two coils, trapping of lines of fluxes can be done by using a common ferromagnetic core.*

Though it might look incredibly trivial to wrap a coil of insulated wire around a loop of iron (ferromagnetic material) to get good coupling, the situation is complicated by the energy losses within the iron core. The effects of hysteresis and eddy currents distort and the current waveform, making it less sinusoidal and altering its phase to be lagging less than 90° behind the applied voltage waveform. Thus both methods of using core or coils with less inductance have their own merits and demerits. This will be discussed in a little more detail in the next section.

Considering a fairly good coupling of two geometrically similar coils is achieved, all the changing magnetic lines coming out of the primary coil would cut the secondary hence eqn(4.4) would give

$$\frac{v_{pri}}{n_{pri}} = \frac{v_{sec}}{n_{sec}}$$

$$v_{sec} = \left(\frac{n_{sec}}{n_{pri}}\right) v_{pri} \tag{4.5}$$

If $n_{pri} = n_{sec}$, a voltage of equal magnitude as the applied voltage will be induced in the secondary coil. However, it is not necessary to keep $n_{pri} = n_{sec}$. If $n_{pri} > n_{sec}$ the voltage across

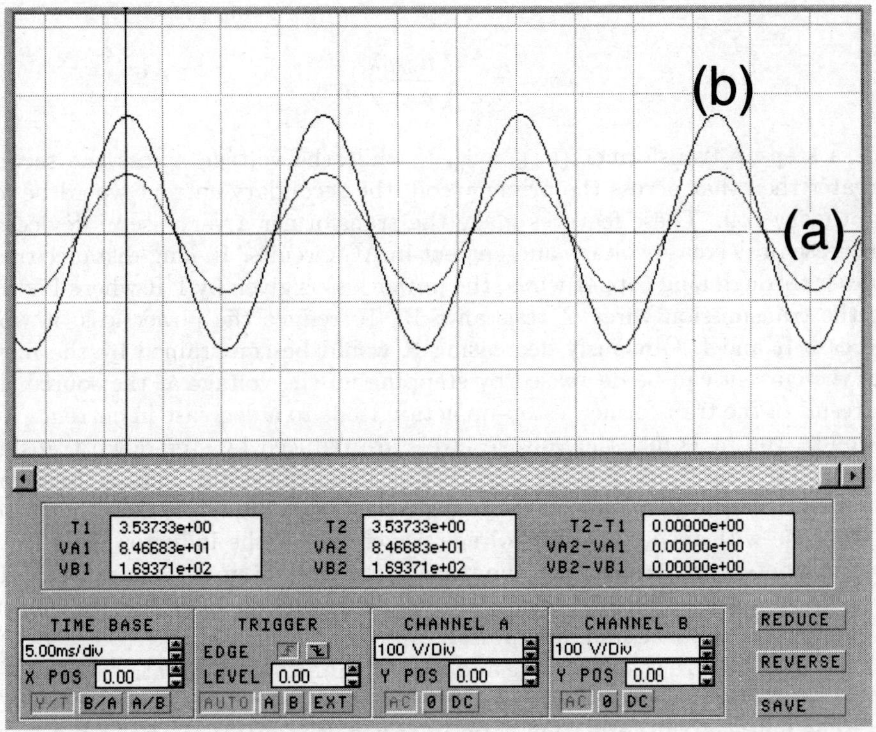

Figure 4.4: *Comparing the (a) input and (b) output voltages fed to a step-up transformer whose transformation ratio is 2.*

the secondary coil would be less then the primary coils (input) voltage. Such transformers having $n_{pri} > n_{sec}$ are called step down transformer. If the reverse is put to use, i.e. $n_{pri} < n_{sec}$, one has a step up transformer, where the output voltage is greater than the input voltage. Figure (4.4) shows the output of a step-up transformer as seen on the oscilloscope. A sinusoidal input voltage of (a) 84.67v becomes (b) 169.37v when fed to a step-up transformer whose transformation ratio (n_{sec}/n_{pri}) is 2. One might jump to the conclusion that the Law of Energy Conservation in physics, namely that energy cannot be created or destroyed (only converted), is being violated. But voltage is not a measure of energy, thus applying the Law of Energy conservation on the above equation is erroneous. However, power is a measure of energy, i.e.

$$p_{pri} = p_{sec}$$
$$v_{pri}i_{pri} = v_{sec}i_{sec}$$
$$v_{pri}i_{pri} = \left(\frac{n_{sec}}{n_{pri}}\right)v_{pri}i_{sec}$$
$$i_{pri} = \left(\frac{n_{sec}}{n_{pri}}\right)i_{sec}$$

or

$$i_{sec} = \left(\frac{n_{pri}}{n_{sec}}\right) i_{pri} \qquad (4.6)$$

Thus, for a step-up transformer ($n_{sec} > n_{pri}$), while the voltage across the secondary coil would be greater then that across the primary coil, the secondary current would be lower than that in the primary coil. These features make the transformer a very useful device, where we can easily increase or decrease voltage and current in AC circuits. In long-distance transmission of electric power through long copper wires, the power loss is given by I^2R where I is the current traveling in the transmission wires of resistance R. To reduce the power loss, it would make sense to decrease R and I. Obviously decreasing R would be constrained by the metal put in use, however the current can be decreased by stepping up the voltage at the source. Increasing voltage as a result of the transformer's step-up action leads to a decrease in current. At the load end (the user like you at home) the voltage levels are reduced by step-down transformers for safer operation. One may think of other devices that can achieve this voltage hike and current reduction, however, view any one side of the transformer as an inductor, then the current would be 90° out of phase with respect to the voltage waveform. If the inductor is perfect (no wire resistance, no magnetic core losses, etc.), the transformer will dissipate zero power. Thus, while other devices might achieve voltage hike and current reduction only the transformer would achieve the same without consuming any power.

Eqn(4.5) and eqn(4.6) are called voltage and current transformation ratio respectively. These transformation ratio of a transformer are related to the number of turns in the two coils of the transformer. The voltage transformation ratio in terms of inductance of the coils is (obtained using eqn 4.2)

$$= \sqrt{\frac{L_{sec}}{L_{pri}}}$$

Another use of transformers in electrical circuits is in its ability to isolate the circuit on it's primary side from the secondary side. Because the two coils are not electrically connected, only the magnetic field between them. Hence, the input and output circuits as connected to the primary and secondary side of the coils are electrically isolated. This is an important use of the transformer.

4.1   Losses in Transformer

We have stated that using ferromagnetic material achieves good coupling. Though, industry uses this method, it is not without flaw. The core introduces energy losses as well as distortions in the output current waveform due to eddy currents and the effect of hysteresis respectively. Let us understand these new terms.

When a electric current flows through a conductor, it generates a magnetic field. This magnetic fields then tries to magnetize the core material. This magnetization force (H) is measured in Amperes per meter (i.e. depends on the current applied). As a result of this magnetization force, the core starts getting magnetized and is reflected by it's developing of core flux density (B). The flux density appearing as a result of the magnetizing force is given as

$$B = \mu H$$

This shows a linear relationship between the magnetization force and the resulting flux density, with μ being the core materials permeability. Higher the materials permeability, the easier it is magnetized by a magnetizing force. Though, the above relationship shows linearity, however, B does not increase to infinity with increasing magnetization force. It in fact saturates, corresponding with the point where all the domains of the magnetic material of the core have aligned itself in one specific direction. This saturation is reflected in the B-H hysteresis curve of the magnetic core material. Thus, if very large current is given in the primary of the transformer, due to the saturation, the increasing flux in the primary is not reflected into the core and in turn to the secondary. This results in a distortion in the secondary output. Core saturation is a very undesirable effect and is avoided through good design. For example, the magnetization force is reduced by controlling the number of winding (n) since,

$$H = \frac{0.4\pi n I}{l}$$

Hence, if the transformer is being designed for large currents, the number of coils in the primary is reduced. Also, the flux density (B) is controlled by proper selection of core material (μ) and also the distortion may be minimized if core is given large cross-sectional area which results in increase in B versus H curves linearity, simply because the magnetization force has more domains to rotate. But this makes the transformer bulky and expensive.

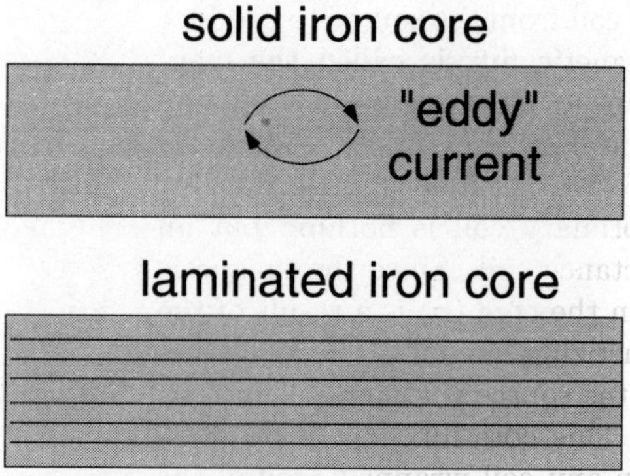

In lieu of good coupling of the primary and secondary coils of the transformer, which was stated as difficult to achieve, earlier we suggested a compromise solution by having transformer coils

Figure 4.5: *Eddy currents are generated in conductors when placed in changing magnetic field. These are reduced by slicing the conductor and insulating neighboring layers by lamination.*

of less inductance. The idea being that less inductance leads to less "leakage" inductance and in turn less loss due to poor coupling. One might ask, "If near-ideal performance can be achieved only by decreasing the leakage inductance, then why worry about coupling efficiency (k)? Since it is near impossible to build a transformer with perfect coupling, but easy to design coils with low inductance, then why not build all transformers with low-inductance coils and have excellent efficiency even with poor magnetic coupling?" To understand this one must realize, with less inductance in the primary winding, there is less inductive reactance (X_L) resulting in much larger primary current and in turn magnetizing force on the core's material. Thus, sending the core easily to saturation. Thus, while a substantial amount of the current through the primary winding merely works to magnetize the core, the useful energy transferred to the secondary winding and load is reduced. Thus, designing a efficient transformer is an involved job for the designer.

Another problem that appears in transformers using cores is the appearance of "Eddy Currents". Eddy currents are induced currents that are set up in a solid conductor (even with a sheet like geometry/ shape) when it is kept in the presence of a changing magnetic field. The current moves in circles as if the solid sheet is made up of numerous closed paths (coils). The plane of the eddy current's circular path is perpendicular to the magnetic field that causes it. Eddy currents lead to the heating of the core and thus acts as a contributor to the loss processes in the transformer. To minimize eddy current loss, the solid conductor is sliced into plates and lamination are placed in parallel plates of the ferromagnetic core material (see fig 4.5). This reduces the area of cross section of the conductor on which the magnetic field is being applied. A reduced cross-sectional area reduces the induced emf generated in the conductor and in turn the induced current.

4.2   Polarity

The current drawn by the primary coil from the source to produce a magnetic flux is called the magnetizing current. This magnetizing current lags the supply voltage by $90°$. This is obvious since, ideally, the primary coil is nothing but an inductance coil. Since the magnetic flux in the core (ϕ) is a result of the magnetizing current, it too lags behind the source voltage waveform by $90°$. This core flux induces a voltage in any coil wrapped around the core. However, what is the phase relationships for voltage and current between primary and secondary circuits of a transformer? While, normally, there is no phase difference between the primary and secondary voltages, transformers are available where they are 180° out-of-phase with respect to each other. If we were to look at an unmarked transformer, we

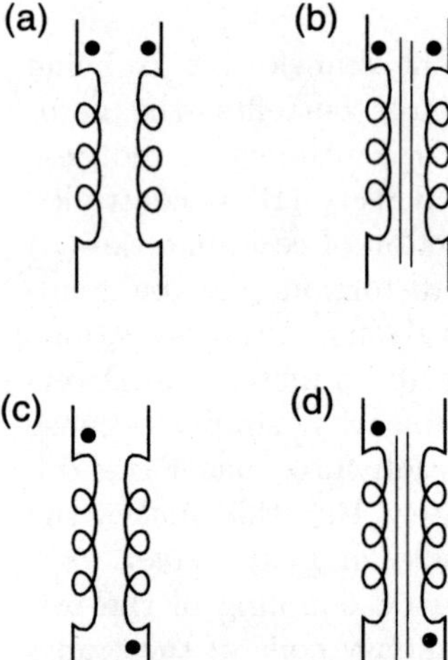

Figure 4.6: *Transformer schematics shown with dot convention, where (a & b) show the input and output voltages to be in phase while (c & d) represents the output to be out of phase with respect to the input by a phase difference of π. The parallel lines in the schematics of (b) and (c) represents coils are wound on a ferromagnetic core, while (a) and (c) represents air core.*

would have no way of knowing which way to hook it up to a circuit to get in-phase (or 180° out-of-phase) voltage and current. Before detailing how to distinguish between the two type of transformers, let us first understand how circuital representations are made to distinguish between the two and how a phase difference may or may not be introduced by a transformer.

Typically transformers schematic diagram labeling follow the "dot convention", where a dot is used for polarity marking showing which end of the winding is which, relative to the other windings. The placement of dots next to the top ends of the primary and secondary winding tells us that whatever instantaneous voltage polarity seen across the primary winding will be the same as that across the secondary winding. In other words, the phase shift from primary to secondary will be zero degrees. Figure(4.6) shows the two possible representations. While, fig(4.6a & b) shows zero phase difference between primary and secondary voltages, fig(4.6c & d) represents a phase difference of 180°. Sometimes dots are omitted (especially in power transformer), and "H" and "X" labels are used to label transformer winding wires, where "H" designations for the higher-voltage winding and "X" designations for the lower-voltage winding. The subscript number is supposed to represent winding polarity. The "1" wires (H1 and X1) represent where the polarity-marking dots would normally be placed.

To understand the physics, consider fig(4.7). To explicitly explain the direction of wire coiling, crosses and dots have been used. While the crosses indicate that the wire is going into the paper, the dot placed along the wire indicates that it is coming out of the paper as seen by the reader (to look up, remember that the coiling is done from top to bottom, i.e. ↓). The primary coil can now be understood to be wound in an anti-clockwise manner (look at coiling around the core's cross-section from the top). Consider at an instant the direction of current flow is as indicated in the figure. The coil would act as a magnet with the direction of magnetic field pointing in the upward direction (as shown in the figure).

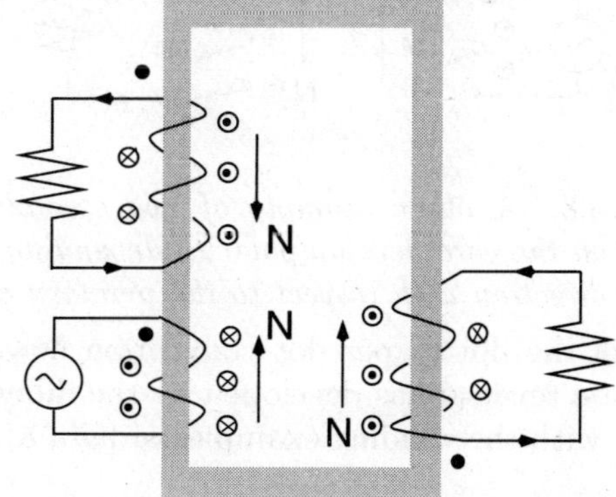

Figure 4.7: *Two secondary coils wound on the core develop polarity depending on their coiling direction with respect to the primary coil.*

The direction being given by the **right hand rule**. Right hand rule states that if one coils the fingers of the right hand in the direction of current flow, the thumb points in the direction of the field. Thus, magnetic lines of force flow in the core from top of the primary coil and enter from the bottom of the primary coil. That is, the coil behaves as a magnet, with the north pole pointing upward and the lower end of the coil acting as the south pole.

There are two possible secondary windings in this example. The example has been selected for simplicity of explanation. Especially, the secondary coil placed on the same arm of the core. An induced current would start flowing in it, in response to the changing magnetic flux. The direction of the induced emf is again determined by the right hand screw rule. While initially we determined the direction of the current and then analyzed the direction of the field as in the case of the primary coil, here for the secondary coil, we determine the direction of the field and then analyze using the right hand rule the direction of the induced current. The direction of the field in the secondary coil should be such that it opposes the very cause of the induced current (Lenz law). Hence, the secondary coil's lower side (fig 4.7) should respond as a magnet's north pole. Now from the right hand rule, one can work out that the current should flow in clockwise

direction in the secondary coil (note that the coils winding is also in clockwise direction). As per the dot convention, a dot is placed at the side which is at a higher potential (+ive of the battery). Since, the primary is a load, the current flows into the dotted side while in case of the secondary, which acts as a source, the current flows out of the dotted side.

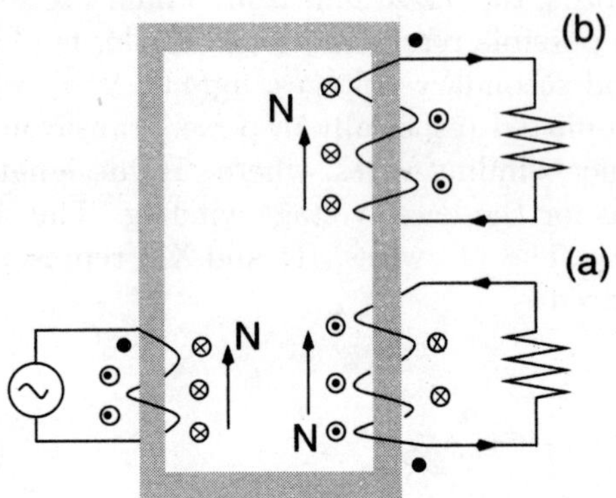

Figure 4.8: *Another example of two secondary coils wound on the core develop polarity depending on their coiling direction with respect to the primary coil.*

Consider the second coil acting as a secondary coil (fig4.7). This coil is wound diametrically opposite to the primary coil on the same core in an anti-clockwise manner. Since, the flow of the flux in the core due to the primary's magnetizing current which makes the primary coil's top end the north pole, Lenz's law demands the induced emf in the secondary coil under consideration should be such that the resulting magnetic field should have it's north pole pointing in the upward direction. By the right hand rule the current should be in counter-clockwise direction (as is the direction of the winding). The position of the dot as per dot convention flows naturally and is indicated in the figure. Note, the phase reversal has developed by the difference in winding direction. Let us investigate this further with the winding examples of fig(4.8).

While the explanation given above example would hold true for the primary and secondary coil labeled (a) (fig 4.8a), the point worth discussing here is response of secondary coil (b) to the flux generated by the primary coil. Due to Lenz law, the responding magnetic field due to the induced current in coil (b) should point in the same direction as that of coil (a), after all the cause of the induced emf in both the coils remains the same. For this, the induced current has to be in a counter clockwise direction, as can be deduced by the right hand rule. For the clockwise coiling of the secondary coil (b), a counter clockwise current can only exist for the shown current direction. This gives a polarity to the secondary coil (b) which is opposite to that obtained in coil (a). Thus, by merely changing the coiling direction of the secondary coil, the polarity of the output can be reversed.

As stated at the start of this section, manufacturers never indicate the polarity of the transformer. The dot representation is restricted to circuital schematics only. Rarely does any user require this information. All calculations and realization of circuits are done with respect to the secondary and not with respect to the input coil, explaining why the polarity is not indicated. However, if for academic interest one is wants to find the same, the following procedure can be used.

Any one of the primary and secondary terminal's wires are connected (same side wires are shorted temporarily) and a small test voltage is applied to the transformer primary. The resultant voltage is measured between the free wires of the primary and secondary coil (see fig 4.9). If the transformer is designed to introduce a phase difference between input and output,

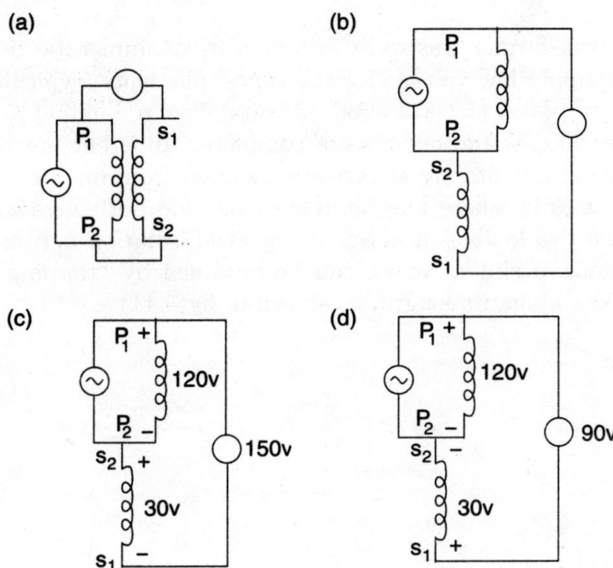

Figure 4.9: *Schematics of the procedure used to identify the transformer's polarity.*

this voltage would be greater than the input voltage. However, for transformers which do not introduce any phase difference this voltage would be less than the input voltage. This is easily understood by comparing the equivalent circuit of the circuital arrangement we have made in fig(4.9c & d). A step down transformer with transformation ratio 0.25 is given an input of 120volts. While for transformers introducing phase difference, the primary and secondary voltages add in series, for the second case they add in anti-series.

4.3   Winding Configurations

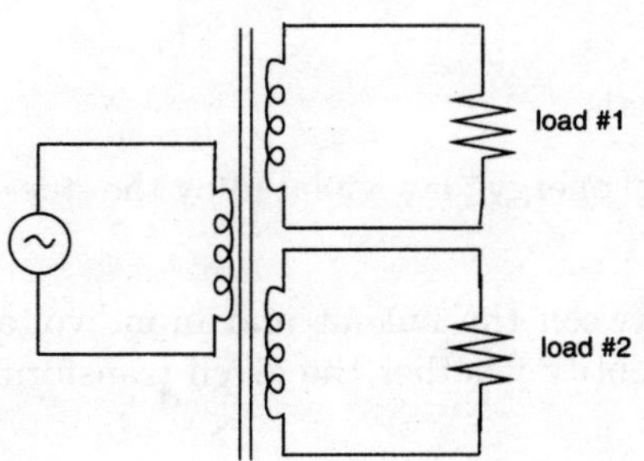

Figure 4.10: *A transformer with two secondary coils.*

Transformers are very versatile devices. The basic concept of energy transfer between mutual inductors explained the functioning of the transformer with a single primary and a single secondary coil, but transformers don't have to be made with just two sets of windings. Consider the transformer shown in fig(4.10).

Here, three inductor coils share a common magnetic core, magnetically "coupling" or "linking" them together. The setup allows for two loads to be driven (given energy to) by two secondary coils.

The relationship of winding turn ratios and voltage ratios seen with a single pair of mutual inductors still holds true here for multiple pairs of coils.

It is entirely possible to assemble a transformer such as the one above (one primary winding, two secondary windings) in which one secondary winding is a step-down and the other is a step-up.

In fact, this design of transformer was quite common in vacuum tube power supply circuits, which were required to supply low voltage for the tubes' filaments (typically 6 or 12 volts) and high voltage for the tubes' plates (several hundred volts) from a nominal primary voltage of 110 volts AC. Not only are voltages and currents of completely different magnitudes possible with such a transformer, but all circuits are electrically isolated from one another. This particular transformer is used in circuits where one intends to provide both high and low voltages and electric isolation between two loads is a must. If electrical isolation between secondary circuits is not of great importance, a similar effect can be obtained by "tapping" a single secondary winding at multiple points along its length, as shown in fig(4.11).

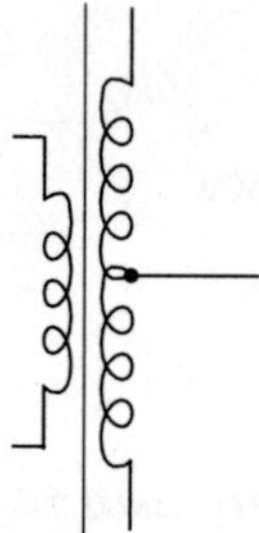

Figure 4.11: *A center tapped transformer where the secondary coil is divided into two equal parts by the tap, a connection made along the length of the secondary coil.*

A tap is nothing more than a wire connection made at some point on a winding between the ends. The winding turn-voltage magnitude relationship of a normal transformer holds true for all tapped segments of windings. The use of center tapped transformers would be seen while studying full wave rectifiers.

Exercise

Q1. Explain why a transformer can not work for DC signals.

Q2. If voltage is a measure of energy, is the "conservation of energy" law violated by the step-up transformer. Explain.

Q3. Does a transformer introduce a phase difference between the output and input voltage signals. Explain how someone can experimentally identify whether the given transformer will introduce a phase difference or not?

Q4. Derive expression for the coefficient of coupling of two magnetically coupled coils.

Q5. Two magnetically coupled coils have a mutual inductance of 32mH. What is the average emf induced in one, if the current through the other changes from 3 to 15mA in 0.004sec. Given that one coil has twice the number of turns in the other. Calculate the inductance in each coil. Neglect leakage.

Chapter 5

Diode

Everyone would be aware of the fact that copper is a metal and is characterized by its shining luster, it's malleability which allows it to be drawn into wires and it's ability to conduct electricity. One reason why copper is used to transmit electricity from the power station to our homes is that for the enormous length of wire (distance from the power station to our homes) the resistance offered by the wire is quite low. This in turn implies that the power loss (I^2R) in transmission is low. Such metallic (marked by low resistance) materials are placed in the first and second group of the periodic table. These elements have either one or two electrons in their outermost orbit, which easily escapes from the atom and freely moves about in the solid to conduct electricity. As you move from group I and II, the conductivity (ease with which current conducts) decreases.

A material of the periodic table's forth group, for example silicon (Si), in its pure state thus does not conduct easily. This is because no free electrons are available for conduction. In fact each Si atom would have four covalent bonding with four immediate Si neighbors. Two electrons would be there in each bond, with one electron being contributed by each neighbor. If the material is heated appreciably, then the covalent bonds would break. The broken bonds would release electrons which would then be free to move about in the solid and conduct electricity. Thus, elements like Silicon do not conduct at room temperature (i.e. behave like insulators). However, on heating their conductivity increases. Such materials are classified as **semi-conductors**.

The conduction of the pure semiconductor may be raised by adding impurities in the pure sample of silicon. Consider an impurity from the periodic tables fifth group (like Arsenic, Phosphorus etc) is added into the sample. Since the material happens to be from the fifth group, the impurity atoms have five electrons in its valence shells (the outer most orbital). Four of these electrons would enter in covalent bonding as if they were Si atoms, leaving the fifth in search of a partner. A slight increase in temperature would release this electron from the Arsenic or Phosphorus atom. The free electron leaves behind it a positively charged massive immovable ion (As^+, P^+). While the free electron would be mobile to move around and contribute to flow of current, the massive ions do not contribute to the current. The controlled addition of fifth group elements give rise to an N type semiconductor. The process of adding impurities to manufacture N type material is called **doping**. The situation becomes interesting when silicon is doped with group three materials like Gallium or Boron. These atoms contribute their three outer most electrons for covalent bonding with silicon atoms. However, the forth electron pair is left without an electron for pairing. This absent charge carrier is called "an *hole*". This acts as a charge carrier whose behavior is opposite to that of electrons. While electrons under the action of an applied electric field moves from lower potential (negative plate) to a higher potential (positive plate), in the opposite direction of conventional current which flows from higher potential to a lower potential, holes move in the same direction of conventional current. One might ask how does a "lack of electron in covalent bond" carry current? Figure (5.1) explains this.

The figure shows a bar of P-type semiconductor. The filled circles represents electrons while the unfilled circles represents the lack of electrons or holes. Carefully note, the representation in figure (5.1) has only one hole for the sake of brevity and is not a comment on the electron/ hole concentration of the doped semiconductor. The first diagram shows the hole is present on the right hand edge of the semiconductor sample. If now an electric field is applied with the positive plate on the right hand edge and the negative plate on the left hand side, the electrons in the sample would move towards the positive plate. The electrons from an adjacent Silicon bonding would escape for conduction, creating or leaving behind a vacancy or a hole. In it's motion towards the positive plate it would be captured by the nearby Silicon bondings which have a hole or vacancy. With this capture, this vacancy would disappear. The net result would be that it appears as if the hole is moving from right to left, towards the negative plate, giving rise to what is called the **"hole current"**. The direction of motion of the holes are opposite to the direction of motion of the electrons. However, since the current due to electrons is in opposite direction to the electron's motion, the net current in a semiconductor is given as

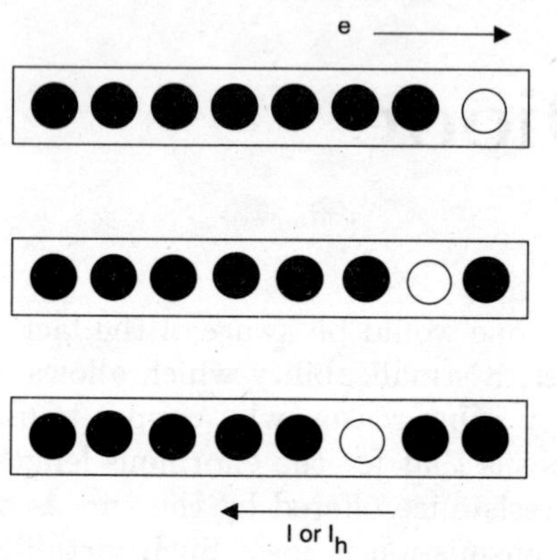

Figure 5.1: *The displacement of a neighboring electron and filling of a vacate bond (hole) leaves a hole at the position where the electron started from. It gives an apparent notion of hole moving in the opposite direction of the electron. Thus, the hole current has the same direction as the conventional current.*

$$I = I_e + I_h$$

In N-type semiconductors, doped with group V elements, the electrons would be in larger number than holes. Hence, I_e would be more dominant than I_h and hence would be called the majority current. Indeed ideally I_h should be zero, however, impurities would make sure I_h does exist. This would be only a small fraction of the majority current and then be aptly called the minority current. In P-type semiconductor, the situation would reverse with I_h becoming the majority current and I_e the minority current.

A junction diode maybe be considered to be formed when a P type semiconductor is bought in contact with a N type material. While holes are the majority carriers in the P type material, electrons constitute the majority in N type material. The concentration gradient present across the junction leads to diffusion of positive carriers into the N-type material and electrons into the

P-type material. The electrons and holes on meeting **recombine** releasing energy. Physically, the recombination process implies the electrons return from the conduction band to the valence band. This is the physicist way of describing a semiconductor. They visualize the electrons of the semiconductor are kept in order within covalent bonds with energies below a restrictive level called the valence level.

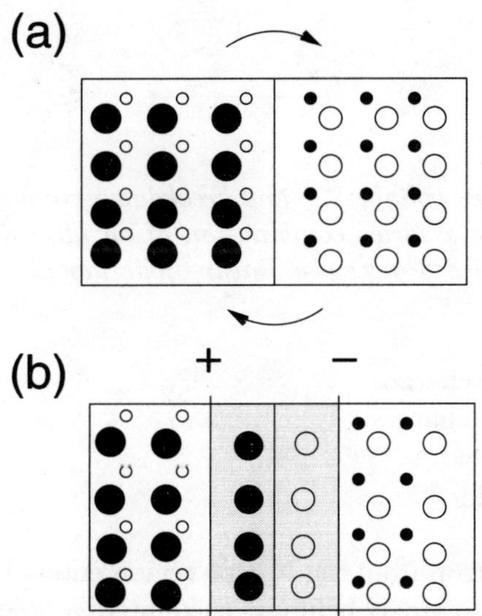

Figure 5.2: *The transfer of mobile charges across the interface leaves behind a collection of massive ions around the interface which is devoid/ depleted of mobile charges. The shaded region represents the depletion region.*

Energy levels that the free conducting electrons have are well above this valence level, so much so that the least mobile free electron's energy marks the start of all possible energy levels that free electrons can have (the least energy level is called the conduction level). Even this conduction level is well above the valence level. The difference between the conduction band and the valence band is like a barrier the electron has to jump (called the band gap). The extra energy required for the jump from valence band to conduction band is either provided by heating or by applying an external electric field (the extra energy has to be greater than the band gap). When an electron jumps out of bonding (valence level) to freedom (conduction level) with this extra energy, the free electron contributes for the conduction (I_e) as also the hole (I_h) that has been created by the electron's escape. So when the free electron enters into the vacancy of some covalent bonding, it returns from the conduction band to the valence band and the hole disappears. The extra energy the electron had acquired to jump into the conduction band is released in its return trip. The magnitude of energy released depends on the energy structure of the semiconductor being used. In normal diodes, the energy released is in the form of heat or in the infra-red range of spectrum. Special diodes are present where the energy released in the visible spectra. Such diodes are called Light Emitting diodes (LED).

Due to the migration of charges from N region to P region and visa verse due to diffusion, the N type material has exposed positive ions while the P type material has negative ions near the junction. A region develops at the junction called the depletion region which is depleted of any mobile free charges. This region is essentially un-doped or just intrinsic silicon. The build up of ions around the junction gives rise to the development of an electric field at the junction or a potential, called the barrier potential which appears as V_K in the IV characteristics. For silicon diodes, the typical barrier potential is 0.7volts while for germanium diodes it is only 0.3 volts. This barrier eventually prevents further diffusion of charges across the barrier and an equilibrium is reached. The equilibrium is achieved with

$$N_A x_p = N_D x_n \tag{5.1}$$

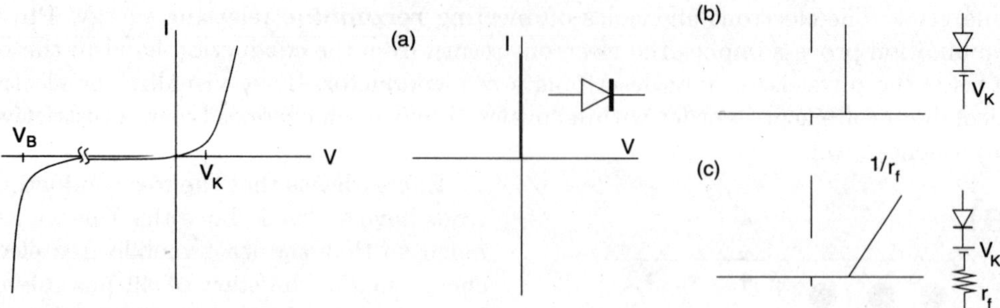

Figure 5.3: *The I-V characteristics of a diode. The knee voltage V_K and breakdown voltage V_B are indicated. A practical diode maybe considered to be a series combination of an ideal diode, a battery of potential V_K and resistance r_f. This is called a piecewise linear diode model.*

where, the depletion width is given by $W = x_p + x_n$

x_p is the extent in the P side which is lacks mobile charges,

x_n is the extent in the N side which is lacks mobile charges,

N_A is the hole carrier concentration in the P side and

N_D is the electron carrier concentration in the N side.

What this equation represents is that when an electron from the N type region enters the P type region, the length (x_p) to which the electron has to travel before it encounters a hole and disappears by recombination depends on the hole concentration (N_A). If the hole concentration is large, the electron would not travel very far (small x_p) and would in a short travel be taken up for recombination. The site of recombination would be left with a massive negative ion which would be immovable and would be depleted of any mobile conducting charge.

So to achieve any flow of current, the diode has to be given an external voltage to overcome the barrier potential (V_K) that has developed across the depletion width (see fig 5.3). Diodes are referred to as non-linear circuit elements because of its non-ohmic I-V characteristics (fig 5.3). The diode's characteristics appear different in the first and third quadrant, i.e. the device is not bilateral. A resistor (ohmic resistance) is bilateral, with the first and third quadrant of the IV characteristics being identical. The multimeter would give the same value of R (resistance), irrespective of which end of the resistor the multimeter's positive probe is kept. This is not the case with diodes. One would have to bear in mind which side of the diode is the positive terminal of the battery/ power supply is being applied. The diode is said to be forward biased when the P side is kept at a higher potential w.r.t. the N side, while it is said to be reverse biased visa verse. The diode does not conduct immediately on application of a forward potential. This is due to the fact that charge carriers can only cross over when they have energy to overcome the barrier potential. Once the barrier potential V_K is overcome, the diode acts as a piece of resistance bar, whose resistance is small. When reversed biased, the diode's depletion width increases. The increase in depletion width is due to electrons of the N region being pulled towards the positive plate of the source and similarly holes of the P region being pulled towards the negative terminal. Thus, more immovable lattice ions are left around the interface giving rise to a enlarged depletion width. A small current however does flow even in reversed biased condition. Present day manufacturing allows to achieve very pure samples of intrinsic (pure Si) and extrinsic (doped P or N type Si) semiconductors, however, impurities remain. The impurity we are talking of here, contribute electrons in the P type region and holes in the N type region. Both these charges obviously would be in minority in their respective region. When the diode is reverse biased, these minority charges see themselves properly biased and move across the boundary. Since the minority charge carriers are very small in number, the current is very small and is fairly independent of the applied reverse voltage. Hence, the reverse current is called reverse saturation current, I_{sat}. Since the number of minority charge carriers in a pure sample would be very small, the current would be in μA or nA, making it very difficult to measure. Increasing, the reverse voltage to a very high value leads to breaking of covalent bonds and the acceleration of released carriers. These carriers have sufficient energy to break more covalent bonds and releasing more carriers. This is very much like an avalanche in mountains, and is rightly called **Avalanche Effect**. The rapid multiplication of free carriers leads to a large current flow. Usually, since the breakdown occurs at a high reverse voltage and is accompanied with large current, the heat dissipation (heat released) in diode destroys the diode. The diode current is given by the equation

$$i_D(v_D) = I_{sat} exp\left(\frac{q v_D}{KT} - 1\right) \tag{5.2}$$

where v_D is the voltage applied across the diode,
I_{sat} is the reverse saturation current of the diode,
K is Boltzmann constant and T is temperature in Kelvin.

The equation kT/q describes the voltage produced within the P-N junction due to the action of temperature, and is called the thermal voltage of the junction. At room temperature, this is about 26 millivolts. For explaining the behavior of many diode circuits we do not need to use the diode equation and may think of an ideal diode has having two regions: a conduction region of zero resistance and an infinite resistance non-conduction region (fig 5.3 a). For many circuit applications, this ideal diode model is an adequate representation of an actual diode and simply requires that the circuit analysis be separated into two parts: forward current and reverse current. The I-V characteristics are exactly like that of a mechanical switch one uses to ON or OFF the light bulb. In subsequent sections of this chapter we introduce the applications of diode making use of this switching behavior. The practical diode can be imagined to be an ideal diode in series with a resistance and voltage source whose magnitude is same as the forward resistance of the diode and the barrier potential, V_K respectively (see fig 5.3 b and c).

$\boxed{\text{Example 5.1:}}$ Analyze and find the current in the load resistance $R_L = 1.8K\Omega$ for the given circuit. The forward resistance of the diode is 200Ω and knee voltage is 0.7v.

Using the piecewise linear diode model, and applying KVL, we have

$$\begin{aligned} V &= V_K + I(R_L + r_f) \\ 10 &= 0.7 + I(1.8 + 0.2) \end{aligned}$$

$$\boxed{\text{Answer I=4.65mA}}$$

5.1 Diode Specifications

In addition to forward voltage drop (V_K), there are many other ratings of diodes important to circuit design and component selection. Since semiconductor manufacturer and appliance manufacturers are different, it is necessary to pass the complete information of the device to the circuit designer and appliance manufacturer. Semiconductor manufacturers provide detailed specifications on their products in publications known as data-sheets. Data-sheets now-a-days are readily available on the Internet. A typical diode data-sheet will contain figures for the following parameters:

1. **Maximum (average) forward current:** The maximum average current the diode is capable to conduct in its forward bias mode. This essentially depends on the diode's thermal limitation, i.e. how much heat can the PN junction handle, Ideally, this figure should be infinite.

2. **Maximum DC reverse voltage:** The maximum voltage that the diode can withstand when reverse-biased on a continual basis. Ideally, this should be infinite.

3. **Typical junction capacitance:** The depletion region acts as a dielectric (insulator) separating the anode and cathode connections, i.e. behaves like a capacitor. The capacitance associated with the depletion region is very small, usually in the range of picofarads (pF).

4. **Peak Inverse Voltage:** The ability of a diode to withstand reverse bias voltages is limited, as seen from the breakdown occurring in practical diodes (fig 5.3). A diode's maximum reverse-bias voltage rating or in other words the maximum reverse voltage for which the diode will not conduct is known as the Peak Inverse Voltage, (PIV). The PIV rating of a diode increases with increased temperature and decreases as the diode becomes cooler. Typically available rectifier diodes have a PIV of around 50 volts at room temperature.

Table 5.1: Important data of Diode 1N4001

Parameter	Symbol	Value
Max forward surge current	I_{FSM}	30A
Max peak reverse voltage	V_{RPM}	50V
Max average value of full wave rectified current (120 Hz)	I_o	1 A
Max DC reverse voltage	V_R	50 V

All these parameters are temperature dependent, hence, the manufacturers usually provide various graphs of the component ratings as a function of temperature. This gives the circuit designer a better idea of how the device would perform. Table 5.1 gives an example of important data of a very common diode, the 1N4001 diode. The data sheet of components are more detailed, where the devices dimension and mechanical features are also listed along with the manufactures address and mode of order placement. Fig(5.4) gives a section of such a data sheet of 1N4001 diode. Note the manufacturer also indicates that the band/ circle indicates the 'n' type or cathode end. The marking made on the diode is also explained.

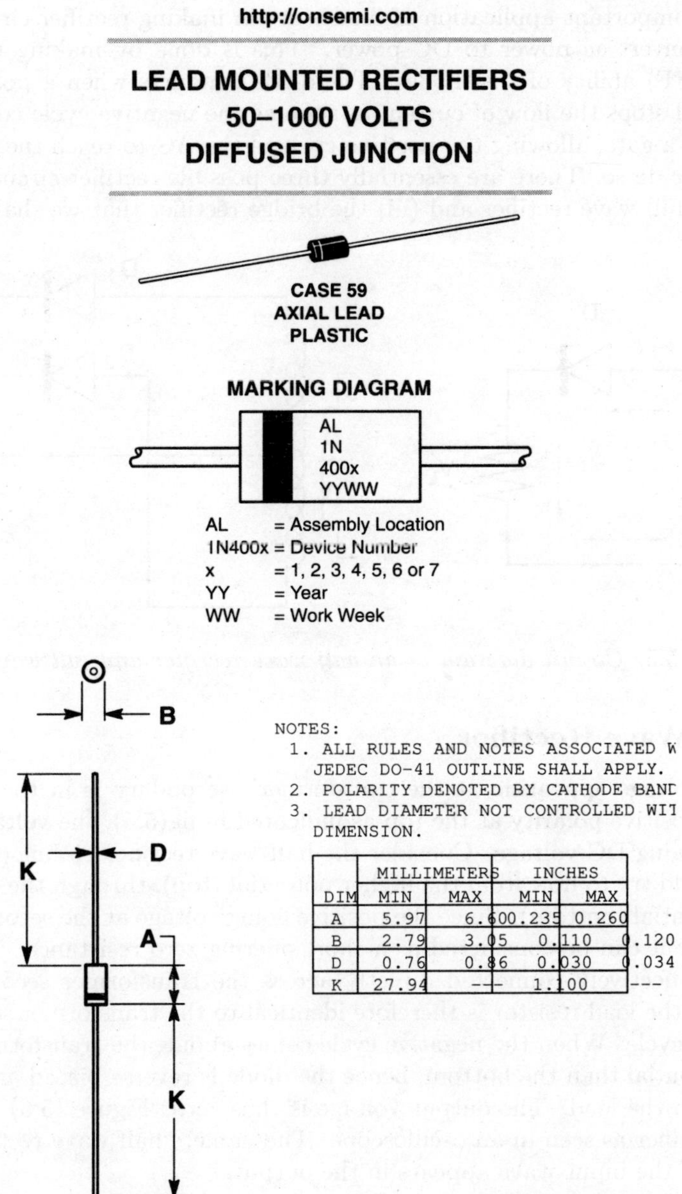

DIM	MILLIMETERS		INCHES	
	MIN	MAX	MIN	MAX
A	5.97	6.60	0.235	0.260
B	2.79	3.05	0.110	0.120
D	0.76	0.86	0.030	0.034
K	27.94	---	1.100	---

Figure 5.4: *Example of a data sheet giving details of non-electrical process parameters.*

5.2   Rectifier

One of the most important application of diodes are in making rectifier circuits. Rectifiers, are circuits which convert ac power to DC power. This is done by making use of the switching (going ON or OFF) ability of the diodes. The diode conducts when a positive signal is given to its 'p' side and stops the flow of current as soon as the negative cycle comes about. That is, the diode acts as a gate, allowing the positive cycle of the AC to reach the load and preventing the negative cycle do so. There are essentially three possible rectifier circuits, (i) the half-wave rectifier, (ii) the full wave rectifier and (iii) the bridge rectifier that we shall consider.

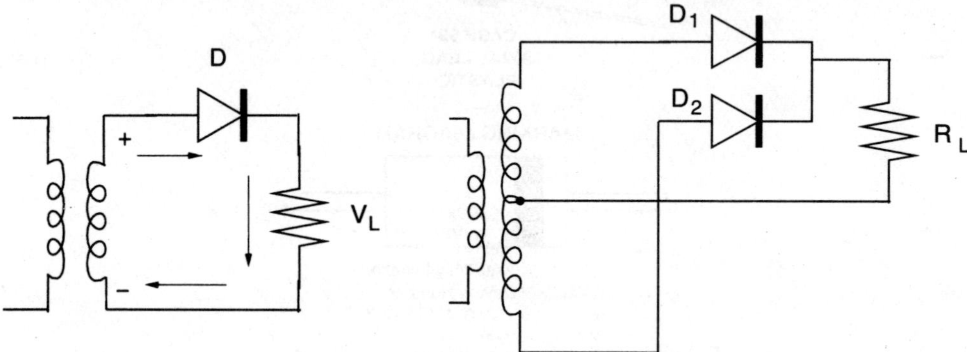

Figure 5.5: *Circuit diagram of an half wave rectifier and full wave rectifier.*

5.2.1   Half Wave Rectifier

When the AC voltage appearing at the transformer's secondary is in the positive half of it's cycle, say with positive polarity at the top as indicated in fig(5.5), the voltage at the secondary looks like a changing DC voltage. Consider the half-wave rectifier circuit (first circuit of figure 5.5), current would try to flow from the higher potential (top), through the diode and resistance to the lower potential (bottom). Since the positive going voltage at the secondary forward biases the diode, ideally it can be considered as a short offering zero resistance. This means that the load resistor is effectively connected directly across the transformer secondary. The voltage waveform across the load resistor is therefore identical to the transformer secondary voltage for the positive half cycle. When the negative cycle comes about, the transformers secondary's top is at a lower potential than the bottom, hence the diode is reverse biased and is open. Thus, no current flows into the load. The output voltage is thus zero. Figure (5.6) shows the output of an half wave rectifier as seen in an oscilloscope. The name, "half wave rectifier" is self evident, since only half of the input wave appears in the output.

   Figure (5.6) shows the oscilloscope in DC mode and AC mode respectively. Without going into the detail of the oscilloscope's construction, one has to understand that in its AC mode, the oscilloscope removes any DC component in the input and only shows the time varying part of the input signal. Consider a line source of 220v is fed to a step down transformer (1:2) of an half wave rectifier. The rectified output when seen in AC and DC mode differ in magnitude, with a shift of $\sim$ 49volts. This would imply, in AC mode, the oscilloscope's internal circuit removed

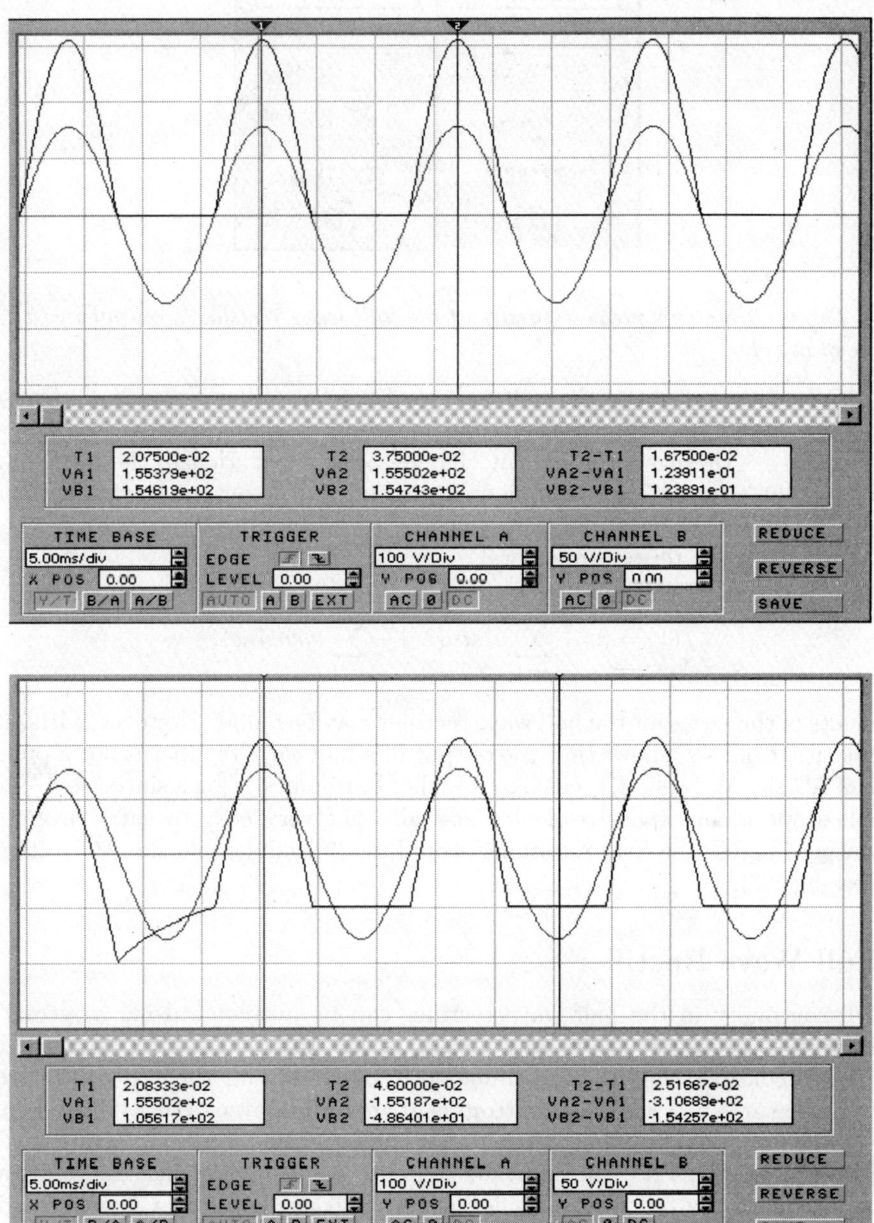

Figure 5.6: *The waveform at the output of a half wave rectifier shown in fig(5.5) is shown when the oscilloscope is in DC and AC modes respectively.*

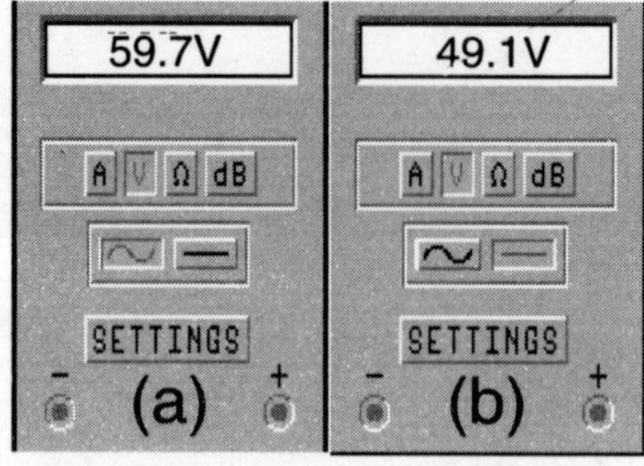

Figure 5.7: *The multimeter's measurements of the half wave rectifier's output in (a) AC and (b) DC mode respectively.*

49volts of DC, and the output of the half wave rectifier has still got some AC! One might have been tempted to think that the rectifier's output is purely DC, since it does not swing from positive to negative. However, the output changes with time. Hence, the rectifier's output is not a direct current, since DC does not change with time. The output waveform is periodic and repeats itself with time. The output can hence be represented by Fourier series, i.e. as a sum of various sine and cosine terms. In general

$$f(x) = a_o + \sum_{i=1}^{n} a_i sin(ix) + \sum_{i=1}^{n} b_i cos(ix) \qquad (5.3)$$

The coefficients of the series for the half wave rectifier may be found. However, without going into the mathematics, Eqn(5.3) shows that the output of a half wave rectifier is not a pure dc (a_o) by also contains AC signals ($a_1 sinx$, $b_1 cosx$, and higher harmonics $\sum_{i=2}^{n} a_i sin(ix) + \sum_{i=2}^{n} b_i cos(ix)$).

An oscilloscope is an expensive device and also not very easy to carry around. The DC voltage measured across the load resistance would be 49.1volts while the AC voltage would be 59.7volts (see fig 5.7).

5.2.2 Full Wave Rectifier

The huge discontinuity in the half wave rectifier can be overcome using a second diode and a center tapped transformer (see second circuit of fig 5.5) giving what is called a full wave rectifier. Again consider that the incoming positive cycle send the top of the transformer's secondary positive at $+V_m$, while the bottom is at zero. Half away along the transformer's coil, the voltage would be $+V_m/2$. Thus, while diode D_1 would be forward biased, the second diode (D_2) would be reverse biased. The equivalent circuit would be just like the half wave rectifier (seen along side with the full wave rectifier). The load current would flow through diode D_1 and load and back into the center tap. As the positive cycle disappears and the negative cycle comes about, the transformer's lower part on secondary side becomes more and more positive w.r.t. the top part. Consider lower part to be at $+V_m$, while the top is at zero volts. Again, at half way, the potential would be at $+V_m/2$. However, under this condition, diode D_2 would be forward biased and conducting while D_1 would remain open. The load current flows from the secondary's lower section through D_2 and load back into the center tap. Thus, one can imagine this circuit to be made up of two half wave rectifiers, each designed so as to work only in one half of the input AC's cycle. For either cycle, the direction of current through the load resistance remains the same. Note, the potential drop across the load resistance would only be half of that appearing across the transformer secondary. Figure 5.8 compares the input and output of a full wave rectifier. Note that the output repeats itself twice for every input AC cycle. Thus, the outputs frequency is twice that of the input wave, however with different waveforms. However, for the same input frequency, the output frequency of the full wave rectifier would be twice that of the half wave rectifier.

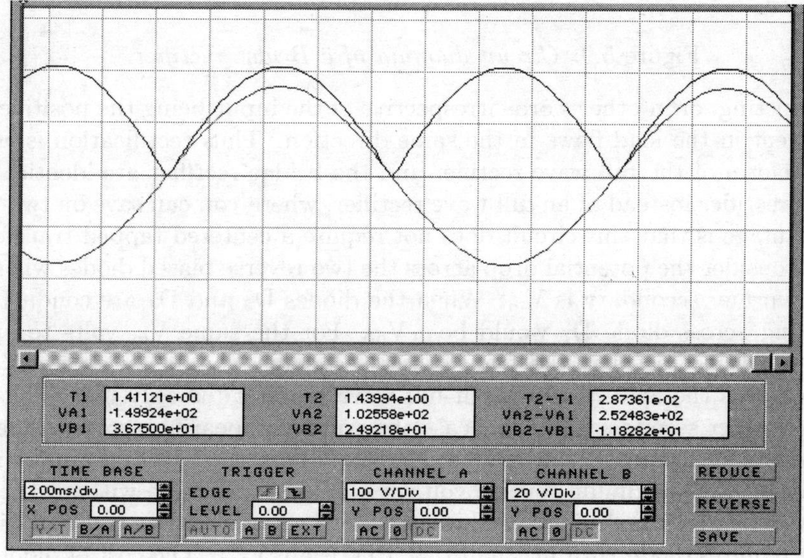

Figure 5.8: *The output waveform of a full wave rectifier shown is compared with the AC input.*

5.2.3   Bridge Rectifier

The bridge rectifier uses four diodes. An immediate advantage seen from the circuit diagram (fig 5.9) is that this circuit does not require a centered tapped transformer. Let us consider that the positive cycle makes the top of the transformer secondary positive, i.e. at a higher potential than the bottom of the transformer. Point 'A' would hence be at the same high potential as the top of the transformer. Consider potential V_A to be greater than potential at points 'B' and 'D'. In this case diode D_2 would be forward biased while D_1 would be reversed biased. Current would flow from 'A' to the load. This current is shown by dotted arrows in fig 5.9. Point 'D' would be at a potential $V_A - V_{load}$, again assuming the diodes to be ideal and $V_D = 0$. Thus, our initial argument that the diode D_1 would be reversed biased is justified. The load current travels through the conducting D_4 diode and back to the transformer.

The onset of the negative cycle sends the bottom of the transformer at a higher potential then the top. This reverses the above picture and makes the diodes D_3 and D_1 forward biased. The path of the current is shown by the dark continuous arrows. Irrespective of which pair of

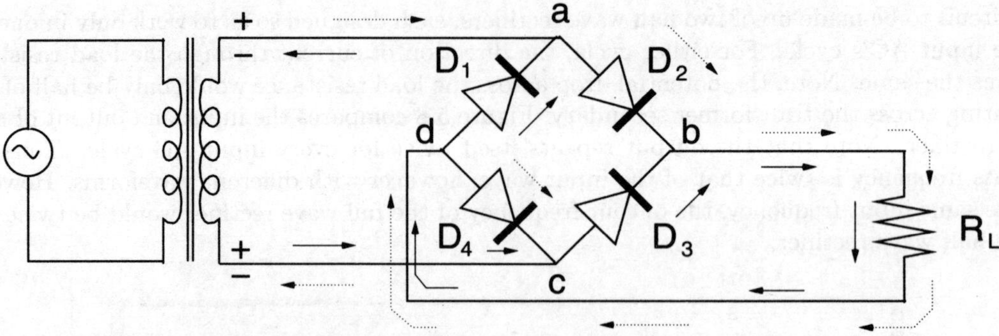

Figure 5.9: *Circuit diagram of a Bridge rectifier.*

diodes are conducting, or in other words irrespective of the input being the positive or negative, the output current in the load flows in the same direction. Thus rectification is achieved with the output waveform of the full wave rectifier and the bridge rectifier are identical. Then why design a bridge rectifier instead of an full wave rectifier, where you can save on two diodes? One immediate advantage is that this circuit does not require a centered tapped transformer.

Let us now consider the potential drop across the two reverse biased diodes when the voltage across the transformer secondary is V_m. When the diodes D_2 and D_4 are conducting then the voltage appearing across diode D_1 would be $-V_m$. For the same V_m volts to appear across the load in a full wave rectifier, the transformer secondary should be given $2V_m$ since, the potential drop across the load resistance in full wave rectifier's is only half of that appearing across the transformer secondary. In that case the voltage appearing across the reverse biased diode D_2 would be $2V_m$. Since $2V_m$ would be the maximum reverse voltage appearing across the diode, the diode's PIV (maximum reverse voltage a diode can handle without breaking down) should be greater or just equal to $2V_m$. However, for the same output considerations a bridge rectifier circuit requires diode with just half that PIV (only V_m). The cost of diodes required is substantially reduced.

There is a third advantage of bridge rectifier's over the full wave rectifiers (the Transformer Utility Factor, TUF). To appreciate this, we require some mathematics.

5.3   Mathematical Analysis

Finding the DC component by evaluating the coefficient a_o using Fourier Series is not trivial. An easier method is available using calculus (average area under the curve). In the following passages we calculate the output DC voltage of various rectifier circuits studied.

For the half wave rectifier

$$
\begin{aligned}
V_{dc} &= \frac{\int_0^{2\pi} V_m sinx dx}{\int_0^{2\pi} dx} = \frac{1}{2\pi}\int_0^{2\pi} V_m sinx dx \\
&= \frac{1}{2\pi}\left[\int_0^{\pi} V_m sinx dx + \int_{\pi}^{2\pi} 0 dx\right] \\
&= \frac{V_m}{\pi}
\end{aligned}
\tag{5.4}
$$

In the example of fig(5.6), the output at the secondary of the transformer is 110v AC (a 1:2 stepdown transformer was used with AC input being 220v). Thus the peak voltage, (V_m), fed to the rectifier is given by

$$V_m = \sqrt{2}V_{ac} = 155.56v$$

The DC voltage can now be calculated using eqn(5.4), which gives 49.1v. This can be seen from the oscilloscope output (fig 5.6). The sine wave at the transformer's secondary swings from +155.5 to -155.5volts, however, after rectification (when viewed on oscilloscope in DC mode) the output swings from zero to +155.5v. While one moves from DC mode to AC mode, the output of the half wave rectifier jumps down by nearly one division (1 div $\sim$ 50v, channel B). This shows the output is nearly 50v (an accurate measurement would give 49.1v).

The output is still not a prefect straight line (constt with respect to time). The waviness or ripples present is due to the remaining AC signal in the output. A figure of merit, i.e. how much waviness is present can be denoted by

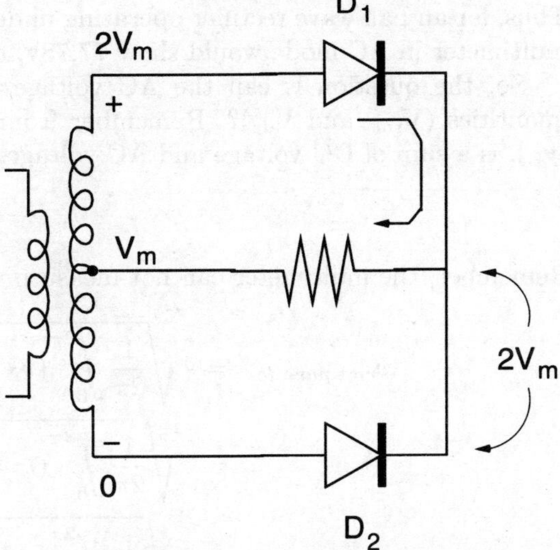

Figure 5.10: *Potential drop across the reverse biased diode D$_2$ of a full wave diode is twice that of diodes in a bridge rectifier which supplies the same to the load.*

ripple factor, a ratio of V_{ac} to V_{dc}. The less the AC present in the output, smaller is the ripple factor (r). But how much AC is left? Let us consider how this measurement is done with a multimeter. The multimeter comes in two modes, AC and DC modes. When the output of an half wave rectifier circuit is measured using a multimeter in AC mode (fig 5.7), it shows the AC output voltage to be 59.7volts. Hence, the ripple factor of an half wave rectifier would be

$$r = \frac{V_{ac}}{V_{dc}} = 1.215 \tag{5.5}$$

However, the problem is in practice, multimeters in AC mode does not report the AC voltage but the **rms** voltage.[1] The rms voltage is the DC equivalent voltage which would result in the same amount of heating in a resistive element. The rms voltage is calculated as

$$V_{rms} = \sqrt{\frac{1}{2\pi}\int_0^{2\pi} V_m^2 sin^2 x\, dx} \tag{5.6}$$

Substituting appropriate values, the V_{rms} for an half wave rectifier is given as

$$V_{rms} = \sqrt{\frac{1}{2\pi}\left[\int_0^{\pi} V_m^2 sin^2 x\, dx + \int_{\pi}^{2\pi} 0\, dx\right]}$$

---

[1]There is also the low end (less costlier) multimeters that are average-type. These have rectifiers in them that convert the sine wave into pulsating DC signals and measure the DC voltage (average voltage of eqn 5.4). The multimeter is calibrated to show corresponding V_{rms} for measured V_{dc}. This calibration is restricted to sine waves of 50Hz.

$$= \frac{V_m}{2} \tag{5.7}$$

Thus, for an half wave rectifier operating under the conditions described (for fig 5.6), a practical multimeter in AC mode would show 77.78v, the RMS voltage in accordance of eqn(5.7).

So, the question is can the AC voltage be expressed in terms of practically measurable quantities (V_{rms} and V_{dc})? Remember from eqn(5.3), it is understood that the load voltage (v_L), is a sum of DC voltage and AC voltage, hence

$$v_{ac} = v_L - V_{dc}$$

Remember, the multimeter can not measure v_{ac} but v_{rms}, hence

$$v_{rms\,pure\,ac} = \sqrt{\frac{1}{2\pi} \int_0^{2\pi} (v_L - V_{dc})^2 dt}$$

$$= \sqrt{\frac{1}{2\pi} \int_0^{2\pi} (v_L^2 + V_{dc}^2 - 2v_L V_{dc}) dt}$$

$$= \sqrt{\frac{1}{2\pi} \int_0^{2\pi} v_L^2\, dt + \frac{1}{2\pi} \int_0^{2\pi} V_{dc}^2\, dt - \frac{1}{\pi} \int_0^{2\pi} v_L V_{dc}\, dt}$$

$$= \sqrt{\frac{1}{2\pi} \int_0^{2\pi} v_L^2\, dt + \frac{V_{dc}^2}{2\pi} \int_0^{2\pi} dt - \frac{V_{dc}}{\pi} \int_0^{2\pi} v_L\, dt}$$

The first and third integrals are given by eq(5.6) and eq(5.4) respectively. Hence, the pure AC's rms contribution would be given as

$$= \sqrt{V_{rms}^2 - V_{dc}^2}$$

The ripple factor is (from eqn 5.5)

$$r = \sqrt{\frac{V_{rms}^2 - V_{dc}^2}{V_{dc}^2}}$$

$$= \sqrt{\left(\frac{V_{rms}}{V_{dc}}\right)^2 - 1} \tag{5.8}$$

Using eqn(5.4), eqn(5.7) and eqn(5.8) we have

$$r = \sqrt{\left(\frac{V_m}{2}\frac{\pi}{V_m}\right)^2 - 1}$$

$$= 1.21$$

To verify this theoretical value, one can use measured values from a circuit. Both V_{rms} and V_{dc} are practically measurable quantities (using a practical multimeter, but in case of oscilloscope

V_{rms} has to be calculated using eqn 5.7), and hence the ripple factor can now be evaluated experimentally. V_{rms} was calculated as 77.78v. Direct substitution of values in eqn(5.8) gives the ripple factor as 1.21 for an half wave rectifier.

Let us gain further experience by applying the above equations for the full wave rectifier. The only difference in mathematics would be the existence of the time varying function between interval π and 2π. The DC output would be

$$
\begin{aligned}
V_{dc} &= \frac{1}{2\pi} \int_0^{2\pi} V_m sinx dx \\
&= \frac{1}{2\pi} \left[\int_0^{\pi} V_m sinx dx + \int_{\pi}^{2\pi} V_m sinx dx \right] \\
&= 2 \times \frac{1}{2\pi} \int_0^{\pi} V_m sinx dx \\
&= \frac{2V_m}{\pi}
\end{aligned}
\tag{5.9}
$$

The symmetrical nature of the two halves of the sine wave allows for the reduction of the integral as seen in the third step above. The time varying AC in output would be expressed in rms terms as

$$
\begin{aligned}
V_{rms} &= \sqrt{\frac{1}{2\pi} \left[\int_0^{\pi} V_m^2 sin^2 x dx + \int_{\pi}^{2\pi} V_m^2 sin^2 x dx \right]} \\
&= \sqrt{\frac{1}{2\pi} \left[2 \times \int_0^{\pi} V_m^2 sin^2 x dx \right]} \\
&= \frac{V_m}{\sqrt{2}}
\end{aligned}
\tag{5.10}
$$

A direct substitution of eqn(5.9) and eqn(5.10) in the expression for ripple factor (eqn 5.8), gives the ripple factor for a full wave rectifier as

$$
\begin{aligned}
r &= \sqrt{\left(\frac{V_{rms}}{V_{dc}} \right)^2 - 1} \\
&= \sqrt{\left(\frac{V_m}{\sqrt{2}} \times \frac{\pi}{2V_m} \right)^2 - 1} \\
&= \sqrt{\frac{\pi^2}{8} - 1} = 0.4834
\end{aligned}
\tag{5.11}
$$

As can be seen, the ripple factor has drastically come down for the full wave rectifier. By including the second half of the input AC's sine wave, the waviness in the output and in turn the AC component in the output has drastically decreased. The primary duty of the rectifier is to convert an AC into a DC signal. By now one must have released that the rectifier circuits do not do their job perfectly, so it would be interesting to compare the circuits efficiency. The rectification efficiency is defined as the ratio of the DC power delivered to the load and the net power delivered to it (i.e. both AC and DC power delivered to the load),

$$\eta = \frac{dc\ power\ delivered}{net\ power\ at\ load}$$

The DC output voltage for both, half and full wave rectifiers have been given (eqn 5.4 and 5.9) from which the DC power delivered is estimated as $V_m^2/\pi^2 R_L$ and $4V_m^2/\pi^2 R_L$ respectively. Also, the rms voltage represents the net power transferred to the load by the rectifier circuit (eqn 5.7 and 5.10), hence the rectification efficiency of a half wave rectifier would be

$$\begin{aligned} \eta &= \frac{V_m^2}{\pi^2 R_L} \frac{4R_L}{V_m^2} \\ &= \frac{4}{\pi^2} \end{aligned} \tag{5.12}$$

For an half wave rectifier, the rectification efficiency expressed in percentage would be 40.6%. Similarly, the rectification efficiency of a full wave rectifier with ideal diodes would be given as

$$\begin{aligned} \eta &= \frac{4V_m^2}{\pi^2 R_L} \frac{2R_L}{V_m^2} \\ &= \frac{8}{\pi^2} \\ &= 81.2\% \end{aligned} \tag{5.13}$$

A rearrangement of eqn(5.8) shows the rectification efficiency is related to the ripple factor

$$\eta = \frac{1}{r^2 + 1}$$

A interesting point one should appreciate from the mathematics of both half wave and full wave rectifiers is that these expressions depend on the nature of the output waveform only. Since the output waveform of a full wave rectifier and a bridge rectifier are identical, the output DC, output rms and the ripple factor of both circuits are identical. While the advantages of the bridge rectifier over the full wave rectifier in terms of PIV and center tap transformer is understood, we investigate whether the bridge rectifier has any other advantages over the full wave rectifier. A figure of merit that can be used to find which rectification circuit is better would be how well does the rectification circuit utilize the input power. For this we define the Transformer Utility Factor (**TUF**, *__how well does the rectifier use the power from the transformer's secondary__*).

5.4   Transformer Utility Factor

The transformer utility factor is defined as how much dc power is obtained from the maximum possible power available at the transformer's secondary, i.e.

$$TUF = \frac{P_{dc}}{P_{ac}(rated)}$$

While the DC power delivered in all three rectifiers are established, the AC power delivered by
the transformer would depend on (a) the rms voltage developed across the secondary (b) the
current drawn by the load circuit. Thus, in the case of a half wave rectifier the AC power drawn
is

$$
\begin{aligned}
P_{ac}(rated) &= V_{ac}I_{rms} \\
&= \frac{V_m}{\sqrt{2}}\frac{V_m}{2R_L} \\
&= \frac{V_m^2}{2\sqrt{2}R_L}
\end{aligned}
\tag{5.14}
$$

Note, if the output of the transformer's secondary is given as $V_m\sin\omega t$, the rms voltage on the
output is $V_m/\sqrt{2}$. The rms current drawn by the half wave rectifier would be

$$
\frac{V_{rms}}{R_L} = \frac{V_m}{2R_L}
$$

Using the above information (eq 5.14), the transformer utility factor for an half wave rectifier
would be

$$
\begin{aligned}
TUF &= \frac{P_{dc}}{P_{ac}(rated)} \\
&= \frac{V_m^2}{\pi^2 R_L}\frac{2\sqrt{2}R_L}{V_m^2} \\
&= \frac{2\sqrt{2}}{\pi^2} = 28.6\%
\end{aligned}
$$

The full wave rectifier's transformer is not effectively used, since only half of the center tapped
transformer comes to play (contributes to the output) when any one of the diodes are conducting.
Hence, truly speaking the full wave rectifier is two half wave rectifiers acting one at a time. Hence,
the TUF of the full wave transformer an be approximated as twice that of the half wave rectifier.

$$
TUF(full\ wave) = 2 \times TUF(half\ wave) = 57.4\%
$$

The AC power at the secondary of the bridge rectifier is given as

$$
\begin{aligned}
P_{ac}(rated) &= V_{ac}I_{rms} \\
&= \frac{V_m}{\sqrt{2}}\frac{V_m}{\sqrt{2}R_L} \\
&= \frac{V_m^2}{2R_L}
\end{aligned}
\tag{5.15}
$$

Hence, the TUF for a bridge rectifier is given as

$$
\begin{aligned}
TUF &= \frac{4V_m^2}{\pi^2 R_L}\frac{2R_L}{V_m^2} \\
&= \frac{8}{\pi^2} = 81\%
\end{aligned}
$$

The high TUF of the bridge rectifier (as compared to the full wave and half wave rectifier) makes the bridge rectifier more attractive commercially. Remember, transformers are costly. Hence, to use it efficiently is important and the transformer utility factor is a figure of merit showing in percentage how effectively the transformer is put into use. As can be seen, the bridge rectifier uses the transformer effectively.

5.5   A Comparative Study

Table 5.2 compares the three rectifier circuits and while the full wave and bridge rectifier out scores the half wave rectifier as far as ripple factor and efficiency is concerned, it is only on the PIV and transformer utility factor (TUF) that the bridge rectifier is seen as a superior circuit for converting ac voltage into DC voltage.

Table 5.2:   Comparison of various rectifiers.

Rectifier Type	V_{dc}	V_{rms}	r	$\eta(\%)$	TUF(%)
Half Wave	V_m/π	$V_m/2$	1.21	40.6	28.6
Full Wave	$2V_m/\pi$	$V_m/\sqrt{2}$	0.482	81.2	57.4
Bridge	$2V_m/\pi$	$V_m/\sqrt{2}$	0.482	81.2	81

| Example 5.2: | Determine the output DC voltage and ripple factor of a full wave rectifier whose output is as shown in fig(5.11).

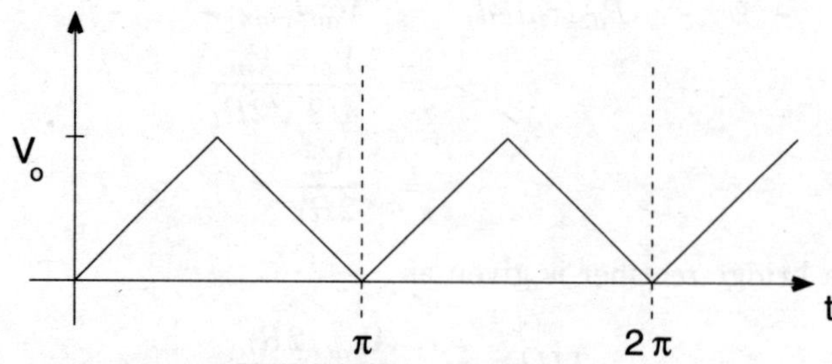

Figure 5.11: *Triangular output of a full wave rectifier whose input was triangular in nature (fig 5.12).*

The desired output is obtained when the full wave rectifier circuit is fed a triangular wave (easily available from function generators) as shown in fig(5.12). The general equation for finding the output DC voltage is given as

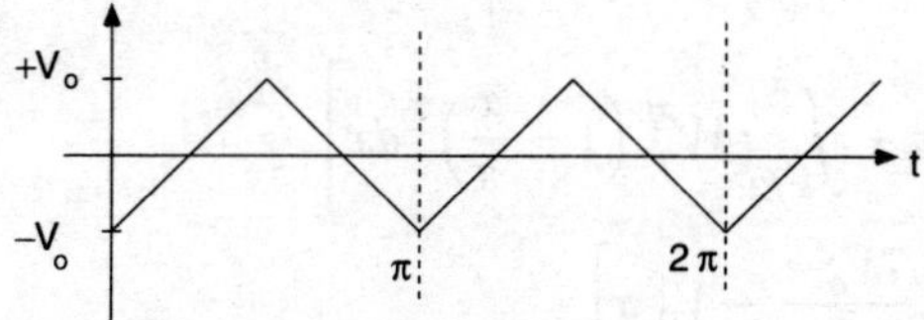

$$V_{dc} = \frac{\int_0^T f(x)dx}{\int_0^T dx}$$

Similarly, the output rms voltage is given as

Figure 5.12: *Triangular input for a full wave rectifier.*

$$V_{rms}^2 = \frac{\int_0^T [f(x)]^2 dx}{\int_0^T dx}$$

where f(x) is the functional form of the output waveform under which we are interested in calculating the area enclosed. For $0 < x < \pi/2$ the equation of the positive slope triangular edge is written as

$$f_1(x) = \left(\frac{2V_o}{\pi}\right) x$$

while for $\pi/2 < x < \pi$ we have

$$f_2(x) = -\left(\frac{2V_o}{\pi}\right) x + 2V_o$$

In one cycle of the input (0 to π), the output repeats itself twice and due to symmetry we may write

$$V_{dc} = \frac{\int_0^{2\pi} f(x)dx}{\int_0^{2\pi} dx}$$

$$= 2 \times \frac{1}{2\pi} \int_0^\pi f(x)dx$$

$$= \frac{1}{\pi}\left[\int_0^{\pi/2} f_1(x)dx + \int_{\pi/2}^\pi f_2(x)dx\right]$$

$$\tag{5.16}$$

Substituting the functional form of the two sides of the triangular wave we have

$$V_{dc} = \frac{1}{\pi}\left[\int_0^{\pi/2} \left(\frac{2V_o}{\pi}\right) xdx + \int_{\pi/2}^\pi \left(2V_o - \frac{2V_o}{\pi}x\right) dx\right]$$

$$= \frac{V_o}{2}$$

$$\boxed{V_{dc} = \frac{V_o}{2}}$$

We have hence found the DC component on full rectification of a triangular wave. For the ripple factor we will have to calculate the rms output also.

$$V_{dc} = 2 \times \frac{1}{2\pi}\left[\int_0^{\pi/2}[f_1(x)]^2 dx + \int_{\pi/2}^{\pi}[f_2(x)]^2 dx\right]$$

$$= 2 \times \frac{1}{2\pi}\left[\int_0^{\pi/2}\frac{4V_o^2}{\pi^2}x^2 dx + \int_{\pi/2}^{\pi}4V_o^2\left(1-\frac{x}{\pi}\right)^2 dx\right]$$

$$= \frac{1}{\pi}\left[\frac{4V_o^2}{3\pi^2}\left(\frac{\pi^3}{8}\right) + 2V_o^2\pi + \frac{7V_o^2\pi}{6} - V_0^2\pi\right]$$

$$= \frac{7}{3}V_o^2$$

Using the two results, we find the ripple factor (use eqn 5.8) as

$$\boxed{r = 2.886}$$

5.6 Other Diode Applications

5.6.1 Clamping Circuits

Sometimes, depending on the demands of the circuit designer, it might be necessary to shift the reference level about which the ac signal swings. With the knowledge on the superposition theorem's validity, all the circuit designer has to do is add an appropriate dc source (level) in series to the ac source. However, batteries are costly and are ideally avoided! A simple alternative method of establishing a DC reference for the output voltage is by using a diode clamp as shown in figure(5.13). Consider the input ac is a square wave varying from $-V_m$ to $+V_m$ is fed to the circuit. The choice of waveform is just from simplicity of explanation, since input has just two states. Consider the negative voltage appearing at the input first. In that case the diode is forward biased. Conducting diode results in the following:

  (i) It shorts the output, giving zero volts across the load during the duration that input is $-V_m$.

  (ii) The conducting diode allows the capacitor to charge to V_m with the capacitor's right hand side plate gaining higher potential w.r.t. the left hand side plate.

If the capacitor is not allowed to discharge (that is the discharge time is made very-very large by proper selection of RC value) the polarity attained by the capacitor is maintained even though the input toggles from $-V_m$ to $+V_m$. So when the positive cycle comes about, it adds to (comes in series to) the V_m stored in capacitor,resulting in $+2V_m$ reverse biasing the diode. The potential $2V_m$ hence appears across the load as shown in the waveform is in fig(5.14).

Consider the same circuit, however, just invert the pn diode (fig 5.15). The capacitor is assumed to be uncharged. This time around we assume the positive cycle to arrive first, which would result in the diode being forward biased. Thus,

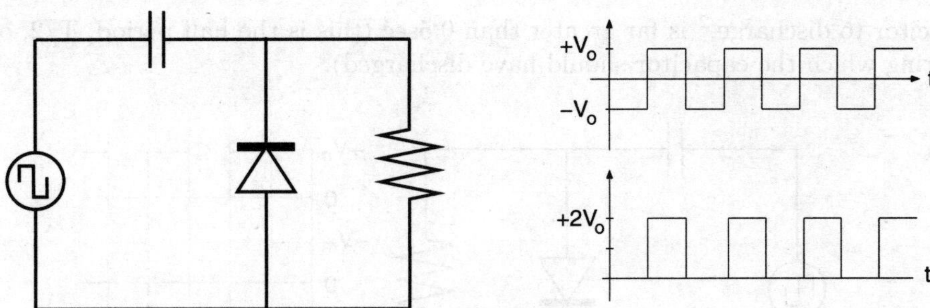

Figure 5.13: *The circuit champs the input voltage by* $+V_m$*. While the input toggles from* $-V_m$ *to* $+V_m$*, the output as a result of the clamping action toggles from ground level (0 volts) to* $+2V_m$*.*

(i) The diode shorts the output and the load output is forced zero on the onset of positive input.

(ii) The conducting diode instantaneously charges the capacitor to V_m, with the capacitor's right hand side plate at higher potential.

During the negative cycle, the net voltage $-2V_m$ appears across the diode and hence, open diode allows net potential $-2V_m$ to appear across the load. Note the output waveforms are ideal, showing that the capacitor did not discharge through 0.5sec while the positive cycle was being fed into circuit (input freq is 1Hz). The RC combination for taking the waveform of

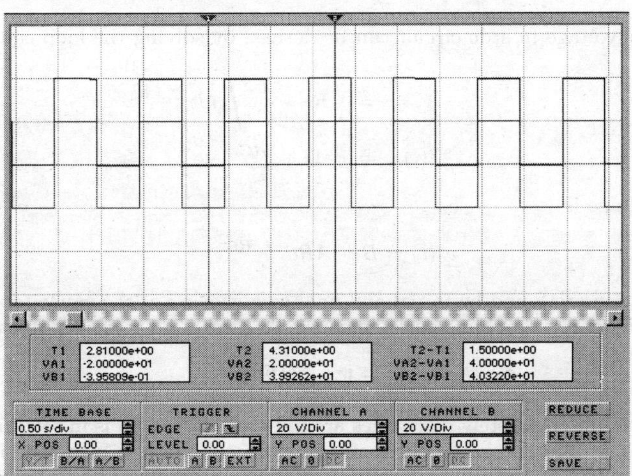

Figure 5.14: *The waveform of the clamping circuit shown in fig(5.13) as seen on the oscilloscope when the output channel is in DC mode.*

fig(5.14) was 100sec. Thus, the RC combination, which is an indication of the time taken for the capacitor to discharge,[2] is far greater than 0.5sec (this is the half period, T/2, of the input cycle during which the capacitor should have discharged).

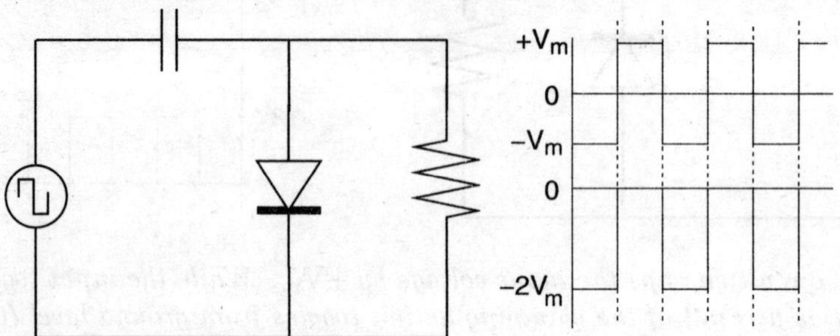

Figure 5.15: *Reverse clamping, clamping at* $-V_m$ *can be achieved by just placing the diode upside down. While the input toggles from* $-V_m$ *to* $+V_m$*, the output as a result of the clamping action toggles from* $-2V_m$ *to 0 volts.*

Consider, what happens be the RC combination is made smaller and smaller (comparable to T/2=0.5sec). Fig(5.16) shows the effect of clamping circuits with RC combination of 1.0 and 0.1sec respectively. As can be seen the discharging and in turn exponential fall of the capacitor voltage are evident. The output voltage is essentially

$$v_{clamped}(t) = -[V_m + v_c(t)]$$
$$= -\left[V_m + V_m(2e^{-t/RC} - 1)\right]$$

---

[2]The falling capacitor voltage in a dc circuit can be derived by solving the loop equation

$$V_m = Ri + \frac{1}{C}\int idt$$
$$i(t) = Ae^{-t/RC}$$

The falling capacitor voltage is

$$v_c(t) = B - ARe^{-t/RC}$$

The constants of integration are found by initial and final conditions, i.e. the capacitor was initial charged [t=0, $v_c(t) = V_m$] and after an appreciably long time the capacitor discharges to zero potential [t = ∞, $v_c(t) = 0$].

$$v_c(t) = V_m e^{-t/RC}$$

as a thumb rule, the capacitor completely discharges around t=5RC (i.e. ∞ is in relation to the RC combination used). However, in our clamping example where steady state condition is attained (i.e. one cycle follows the other and this has been going on and on), the initial conditions are different and hence expression differs. The capacitor was initial charged [t=0, $v_c(t) = V_m$] and after an appreciably long time the capacitor charges to the opposite polarity [t = ∞, $v_c(t) = -V_m$]. The voltage across the capacitor varies as

$$v_c(t) = V_m(2e^{-t/RC} - 1)$$

For the first example shown in fig(5.16), at the end of the half cycle (t=0.5sec), the voltage appearing at the load is

$$v_{clamped}(t) \;=\; -\left[V_m + V_m(2e^{-0.5} - 1)\right]$$
$$\sim\; -1.2V_m$$

Extending this to the second example of fig(5.16), i.e. for RC combination of 0.1sec, at the end of the half cycle (t=0.5sec) we have

$$v_{clamped}(t) \;=\; -\left[V_m + V_m(2e^{-5} - 1)\right]$$
$$\sim\; -0.012V_m$$

Again a thumb rule used for designing a proper clamping circuit is that the load resistance and capacitor should meet the condition, RC $>$ 10T/2 or $>$ 5T.

Various different clamping levels can be attained by varying the peak voltage of the input voltage supply, since the capacitor charges to in peak input voltage, as was seen from previous examples. This also makes the output voltage to toggle either only on the positive side (0 to $2V_m$) or the negative side (0 to $-2V_m$). These are restrictions and though, we started with the demand that we do away with batteries, if clamping is to be done at voltage levels other than the input ac peak value, we are forced to use a battery source (after all, you can never tell what the user wants).

Consider the circuit shown in fig(5.17). A dc source V_{dc} has been kept in series with the diode. When the negative pulse arrives, the diode is forward biased by a potential $(V_m - V_{dc})$ and conducts. The conducting diode brings the dc supply across the load resistance sending the output to $-V_{dc}$. The capacitor charges to $(V_m - V_{dc})$, with the right hand side plate at a higher potential. Thus, on the arrival of the positive pulse from the input supply, the two potentials add in series, $V_m + (V_m - V_{dc})$. This voltage appears across the load resistance, since the diode is reverse biased by $2(V_m - V_{dc})$.

A check on whether you have predicted the output of a clamping circuit properly is by noting, whatever might be the output's swing (say from V_1 to $-V_2$), the peak-to-peak of the output must be identical to the input's peak-to-peak. In our case the input voltage swings from $+V_m$ to $-V_m$, i.e. a swing of $2V_m$. Let us now investigate the swing in the output

$$V_1 - (-V_2) \;=\; V_1 + V_2$$
$$=\; (2V_m - V_{dc}) + V_{dc} = 2V_m$$

As you can see, the output peak-to-peak is same as the input's peak-to-peak. Keeping this check in mind should give one confidence to analysis various clamping circuits.

5.6.2 Clipping Circuits

Another application of diodes are in clipping circuits where the output voltage can be limited at a predecided level. Thus, their function is to prevent the output voltage from exceeding a given voltage. The half wave rectifier circuit is a chipping circuit, since the negative part of the input cycle is **clipped** from the output.

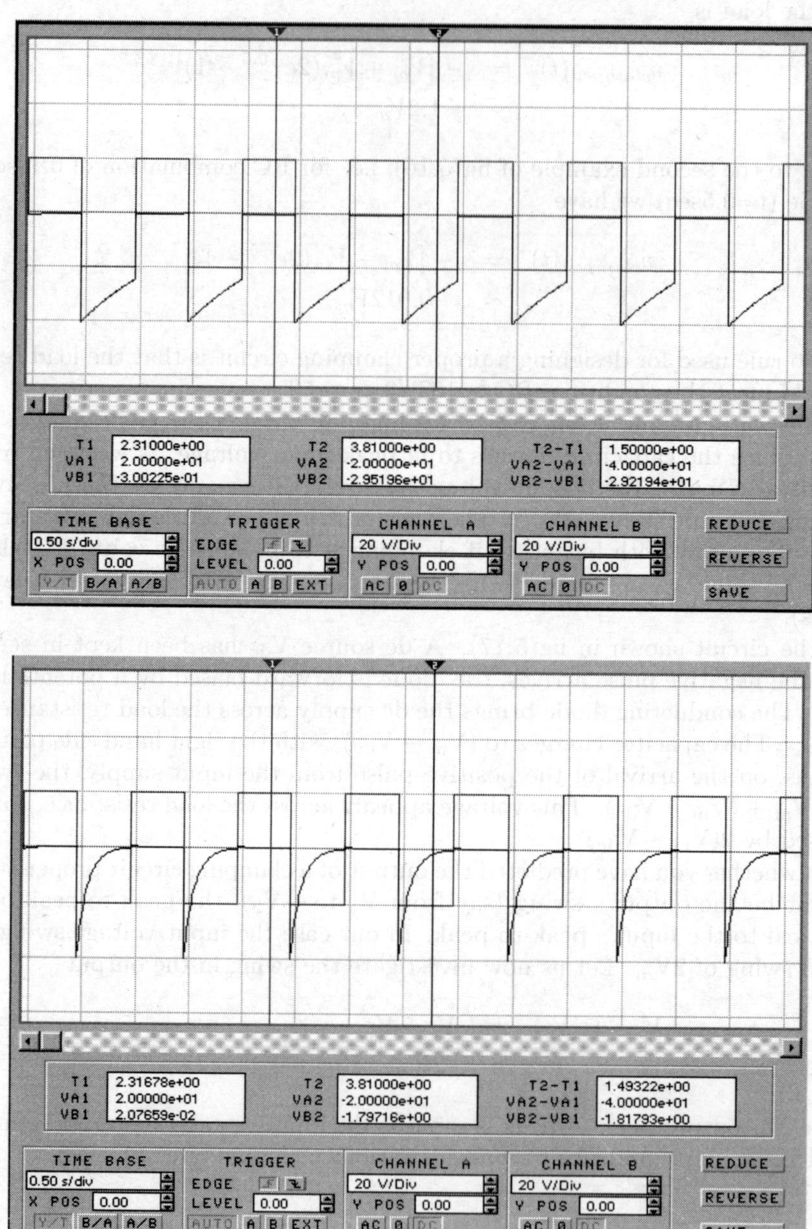

Figure 5.16: *The waveform of a clamped output, where the RC combination in circuit is comparable to the half cycle of the input.*

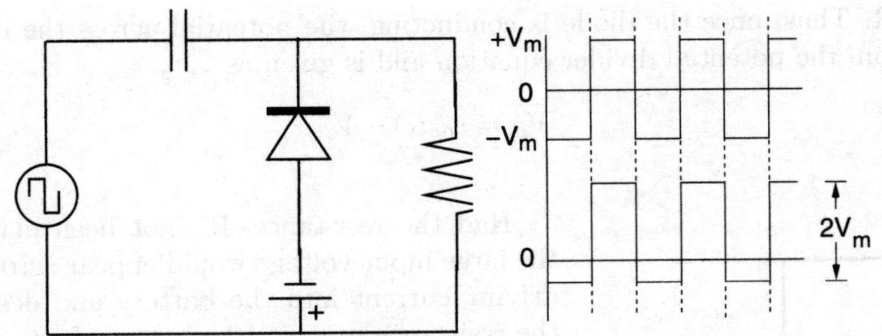

Figure 5.17: *Another example of a clamping circuit.*

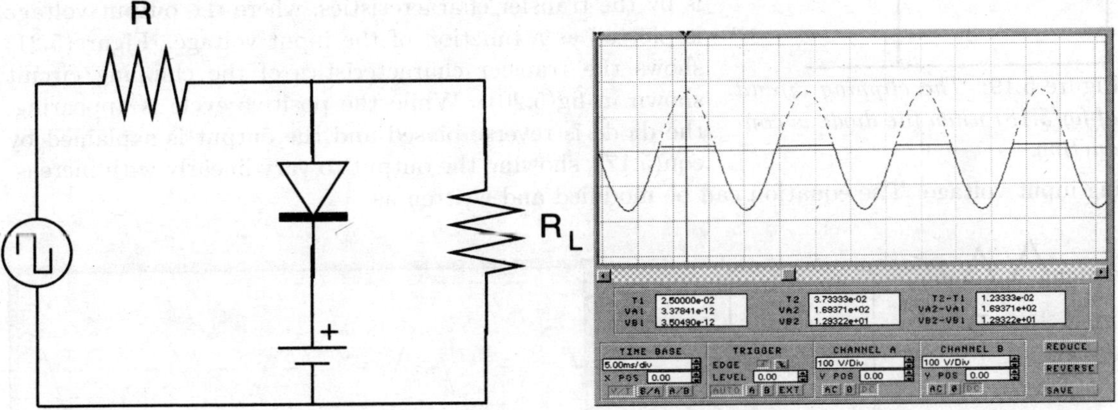

Figure 5.18: *A clipping circuit along with the output seen on oscilloscope is shown.*

Consider the clipping circuit shown in fig(5.18). As the positive cycle appears, the diode is reverse biased till $v_{ac}(t) < V_{dc}$, where V_{dc} is the drop across the battery. Since, the diode is reverse biased, the circuit reduces to that of a potential divider and the output across the load is given by the potential divider equation

$$v_o(t) = \frac{R_L}{R_L + R} v_{ac}(t) \tag{5.17}$$

By selecting the load resistance to be far larger than R, the output can be made to follow the input [i.e. $v_o(t) = v_{ac}(t)$]. The condition is the same when the negative cycle of the input is reverse biasing the diode. The output follows the input. However, as the input voltage increases, the diode becomes forward biased by which the battery is brought parallel to the load. The load is hence, *pegged* to the same potential as the battery, irrespective of the input. Thus, the circuit clips voltages above V_{dc} from the output. One might be tempted to ask of what use is the resistance 'R' in the circuit? Fig(5.19) shows the equivalent circuit when the diode is conducting. The output is chipped and the potential drop across the load is V_{dc}, the drop across the battery. Then where is the input voltage disappearing to? It appears across this resistance R. Thus, once the diode is conducting, the potential across the resistance R is not obtained from the potential divider equation and is given as

$$V_R = v_{ac}(t) - V_{dc}$$

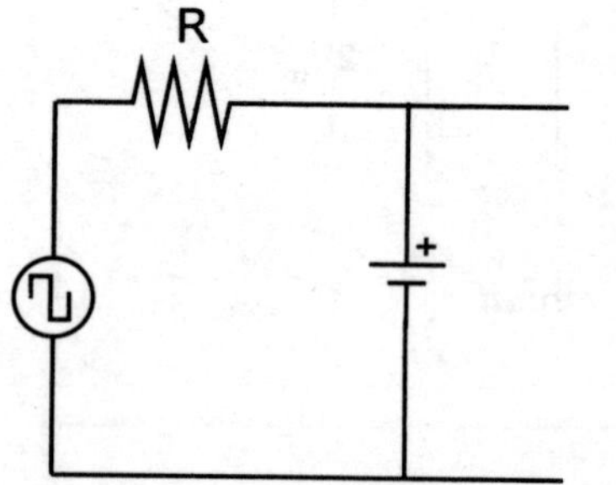

Figure 5.19: *The clipping circuit of fig(5.18) when the diode is conducting.*

Had the resistance 'R', not been placed in the circuit the large input voltage would appear across the DC battery, driving current into the battery and destroying it. Thus, the resistance protects the battery from damage.

Figure(5.20) shows the chipping action taking place at $-V_{dc}$. Another method of representing the chipping action is by the transfer characteristics, where the output voltage is plotted as a function of the input voltage. Figure(5.21) shows the transfer characteristics of the chipping circuit shown in fig(5.20). While the positive cycle is appearing, the diode is reverse biased and the output is explained by eqn(5.17), showing the output to vary linearly with increasing input voltage. The equation can be modified and written as

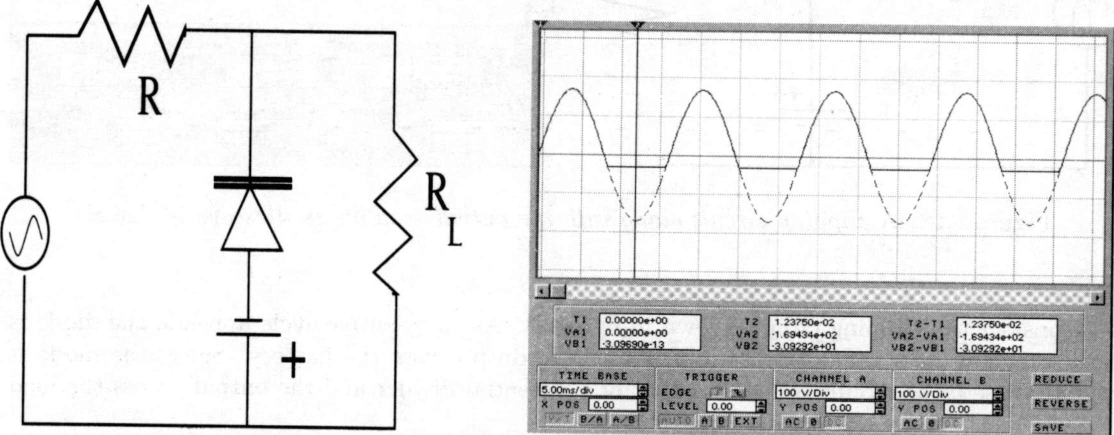

Figure 5.20: *A clipping circuit along with the output seen on oscilloscope is shown.*

$$\frac{v_o(t)}{v_{ac}(t)} = \frac{R_L}{R_L + R}$$

which gives the slope of the line. For the designing condition that we have set, $R_L \gg R$, we have unit slope on the transfer characteristics. This slope is maintained till the input voltage is $-V_{dc}$, which keeps the diode reverse biased. Beyond $-V_{dc}$, the diode is forward biased and hence $-V_{dc}$ appears across the load, which is maintained whatever is the input voltage. This is shown by the horizontal line on the transfer characteristics, with the point of discontinuity at $-V_{dc}$.

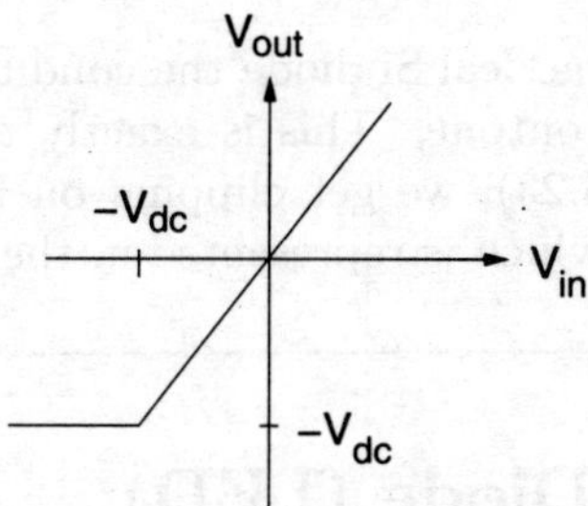

Figure 5.21: *The transfer characteristics, a plot between the input and output voltages, of the clipping circuit discussed in fig(5.20).*

Example 5.3: Determine the chipping circuit whose output characteristics is as shown in fig(5.22).

Since, no mention is made about the slope of the line segment enclosed between $(-V_{dc2}, -V_{dc2})$ to $(+V_{dc1}, +V_{dc1})$, we do not have to calculate the size of the load resistance (R_L) with respect to the resistance kept in series to protect the batteries.

Obviously, since clipping takes place at two different voltage levels, two diodes are required. While the clipping shown in the transfer characteristics third quadrant can be achieved by the circuit discussed in fig(5.20), the first quadrant requires the diode to be in it's blocking state (reverse biased) for

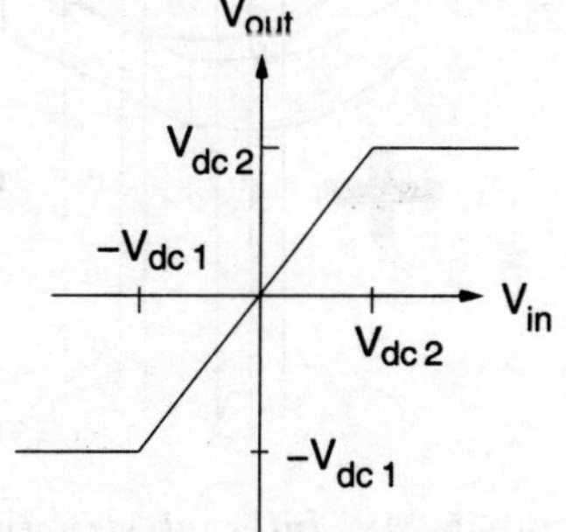

Figure 5.22: *Transfer characteristics for an unknown clipping.*

$$v_{ac} < V_{dc2}$$

The reverse biased diode would act as an open circuit and would not bring the battery into the output. Hence,

$$v_{out} = V_{ac}$$

With increasing AC voltage, when the diode is forward biased, i.e.

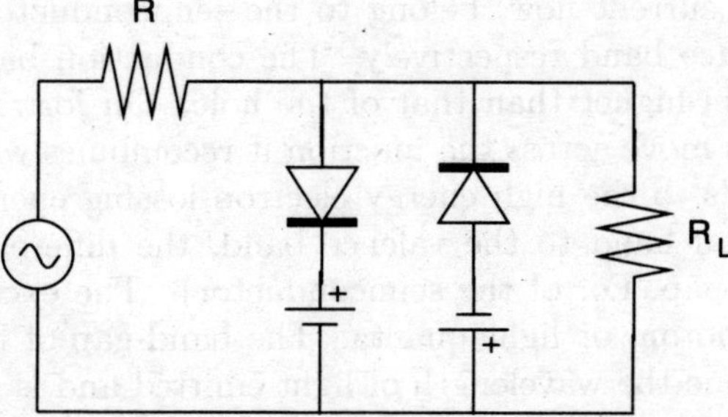

Figure 5.23: *Circuit that achieves clipping on both the positive and negative cycle of the input AC.*

$$v_{ac} \geq V_{dc2}$$

(this is for an ideal diode, for a practical Si diode the condition would be $v_{ac} \geq V_{dc2} +0.7$) the battery is bought parallel to the output. This is exactly the condition achieved in fig(5.18). Combining the two circuits (fig 5.23), we get clipping on both the sides (positive going and negative going) of the input AC, which is represented in the transfer characteristics (fig 5.22).

5.7   Light Emitting Diode (LED)

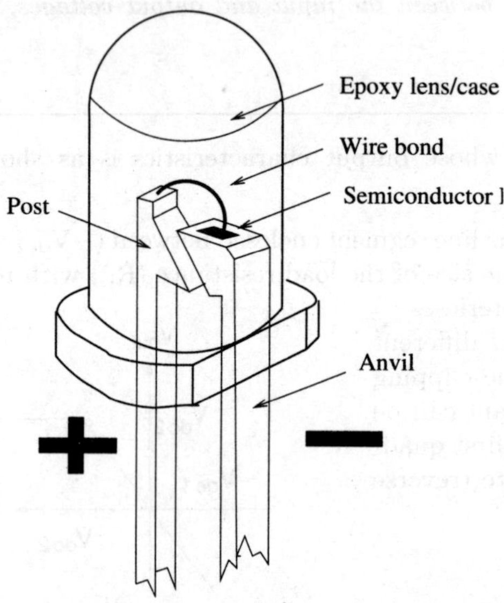

Figure 5.24: *Internal structure of Light Emitting Diode.*

Light emission from an incandescent lamp takes place when its filament is heated to a high temperature. They are not very efficient with very small amount of energy being converted to light and the remaining being lost as heat. The discovery of LED or Light Emitting Diode (fig 5.24) in 1907[3] gave hope of an efficient method of generating light. As the name suggests, Light Emitting Diodes are pn junction diodes. Infact their IV characteristics and behaviour is exactly the same as the normal diode. Fig 5.24 shows the circuit representation is essentially that of a diode with arrows representing light being emitted. Normal diodes also emits light but the light is not visible since its wavelength is in the Infra-Red (IR) range. Any body mass mildly hot with temperatures above its surrounding would be emitting IR wavelengths.

The free electron in the 'n' region of the diode and free hole in the 'p' region contributing to the current flow, belong to the semiconductor's conduction band and valence band respectively. The conduction band electron has an energy level higher than that of the hole. On forward biasing, when the electrons move across the junction it recombines with the holes. This corresponds to the high energy electron lossing energy (falling from the conduction band to the valence band, the difference of which we call the band-gap, E_g, of the semiconductor). The excess energy is given out as a photon, or light quanta. The band-gap of the semiconductor will determine the wavelength of light emitted and is given by the relation

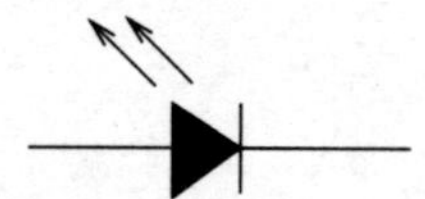

Figure 5.25: *Circuital symbol of LED.*

$$\lambda\,(nm) = \frac{1240}{E_g}$$

---

[3]It was the Russian scientist Oleg V Lossev who first reported emission of light from SiC pn diodes. It took another 90 years to go from red LEDs to blue LEDs. Shuji Nakamura obtained blue emission using GaN

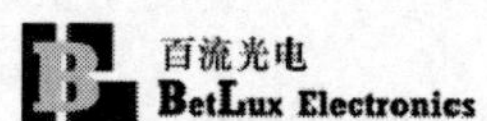

Bullet Type LED lamp
BL-L324

■ **Features:**
➢ 3.0mm Round Bullet Type LED Lamps
➢ Ultra brightness.
➢ Choice of various viewing angles.
➢ Diffused, Transparent and Water clear lens.
➢ IC compatible /Low current capability.
➢ RoHs Compliance

■ **Electrical-optical characteristics: (Ta=25℃)** (Test Condition: IF=20mA)

| Part Number | Chip | | | Lens Type | Forward Voltage(VF) Unit:V | | Luminous Intensity (Iv) Unit:mcd | | Viewing Angle 2θ1/2(deg) |
	Emitted Color	Material	λ_P (nm)		Typ	Max	Min.	Typ.	
BL-L324SRC	Hi Red	AlGaAs,SH	660	Water Clear	1.85	2.20	100	440	13
BL-L324LRC	Super Red	AlGaAs,DH	660		1.85	2.20	400	1000	
BL-L324URC	Ultra Red	AlGaAs,DDH	660		1.95	2.20	700	1800	
BL-L324UEC	Ultra Orange	AlGaInP	630		2.10	2.50	1200	2500	
BL-L324UYC	Ultra Yellow	AlGaInP	590		2.10	2.50	500	2000	
BL-L324UGC	Ultra Green	AlGaInP	574		2.20	2.50	800	1800	
BL-L324PGC	Ultra Pure Green	InGaN	525		3.80	4.50	2000	5000	
BL-L324BGC	Ultra Bluish Green	InGaN	505		3.80	4.50	1200	4500	
BL-L324BC	Blue	InGaN	430		3.80	4.50	700	1800	
BL-L324UBC	Ultra Blue	InGaN	470		2.70	4.20	2000	4000	
BL-L324VC	UV	InGaN	405		3.80	4.50	150	220	
BL-L324UWC	Ultra White	InGaN			2.70	4.20	3000	10000	

Figure 5.26: *Example of a data sheet giving details composition of semiconductor and corresponding wavelength emitted.*

One point that should be clear is that photons are only emitted from the junction where electrons and holes recombine. The photon thus generated would have to travel through either the n-type or p-type layer to escape and be usable light. For this the layer must not absorb the released photon. This would require the semiconductor layers to be very thin. The choice of the semiconductor materials is critical. For a good LED, the semiconductor materials should be transparent or non-absorbing for the wavelength it is being made for. Figure 5.26 shows the varius semiconductor compositions used and the wavelengths emitted by the LEDs thus fabricated.

The amount of photons emitted would depend on the current flowing in the LED. The current however should be limited to that which the diode can sustain. Practically, while using a diode it is safe to put a 330Ω resistance in series with the LED.

Exercise

Q1. What is an ideal diode?

Q2. Discuss the IV characteristics of the ideal diode and compare it with that of a practical diode, under forward bias and reverse bias conditions?

Q3. Find the incremental (dynamic) resistance of the diode using the diode equation.

Q4. Describe the piecewise diode model for circuit analysis.

Q5. How does a rectifier work?

Q6. Can the knee potential or the barrier potential (V_K) of a diode be measured using a multimeter. Discuss (It would be worthwhile to go into a lab and try doing this).

Q7. What is the difference between intrinsic and extrinsic semiconductor. Differentiate between an N-type semiconductor and a P-type semiconductor.

Q8. If a voltage $V_o\sin\omega t$ is placed across the terminals of a digital multimeter, what value does the multimeter show in AC mode (V_o= 5v).

Q9. Fig(5.27) shows various clamping circuits. Verify and confirm the outputs shown along with the circuits are correct.

Q10. Consider that a DC voltage of V=10v is applied to a potential divider circuit, with each resistor having a value of R = 1KΩ. Assume the output voltage is measured using an oscilloscope. What is the measured value of output voltage? Would this value be different if read by a multimeter (in DC mode)? What would be the situation if the DC source is replaced with an AC source which gives an input of 10v peak-to-peak?

Q11. If a current of 5mA is flowing through the diode when forward biased, what is the voltage drop across the diode?

Q12. A diode is operated at room temperature with I_{sat}=1nA.

    (a) What is the diode current, if the voltage across the diode is 1v?

    (b) What is the diode current if the diode is reverse biased by 1v?

    (c) At what voltage would the current in the diode be $200\mu A$?

Q13. A periodic wave has the following Fourier series: $v(t) = -1 + \frac{1}{2}\cos(100t) - \frac{1}{4}\cos(200t) + \frac{1}{6}\cos(300t) - \frac{1}{8}\cos(400t)$ Volts

    (a) What is the dc value of this waveform?

    (b) What is the fundamental frequency?

    (c) What is the frequency of the second harmonic (Hz)?

    (d) What is the amplitude of the 3rd harmonic?

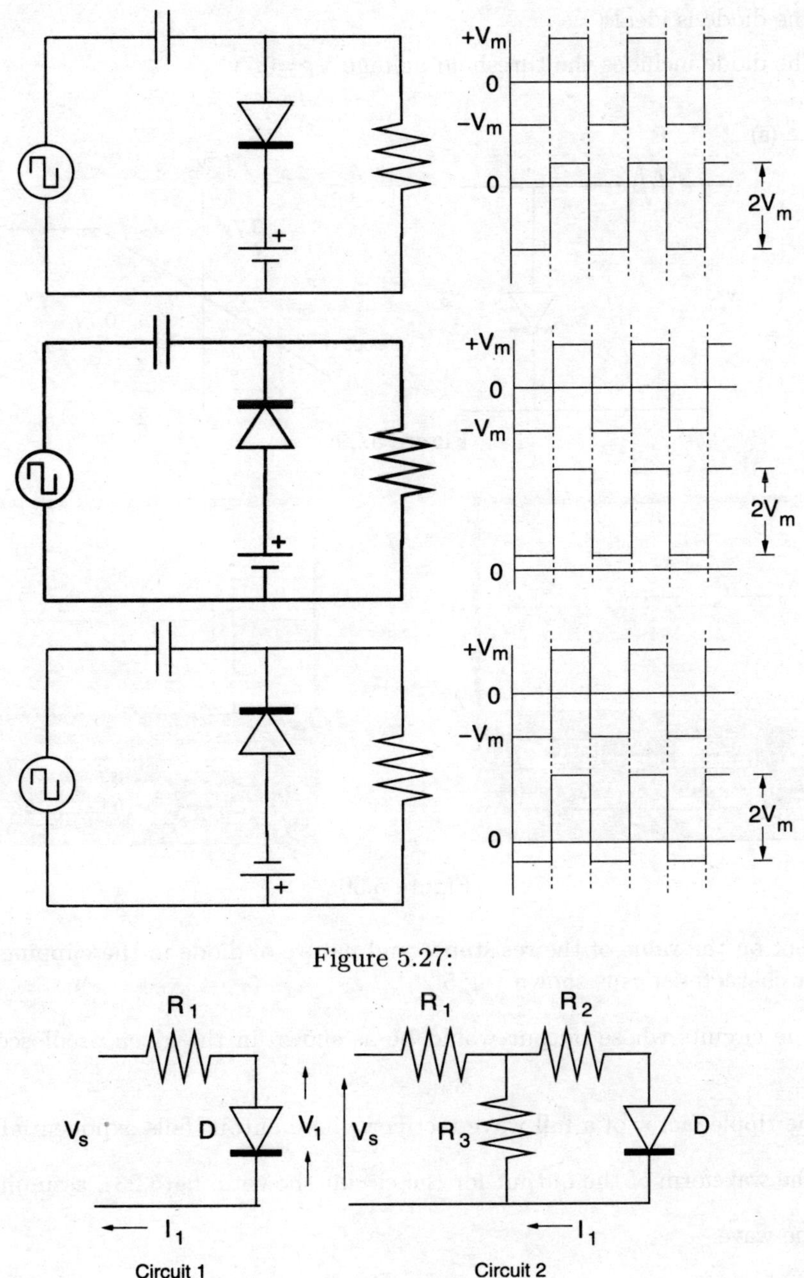

Figure 5.27:

Figure 5.28:

Q14. For the circuits shown in fig(5.28), V_s=5V, $R_1 = 1K\Omega$, $R_2 = 2K\Omega$ and $R_3 = 1.5K\Omega$. Calculate the current I_1 and the voltage V_1 assuming

(a) The diode is ideal.

(b) The diode includes the threshold voltage $V_K=0.7v$

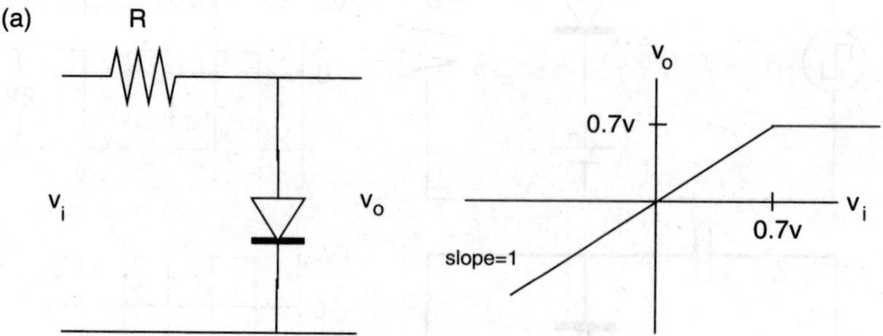

Figure 5.29:

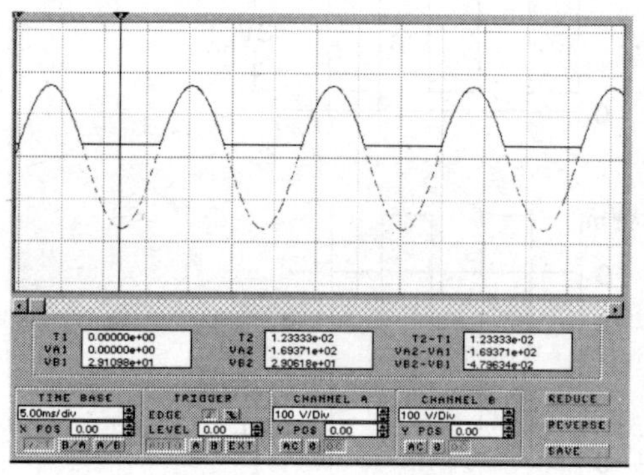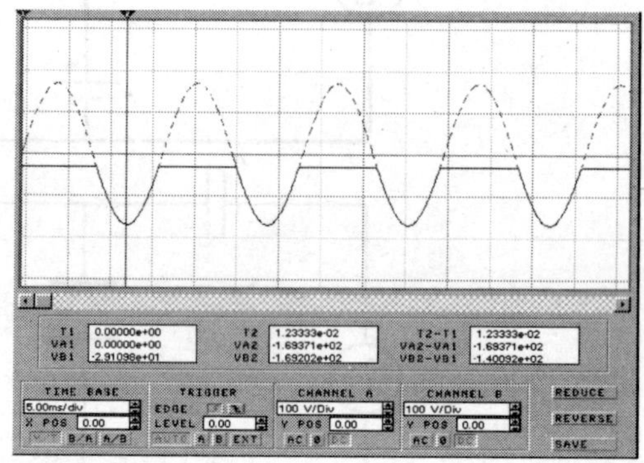

Figure 5.30:

Q15. Comment on the value of the resistance and nature of diode in the clipping circuit whose transfer characteristics is shown (fig 5.29).

Q16. Draw the circuits whose output would be as shown in the given oscilloscope traces (fig 5.30).

Q17. Find the ripple factor of a full wave rectifier whose output falls exponentially.

Q18. Draw the waveform of the output for the circuit shown in fig(5.23), assuming input to be

   (a) sine wave,

   (b) square wave.

Q19. What is the diode's PIV?

Q20. If non-ideal Si diodes are being used to rectify an 4v peak-to-peak ac voltage, which rectifier circuit would you use, full wave or bridge rectifier. Justify your selection.

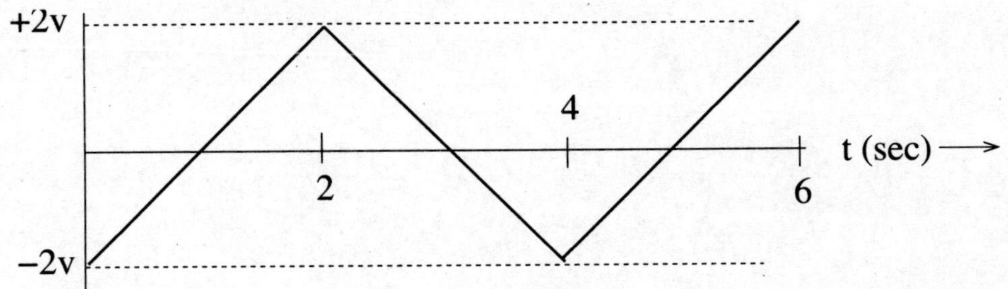

Figure 5.31:

Q21. Explain what is a depletion region and why it forms about a PN junction.

Q22. For a given waveform (fig 5.31), draw the waveform of a

(a) Half wave rectifier circuit with ideal diode.

(b) Full wave rectifier circuit with ideal diode.

(c) Half wave rectifier circuit with Silicon diode having negligible forward resistance.

(d) Full wave rectifier circuit with Silicon diode having negligible forward resistance.

(e) Consider the Silicon diode being used to deliver power to a load of 1KΩ, while the diode itself has a forward resistance of 1KΩ. Draw and explicitly explain the output waveform.

Chapter 6

Filters

As seen from the last Chapter, the output of a rectifier circuit has both DC component as well as AC components. Thus, we have not yet achieved the objective of obtaining a pure DC output and in turn developing our home made power supply. To remove the AC ripples present in the rectifier's output, we require an additional block of circuitry. The circuits used to remove any residual AC components are called filter circuits. The objective is to provide the AC a lower impedance path, which would however provide a higher impedance path to the DC component. Thus, the filter circuit would be made of circuital elements which have frequency dependent impedance. Since, capacitors and inductors provide frequency dependent impedance, they are used to build filter circuits.

6.1   The Shunt Capacitor Filter

This is the simplest possible filter circuit. A capacitor is placed parallel to the load. As the rectified current flows from the rectifier circuit, it is presented with two possible path, the load and the capacitor. The capacitor provides a lower impedance path to the AC ripples, while an infinite impedance is offered to the DC component. Thus, the load impedance is the only path available for the DC component.

As the AC current flows into the capacitor, the capacitor is being charged. This results in a voltage developing across the capacitor. Since the diode is conducting it brings the capacitor in parallel to the secondary of the transformer, as is the case with the load. If the diode is considered ideal the voltage drop across the capacitor is the same as that of the transformer's secondary, i.e.

$$v_c(t) = v_i(t) = V_m sin\omega t = v_r(t) \tag{6.1}$$

The net current flowing from the rectifier can be found using Kirchhoff's current law at the node. The input current divides into two, both of which can be deduced from the potential difference generated due to their flow in respective impedance.

$$i_r(t) = \frac{V_m sin\omega t}{R_L} \tag{6.2}$$

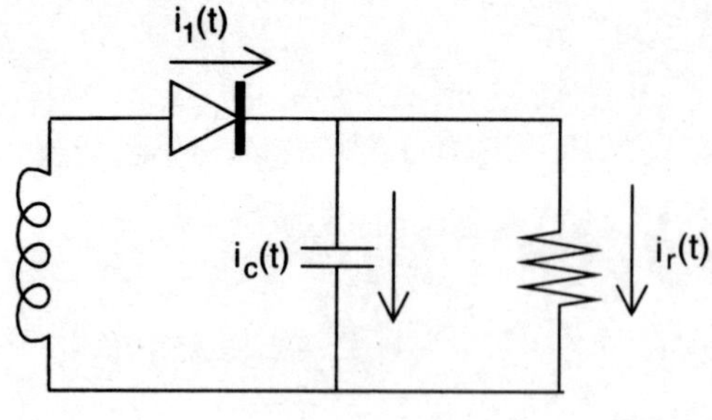

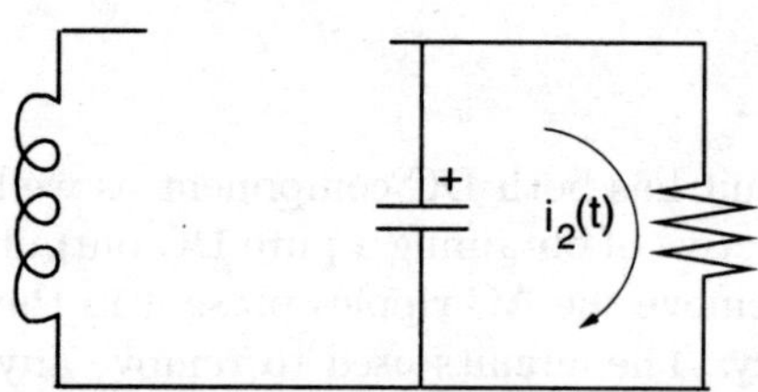

Figure 6.1: *Circuit diagram of an half wave rectifier with shunt filter. The various currents and their directions are indicated. Second circuit shows an open circuit since diode is reverse biased. The capacitor discharge current flows into resistance in the direction shown.*

and

$$i_c(t) = C\frac{dv_c}{dt} = \omega C V_m cos\omega t \qquad (6.3)$$

The input (diode) current from KCL is

$$
\begin{aligned}
i_i(t) &= i_r(t) + i_c(t) \\
&= \frac{V_m sin\omega t}{R_L} + \omega C V_m cos\omega t \\
&= aV_m sin(\omega t + \phi) \qquad (6.4)
\end{aligned}
$$

where

$$
\begin{aligned}
a &= \sqrt{\frac{1}{R_L^2} + \omega^2 C^2} \\
\phi &= tan^{-1}(\omega C R_L) \qquad (6.5)
\end{aligned}
$$

The voltage drop across the capacitor follows the AC input (transformer's secondary) till V_{max}, after which the rate of fall of AC input is more rapid than the capacitor's voltage fall via it's discharge. Thus, the rectifier (or rather the rectifying diode) is reversed biased. Once the capacitor forces the rectifier diodes in to its reverse bias state, the diodes are open (see the difference between the conducting and non-conducting cases shown in fig 6.1).

This disconnects the input (transformer's secondary voltage) from the load, hence, the input current goes to zero. From eqn(6.4) we can compute the instant that the input current goes zero.

$$
\begin{aligned}
i_i(t) = aV_m sin(\omega t + \phi) &= 0 \\
\omega t + \phi &= n\pi \qquad (6.6)
\end{aligned}
$$

Physically, only the n=1 solution is of interest, hence, the input gets disconnected from the load at instant t_2, given by the expression

$$\omega t_2 = \pi - tan^{-1}(\omega C R_L) \qquad (6.7)$$

The current represented by eqn(6.4) is the current flowing through the diode while it conducts from the onset of the positive cycle to the instant when the diode is reversed biased by the capacitor. Using eqn(6.5) and eqn(6.6) in eqn(6.7), this diode current can be represented as

$$i_i(t) = V_m\left(\sqrt{\frac{1}{R_L^2} + \omega^2 C^2}\right) sin[\omega(t_2 - t)] \qquad (6.8)$$

It should be remembered the above expression is valid only for the time interval $0 < t \leq t_2$, where t=0 represents the onset of the positive cycle at the input. After the instant t_2, the current flowing in the load is due to the discharging capacitor. The loop equation is set up as a differential equation from which the time varying current may be evaluated as

$$v_r(t) + v_c(t) = 0$$

$$iR_L = -\frac{1}{C}\int i\,dt$$

$$\frac{di}{i} = -\frac{dt}{R_L C}$$

The standard solution of this first order differential equation is given as

$$i_2(t) = Be^{\left(-\frac{t}{R_L C}\right)} \tag{6.9}$$

The current falls exponentially with time, t. However, since the input varies sinusoidally, instead of talking in terms of time, it would be better to write eqn(6.9) in terms of an angle or angular time (ωt).

$$i_2(t) = Be^{\left(-\frac{\omega t}{tan\phi}\right)} \tag{6.10}$$

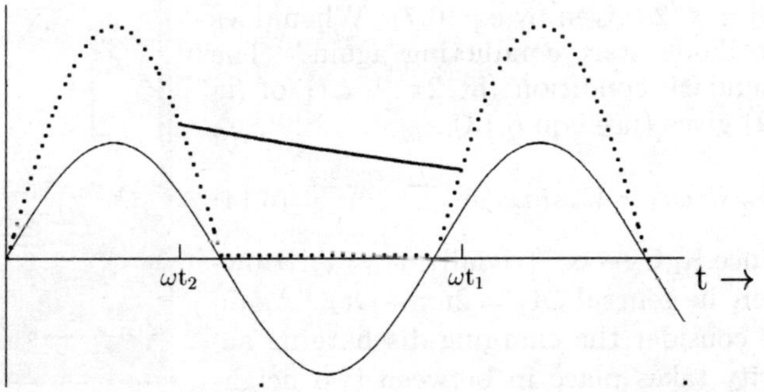

Figure 6.2: *The capacitor discharge current gives a continuous current in the output ($t_2 < t < t_1$). Input AC waveform is not to scale.*

This current would flow till the potential drop across the capacitor becomes less then the input voltage. This forwarded biases the diode which until this point was not conducting due to the reverse bias it experienced. Under such circumstances, the input/ diode current (eqn 6.8) would start flowing again, charging the discharged capacitor. For the input voltage to be greater then the capacitor voltage, the positive cycle should reappear at the input. This would only happen at an instant $\geq 2\pi$. We assume the instant to be represented by $2\pi + \omega t_1$ (see fig 6.2). Thus, the load voltage for the time ($\omega t_2 < \omega t < 2\pi + \omega t_1$) would be given as

$$v_2(t) = BR_L e^{\left(-\frac{\omega t}{tan\phi}\right)} \tag{6.11}$$

Now to determine the coefficient 'B', we use the boundary condition that at ωt_2, the magnitude of load voltage given by $v_2(t)$ and $v_i(t)$ are equal

$$V_m sin(\omega t_2) = BR_L e^{\left(-\frac{\omega t_2}{tan\phi}\right)} \tag{6.12}$$

Thus, the falling voltage can be expressed as

$$v_2(t)_{\omega t_2 < \omega t < 2\pi + \omega t_1} = V_m sin\omega t_2 \, e^{-\frac{\omega(t-t_2)}{tan\phi}} \tag{6.13}$$

To make a prefect DC output, we require the output voltage to be independent of time. This would require the terms of the exponential, $\frac{\omega(t-t_2)}{\tan\phi}$ to be equal to zero. This can only be achieved when $R_L C \to \infty$. Fig(6.3) shows the output of two half wave rectifier circuits which have a shunt capacitor filter. The second circuit has an $R_L C$ combination ten times larger then that used in the first circuit. A larger combination would achieve a perfectly flat line. The practical problem in increasing the capacitance is that the physical size of the capacitor increases making the circuit bulky. Increasing load resistance decreases the current flowing into the load. Thus, the shunt capacitor filter circuit is ideally suited for circuits whose current demand is low.

Other physical implications are there on making $R_L C \to \infty$. The instant when the diode reverse biases would be when the input voltage attains its maximum, i.e. at $\omega t_2 = \pi/2$ (given by eqn 6.7). When does the diode start conducting again? The boundary condition (at $2\pi + \omega t_1$ of fig 6.2) gives (use eqn 6.13)

$$V_m sin\omega t_1 = V_m sin\omega t_2 e^{-\frac{\omega t_1 - 2\pi + \omega t_2}{\omega C R_L}} \qquad (6.14)$$

Since $R_L C \to \infty$, trivially, $t_1 = t_2$. However, in general $\omega t_1 = 2n\pi + \omega t_2$. Again we consider the charging-discharging activity takes place in between two neighboring peaks, hence n=1. Thus we have $\omega t_1 = 2\pi + \omega t_2$. We have shown for $R_L C \to \infty$, $\omega t_2 = \pi/2$. Hence, $\omega t_1 = 2\pi + \omega t_2 = 5\pi/2$. This would mean in steady state, the instant of capacitor starting to charge (ωt_1) would coincide with the instant the capacitor starts discharging (ωt_2). The diode hence conducts only for a small time and surges the capacitor with power during that instant. Since, all this occurs at the instant when V_m, the maxima of the input signal occurs. The waviness is bound to decrease, decreasing the ripples.

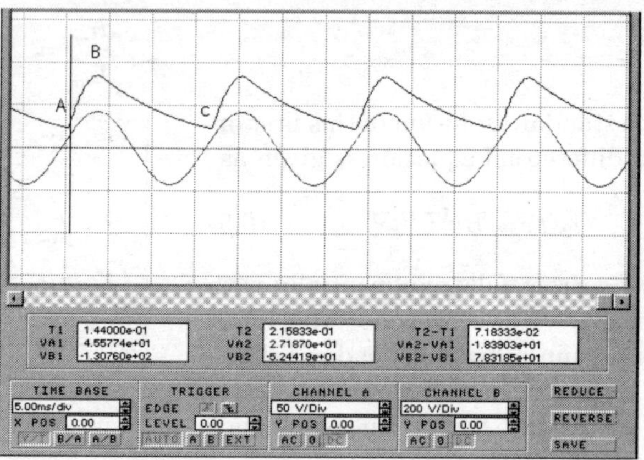

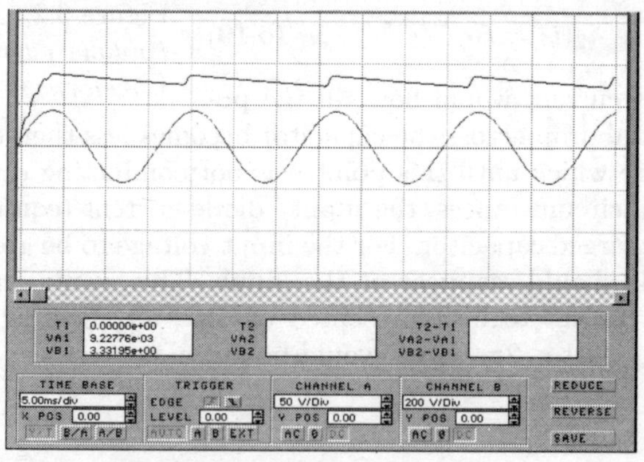

Figure 6.3: *The shunt filtered output of an half wave rectifier is shown w.r.t. the input AC. Higher removal of AC is achieved by increasing the $R_L C$ combination by ten times.*

6.1.1 Ripples in Filtered Half Wave Rectifier

We now calculate the waviness in a shunted filtered half wave rectifier, which in turn implies we are calculating the ripple factor (as given in eqn 5.8). For this we require to know the output rms voltage along with the DC output. For calculating both these quantities a functional form of the output waveform should be known. However, an examination of fig(6.3) would tell you that a trivial mathematical function does not exist to describe the waveform. However, a triangular waveform as shown in fig(6.4a) can approximate the filtered output waveform. A further idealization of the output would be a sawtooth waveform shown in fig(6.4b). The mathematical function used to describe the output would be that of a straight line. This straight line is symmetrical about the average/ DC output. Hence, the co-ordinates for a half-wave rectifier's output would be $(\omega t_2, V_{dc} + V_r/2)$ and $(2\pi + \omega t_1, V_{dc} - V_r/2)$. From our knowledge $\omega t_1 = \omega t_2$, we can shift the waveform of fig(6.4b) such that ωt_2 coincides with the origin, (i.e. $\omega t_2 = 0$). This gives the two co-ordinates of our idealized output waveform as $(0, V_{dc} + V_r/2)$ and $(2\pi, V_{dc} - V_r/2)$. Using this, the equation of the straight line can be setup using which we compute the required terms for ripple factor:

$$
\begin{aligned}
V_{rms}^2 &= \frac{1}{2\pi} \int_0^{2\pi} \left[-\frac{V_r}{2\pi} x + \left(V_{dc} + \frac{V_r}{2} \right) \right]^2 \\
&= \left[\frac{V_r^2}{3} + \left(V_{dc} + \frac{V_r}{2} \right)^2 - V_r \left(V_{dc} + \frac{V_r}{2} \right) \right]
\end{aligned} \tag{6.15}
$$

The ripple factor is the measure of AC voltage present in the output and is given by eqn(5.8). Using eqn(6.15) we can find the ripple factor for this saw tooth output waveform

$$
r = \frac{1}{V_{dc}} \sqrt{V_{rms}^2 - V_{dc}^2} = \frac{1}{2\sqrt{3}} \frac{V_r}{V_{dc}} \tag{6.16}
$$

The ripple factor is in terms of V_{dc} and V_r, both are at present unknown and hence, we require a relation between the two voltages. Also, since the filter circuit is to be designed using capacitors, it is necessary to express the ripple factor in terms of C and R_L. For this, we require to find I_{dc} for getting $V_{dc} = I_{dc}R_L$. Since, the discharging capacitor provides the current, from the saw-tooth waveform we have

$$
\begin{aligned}
i(t) &= C \frac{dv}{dt} \\
&= C \frac{V_r/2 - (-V_r/2)}{(2\pi/\omega + t_1) - t_2}
\end{aligned}
$$

for, the prefect saw-tooth assumption $t_1 = t_2$, hence

$$
i(t) = \omega C \frac{V_r}{2\pi} \tag{6.17}
$$

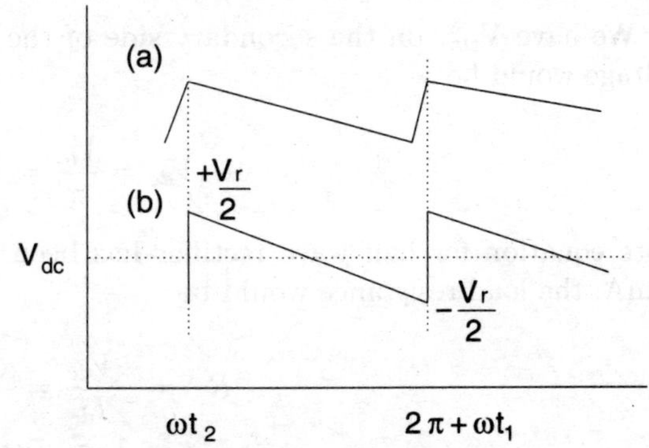

Figure 6.4: *The filtered output can be approximated by a sawtooth waveform, where triangular ripples (V_r) may be considered riding on a DC voltage (V_{dc}).*

As can be seen, the R.H.S. independent of time and hence the current we are talking of is a DC current, hence

$$
I_{dc} = \omega C \frac{V_r}{2\pi} \tag{6.18}
$$

giving the output DC voltage as

$$V_{dc} = R_L I_{dc} = \omega R_L C \frac{V_r}{2\pi} \tag{6.19}$$

Substituting eqn(6.19) in eqn(6.16) gives us the ripple factor in terms of circuital elements

$$r = \frac{\pi}{\sqrt{3}} \frac{1}{\omega R_L C}$$

$$= \frac{1}{2\sqrt{3}} \frac{1}{f R_L C} \tag{6.20}$$

The waviness of the output or the ripples can be minimized by,

(a) Increasing the value of C

(b) Increasing the load resistance

(c) Increasing the input frequency

Example 6.1: Design a shunt capacitor filter for a half-wave rectifier which delivers a load current of 10mA. The ripple factor should be 0.4, also the V_{rms} on the secondary side of the transformer is 110 volts. AC frequency is 50Hz.

We have V_{rms} on the secondary side of the transformer as 110 volts. Hence, the load DC voltage would be

$$V_{dc} = \frac{V_m}{\pi} = \frac{\sqrt{2} V_{rms}}{\pi}$$

Note equation for half wave rectifier has been used. Since we demand a DC load current of 10mA, the load resistance would be

$$R = \frac{V_{dc}}{I_{dc}} = \frac{\sqrt{2} V_{rms}}{\pi I_{dc}}$$

$$= \frac{\sqrt{2} \times 110}{10\pi} K\Omega = 5.5 K\Omega$$

The ripple factor is given as

$$r = \frac{1}{2\sqrt{3} f RC}$$

Giving C as $2.91 \mu F$.

$$\boxed{\text{Answer } R = 5.5 K\Omega \text{ and } C = 2.9 \mu F.}$$

| Example 6.2: | A simple power supply consists of a transformer, a half wave rectifier and a shunt capacitor filter with C=1000μF. What is the load current that can be drawn if the peak-to-peak ripple voltage (V_r) is to be no greater than 5V?

Refer to eqn(6.18), the current is given as

$$I_{dc} = \omega C \frac{V_r}{2\pi}$$
$$= fCV_r$$

A direct substitution gives

| Answer I_{dc}=250mA |

(for line frequency 50Hz.)

Problems with Multimeters

A practical shunt filter circuit was designed in an under-graduate lab and the measurements were made both with an oscilloscope and a multimeter. The observations made were as follows:

with oscilloscope

$$V_r = 0.9v$$
$$V_{dc} = 5v$$

giving ripple factor, r=0.052 (using eqn 6.16). with multimeter

$$V_{dc} = 4.95v$$
$$V_{rms} = 10.9v??$$

giving ripple factor, r=1.87 ??(using eqn 6.16).

What went wrong? As stated in the last chapter, in an under-graduate lab the multimeters available for measuring V_{rms} are of the average-type, i.e. they measure the average voltage (V_{dc} of eqn 5.4). They are however, calibrated to display V_{rms} for the corresponding V_{dc}. The multimeter calibration is restricted to sinusoidal waves of frequency 50Hz. The output of a filtered circuit is not sinusoidal and hence, the calibration would fail. The multimeter is bound to give errors. What is required is true-rms based multimeters which measure the "heating" potential of an applied voltage.

Another precaution while measuring ripple factor with multimeters (even with true-rms type)is that when measuring voltages less then 100mV, the measurements are susceptible to errors introduced by noise. Thus, ripple factors of filtered circuits are best studied with oscilloscopes.

6.1.2   Ripples in filtered Full Wave Rectifier

It is easily appreciated from the figures comparing the output of half and full wave rectifiers (fig 5.6 and fig 5.8), that for the same input AC frequency, the discontinuity (time interval for which the output signal goes below V_m[1]) is more in half wave rectifiers. Thus, the capacitor to be used has to be large so that it can release energy for a longer time.

However, an expression for the ripple factor of a shunt filtered full wave rectified output can be easily obtained by following the saw-tooth approximation and the above mathematics. Care should be taken for the limits. While in half rectifiers only the positive signal appears for very incoming AC cycle (limits hence, was from 0 to 2π), in full wave rectifiers there are two positive going signals for every incoming AC cycle. Thus, the limits to be used in this example, for the corresponding equation of eqn(6.15), are 0 to π.

Without going through the involved mathematics of the previous section, one can still compute the ripple factor of the shunt filtered full wave rectifier. The output frequency of a full wave rectifier is twice that of a half wave rectifier. Hence replacing term 'f' of eqn(6.20) by 2f, we get the ripple factor of a full wave rectifier, expressed as

$$r = \frac{1}{4\sqrt{3}} \frac{1}{f R_L C} \tag{6.21}$$

6.2   Choke Filter

The last section talked of a shunt capacitor filter, where the capacitor stores electrical energy, which is slowly discharged to the load when the input stops providing an output signal. This alternation from input source to energy from the discharging capacitor maintains a continuous current in the output which leads to a reduction in the ripples. So any device which is capable of storing energy should be an able filtering circuit. The inductor stores electrical energy in the form of magnetic energy. Thus, the inductor should also be capable of removing ripples from the output. Inductor filters are popularly called *"choke filters"*. We now study the output of an half-wave rectified and fully rectified output with a choke filter in series to the load.

6.2.1   For Half Wave Rectifier

Figure (6.5) shows how the inductor is used in a half wave rectifier circuit to remove ripples. The output is taken across the load resistance. Initially, while the positive AC cycle appears across the transformer's secondary, the inductor stores magnetic energy. At the onset of the negative cycle the current falls, (i.e. +di/dt goes -di/dt) hence, the inductors polarity flips direction. Feeding current into the load at expense of the stored magnetic energy, i.e. magnetic energy is being converted to electric energy. The circuit is complete and current will continue to flow while the voltage across the inductor is greater than the negative going AC signal, thus keeping the diode forward biased. The moment the AC voltage goes more negative then the voltage across the inductor, the diode is reverse biased and no current flows in the load.

---

[1]some authors judge discontinuity from interval for which the output signal goes zero.

Assuming the diode to be conducting and ideal, no voltage drop would exist across it. The loop equation (using KVL) can be written as

$$V_m sin\omega t = L\frac{di}{dt} + Ri$$

Selecting a new variable 'ωt' the above equation reduces to

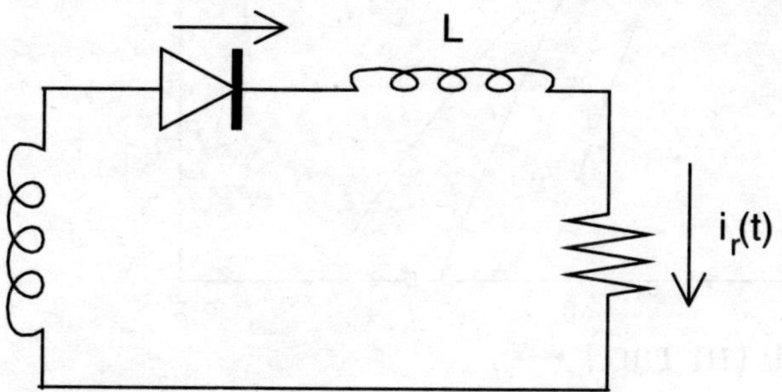

$$V_m sin\omega t = \omega L\frac{di}{d\omega t} + Ri$$

$$V_m sinx = \omega L\frac{di}{dx} + Ri$$

The solution[2] of this equation is

$$i(t) = \frac{V_m}{\omega L}e^{-ax}\int e^{ax}sinxdx$$

Figure 6.5: *The inductor is kept in series to remove ripples from the output of an half wave rectifier circuit. Primary of the step-down transformer is not shown.*

$$i(t) = \frac{V_m}{\sqrt{R^2 + \omega^2 L^2}}sin(\omega t - \phi) + Ce^{-Rt/L}$$

where

$$\phi = tan^{-1}\left(\frac{\omega L}{R}\right) \tag{6.22}$$

When the circuit starts conduction (t=0), the load current is zero. Applying this initial condition for calculating the integral constant, we have

$$C = \frac{V_m}{\sqrt{R^2 + \omega^2 L^2}}sin\phi$$

---

[2]The loop equation for the half wave rectifier circuit with choke filter is of the form:

$$\frac{dy}{dx} + Py = Q(x)$$

multiply equation by e^{Px}

$$e^{Px}\frac{dy}{dx} + Pe^{Px}y = Q(x)e^{Px}$$

$$\frac{d}{dx}(ye^{Px}) = Q(x)e^{Px}$$

$$y(x) = e^{-Px}\int Q(x)e^{Px}dx$$

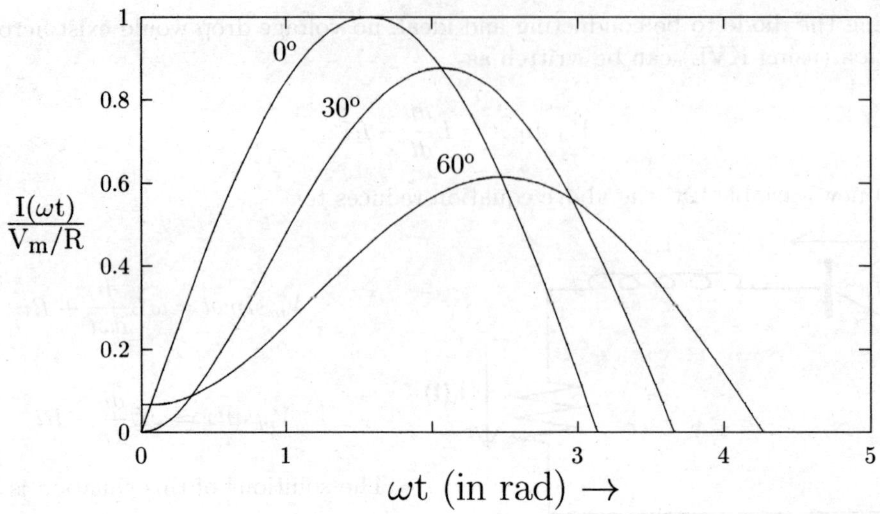

Figure 6.6: *The variation of load current with increasing inductance ($\phi = \tan^{-1}(\omega L/R)$) is shown. The plot is between $\frac{i(\omega t)R}{V_m}$ ($= \cos(x)\left[\sin(x-\phi) + e^{-x/\tan\phi}\sin(\phi)\right]$) and x = ωt. As can be seen even for large x = ωt the output current falls zero even before the next positive cycle appears in the output ($2\pi \sim 6.28$ rad).*

The current through the load, hence, can be represented as

$$i(t) = \frac{V_m}{\sqrt{R^2 + \omega^2 L^2}}\left[sin(\omega t - \phi) + e^{-Rt/L}sin\phi\right]$$

or completely in terms of ϕ and ωt, the above eqn can be written as

$$i(t) = \frac{V_m}{R}cos\phi\left[sin(\omega t - \phi) + e^{-Rt/L}sin\phi\right] \tag{6.23}$$

The variation of output current for various combinations of inductors (in terms of ϕ) is shown in fig(6.6). As can be seen, even with a large inductor, or small enough load resistance, the output current falls zero even before the start of the second cycle at 2π (or 6.28). To obtain a continuous output current which would reduce the ripples we demand that the falling current should never fall equal to zero. For the worst possible case we allow the current to fall zero at 2π at which instance the second positive input comes in for the case of half wave rectifiers. We ask what should be the value of load resistance and filtering choke to achieve this condition. To compute this we re-arrange eqn(6.23)

$$i(t) = \frac{V_m}{R}cos\phi\left[sin(\omega t - \phi) + e^{-Rt/L}sin\phi\right] = 0$$

and ask at what instant (say $\omega t = \phi_c$) does the load current go zero?

$$0 = cos\phi\left[sin(\phi_c - \phi) + e^{-R\phi_c/\omega L}sin\phi_c\right]$$

ideally ϕ_c should be $\geq 2\pi$. Re-arranging the above equation we have

$$sin\phi_c cot\phi = cos\phi_c - e^{-R\phi_c/\omega L} \tag{6.24}$$

This equation can not trivially solved to obtain ϕ_c. Hence we ask for the requirement to achieve the worst case condition (i.e. $\phi_c = 2\pi$), we have

$$1 - e^{-2\pi R/\omega L} = 0$$
$$e^{-2\pi R/\omega L} = 1$$

This condition can only be achieved for $R \to 0$ or $L \to \infty$. Both these are impractical, thus, under no circumstances can we achieve a continuous current at the output using a choke filter for an half wave rectifier. Obviously this results in high ripples in the output. However, in a full wave rectifier where the second peak starts at π itself, one can expect to get continuous output with a choke filter.

6.2.2   For Full Wave Rectifier

Using Fourier series, the output of the full wave rectifier can be written as[3]

$$v(t) = \frac{2V_m}{\pi} - \frac{4V_m}{\pi}\left[\frac{1}{3}cos2\omega t + \frac{1}{15}cos4\omega t + \ldots\ldots\right] \tag{6.25}$$

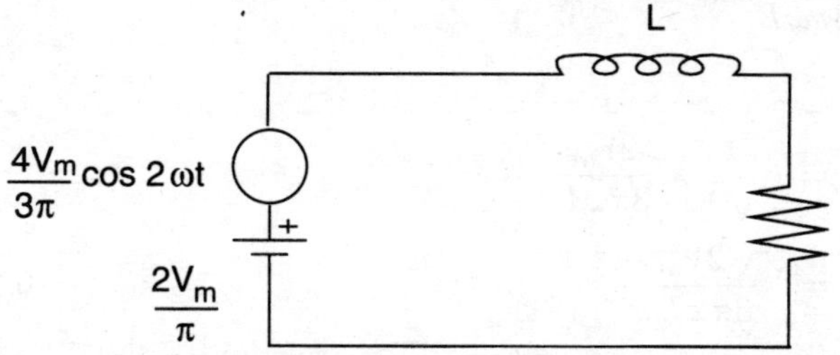

Figure 6.7: *The output of the full wave rectifier can be approximated as a DC source and AC source feeding power into the output.*

Note the time independent term of the series above is the DC output (V_{dc}) for a full wave rectifier (see table 5.5) which was evaluated by finding the average/ area under the curve using calculus. The output is across the resistance, hence the inductor resistance combination acts as a potential divider circuit. The output would be small if the impedance offered by the inductor is greater than that offered by the resistance. Since, the impedance offered by the inductor is proportional to the AC frequency, higher harmonics see the inductor offering larger and larger impedance as compared to the resistance. Hence, the higher harmonics do not appear across the resistance, i.e. at the output. Thus, only the DC term and the first harmonics is of any significance in other understanding of the choke filters response on the full wave rectifier circuit. Figure(6.7) shows the equivalent circuit of the full wave rectifier with choke filter. The transformer's secondary and diode can be replaced by two potential sources, a DC and AC source. The net voltage appearing in the circuit is

$$v(t) = \frac{2V_m}{\pi} - \frac{4V_m}{3\pi}cos2\omega t$$

---

[3]The derivation of this is left as a Mathematical exercise for students of physics. For the remaining audience, this expression is best treated as a black box. However, a point to appreciate is that any period (repeating) waveform can be mathematically represented as an infinite sum of sine and cosines terms.

The AC and DC current flowing in the output can be calculated individually. The net current in the circuit then is a linear sum of the two, as understood by superposition theorem. To obtain the AC current solve the voltage loop equation:

$$L\frac{di}{dt} + Ri = -\frac{4V_m}{3\pi}cos2\omega t$$

The differential equation is similar to the equation we met in the last section, and the solution gives

$$i(\omega t) = -\frac{4V_m}{3\pi}\frac{cos(2\omega t - \phi)}{\sqrt{R^2 + 4\omega^2 L^2}} - Ke^{-Rt/L}$$

where

$$\phi = tan^{-1}\left(\frac{2\omega L}{R}\right)$$

The expression is quite long and is best to approximate for simplification. We assume the load resistance is small compared to the inductors impedance. This is consistent with the requirements required to ignore the higher harmonics. In that case the above equation reduces to

$$i(\omega t) = -\frac{2V_m}{3\pi\omega L}cos(2\omega t - \phi)$$

The (rms) AC current flowing in the load is

$$\begin{aligned}|i_{ac}| &= \frac{1}{\sqrt{2}} \times \frac{2V_m}{3\pi\omega L} \\ &= \frac{\sqrt{2}V_m}{3\pi\omega L}\end{aligned} \qquad (6.26)$$

Thus, we have obtained the AC current flowing in the load. This was calculated using superposition theorem, where sources are seen to act independent of each other. Similarly, we can calculated the DC current by removing the AC source. Since the inductor does not provide any impedance to DC sources, the DC current is

$$i_{dc} = \frac{2V_m}{\pi R}$$

Unlike in the previous cases where we were not able to resolve the pure AC and pure DC part, leading to the requirement of eqn(5.8), we have both the pure AC and pure DC flowing in the load. Hence, we can directly calculate the ripple factor from eqn(5.5), giving

$$\begin{aligned}r &= \frac{\sqrt{2}V_m}{3\pi\omega L} \times \frac{\pi R}{2V_m} \\ &= \frac{R}{3\sqrt{2}\omega L}\end{aligned} \qquad (6.27)$$

The ripple factor can hence be reduced by

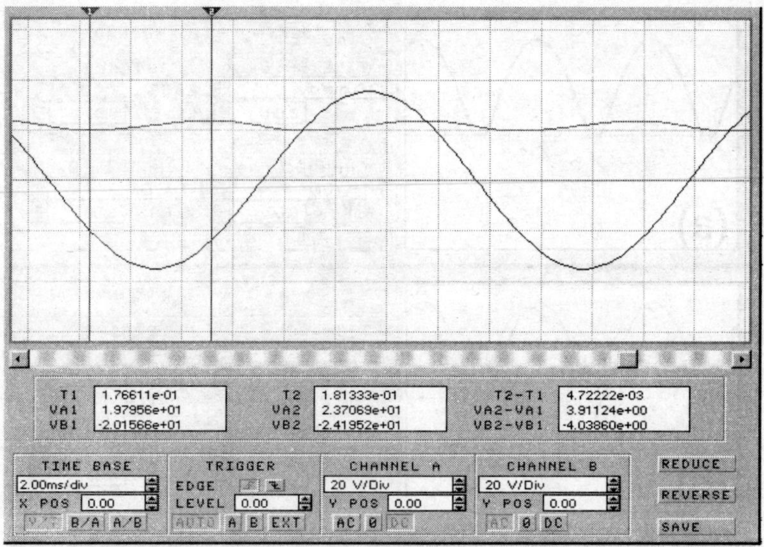

Figure 6.8: *The output of the full wave rectifier after ripple reduction using a choke filter.*

(i) Reducing the load resistance. This would result in large load current. Thus if the load demands a large DC current, the choke filter is best suited for removing ripples unlike the shunt capacitor which feeds low current to the load for small rippled output.

(ii) Making the inductance large. Obviously, larger the inductance, more magnetic energy is stored to provide continuous current to the output. The practical problem is of course the bulky size of the inductor as larger and larger coil is taken up.

(iii) Finally, increasing the input frequency.

Also, examine fig(6.8) and note the difference in the shunt filtered output and that obtained by the choke filter. In the case of the shunt filter, V_{dc} is equal to peak voltage level of the AC signal from the transformer's secondary coil (i.e. V_m). A little bit of imagination and eqn(6.13) would show this statement to be true. In the case of choke filtered output, the DC level is well below the peak voltage level. Usually, it is convenient for designing a shunt filter when there is a specific demand for the output DC level as the peak voltage level is known before hand and no mathematics is required. For the choke filter output DC level is given as $2V_m/\pi$.

Examining fig(6.8) again, one would notice the output signal oscillating around this DC level. The inductor obviously does not remove all the AC signal and hence some AC signal does travel into the resistance. The DC contribution to the output is known while the AC would be different. The AC output would be given as

$$v_{ac} = i_{ac}R$$

From eqn(6.26), we have i_{ac}. Thus, the AC output would be

$$v_{ac} = \frac{\sqrt{2}V_m}{3\pi\omega L} \times R$$

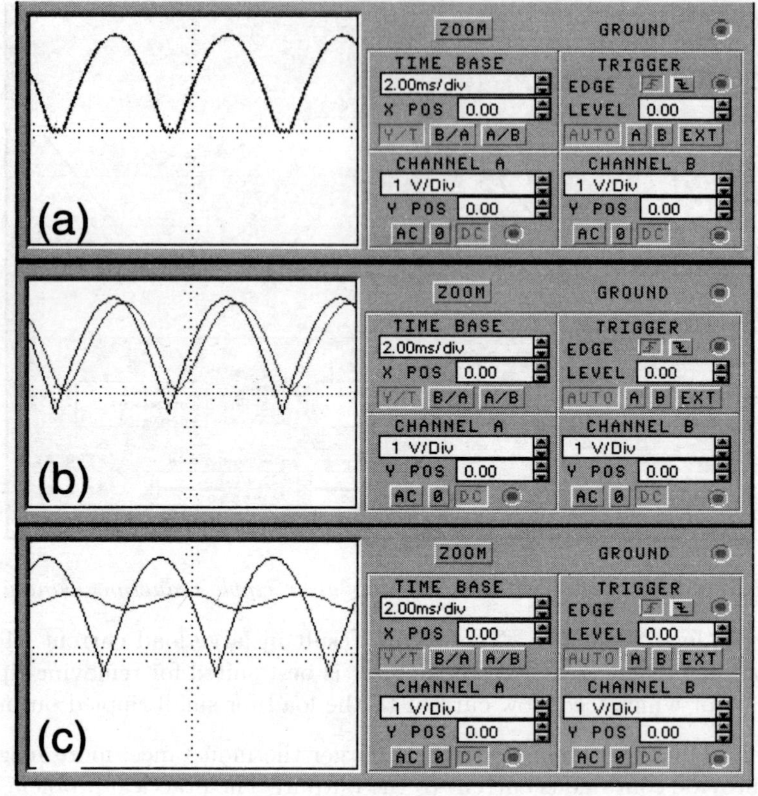

Figure 6.9: *The three outputs are of the full wave rectifier after ripple reduction using a choke filter. In all cases the load resistance is of* $1K\Omega$ *while (a) L=0.1H* $(L << \frac{R}{3\omega})$*, (b) L=1H* $(L = \frac{R}{3\omega})$ *and (c) L=10H* $(L \gg \frac{R}{3\omega})$*.*

this is the output rms (AC) voltage. While designing, a precaution to be taken is that the AC voltage in it's negative cycle (maxima in negative side $V_m = \sqrt{2}v_{rms}$) should not be greater then the output DC voltage, i.e

$$\frac{2V_m}{\pi} - \frac{2V_m R}{3\pi\omega L} \geq 0$$

Simplifying, this gives us the designing precaution to be taken

$$L \geq \frac{R}{3\omega} \; (in \; Henry) \tag{6.28}$$

Figure(6.9a) shows the situation where the condition stated in eqn(6.28) is not satisfied. Note the discontinuous output waveform. This discontinuity is just removed when $L = \frac{R}{3\omega}$, (fig 6.9b). The ripple is minimized when $L > \frac{R}{3\omega}$, (fig 6.9c). This condition is restrictive and calls for precautions from the designer. The designer can employ two possible solutions (for this given circuit).

In the first case, the designer can place a resistance parallel to the load resistance. This resistance is called a **bleeder resistance**. The choice of the bleeder resistance is such that it should satisfy the condition given in eqn(6.28). It would obviously be a small entity. The load resistance kept would usually be larger then the bleeder resistance, however, the effective resistance of their parallel combination would be dictated by the value of the bleeder resistance. Hence the condition defined by eqn(6.28) would be satisfied preventing a discontinuous output abide condition user maintains $R > R_B$.

The second technique that the designer can employ is to use a swinging inductor (also called a saturable inductor). The inductance offered by this coil depends on the direct current (DC) flowing in it ($L \propto 1/I_{dc}$). As the load resistance increases, I_{dc} decreases. This in a swinging inductor increases the value of the inductance. The condition given in eqn(6.28) is thus maintained for whatever load the user might care to use.

6.3   The LC Filter

To improve the removal of ripples from the rectified output we now study the third example of filtered circuits. This circuit uses features of both a choke filter and a shunt filter in the same circuit. An inductor is kept in series of the load with a capacitor in parallel with the load. This parallel placed capacitor prevents AC current in the resistance by presenting an alternative path, thus not calling for the problem discussed in the last chapter. Considering a full wave rectifier's output is to be filtered using this LC filter, the equivalent circuit would look like that shown in figure(6.10). This again is based on the Fourier series that represents the output of our full wave rectifier.

$$V = \frac{2V_m}{\pi} - \frac{4V_m}{\pi}\left[\frac{1}{3}cos2\omega t + \frac{1}{15}cos4\omega t + \ldots\ldots\ldots\right]$$

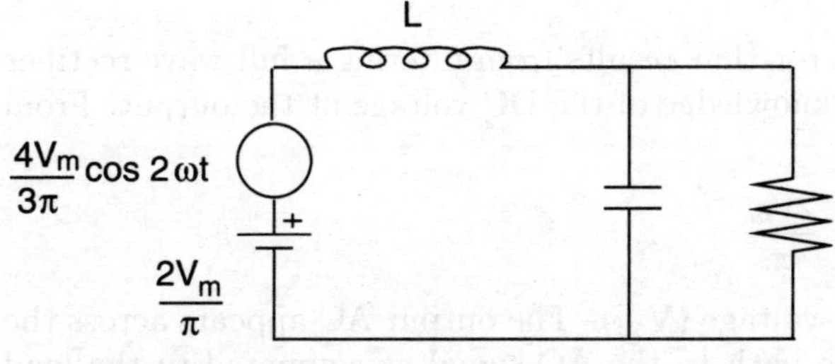

Figure 6.10: *The inductor and capacitor kept in an 'L' shape constitutes an 'LC' filter or just 'L' section filter.*

Notice the higher harmonics are neglected in our equivalent circuit. This is due to their negligible contribution to the load current. Trivial mathematics can be used to compare the contribution of the first and second harmonics (the first two terms in the bracket of the Fourier series, the time dependent AC terms) to the load current. Contribution from the first harmonics

$$V_{rms} = \frac{1}{\sqrt{2}}\frac{4V_m}{3\pi}$$

To calculate the current we require to know the impedance offered by the series and parallel combination of the inductor, capacitor and load resistance.

$$Z_L = X_L + \frac{R_L X_C}{R_L + X_C}$$

Assuming $X_C \ll R_L$ and $X_C \ll X_L$ we have the impedance offered by the circuital elements as

$$= X_L + X_C$$
$$= X_L$$

The magnitude of the impedance offered by the inductor depends on the frequency of the AC signal. An inspection of the Fourier series show that the first harmonics has a frequency of '2f', where 'f' is the frequency of the AC signal. The full wave rectifier's output has a frequency of '2f' as discussed earlier. Hence, the magnitude of the impedance offered by the inductor is

$$Z_L = 2j\omega L$$
$$|Z_L| = 2\omega L$$

Using the information analyzed above, the current due to the first harmonics is

$$I_{rms} = \frac{V_{rms}}{|Z_L|} = \frac{\sqrt{2}V_m}{3\pi\omega L}$$

(The results of eqn 6.26 can be obtained following the same steps) Similarly, the current due to the second harmonics is

$$I_{rms2} = \frac{V_{rms}}{|Z_L|} = \frac{V_m}{15\sqrt{2}\pi\omega L}$$

As you can see the contribution of the second harmonics is only one tenth that of the first. Thus, effectively only the first harmonics is going to contribute, justifying our equivalent circuit.

$$\frac{I_{rms2}}{I_{rms}} = \left(\frac{V_m}{15\sqrt{2}\pi\omega L}\right)\left(\frac{3\pi\omega L}{\sqrt{2}V_m}\right)$$
$$= \frac{1}{10}$$

Our next step is to compute the ripple factor that results from filtering a full wave rectifier with a LC filter. For this we first require the knowledge of the DC voltage at the output. From the Fourier series we have

$$V_{dc} = \frac{2V_m}{\pi}$$

Second we require to know the AC output voltage (V_{ac}). The output AC appears across the capacitor since it provides a lower impedance path to the AC signal as compared to the load resistance (or mathematically, in a parallel combination of X_c and R, the effictive impedance is X_c considering the fact we have already demanded $X_c \ll R$). Hence,

$$v_{ac} = X_C I_{rms}$$
$$= \frac{1}{2\omega C} \times \frac{\sqrt{2}V_m}{3\pi\omega L}$$
$$= \frac{V_m}{3\sqrt{2}\pi\omega^2 LC}$$

The ripple factor now can be calculated

$$r = \frac{v_{ac}}{V_{dc}} = \frac{V_m}{3\sqrt{2}\pi\omega^2 LC} \times \frac{\pi}{2V_m}$$

$$= \frac{1}{6\sqrt{2}\omega^2 LC} \tag{6.29}$$

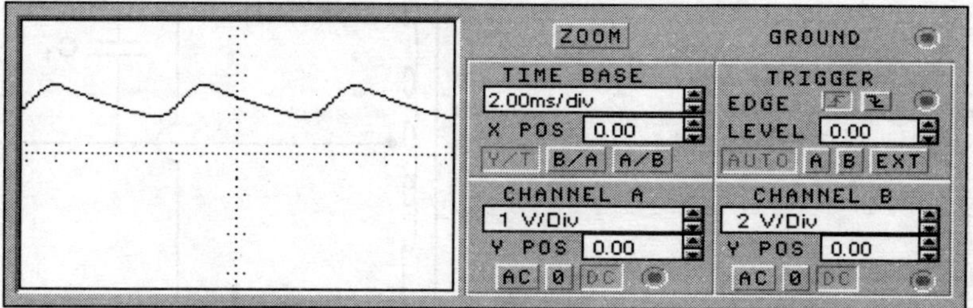

Figure 6.11: *The output of an 'LC' filter. In the L section filtered output of fig(6.9a), a 10μF capacitor across the load improves the removal of ripples even though the condition of eqn(6.28) has not been attained.*

The result of cqn (6.29) suggests that the ripple factor of this circuit is independent of the load resistance. Unlike the shunt capacitor filter which required large resistance and the choke filter which required small load resistance, eqn(6.29) does not seem to impose such restrictions. This is not totally true, since to arrive to the result given by eqn(6.29) we imposed the condition $X_L >> R$ and $R >> X_C$.

Figure(6.11) shows the output of an LC filter. Infact a 10μF has been added to the choke filter circuit of load 1KΩ and L=0.1H. The output without the capacitor is shown in fig(6.9a). The output had large ripples and was discontinuous. How the designer comes out of the problem of discontinuous output in designing a choke filter was discussed in the previous section. However, a simpler solution is to opt for an LC filter and add a capacitor.

6.4   The π Filter

The π filter cascades the LC filter to a shunt capacitor filter (see fig 6.12). As can be seen from the earlier analysis, the mathematics gets more and more involved with increased filtering elements. We hence fall back to the Fourier series. Here we consider the Fourier series of the sawtooth waveform[4] of fig(6.4) which acts as the input signal for the LC filter that follows. Here it should be noted that the Fourier series of eqn(6.30) is for a full wave rectified output.

$$v(t) = V_{dc} - \frac{V_r}{\pi}\left[sin2\omega t - \frac{sin4\omega t}{2} + \frac{sin6\omega t}{3} - \ldots\ldots \right] \tag{6.30}$$

where V_r is the peak to peak of the AC ripples riding on the DC output, given as

$$V_r = \frac{\pi I_{dc}}{\omega C_1}$$

---

[4]Can you analyze the shunt filter using this Fourier series and not all the mathematics used earlier?

The method of calculating V_r is already discussed for the half wave rectifier in section on shunt capacitor filters. It can be extended to full wave rectifiers. However, to brush our memory, V_r can be calculated in a couple of steps. The output of a shunt capacitor is idealized as a sawtooth waveform (see fig 6.4). In half the cycle, the waveform is described by the equation of a line from which the slope and hence the capacitor current can be computed as

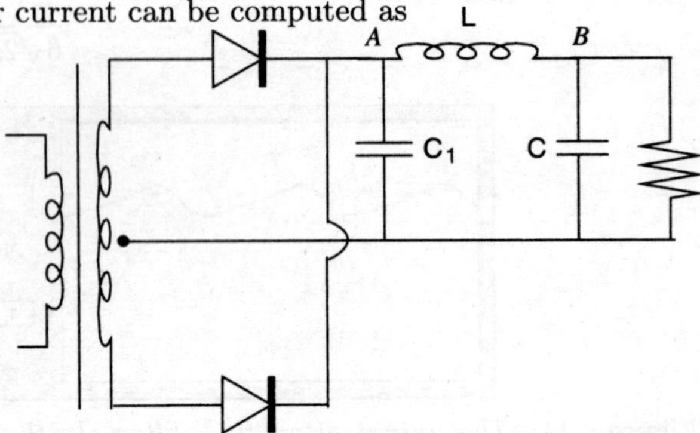

$$\begin{aligned}
i_c(t) &= C_1 \frac{d}{dt}[v_c(t)] \\
&= \omega C_1 \frac{V_r}{\pi - \omega(t_2 - t_1)} \\
&= \omega C_1 \frac{V_r}{\pi} = I_{dc}
\end{aligned}$$

giving the required form of V_r. The V_{rms} of the AC ripples, whose peak to peak is V_r is given as

$$V_{rms} = \frac{1}{\sqrt{2}} \frac{V_r}{\pi}$$

Figure 6.12: *The capacitor-inductor and capacitor kept in an 'π' shape achieves high filtration and can be imagined as an 'L' section filter smoothing the output of the shunt capacitor filter.*

Substituting the value of V_r, we have

$$V_{rms} = \frac{I_{dc}}{\sqrt{2}\omega C_1}$$

in LC filter, for $Z_L > X_C$ and $R > X_C$, the load impedance is $Z_L = 2\omega L$, giving

$$\begin{aligned}
I_{rms} &= \frac{V_{rms}}{2\omega L_1} \\
&= \frac{I_{dc}}{2\sqrt{2}\omega^2 L_1 C_1}
\end{aligned}$$

The AC voltage is across the capacitor ('C') as we have already assumed $X_c \ll R$

$$\begin{aligned}
v_{ac} &= I_{rms} X_C \\
&= \frac{I_{dc}}{4\sqrt{2}\omega^3 L_1 C_1 C}
\end{aligned}$$

With the knowledge of the AC signal present in the output, we can now compute the ripple factor for this circuit.

$$\begin{aligned}
r = \frac{v_{ac}}{V_{dc}} &= \frac{I_{dc}}{4\sqrt{2}\omega^3 L_1 C_1 C V_{dc}} \\
&= \frac{\sqrt{2}}{8\omega^3 R_L L_1 C_1 C}
\end{aligned}$$

Note, as we proceeded in our discussion on filters, the ripple factor of successive circuits had higher order of ω in the denominator. This made sure that the ripple in the output was being smoothed out faster and faster.

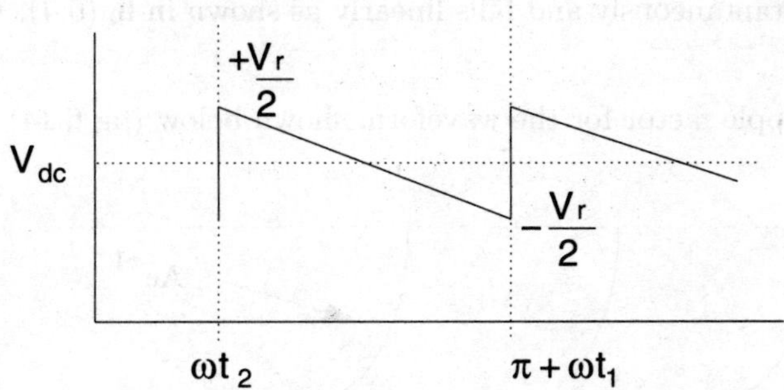

Figure 6.13: *The idealized sawtooth waveform that approximates the output of a shunt capacitor filtered full wave rectifier's output.*

Exercise

Q1. What is the purpose of the filter capacitor in the rectifier?

Q2. Describe effect of selected value of filtering capacitor on various parameters like the desired ripple voltage, load, ripple frequency, and peak voltage.

Q3. What is the rate of discharge of a capacitor in an RC filter?

Q4. Calculate the ripple voltage (V_r) measured at the load of a shunt capacitor filtered rectifier (both for half wave and full wave).

Q5. Compare the DC voltage level with the peak voltage (V_p, maxima of the input sine wave) of a shunt capacitor filtered circuit which achieves a prefect dc output.

Q6. Draw a half wave rectifier circuit with a 10 V_{rms}, 50Hz supply and a 100Ω load resistance. Draw a graph of the input and output voltages. Calculate the peak voltage for the output waveform.

Q7. Add a 5000μF capacitor in parallel with the resistor load of a half wave rectifier circuit with a 10 V_{rms}, 50Hz supply and a 100Ω load resistance.. Calculate the time constant of the RC combination. How does it compare with the period of the rectified waveform? Draw an approximate graph of the output voltage waveform.

Q8. Will a smaller capacitor (for the above question) increase or decrease the time constant? Will the output voltage have reduced ripples?

Q9. If a capacitor were chosen such that the change in output voltage (ripple) was 5% of the average voltage value, what would be the value of the average voltage. Assume that the voltage rises instantaneously and falls linearly as shown in fig(6.4). Calculate the value of C.

Q10. Calculate the ripple factor for the waveform shown below (fig 6.14)

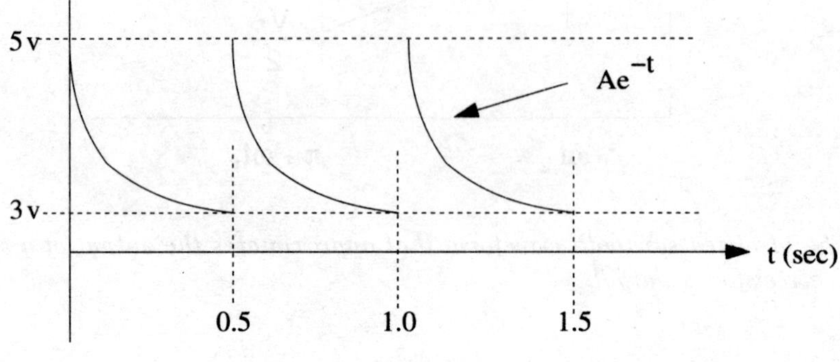

Figure 6.14:

Q11. Discuss the problems of choke filter in (a) half wave rectifiers and (b) full wave rectifiers.

Q12. What is a bleeder resistance? What is it's use.

Q13. How is a swing inductor useful in designing a choke filter?

Q14. Compare the performance of a shunt filter with a choke filter when used with a half wave rectifier.

Q15. Compare the performance of a shunt filter with a choke filter when used with a full wave rectifier.

Q16. Design a power supply with a full wave rectifier using a shunt filter so as to provide 10v (DC) at 150mA with maximum 3% ripple for f=50Hz.

Chapter 7

Regulation

The common energy source that students must have seen is the battery. Most of them must have seen these DC source in school laboratories or in automobiles like car, buses and trucks. The smaller versions, popularly known as "cells" are used in Walkmans, torches etc. These provide energy from chemical reactions taking place in them. Easily understood and also experienced, is that with time their ability to provide energy diminishes. Thus, the best way is to convert the main supply's AC into DC. Circuits achieving this are called DC supplies or power supplies. The essential circuitry required for making a power supply is shown as blocks in fig(7.1).

The first three blocks have been discussed in the preceding chapters. The selection of the transformer is based on how much the final output voltage is required. For an example, on designing a +30v DC power supply, the transformer best suited would be the 16-0-16 center tapped transformer. This is the most easily available transformer for the particular output that is required. Without using the center tap, the secondary side would give 32v when given line voltage. The second important point which the designer would have to keep in mind is, *"how much current would the output draw?"* The wire or more precisely the wire gauge of the transformer secondary and primary is determined by the current demand. Obviously the cost of the transformer escalates with increasing current demand.

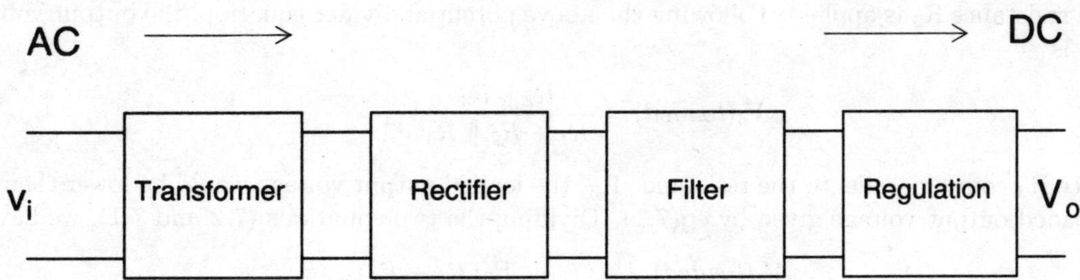

Figure 7.1: *Block diagram of an DC power supply.*

Depending on the output of the transformer's secondary one has to decide on using a full wave rectifier or a bridge rectifier. If the output of the secondary is low, say 3v, then using a bridge rectifier consisting of Si diodes would mean a loss of 1.4v, which would be required to send the diodes into conduction. Assuming no further losses etc, the output voltage would only be 1.6v for the given example. Since only one diode goes into conduction for each output cycle in the full wave rectifier, it would be ideal for this particular example. However, if the voltage considerations are large, then a bridge rectifier would be more suitable. Since the two reverse biased diodes share the negative voltage appearing across them, it works out cheaper to work with diodes requiring only half the PIV's that the full wave rectifier's diode would require while working between the same voltages.

The third block of the filter circuit depends on whether the rectifier is an half wave rectifier or a full wave rectifier. If its an half wave rectifier, then the choke filter is immediately ruled out. Next criteria would depend on the load requirement. If the load requires small current the shunt filter would be enough, however if the output load's current demand is large than one thinks in terms of a choke filter.

7.1　Loading and Regulation

The forth block is new to the readers of this book. It is required to achieve voltage regulation for the output load. Before defining regulation, it is desirable to understand the phenomenon of *"loading"*. Ideally when a power supply is designed for an user, the manufacturer designs it as a variable voltage source. A variable voltage is obtained by applying the load across a potential divider. Resistance R_1 and R_2 (fig 7.2) constitute a potential divider. The voltage drop across R_2 can be calculated trivially using Ohms' law or more fancifully by Kirchhoff's law, and is given by

$$V_o = \frac{R_2}{R_1 + R_2} V_{in}$$

$$V_o = \frac{1}{1 + R_1/R_2} V_{in} \tag{7.1}$$

Thus, the output voltage can be varied by varying R_2 or the ratio of R_1 and R_2. The varying voltage across R_2 appears across the load. Now let us analysis the situation when the load resistance R_3 is applied. Following the above potential divider equation, the output voltage now is

$$V_o(loaded) = \frac{R_2 \parallel R_3}{R_1 + R_2 \parallel R_3} V_{in} \tag{7.2}$$

Since R_3 comes parallel to the resistance R_2, the loaded output voltage would be lower then the unloaded output voltage given by eq(7.1). Dividing the two equations (7.2 and 7.1), we have

$$\frac{V_o(loaded)}{V_o} = \frac{R_3(R_1 + R_2)}{R_2 R_3 + R_1(R_2 + R_3)}$$

$$= \frac{1}{1 + \frac{R_1 R_2}{R_3(R_1 + R_2)}} \tag{7.3}$$

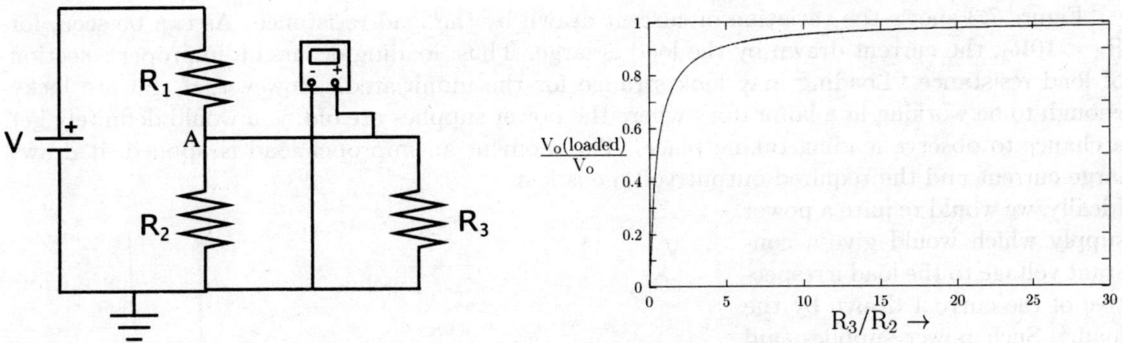

Figure 7.2: *Circuit diagram of a variable power supply used to give different voltages to the load R_3. The variation is obtained by altering the value of R_2. Shown alongside is the variation of the output voltage with varying load resistance. Note how loading takes place for small loads.*

If R_3 is made very large, the output voltage of the loaded circuit would be the same as that of the open circuit or unloaded circuit. However, if the load resistance is small, V_o(loaded) would be less than the unloaded output V_o. That is, the load resistance is effecting the output voltage. The resistance R_3 is said to be *"loading"* the power supply. The graph of figure 7.2 was plotted by assuming $R_1 = R_2$ in eqn(7.3).

When load resistance R_3 is small, it nearly shorts the resistance R_2, and the output voltage is small. As R_3 increases, the output voltage increases. Thus, the value of R_3 is determined in terms of resistance R_2. As a thumb rule the circuit designers select R_3 to be 10 times greater than R_2. This is a thumb rule only, and on maintaining this rule, the output voltage is $\geq 0.95 V_o$. There is another way of conveying what happens during loading. The common description one gets to hear is that "the load is drawing too much current from the supply". To understand this statement, let us analyse the circuit again. The circuit shown in fig(7.2) has two loops.

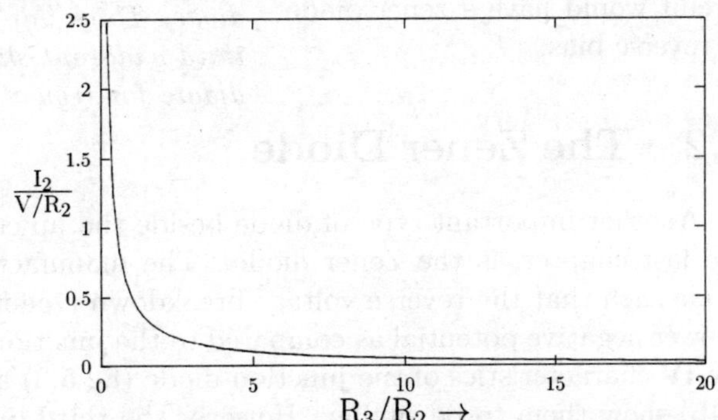

Figure 7.3: *Corresponding to the region of loading shown in fig 7.2 (in terms of voltage), the current drawn by the load is large. Thus, if the load is drawing too much current, it is loading the power supply.*

The current in R_3 can be determined by solving the simultaneous equation

$$V = R_1 I_1 + R_2(I_1 - I_2)$$
$$0 = R_3 I_2 + R_2(I_2 - I_1)$$

Again assuming $R_1 = R_2$ and solving for I_2 we have

$$\frac{I_2}{V/R_1} = \frac{1}{2R_3/R_2} \tag{7.4}$$

Figure 7.3 shows the variation of current drawn by the load resistance. As can be seen, for $R_3 < 10R_2$, the current drawn by the load is large. Thus, loading occurs on improper selection of load resistance. Loading may look strange for the uninitiated. However, if you are lucky enough to be working in a laboratory where the power supplies are old, you would definitely get a chance to observe loading taking place. The moment an improper load is applied, it draws large current and the required output voltage is lost.

Ideally, we would require a power supply which would give a constant voltage to the load irrespective of the current drawn by the load.[1] Such power supplies said to regulated. For prefect regulation, the regulating device/ circuit should have an I-V characteristics shown in figure (7.4a). The zener diode in reverse biased state as a very similar I-V characteristics. The forth block in a power supply, shown in fig(7.1), is for regulation. The simplest circuit would have a zener diode in reverse bias.

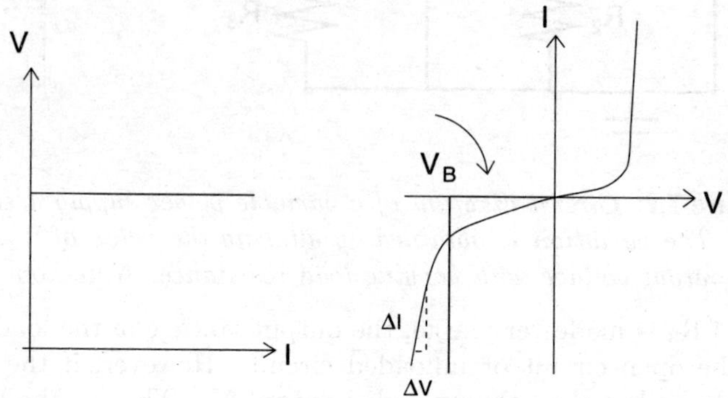

Figure 7.4: *(a) I-V characteristics of an ideal or regulated voltage source, and (b) the I-V characteristics of a zener diode. Thick line indicated by arrow shows ideal nature. The third quadrant shows that the zener diode is a suitable candidate for regulation.*

7.2   The Zener Diode

Another important type of diode beside the junction diode, discussed in the last chapter, is the Zener diode. The manufacturer designs the zener diode such that the reverse voltage breakdown (zener breakdown) occurs at a lower negative potential as compared to the junction diode. Comparison of the IV characteristics of the junction diode (fig 5.3) and the zener diode (fig 7.4b) show them to be similar. However the third quadrant shows that the breakdown in zener diodes occur at a lower negative voltage. Diode breakdown due to avalanche breakdown, which occurs when the minority carriers are accelerated by the electric field to sufficient energies that they can collide and excite electrons from covalent bonding of the semiconductor. Zener breakdown is different and occurs when the electric field near the junction becomes large enough to bend energy levels and excite valence electrons directly into the conduction band. Further explanation of the phenomenon of energy level bending etc. is beyond the scope of this book. Streetman has explained zener breakdown in detail. An important point to note is that the break-over of the zener is not destructive as is the case in junction diodes, in fact the zener finds its application in its breakdown region.

Figure 7.5: *Circuital representation of a zener diode.*

---

[1]The output voltage can also vary resulting from the AC line voltage not being constant. The AC voltage can vary. This means that the peak AC voltage, V_m, to which the rectifier responds can vary. The would result in a change in the DC output voltage. This requires for line regulation. The regulation that we are going to discuss here is the load regulation, i.e. maintaining the DC output for various load applied.

7.3 The Regulating Zener

Figure (7.4b) and fig(7.5) shows the current-voltage characteristic of a zener diode and its schematic symbol. The zener diode acts as a regulator when it is reversed biased beyond its break-over region. How does the zener diode achieve regulation? Trivially one can argue that the resistance of the zener diode beyond its breakdown is zero (from the ideal zener's IV character). Hence, any load would be greater than this zener resistance, satisfying the condition $R_3 > 10R_2$ (zener diode plays the role of R_2). This results in no loading of the power supply. Every circuit designer tries to achieve this condition. He thrives to make the output impedance of his circuit zero (as in case of the ideal zener) while he makes the circuit's input impedance (R_3) infinite. This ensures cascading/ coupling various blocks of circuits would not result in loading.

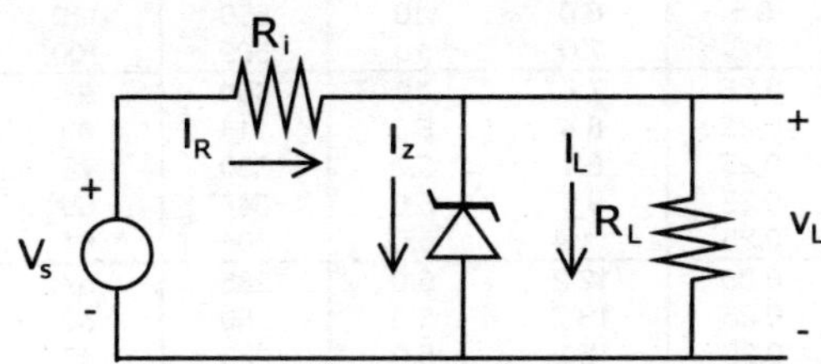

Figure 7.6: *Circuit showing how a zener diode is used to regulate the output voltage.*

However, practical zener diode's characteristics deviate from that of the ideal zeners. The resistance R_2 of the circuit shown in figure (7.2) is replaced by a zener diode such that it is reverse biased. Let us consider that a 12v supply is connected to a 4.7v zener in series to a 1KΩ resistance (R_i). In an open load condition, the zener would maintain a voltage drop of 4.7v across it and the current in the circuit would be given as

$$12 = 4.7 + 1 \times 10^3 I$$
$$I_R = \frac{12 - 4.7}{1 \times 10^3} = 7.3mA$$

If now a load of 1kΩ is connected across the zener (see fig 7.6), the zener would maintain the same voltage across it. Hence the current in the load would be 4.7mA (=4.7/1000). The current through the resistance R_1 remains the same, hence the current through the zener diode would be

$$I_z = 7.3 - 4.7 = 2.6mA$$

Thus, the zener diode regulates the output voltage by varying the current allowed to pass through it. This obviously is done by having a variable resistance. In the above example the zener diodes resistance varied from 640Ω to 1800Ω depending on whether circuit is unloaded or loaded. From fig(7.4b) one can see the resistance of the zener diode is larger near the 'x' axis (small current) and decreases as the current increases. Also, note the power dissipated in the zener in the no load and loaded condition would be different. In the no load condition, the power dissipated is 35mW (4.7v × 7.3mA), while when loaded since the current falls, power dissipation falls to 12mW. A designer hence has to select the value of the series resistance, R_i very carefully such that the power dissipated across the unloaded zener is not more than the maximum power handling capacity of the zener diode. Fig(7.7) lists information of some of the popularly available zener diodes.

A commonly used *figure of merit* for a power supply is its **PERCENT OF REGULA- TION**. The figure of merit gives us an indication of how much the output voltage changes over

Electrical Characteristics

$T_A = 25°C$ unless otherwise noted

Device	V_Z (V)	Z_Z (Ω) @	I_{ZT} (mA)	Z_{ZK} (Ω) @	I_{ZK} (mA)	V_R (V) @	I_R (µA)	I_{SURGE} (mA)	I_{ZM} (mA)
1N4728A	3.3	10	76	400	1.0	1.0	100	1,380	276
1N4729A	3.6	10	69	400	1.0	1.0	100	1,260	252
1N4730A	3.9	9.0	64	400	1.0	1.0	50	1,190	234
1N4731A	4.3	9.0	58	400	1.0	1.0	10	1,070	217
1N4732A	4.7	8.0	53	500	1.0	1.0	10	970	193
1N4733A	5.1	7.0	49	550	1.0	1.0	10	890	178
1N4734A	5.6	5.0	45	600	1.0	2.0	10	810	162
1N4735A	6.2	2.0	41	700	1.0	3.0	10	730	146
1N4736A	6.8	3.5	37	700	1.0	4.0	10	660	133
1N4737A	7.5	4.0	34	700	0.5	5.0	10	605	121
1N4738A	8.2	4.5	31	700	0.5	6.0	10	550	110
1N4739A	9.1	5.0	28	700	0.5	7.0	10	500	100
1N4740A	10	7.0	25	700	0.25	7.6	10	454	91
1N4741A	11	8.0	23	700	0.25	8.4	5.0	414	83
1N4742A	12	9.0	21	700	0.25	9.1	5.0	380	76
1N4743A	13	10	19	700	0.25	9.9	5.0	344	69
1N4744A	15	14	17	700	0.25	11.4	5.0	304	61
1N4745A	16	16	15.5	700	0.25	12.2	5.0	285	57
1N4746A	18	20	14	750	0.25	13.7	5.0	250	50
1N4747A	20	22	12.5	750	0.25	15.2	5.0	225	45
1N4748A	22	23	11.5	750	0.25	16.7	5.0	205	41
1N4749A	24	25	10.5	750	0.25	18.2	5.0	190	38
1N4750A	27	35	9.5	750	0.25	20.6	5.0	170	34
1N4751A	30	40	8.5	1,000	0.25	22.8	5.0	150	30
1N4752A	33	45	7.5	1,000	0.25	25.1	5.0	135	27

V_F Foward Voltage = 1.2 V Maximum @ I_F = 200 mA for all 1N4700 series

Figure 7.7: *Details of popularly available zener diodes.*

a range of load resistance values. The percent of regulation is determined by the equation:

$$POR = \frac{(V_{nl} - V_{fl})}{V_{fl}} \times 100$$

Where POR is the percent of regulation and V_{nl} is the output voltage when no load is applied, or in other words the output is open (**no load does not mean zero resistance load and hence short**). V_{fl} is the output voltage when full load is applied. That is, this equation compares the change in output voltage at the two loading extremes. For example, assume that a power supply produces 15 volts when the load current is zero, i.e. output is open. If the output voltage drops to 10 volts when full load current flows, then the percent of regulation is

$$POR = \frac{(15 - 10)}{10} \times 100$$
$$= 50\%$$

Ideally, there should be no change over the full range of the power supply's operation. That is, the 15 volt power supply should produce 15 volts at no load, at full load, and for any load in between. In that case, the percent of regulation of an ideal power supply or perfectly regulated supply would be

$$POR = \frac{(V_{nl} - V_{fl})}{V_{fl}} \times 100$$

$$= 0\%$$

Zero percent of regulation implies that the output voltage is constant under all load conditions. In practical circuits it is near impossible to obtain zero percent in regulation and the designer must settle for something less then ideal.

Let us visit fig(7.6) again. In a practical situation the load resistance, R_L, would vary with usage resulting in a variation of the load current, I_L. As discussed V_z by nature of the zener would remain constant thus achieving regulation. How good is this regulation? To investigate this we shall do some calculations and for developing the mathematics we assume that the input voltage, 'V_s', remains constant throughout. Then, we have (using KCL)

$$I_R = I_z + I_L$$

Which when expressed as nodal voltages is given as

$$\frac{V_s - V_z}{R_i} = I_z + \frac{V_L}{R_L}$$

$$\frac{V_s - V_z}{R_i} = I_z + \frac{V_z}{R_L} \tag{7.5}$$

Since, a sequence of event is set off by changing R_L, mathematically we investigate this by differentiating the above equation w.r.t. R_L. On differentiating we have

$$\frac{1}{R_i}\left(\frac{dV_s}{R_L}\right) - \frac{1}{R_i}\left(\frac{dV_z}{R_L}\right) = \frac{dI_z}{dR_L} - \frac{V_z}{R_L^2} + \frac{1}{R_L}\left(\frac{dV_z}{R_L}\right)$$

Since the starting assumption is that the input voltage will not change, $dV_s/dR_L = 0$. Also, say, how good the zener regulates is given by

$$\eta_o = \frac{dV_z}{dR_L}$$

the subscript 'o' to denote that regulation due to changes on the output/ load resistance, the above equation can now be written as

$$-\frac{1}{R_i}\eta_o = \frac{dI_z}{dR_L} - \frac{V_z}{R_L^2} + \frac{1}{R_L}\eta_o$$

$$-\frac{1}{R_i}\eta_o = \frac{dI_z}{dV_z}\frac{dV_z}{dR_L} - \frac{V_z}{R_L^2} + \frac{1}{R_L}\eta_o$$

$$-\frac{1}{R_i}\eta_o = \frac{1}{R_z}\eta_o - \frac{V_z}{R_L^2} + \frac{1}{R_L}\eta_o$$

where $\frac{dV_z}{dI_z}$ is the resistance of the zener diode in the breakdown region. Re-arranging, we have

$$\frac{V_z}{R_L^2} = \left(\frac{1}{R_z} + \frac{1}{R_i} + \frac{1}{R_L}\right)\eta_o$$

$$\eta_o = \frac{R_i R_z V_z}{R_L(R_i R_L + R_i R_z + R_L R_z)} \tag{7.6}$$

From the equation it becomes clear that the zener resistance at breakdown if is zero, $R_z = 0$

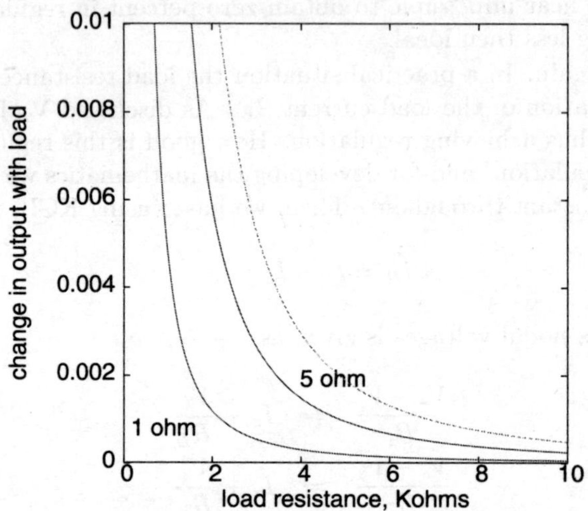

Figure 7.8: *Variation of $\frac{dV_z}{dR_L}$ with varying load resistance for three possible R_zs, namely 1Ω and 5Ω.*

then the POR would match the ideal case and the output would remain equal to V_z irrespective of the load resistance being used. Fig 7.8 clearly shows that for lower zener resistance, the output remains constant over larger range of load resistance.

$\boxed{\text{Example 7.1:}}$ What regulation does a zener achieve for case of varying input voltage. (Assume the load R_L remains constant)

Starting with eqn(7.5) and differentiating w.r.t. V_s, we have

$$\frac{1}{R_i} - \frac{1}{R_i}\left(\frac{dV_z}{dV_s}\right) = \frac{dI_z}{dV_s} + \frac{1}{R_L}\left(\frac{dV_z}{dV_s}\right)$$

$$\frac{1}{R_i} - \frac{1}{R_i}\left(\frac{dV_z}{dV_s}\right) = \frac{dI_z}{dV_z}\frac{dV_z}{dV_s} + \frac{1}{R_L}\left(\frac{dV_z}{dV_s}\right)$$

$$\frac{1}{R_i} - \frac{1}{R_i}\eta_i = \frac{1}{R_z}\eta_i + \frac{1}{R_L}\eta_i$$

or

$$\left(\frac{1}{R_z} + \frac{1}{R_L} + \frac{1}{R_i}\right)\eta_i = \frac{1}{R_i}$$

giving

$$\eta_i = \frac{R_L R_z}{R_i R_L + R_i R_z + R_L R_z} \tag{7.7}$$

The expressions (eqn 7.6 and eqn 7.7) indicate that for good POR, the regulator's resistance should be zero or atleast near zero. Also, notice for non-zero R_z, the resistances in the circuit also play a role in deciding the ability of regulation. The modern IC regulators have percent of regulation that are very low.

7.4   Regulating ICs

The final block of a DC power supply is the regulator (fig 7.1). The voltage regulator makes sure that the output voltage has little or ideally no variation irrespective of the load applied etc. In the last section we saw the zener diode is capable of achieving this, where the zener diode senses any change in output voltage and compensate for that change. Now-a-days IC regulators are present which achieve better regulation then the zener diode. Commonly and popularly used IC regulators are the 78xx and 79xx series. The '78' stands for the fact that the regulated output is positive while '79' implies that the output is below the ground potential. The last two numbers 'xx' represents the regulated voltage levels. For example IC7805 would give an regulated +5v output, while IC7912 would give an regulated -12v DC output. Figure (7.9) shows the pin configuration of IC78xx and IC79xx. Figure (7.10) shows the usage of the IC78xx regulator in a practical power supply.

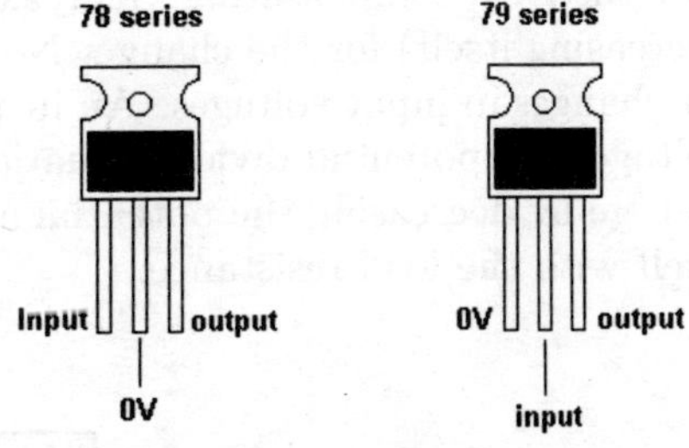

Figure 7.9: *A three pin IC voltage regulator IC7805 which gives a regulated +5v DC output along with the negative counterpart, IC7905.*

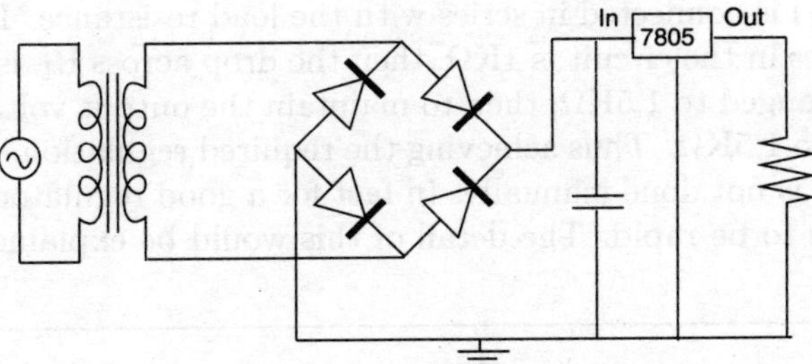

Figure 7.10: *A complete power supply with full wave rectifier circuit followed by a shunt capacitor filter and finally an IC regulator (IC7805) which gives a regulated +5v DC output.*

To understand the working of the IC regulator we must first understand how the transistor works, hence IC regulators will be discussed in chapter 8. However, the basic principle behind its working can be understood by assuming a trivial circuit containing only resistances. Just as in the case of the zener diode acting as a regulator by varying its resistance, the IC would have a regulating element which either varies it's resistance or controls the current through it as the zener does. Depending on the position of the regulating element in relation to the load resistance, voltage regulators are classified as either SERIES or SHUNT voltage regulators.

Figure(7.11) shows the basic circuit for a shunt-type regulator. As can be seen the regulating device, represented by the variable resistance, is connected in parallel with the load resistance. On comparison, the zener diode would also appear to you as a shunt regulator, that maintains the desired output voltage by sensing the current change in the parallel load resistance and then by compensating (increasing or

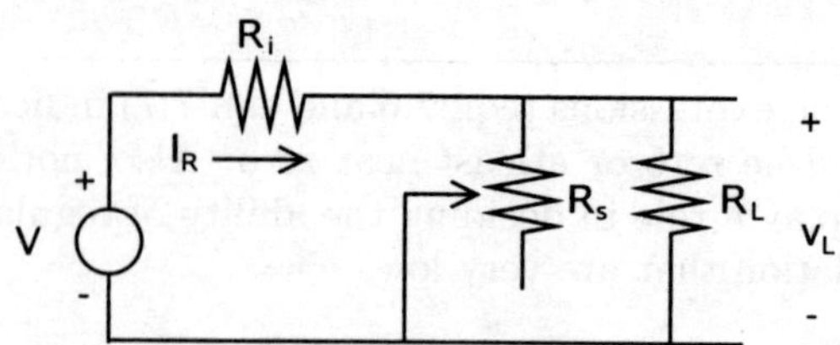

Figure 7.11: *An idealized voltage regulator using a resistance kept in shunt with the load resistance.*

decreasing itself) for the change. Now consider how the voltage regulator works to compensate for changes in input voltages. An increase in input voltage would result in increase in the output voltage (use potential divider equation and verify). Resistance R_s decreases to maintain output voltage by decreasing the potential across the effective resistance of the parallel combination of itself with the load resistance.

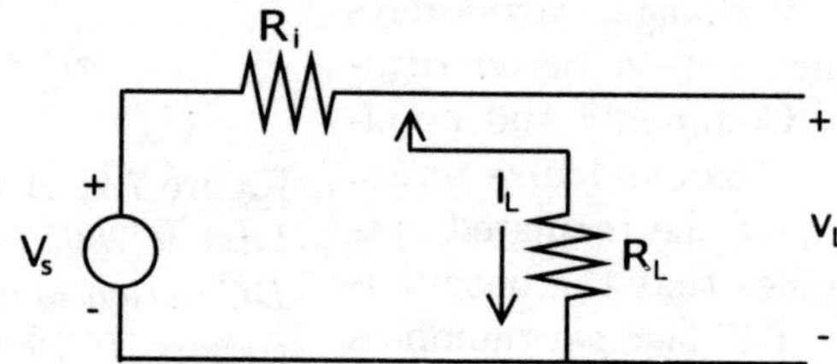

Figure 7.12: *An idealized voltage regulator using a resistance kept in series with the load resistance.*

Figure (7.12) illustrates the series voltage regulation. The circuit diagram shows the regulating device (resistance R_i) is connected in series with the load resistance. If the input DC supply is 10v and the resistances in the circuit is $1K\Omega$, then the drop across R_L is the regulated +5v. If the load resistance is changed to $1.5K\Omega$, then to maintain the output voltage at +5v the control resistance, R_i, changes to $1.5K\Omega$. Thus achieving the required regulation. Obviously, the change in the control resistance is not done manually. In fact for a good regulator the response to input and output voltages has to be rapid. The detail of this would be explained in Chapter 8.

Exercise

Q1. Explain how is the zener diode different from the 'pn' diode.

Q2. Explain the phenomenon of loading.

Q3. The zener of 1W and 1v, is used to regulate the output for a 1KΩ load. The input voltage is 500v and the series resistance used is of 100Ω. Comment whether the zener would be burnt out or not.

Q4. How does a voltage regulator work?

Q5. Describe the variable resistance model for the zener diode used for voltage regulation.

Q6. What is the definition of line regulation?

Q7. Define load regulation?

Q8. What are the output voltages for the following regulators: 7805, 7812, 7905 and 7912?

Q9. Draw a circuit of a power supply with bridge rectifier and an IC regulator that can provide +5V.

Q10. Experimentally verify the results of eqn 7.6 and 7.7.

Chapter 8

Transistor

The point-contact transistor was invented at Bell Laboratories in December 1947 by John Bardeen[1] and Walter Brattain. Bell Labs kept their discovery quiet and announced it only in June 1948. Very few people realized its significance at that time, and the discovery did not even make the front page of newspapers. However, today the transistor has changed the way that we live in fifty years of its existence. The transistor we study in this chapter was however developed by Shockley and is called the junction transistor. This transistor came nearly three years after the point contact transistor.

8.1   The Transistor

Recall from the chapter on diodes, a forward-biased diode is comparable to a low-resistance circuit element. In turn, a reverse-biased diode is comparable to a high-resistance circuit element. Trivially it is understood, for the same current, the power developed across a high resistance is greater than that developed across a low resistance (from $P = I^2R$). Thus, if a device were to contain two pn junctions, one forward-biased forming the input and the other reverse-biased output, a low-power signal could be given/ injected into the input to produce a high-power signal at the output. In this manner, a power gain would be obtained across the device. Current in both the input and output side is the same and amplification is taking place due to the current being *trans*ferred from lower input resistance to an higher output re*istor*, giving the device the name **transistor**.[2] This is the basic theory behind how the transistor amplifies, we now proceed to learn

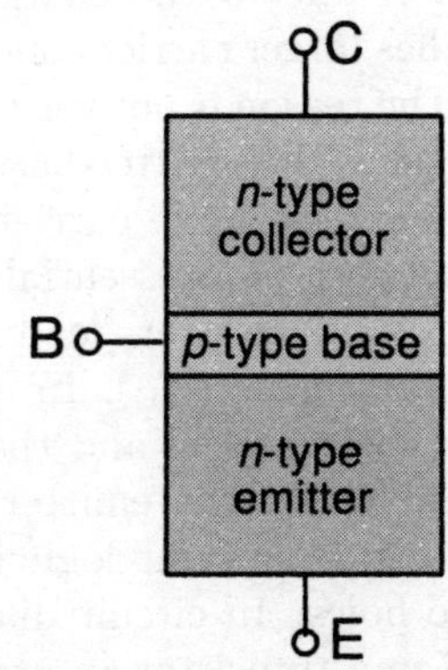

Figure 8.1: *A sandwich of 'p' layer between two 'n' layers giving a NPN junction transistor.*

---

[1]Bardeen won two Nobel prizes in physics, one for this invention and the other for his theory on superconductivity, more famously called as BCS theory.

[2]This example is best suited for transistor in common- base (CB) configuration, of which you will read in the following pages. Also see worked example

more about transistors.

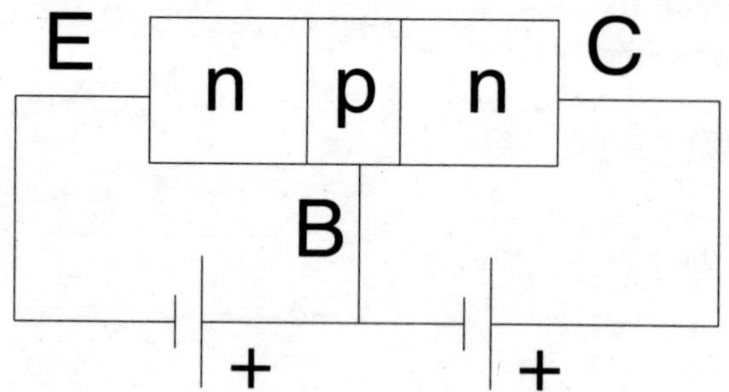

Figure 8.2: *The two junctions of the transistors are biased by two external voltage sources.*

From the geometry, we have demanded for the transistor, the device requires a pn diode placed along side with an np diode. The two diodes brought in close proximity would appear as a sandwich of a 'n' layer between two 'p' layers. A forward and reversed biased structure can also be made by "*sandwiching*" a 'p' layer between two 'n' layers. Figure(8.1) shows an NPN transistor with a 'p' layer between two 'n' layers.

The three layers of the transistor give rise to two depletion regions as in the case of the pn junction diode, the 'n' material containing free electrons while the 'p' section contains holes. To use the transistor, one junction has to be forward biased, while the second has to be reverse biased by some external potential sources or technically more appropriate the transistor has to be properly biased. By the symmetrical nature of the transistor represented in fig(8.1), one might argue any one junction can be forward biased as the user may deem fit. However, this is not the case.

The transistor's emitter-base junction is forward biased (see fig 8.2), while base-collector junction is biased in the reverse direction. This uniqueness or preferred selection of junction is due to the charge carrier concentration. The emitter has larger carrier concentration than that of the collector. The reason is not very difficult to appreciate. On forward biasing the emitter-base junction current starts flowing. Due to the large carrier concentration of the emitter layer, the current is essentially due to the majority charge carriers of the emitter, leading to naming this current as the emitter current (I_E). In an NPN transistor the emitter is a 'n' type material and the majority charge carriers are electrons, hence, the emitter current is essentially due to electrons. By the same logic in a PNP, the emitter current is due to holes. In circuit diagrams to distinguish between npn and pnp transistor an arrow is drawn on the emitter in the direction of the conventional current[3] (see fig 8.3). On forward biasing, electrons of the n layer move into the base

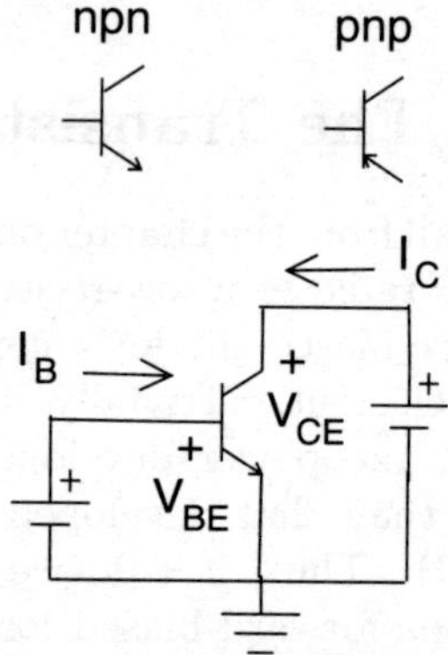

Figure 8.3: *Symbols for the npn and pnp transistor. The npn transistor is shown to be properly biased with two external voltage sources.*

in an npn transistor, i.e. the current flows from base to emitter and as seen in the schematic for npn transistor, the arrow points from base to emitter. Following the same logic, one can understand the direction of the arrow in the schematic for the pnp transistor.

---

[3]The direction of conventional current is opposite to the direction of flow of electrons in the circuit, which conversely implies that the direction of conventional current is the same as the direction of motion of holes in the semiconductor.

NPN transistor

Popularly, the working of a transistor is always done using the NPN as an example. We shall also follow the convention, and restrict our discussion of the transistor working to the npn transistor. However, comparison with pnp transistor will be made to show that the equations/ mathematics would remain identical.

If the electrons *emitted* from the emitter of the npn transistor can cross the base without being lost via recombination, they enter the collector crossing the reverse biased base-collector junction. The current on reaching the collector is called the collector current (I_C). If no loss of charge carriers take place in the base, $I_E = I_C$. Thus, we achieve the objective we had demanded of transferring a current from a region of low resistance to a region of high resistance! But how does one make sure recombination does not take place? By (i) making the carrier concentration of holes in the base very low and (ii) making the base thin, so that the electrons cross over to the collector with lower probability of meeting a hole. By making the base region very thin other than reducing the opportunity for an electron to recombine with a hole and be lost, the electrons that move into the base region come under the influence of the large collector reverse bias. This bias acts as forward bias for the minority carriers (electrons) in the base and accelerates them through the base-collector junction and on into the collector region. These two conditions explain why you can not make a transistor by connecting two diodes back to back. In practical transistor there is always a finite possibility of recombination, which prevents all of the electrons starting from the emitter from reaching the collector. That is, only a fraction of electrons reach the collector, giving

$$I_C = \alpha I_E \tag{8.1}$$

α is the called collector-emitter current gain. Where does the remaining electrons go too? The remaining electrons (proportional to $1 - \alpha$) that move into the base recombine with available holes. For each electron that recombines, another electron moves out through the base lead/ connector as base current I_B, creating a new hole for eventual combination and returns to the forward biasing supply battery, completing the loop motion of current. The electrons that recombine are obviously unavailable for the collector. The collector is made physically larger than the emitter and base for the following two reasons, (i) to increase the chance of collecting carriers that diffuse to the collector as well as those drifting across the base region, and (ii) to enable the collector to handle more heat without damage. The currents of the transistor are related by applying Kirchhoff's current law where the device itself is treated as a node in the circuit (schematic of npn transistor in fig 8.4).

$$I_E = I_C + I_B \tag{8.2}$$

The directions of the current in a pnp transistor would be opposite to the current due to electrons of the npn transistor, hence the current relationship would be the same, i.e.

$$
\begin{aligned}
-I_E &= -I_C - I_B \\
I_E &= I_C + I_B
\end{aligned}
$$

From eqn(8.1), we have

$$I_E = I_C + I_B$$

$$I_E = \alpha I_E + I_B$$
$$I_B = (1 - \alpha)I_E \tag{8.3}$$

This is was already mentioned that the electrons which do not reach the collector (proportional to 1- α) move into the base and recombine with available holes, contributing to the base current.[4] Note the currents are represented in capital letters. This is essentially done to emphasis the fact that the discussions made here are for dc currents. AC currents shall be represented in small letters. Eqn(8.1) can also be manipulate as

$$I_E = I_C + I_B$$
$$\frac{I_E}{\alpha} = I_C + I_B$$
$$I_C = \left(\frac{\alpha}{1 - \alpha}\right)I_B$$
$$I_C = \beta I_B \tag{8.4}$$

β is the collector-base current gain. In summary, for the proper transistor action, by making the base thin and lightly doped, the collector -emitter gain, α can be made large. Typical value of α is 0.98-0.99 or 98-99%. Thus, the collector current is nearly equal to the emitter current. On the other hand from eqn (8.2) and eqn (8.3), the base current is 1-2% of the emitter current, while the collector current is 49-99 times the base current. The amount of current leaving the emitter is solely a function of the emitter -base bias, and because the collector receives most of this current, a small change in emitter-base bias will have a far greater effect on the magnitude of collector current than it will have on base current. In conclusion, the relatively small emitter-base bias controls the relatively large emitter-to-collector current.

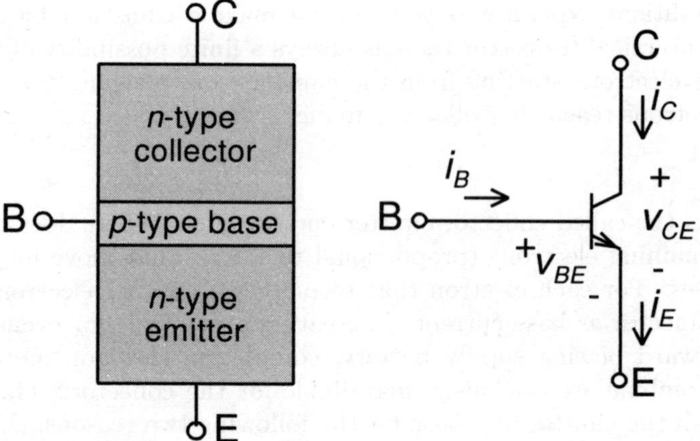

Figure 8.4: *The npn transistor with it's schematic circuit representation.*

---

[4]This is another example of conservation of matter

Example 8.1: A transistor has β of 62. What is it's alpha?

$$\alpha \;=\; \frac{\beta}{\beta + 1}$$

$$=\; \frac{62}{(1 + 62)} = 0.984$$

Answer $\alpha = 0.984$

Obviously, a transistor purchased from the market (fig 8.5) will not look like the circuital representation we have been talking of. It then becomes impossible to identify the emitter, base and collector leads of the transistor. However notice a lot of markings made on the transistor package. The markings have meaning, which allow for identifying the type of transistor available. Each type of transistors are named by a unique combination of letters and numbers. The first letter identifies the semiconductor type:

A = Germanium
B = Silicon
C = Gallium Arsenide
D = other compound semiconductor material

The second letter indicates the intended use:
A = Small signal diode
B = Varicap diode
C = Small signal Low Frequency (LF) transistor
D = LF power transistor
E = Tunnel (Ersaki) diode
F = RF small signal transistor
K = Hall effect device
L = RF power device
N = Optocoupler
P = Radiation sensitive device (e.g photo transistor)
Q = Radiation emitting device (e.g LED)
R = Low power SCR
T = High power SCR or triac
U = High voltage switching transistor
Y = Rectifier diode
Z = Zener diode
Sometimes a third letter is used which indicates the device's professional and industrial use. A suffix letter after the number like A, B or C indicates the order of the gain one can achieve with the device (A = low gain, B = mid gain, C = high gain). An "L" suffix, like BC183L means

the lead configuration is ECB (Emiter-Collector-Base) instead of the usual EBC (reading right to left).

Another equation that can be setup, is by applying Kirchoff's Voltage law on the transistor shown in fig(8.4), i.e.

$$V_{BC} = V_{BE} - V_{CE}$$

You have a three pin device which has to be put in a circuit which requires two pins as the input and two pins which are the output. This is physically possible only if one of the three pins are common on both input and output sides. This gives you three possibilities, (i) emitter pin is common on both input and output side (called the common emitter configuration or CE for short), (ii) base pin is common on both sides (common base configuration or CB) and (iii) collector pin is common on both sides (common collector configuration or CC).

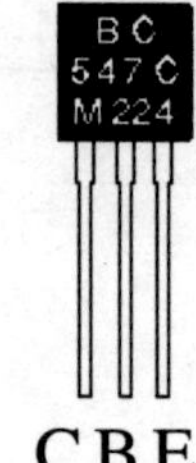

Figure 8.5: *Plastic packed Transistor available in markets.*

8.2   Common Emitter Configuration

Input Characteristics and Input Impedance

The input signal is given between the base-emitter terminals while the output is taken across the collector-emitter terminals (fig 8.6). The input characteristics is the variation of the base current with increasing base-emitter voltage (V_{BE}). For the proper functioning of the transistor, the base-emitter is forward biased, hence the input *I-V* characteristics is the same as any pn diode in it's first quadrant, i.e. when it is forward biased (see fig 8.7). The *I-V* characteristics of the transistor in CE mode can be experimentally determined by proper placement of voltmeters and ammeters in circuit shown in fig(8.3).

The graph of figure(8.7) is essentially for the case when the output side is grounded, that is the collector-emitter voltage (V_{CE}) is set equal to zero. A family of input characteristic curves can be obtained for various collector-emitter voltage. On increasing V_{CE}, the collector-base junction becomes more and more reverse biased. This results in the depletion width growing. Since the doping profile is not symmetrical, i.e. the amount of doping is not the same in the collector and base, the depletion width grows more into the the region with lower doping.[5] In the transistor's case the depletion region extents more into the base region. Remember, a depletion region is a region devoid of free mobile charges. Hence a depletion region

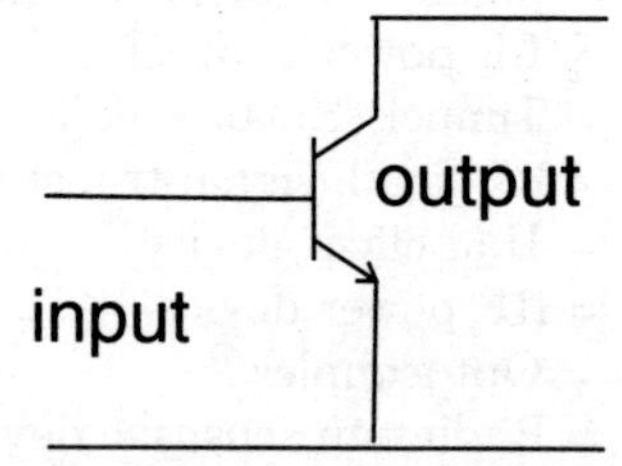

Figure 8.6: *The transistor in CE mode.*

in the base implies less holes available for recombination. With the probability of recombination decreasing α, the fraction of emitter current reaching the collector increases. It logically follows that the base current hence comes down. The base current was after all due to recombination, the number of available holes comes down, recombination comes down. This is also understood

---

[5]Can be explained with help of eqn(5.1) introduced in chapter on diodes.

mathematically from eqn(8.3). Thus, for the same value of forward biasing V_{BE}, the base current decreases with increasing reverse bias of collector-base junction by V_{CE} (see fig 8.8).

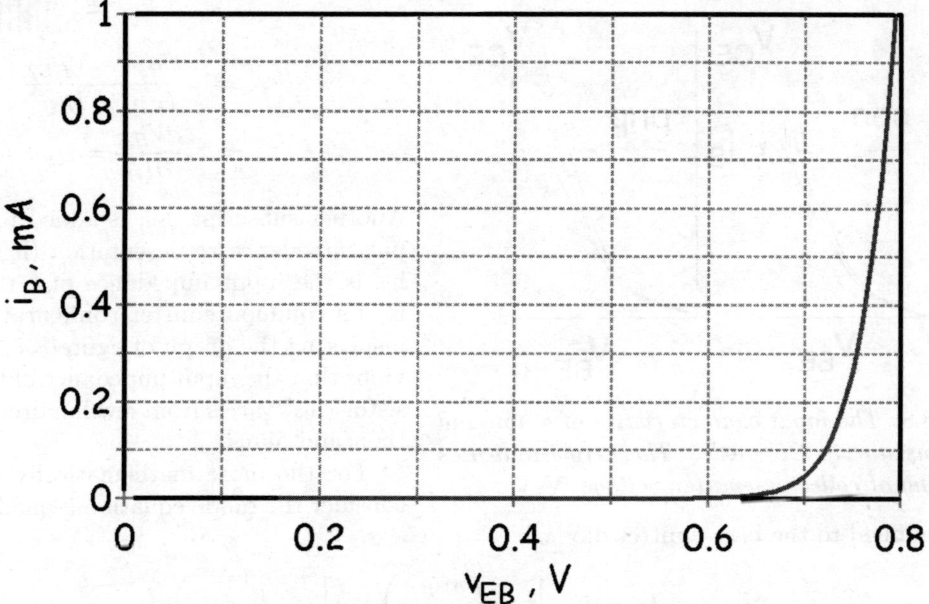

Figure 8.7: *The input characteristics of a transistor in CE mode.*

As expected the input characteristics of the pnp transistor is similar to the npn transistors characteristics. Difference if looked with respect to the npn would be in the labeling of the axis. The polarity of the batteries have to be reversed (as compared to the polarities shown in fig 8.6) to bias the pnp transistor. Hence, with respect to the labelling of the npn transistor, the pnp transistor characteristics label should read $(-)I_B$ versus $(-)V_{BE}$. The family of graphs for the npn transistor can be considered for V_{CE}=0, 2 and 4volts. Since V_{CE} would also be reversed for the pnp transistor, the family of graphs for the pnp transistor would be for V_{CE}=0, -2 and -4volts, explaining the reversal of the arrow.

The input resistance of the the transistor can be found from the slope of the curve of fig(8.7). The input resistance would be given as

$$r_{in} = \frac{\Delta V_{BE}}{\Delta I_B} = \frac{1}{slope}$$

The curve is typically not linear. However, the curvature is not very pronounced and in small sections we may assume the curve to be perfectly linear. Thus, for operational designing, one restricts operation on such a linear region, of whose resistance is easily established. Any input AC signal would add or superimpose with the biasing DC voltage (say V_{BB}), the net voltage appearing across the base-emitter junction would swing between '$V_{BB} - v_{max}$' and '$V_{BB} + v_{max}$'. The swinging voltage would result in the current changing from '$I_{BB} - i_b$' to '$I_{BB} + i_b$'. Both '$V_{BB} - v_{max}$' and '$V_{BB} + v_{max}$' may be considered as two dc voltages, V_{BE2} and V_{BE1} respectively, i.e. the swinging AC signal maybe considered as varying DC bias voltage. The variation maybe restricted along the linear region of the curve. In actual conditions we are able to achieve this by restricting the varying input (v_{max}) to be small, hence as far as the input signal is concerned it swings along a linear curve. The AC input resistance would be

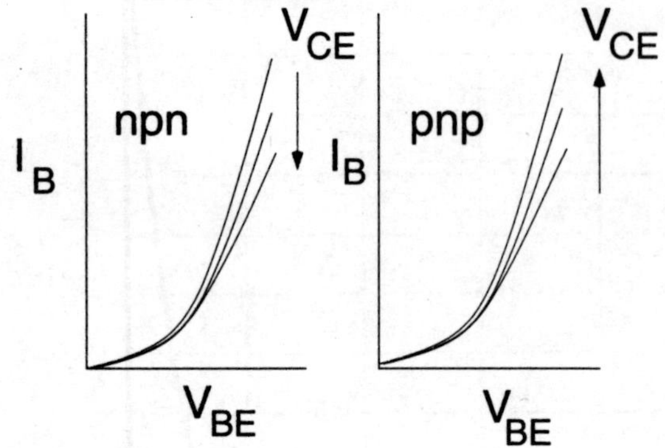

Figure 8.8: *The input characteristics of a npn and pnp transistor in CE mode. The arrow indicates increasing of collector-emitter voltage, V_{CE}.*

$$h_i = \frac{V_{BE1} - V_{BE2}}{I_{BB1} - I_{BB2}}$$
$$= \frac{\partial V_{BE}}{\partial I_B} = r_{in}$$

Another subscript, 'e', is added to the input impedance representation (h_i). Thus, h_{ie} is the input impedance of a transistor in it's common emitter configuration. On inspecting the graph of figure(8.7) it is obvious that the input impedance of the transistor (h_{ie}) varies from few hundreds to few thousand ohms.

For the more mathematically inclined, consider the diode equation (eqn5.2) which

can be applied to the base-emitter diode.

$$I_B = I_{sat}\left[exp\left(\frac{qV_{BE}}{KT}\right) - 1\right] \tag{8.5}$$

where the terms v_D and i_D have been replaced in V_{BE} and I_B respectively, terms applicable for the transistor's base-emitter diode. Differentiating the above equation we have

$$\frac{dI_B}{dV_{BE}} = \frac{qI_{sat}}{KT}exp\left(\frac{qV_{BE}}{KT}\right)$$

The exponent term on the left hand side can be replaced by terms of I_B using the eqn(8.5), giving the transistor's input impedence as

$$h_{ie} = \frac{KT}{q(I_B + I_{sat})}$$
$$= \frac{26mV}{(I_B + I_{sat})}$$

Remember at room temperature KT/q is 26 mV. Also, remember I_{sat} is the current following when the diode is reverse biased. In well fabricated diodes I_{sat} is of the order of nano-amperes. Obviously, far less than the micro- to milli-amperes base current, I_B. Hence, the transistor's input impedance in common emitter mode is

$$h_{ie} = \frac{26mV}{I_B} \tag{8.6}$$

Around a base current of about 0.2mA, the input resistance would be 130Ω.

ELECTRICAL CHARACTERISTICS (Ta=25 deg C Unless Otherwise Specified)			BC107XX		
DESCRIPTION	**SYMBOL**	**TEST CONDITION**	**MIN**	**MAX**	**UNIT**
Collector Knee Voltage	VCE (K)	IC=10mA, IB=The Value for Which IC=11mA, @ VCE=1V	-	0.60	V
Transition Frequency	ft	VCE=5V,IC=10mA, f=100MHz	150	-	MHZ
Noise Figure	NF	VCE=5V, IC=0.2mA Rg=2kohms,			
		F=30Hz to 15 KHz BC109	-	4.0	dB
		F=1kHz, B=200Hz BC109	-	4.0	dB
		BC107/108	-	10	dB
Output Capacitance	Cobo	VCB=10V, f=1MHz	-	4.5	pF
		ALL f=1kHz			
Small Signal Current Gain	hfe	IC=2mA,VCE=5V BC107	125	500	
		BC108	125	900	
		BC109	240	900	
		A Group	125	260	
		B Group	240	500	
		C Group	450	900	
Input Impedance	hie	IC=2mA,VCE=5V A Group	1.6	4.5	Kohms
		B Group	3.2	8.5	Kohms
		C Group	6.0	15	Kohms
Output Admittance	hoe	IC=2mA,VCE=5V A Group	-	30	umhos
		B Group	-	60	umhos
		C Group	-	110	umhos

Figure 8.9: *Some typical information about a transistor given in it's data-sheet.*

Figure (8.9) is an example of the transistor manufacturer's datasheet, giving details of various parameters. The datasheet shows the input impedance of BC107 transistor which is in $K\Omega$ (1.6KΩ). This looks higher than what we calculated above, however, read the fine print of the test condition (third column), which says the measurements were made at collector current (I_C) of 2mA. Thus eqn(8.6) has to be written in terms of the collector current. Using eqn(8.4), we have

$$h_{ie} = \beta \frac{26mV}{I_C} \tag{8.7}$$

Assume collector-emitter current gain (β) to be 125, the input impedance at 2mA works out to be 1.6KΩ[6] as stated in the datasheet.

Equation (8.1) is not the complete story. Since the collector-base junction is reverse biased, an reverse leakage current would be flowing due to the minority charge carriers of the collector and base layers, just like in an ordinary diode. Hence, the two currents contribute to the collector current, which is now written as

$$I_C = \alpha I_E + I_s$$

The reverse leakage current is across the collector-base junction, hence the subscript is modified to reflect this (remember a transistor has two junctions).

$$I_C = \alpha I_E + I_{CB}$$

---

[6]When we calculated the input impedance with eqn(8.6) at I_B=0.2mA, the corresponding collector current was 25mA.

The reverse leakage current in the reverse biased collector-base is measured with the emitter open (i.e. $I_E=0$), hence a third subscript 'O', is added to show the emitter is open. The above equation is now written as

$$I_C = \alpha I_E + I_{CBO} \tag{8.8}$$

Using the current relation $I_E = I_C + I_B$ we have

$$\begin{aligned} I_C &= \frac{\alpha}{1-\alpha} I_B + \frac{I_{CBO}}{1-\alpha} \\ &= \beta I_B + (1+\beta) I_{CBO} \end{aligned} \tag{8.9}$$

Usually the above equation is written as

$$I_C = \beta I_B + I_{CEO} \tag{8.10}$$

where I_{CEO}[7] is the reverse leakage current, when base is left open and collector-base junction is reverse biased by voltage applied across the collector-emitter. This obviously can only be possible in common emitter configuration. Even when the base-emitter is not given any biasing by a source connected across it, the base-emitter junction can be forward biased by the source V_{CC} applied across the collector-emitter. As can be seen from the figure 8.10, there is potential drop across each layer of the transistor, with $V_C = V_{CC} > V_B > V_E$, the collector-base junction is reverse biased while the base-emitter is forward biased.

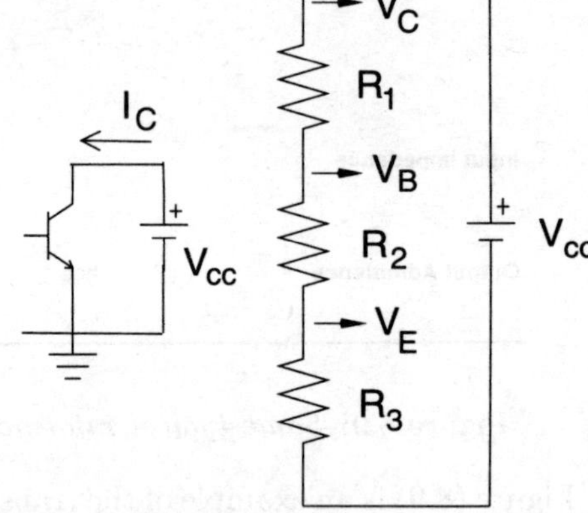

Figure 8.10: *The transistor can be looked upon as three resistances in series, acting as a potential divider when used with the DC supply.*

Example 8.2: A common emitter circuit has a beta of 120, a collector current of 68mA and a base current of 542μs. What is it's reverse leakage current?

Using eqn(8.9), we have

$$\begin{aligned} I_C &= \beta I_B + (\beta + 1) I_{CO} \\ I_{CO} &= \frac{I_C - \beta I_B}{\beta + 1} \\ &= \frac{68 - 120 \times 0.542}{120 + 1} \end{aligned}$$

giving

Answer $I_{CO} = 24.5\mu$A

---

[7]$I_{CEO} = (1 + \beta)I_{CBO}$, I_{CBO} is in nano-amperes, and considering β is in hundreds (100-200), I_{CEO} is in micro-amperes.

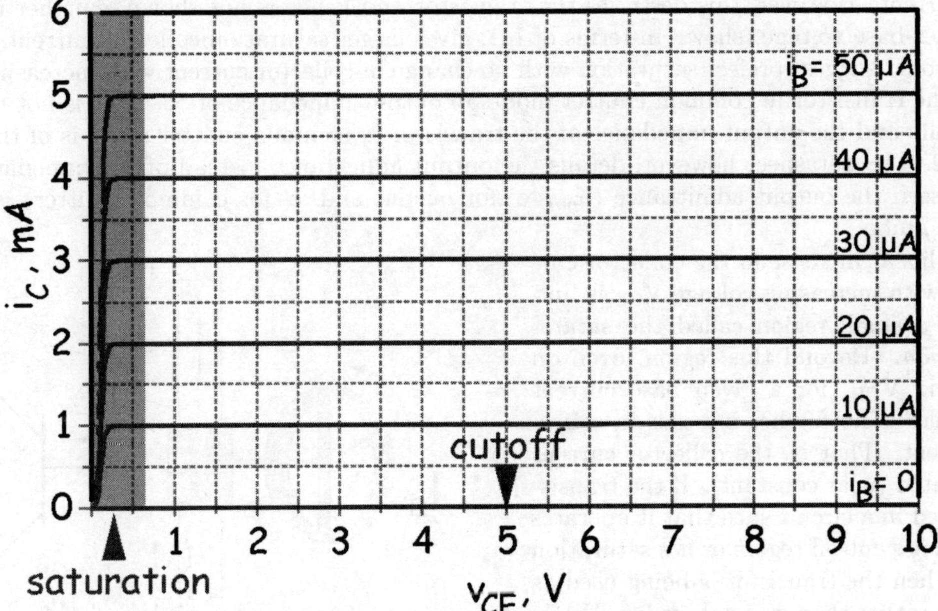

Figure 8.11: *The output characteristics of a npn transistor in it's common emitter mode.*

Output Characteristics and Output Impedance

Figure (8.11) shows the output characteristics of the npn transistor in common-emitter configuration. If there is no source applying voltage across the base-emitter, I_B would be zero. However, as explained by eqn(8.10) the reverse biased collector-base junction would have a current flowing across it due to the movement of the minority charge carriers. However, since the magnitude of this current is very small. It gives a curve that runs parallel and near the x-axis. It envelopes a region called the **cut-off region** where irrespective of the voltage applied across the collector-emitter, there is no current flowing into the collector circuit, i.e. the transistor fails to conduct current.

This gives us a model to compare the transistor with a gate valve placed along a water pipe to control the flow of water (see fig 8.12). Increasing base current, I_B is similar to opening the valve. When the valve is open water flows in the pipe, while in the case of the transistor, collector current starts flowing. For base current going zero, the situation is similar as closing the gate valve.

On increasing the forward bias of the base-emitter junction, emitter current starts flowing. This in turn contributes to the collector current. Increasing the collector-emitter voltage (V_{CE}), increases the kinetic energy of the electrons starting at the emitter. Thus, more electrons are able to cross over into the collector region (as good as saying α increases) increasing collector current. Obviously, there is a limit to the number of electron's that can successfully cross over. Hence, any further increase in the collector-emitter voltage (V_{CE}) leads to a saturation in collector current. At a very large voltages, the collector-base junction can break-down and give large current. However, this destroies the transistor and hence is not shown. Further increase in emitter-base voltage (shown in terms of I_B), gives larger saturation collector current. Figure (8.11) would suggest prefect saturation with no change in collector current with increasing V_{CE}, giving the transistor in common emitter mode an output impedance of ∞. This is not the case practically and the output impedance of the transistor in common emitter mode is of the order of 10KΩ. The datasheet however, details the output admittance instead of out impedance. As can be seen the output admittance (h_{oe}, 'o' for output and 'e' for common emitter) is of the order of μmho.

The linear increase in the collector current I_C with increasing voltage V_{CE} is limited for a small region called the *saturation region*. Beyond this region, even on increasing V_{CE}, for a given base current (I_B), there is no further increase in collector current. That is, the collector current is saturated (near constant). If the transistor is used in a circuit such that it operates either in it's cut-off region or it's saturation region, then the transistor is being used as a switch rather than as a gate valve. While the gate-valve allows water to flow at different rates, the switch either allows current to flow or stops it completely. This feature allows transistors to be used in digital circuits which toggles between two possible states 0 and 1. However, in this book we are more interested in understanding the

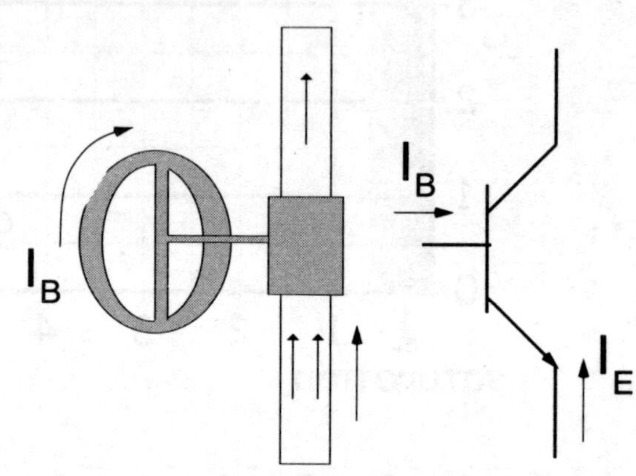

Figure 8.12: *The transistor is like a gate valve which controls the flow of water, the base current controls the collector current flowing in the transistor.*

transistor's action as an analog device, i.e. the working of the transistor when it is biased to work in the region between the cut-off and saturation region. This region is also called the transistor's *active region* .

Current Gain

The current gain of a transistor in common emitter mode is the change in the output AC current (collector current) for the changing input AC current (base current). That is,

$$h_{fe} = \frac{i_c}{i_b}$$

The subscript 'e' signifies the fact that we are talking of the transistor in it's common emitter configuration. A DC base current and collector current exists in the transistor for properly biasing it. The AC is superimposed on this DC currents, i.e.

$$h_{fe} = \frac{I_C + i_c}{I_B + i_b}$$

while the positive cycle is coming, or

$$h_{fe} = \frac{I_C - i_c}{I_B - i_b}$$

for the negative cycle. The AC currents may be modeled as time varying DC current levels, in that case

$$h_{fe} = \frac{I_C - I_{C1}}{I_B - I_{B1}} \tag{8.11}$$

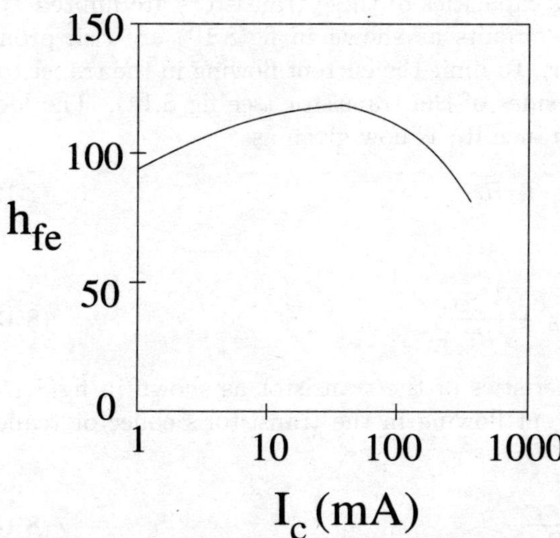

Figure 8.13: *The variation of the current gain with the transistor's collector current shows that it is not a constant.*

In other words h_{fe} is the slope of a graph between the collector current and base current, i.e.

$$h_{fe} = \frac{\Delta I_C}{\Delta I_B}$$

Applying the DC current gain ($I_C = \beta I_B$, this is not a slope [8]) in eqn(8.11), we have

$$\begin{aligned} h_{fe} &= \frac{I_C - I_{C1}}{I_B - I_{B1}} \\ &= \frac{\beta I_B - \beta I_{B1}}{I_B - I_{B1}} \\ &= \beta \end{aligned}$$

As can be seen, the transistor's AC current gain is equal to it's DC current gain and some-times h_{fe} and β are used indistinguishably. Now consider that the DC current gain is not constant (see fig 8.13),

i.e. for collector current I_C it is β while for collector current I_{C1} it is $\beta - a$ (assuming I_{C1} is less than I_C), then we have

$$\begin{aligned} h_{fe} &= \frac{I_C - I_{C1}}{I_B - I_{B1}} \\ &= \frac{\beta I_B - (\beta - a)I_{B1}}{I_B - I_{B1}} \\ &= \beta + \frac{a I_{B1}}{I_B - I_{B1}} \end{aligned}$$

Thus in the region of I_C=1 to 10mA, the AC current gain is greater than the DC current gain (i.e. $h_{fe} > \beta$). By the same argument, in the region of I_C=100mA to 1A, the AC current gain is less than the DC current gain (i.e. $h_{fe} < \beta$). In the usual operating range of I_C=10 to 50mA, β is nearly constant and hence the approximation of the transistor's AC current gain equal to it's DC current, i.e. $h_{fe}=\beta$, is justified. The value of h_{fe} lies between 100-150 (see fig 8.13).

---

[8] Consider line y=2x+1, the slope is 2, however, the ratio (y/x) is not the equal to two at any x.

Graphically, note from fig(8.11), that in the active region of the transistor for an increase in base current of $10\mu A$, there is a corresponding increase in collector current of 1mA. Hence, h_{fe} works out to be 100.[9]

Voltage Gain

The transistor's are essentially used as amplifiers used to amplify the input AC signals. Before understanding how much amplification or voltage gain the transistor can achieve in common emitter mode, it is important to understand that transistor's of the genre BC107 are small signal transistors. That is, the current handling capacities of these transistors are limited and usually of the order of milli-Amperes. Hence, circuits as shown in fig(8.10) are risk prone, where large currents may burn out the transistor. To limit the current flowing in the transistor, resistances are placed in the input and output sides of the transistor (see fig 8.14). The loop equation of the output on placing the load resistance R_C is now given as

$$V_{CC} = V_{CE} + I_C R_C \tag{8.12}$$

Re-writing this equation in the form

$$I_C = -\frac{1}{R_C}V_{CE} + \frac{V_{CC}}{R_C} \tag{8.13}$$

This line can be plotted on the output characteristics of the transistor as shown in fig(8.15). When V_{CE} is zero, we have the maximum current flowing in the transistor's collector (called the saturation current I_{sat})

$$I_{sat} = \frac{V_{CC}}{R_C} \tag{8.14}$$

From eqn(8.13), we can see that the slope of the line is inversely proportional to the load resistance. Thus, both the slope of the line and it's intercept depends on the magnitude of the load resistance, hence the line is called the *load line*.

The output of the amplifier is taken between the collector and the ground (here emitter is grounded). The variation in the output voltage is obviously the variation introduced by an AC signal "riding on" or super-imposed on the DC biasing level. As explained, the AC variation can be treated as a DC voltage level varying with time. Thus, using the load line equation (eqn 8.12) we have

$$v_o = \Delta V_{CE} = -\Delta I_C R_C$$

The negative sign implies that the collector-to-emitter voltage, V_{CE} and in turn output voltage increases with decreasing collector current and visa versa. More would be appreciated from the graphical interpretation of the transistor's amplification action.

The output is obtained by an amplification of an input voltage given at the base, which results in varying base current

$$v_{in} = \Delta I_B R_B$$

---

[9]Note since h_{fe} is a ratio of currents, it is dimensionless

Thus, the voltage gain of the transistor in CE mode is

$$A_v = \frac{v_o}{v_{in}} = \frac{-\Delta I_C R_C}{\Delta I_B R_B}$$

$$= -h_{fe}\frac{R_C}{R_B} \tag{8.15}$$

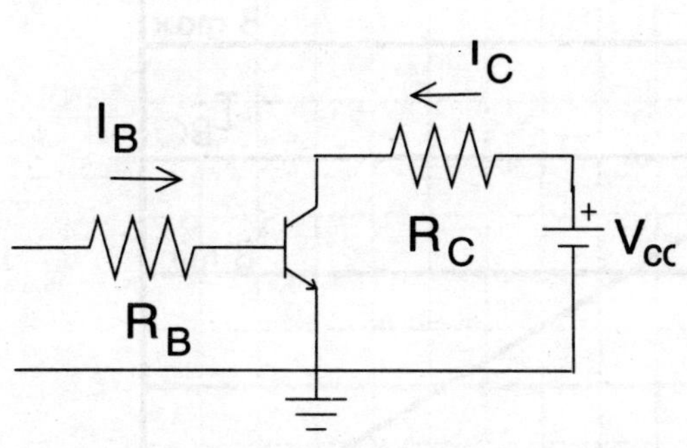

Figure 8.14: *Transistors are used in circuits with resistances to control current etc., this changes the transistor's parameters like input impedance, gain etc.*

Eqn(8.15) is not the true story. There is more to be discussed before getting the formula required to design an amplifier, but that will have to wait till we have learnt more. The complete picture will only emerge after understanding the small signal model of the transistor. However, eqn(8.15) does illuminate the following features: (i) The voltage gain of the transistor in CE mode is directly proportional to it's current gain.

(ii) The negative sign of eqn(8.15) implies that the output voltage is out of phase with respect to the input voltage (phase difference of π). The output voltage being out of phase with respect to the input voltage, as will be seen on comparing with the other modes, is an unique feature of the transistor in CE mode.

Example 8.3: A common emitter circuit has an input resistance of 2KΩ, a load resistance of 25KΩ and a beta of 92. What is it's power gain?

The power gain is a product of the voltage and current gain, where the current gain is given by h_{fe} or β. The voltage gain is given by eqn(8.15), i.e.

$$|A_v| = \beta\frac{R_C}{R_B}$$

$$= 92\left(\frac{25}{2}\right) = 2300$$

giving power gain as

$$|A_p| = |A_i| \times |A_v|$$
$$= 92 \times 2300 = 106,000$$

$$\boxed{\text{Answer } |A_p| = 106,000}$$

The base resistance R_B selects the DC base current I_B, which is represented by one of the family of curves of the output characteristics. The point of inter-section of the load line and this

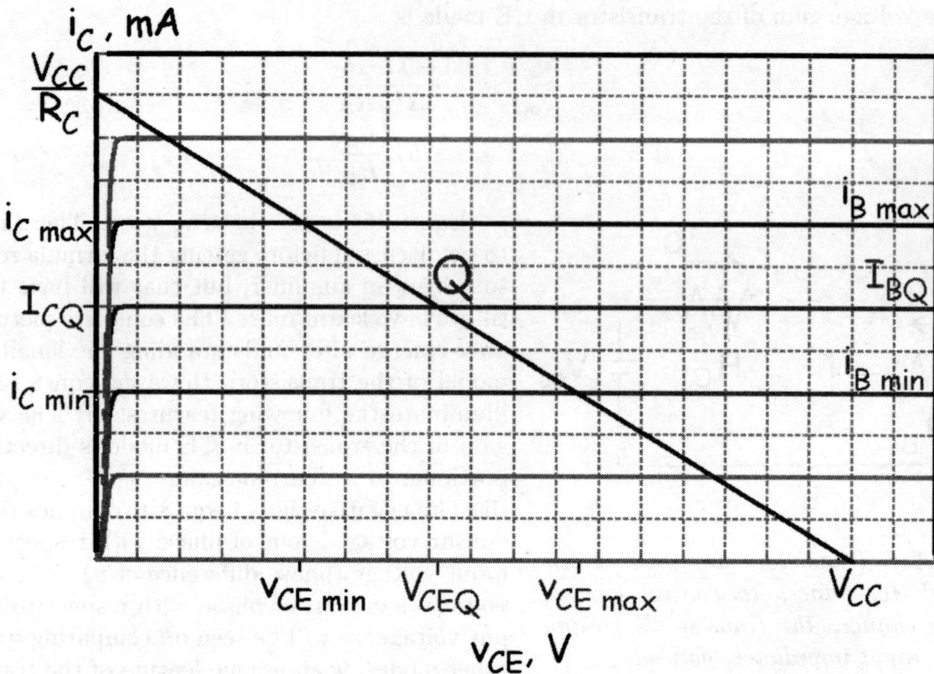

Figure 8.15: *The transistor is like a value whose performance is effected by the base current. The base current is controlled by the base resistance, R_B. This along with the collector resistance, R_C fixes the transistor's operating point (or Q point) on the IV characteristics.*

family of curve completely defines the DC currents flowing and voltages appearing across the transistor. This point is called the transistors *"operating point"* or the transistor's **'Q' point** (see figure 8.15). The point on the output IV characteristics is represented as (V_{CEQ}, I_{CQ}), and for proper functioning is usually selected at the center of the active region of the transistor.

Now let us see what happens when a small AC signal is given as an input to the base of the CE amplifier circuit. Carrying our comparison of the transistor with a gate valve further, a selection of the 'Q' point in the center of the active region is as good as saying that the gate valve is partially open. As the positive AC cycle is inputted, the base current increases from I_B to $I_B + i_B$ to a possible $I_B + i_{Bmax}$, where i_{Bmax} is the maximum of the input AC current. The 'Q' point moves along the load line in an upward direction (see figure 8.16) leading to an increase in collector current ($I_{CQ} + \Delta I_C$) but decrease in collector-emitter voltage ($V_{CEQ} - \Delta V_{CE}$). *As the 'Q' point moves up, it indicates that the valve is moving towards an more open position and the water flow is increasing. The decrease in the collector-emitter voltage is equivalent to the decrease in pressure between the input and output side of the gate valve. The greatest pressure difference takes place when the valve is completely closed, water coming at the input but no water at the output as the valve is closed.* The negative cycle of the AC input decreases the net base current and is represented by the 'Q' point moving down along the load line. It is as good as closing the gate valve decreasing the water flow (collector current) and increasing the water pressure (the collector-emitter voltage), see fig(8.16).

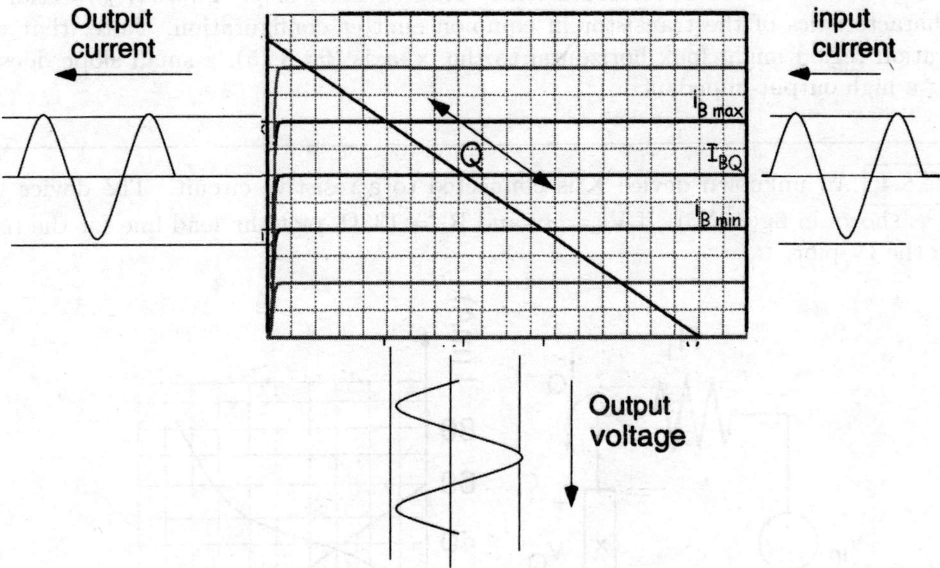

Figure 8.16: *Graphical representation explaining how the input is inverted, i.e. the output is out of phase by π as compared to the input.*

While the decrease in collector current with decreasing base current can be appreciated from the graph or from the mathematical relation $I_C = \beta I_B$, the **base-width modulation effect** explains the physical changes occurring in the transistor that helps to understand the relation between the two currents. As the voltages applied to the base-emitter and base-collector junctions are changed, the depletion layer widths also vary. An positive voltage at the collector, increases the reverse bias between the collector-base junction. This causes an increase in the depletion width at this junction. As discussed in Chapter 4, eq(5.1), since the base has lower charge carrier concentration as compared to the collector region, the depletion width grows more into the base region. This causes the collector current to increase as the probability of recombination in the base region decreases and more and more emitter released carriers cross over into the collector region (i.e. α increases, study eqn 8.8). This is referred to as the **Early effect**. Infact the collector-base junction can be reverse biased to such a large extent that the whole base region is encompassed by the depletion width. This is called *punch through*.

The important aspect to note is that as the input grows, the output decreases for a CE mode transistor amplifier. Thus when the input reaches its maxima, the output reaches its minima. In other words the input and output are out of phase by π or the transistor amplifier introduces a phase shift of π between the input and output.

The reverse biasing of the collector-base junction, as explained (with reference to fig 8.10), would also take place with the applied collector-emitter voltage. An increase in the collector-emitter voltage (V_{CE}) would give rise to an increase in depletion width at the collector-base junction as well as increase the forward biasing of the base-emitter junction. This leads to an increase in emitter current (I_E). As can be seen from eqn(8.8), an increase in α and I_E leads to an strong increase in the collector current. Thus, a finite slope (dI_c/dV_{CE}) exists in the output characteristics of the transistor in common emitter configuration. Note, that though the saturation region might look horizontal to the 'x'-axis (fig 8.15), a small slope does exist, indicating a high output impedance.

| Example 8.4: | An unknown device X is connected to a resistive circuit. The device has an IV curve as shown in fig(8.17b). If V_{in} = 8v and R_1 = 133Ω, plot the load line for the resistive circuit on the IV plot.

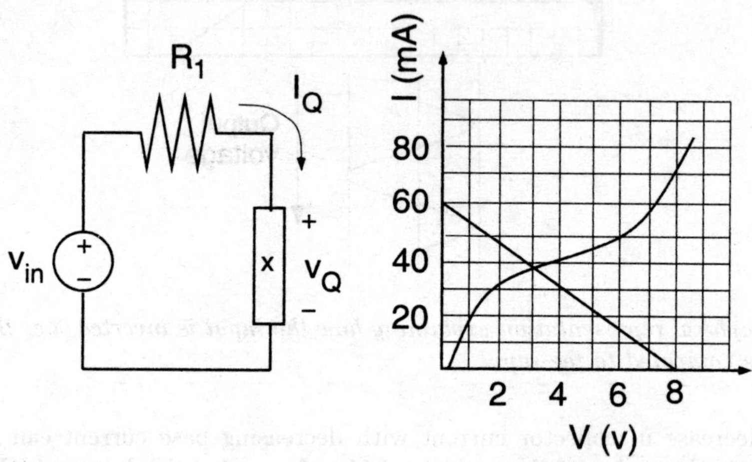

Figure 8.17:

This problem provides a good insight to the problem of finding the operating point of a device graphically from the IV characteristics. First the saturation current is found by basically short circuiting the active device and calculating the current. The loop equation is

$$V_{in} = V_R + V_X$$

Shorting the active device, i.e. V_X=0, the saturation or short-circuit current is

$$I_{sc} = \frac{V_{in}}{R_1}$$
$$= \frac{8}{133K} = 60mA$$

This gives the 'y' axis intercept. The 'x' axis intercept ($I = 0$) is the maximum voltage available in the circuit, $V_{in} = 8V$. Joining the two intercepts we get the *load-line*.

To estimate the operating point of the circuit (I_Q and V_Q), we look for the point of intersection of the load line and the devices IV characteristic curve. We have the operating point as

| Answer I_Q=40mA and V_Q=2.5v |

8.3   Common Base Configuration

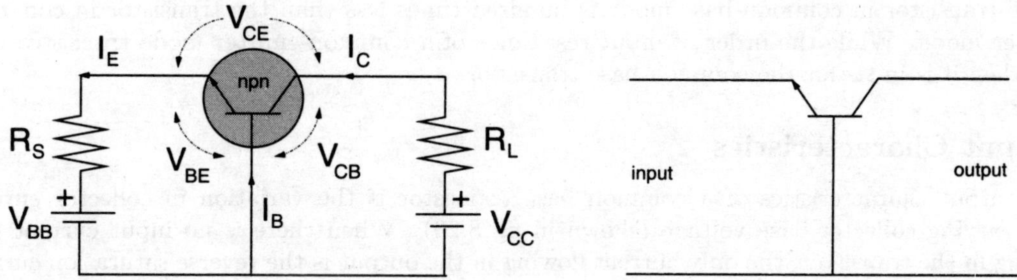

Figure 8.18: *An NPN transistor in common base configuration. The second figure exhibits the two pins of the input and output ports of the common base transistor amplifier (biasing circuits etc not shown).*

In this configuration, the input signal is given between the emitter and base. The base is common on both input side and the output side, hence output is taken across the collector-base terminals (see fig 8.18). The transistor in this mode presents different characters as compared to that when it would be in common emitter mode.

Input Characteristics and Input Impedance

The emitter-base pins forms the input port of the transistor in common -base mode. The input characteristics is the variation of emitter current (I_E) with increasing emitter -base voltage (V_{EB}). Again this is (base-emitter) a pn diode, and the input characteristics is the same as the IV character of a pn diode (fig 8.19).

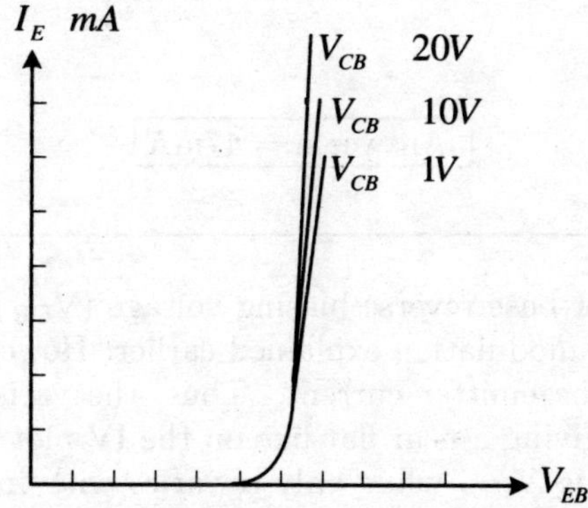

Figure 8.19: *The input IV characteristics of a transistor in common base configuration.*

The input resistance of the transistor in this mode is found just as is done for diodes and the transistor in CE mode. Infact one would be tempted to straight away state that the input resistance of the transistor in CE and CB modes are equal. However, note while the input current for the common-emitter transistor was I_B, the input current for a transistor is common-base mode is I_E. While the base current in the transistor is in μA, the emitter current is at least hundred times stronger and in the range of milli-Amperes. Thus, typically, the input impedance of the transistor in common-base mode is hundred times less than the transistor in common-emitter mode. While the order of input resistance of a common-emitter mode transistor is in $K\Omega$, then it is in Ωs for the common-base transistor.

Output Characteristics

The output characteristics of a common base transistor is the variation in collector current with varying collector-base voltage (shown in fig 8.20). When there is no input current (I_E) flowing in the transistor, the only current flowing in the output is the reverse saturation current I_{CBO}, resulting from the minority charge carriers crossing over the reverse biased collector-base junction. Even this current falls to zero when the collector-base junction is not given any biasing, i.e. $V_{CB} = 0$. This curve ($I_E = 0$) on the IV characteristics bounds the cut-off region of the transistor.

An increase in the forward biasing at the input, leads to flow of emitter current (I_E). The net collector current or the output current given by eqn(8.8), i.e.

$$I_C = \alpha I_E + I_{CBO}$$

Example 8.5: A common base circuit has an alpha of 0.941, an emitter current of 50mA and a reverse saturation current of 5μA. What is it's collector current?

Using the above equation we have

$$\begin{aligned} I_C &= \alpha I_E + I_{CBO} \\ &\approx 0.941 = 47mA \end{aligned}$$

The collector current is

Answer $\alpha = 47$mA

An increase in the collector-base reverse biasing voltage (V_{CB}) increases α which is readily understood by the base-width modulation explained earlier. However, this increase in V_{CB} will not have a strong effect on the emitter current. Thus, the variation in output current with output voltage is very small, giving a near flat line on the IV plot. Since, in the CE mode both α and I_E varies with increasing V_{CE}, while only α varies with increasing V_{CB}, the CE mode output characteristics show a finite abide small slope compared to near zero slope of the CB mode.

The dissimilarity between the two modes does not end here. As explained above, if the reverse biasing voltage (V_{CB}) of the transistor is removed, the reverse leakage current would also fall zero. However, the output would still have a current ($I_C = \alpha I_E$) contributed by the highly kinetic

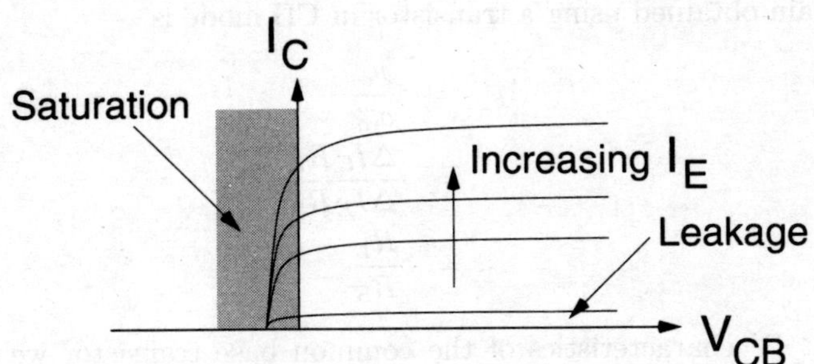

Figure 8.20: *The output IV characteristics of a transistor in common base configuration.*

electrons originating from the emitter. They reach the collector without recombination due to (a) the base being very thin and (b) the base having low carrier concentration and hence less chances of recombination.

Thus, to reduce the output current to zero, one would have to forward bias the collector-base junction, giving a current in the opposite direction to I_E. The near flat output characteristics implies whatever might be the output voltage, the output current remains the same. This is what a constant current source is supposed to do. Hence, the CB configuration of a transistor finds it's use in circuits of constant current sources.

Current and Voltage Gain of transistor in CB mode

With both the input and output currents being of the same magnitude ($I_E = I_C$), the current gain of the common base configuration is always unity, i.e.

$$A_i = \frac{I_C}{I_E}$$
$$\approx 1$$

The output equation or load line equation for the common base mode is

$$V_{CC} = V_{CB} + I_C R_L$$

As in the case of the CE mode, the AC variation is treated as a DC voltage level varying with time, with the output AC being super-imposed on the "operating point" DC levels. Hence,

$$v_o = \Delta V_{CB} = -\Delta I_C R_L$$

Applying Kirchhoff's current law on the input side, we have (see fig 8.18)

$$V_{BB} = V_{EB} + I_E R_S$$

From which a superimposed AC signal signal, which on being treated as a time varying DC signal, maybe written as

$$v_{in} = -\Delta I_E R_S$$

Thus, the voltage gain obtained using a transistor in CB mode is

$$A_v \;=\; \frac{v_o}{v_{in}}$$

$$=\; \frac{\Delta I_C R_L}{\Delta I_E R_S}$$

$$=\; \frac{R_L}{R_S}$$

From the output IV characteristics of the common base transistor we understand that it presents itself as an ideal candidate for being used as a current source (output impedance $R_o \sim \infty$). Also, the output voltage is in phase with the input voltage (negative sign is absent in the voltage gain expression). However, its current gain is only unity. Thus, however much the voltage gain might be the transistor in this mode would have a smaller power gain. The power gain of the transistor in CE mode is proportional to β^2 and hence quite appreciable. This, hence restricts the utility of the transistor in common base mode.

Example 8.6: A common base circuit has a power gain of 4.8, an input resistance of 3KΩ, a load resistance of 15.6KΩ. What is the alpha of the transistor?

The current gain in common base is equal to alpha, i.e.

$$|A_i| = \alpha$$

and the voltage gain is

$$|A_v| = \alpha \left(\frac{R_L}{R_S} \right)$$

Therefore, the power gain is given as

$$|A_p| \;=\; \alpha^2 \left(\frac{R_L}{R_S} \right)$$

$$4.8 \;=\; \alpha^2 \left(\frac{15.6}{3} \right)$$

giving alpha of the transistor as

Answer $\alpha = 0.965$

8.4   Common Collector Configuration

The third configuration possible using the three pin BJT is the common collector configuration. As the name suggests, the collector is common to both the input and output side. Figure(8.21a) shows how a BJT is used as a common collector. The base-collector pins are the input pins while the emitter-collector pins forms the output. However, in practical circuits, the common collector amplifier would look as shown in fig(8.21b). This at first look would appear like a transistor in it's common-emitter mode. This circuit is infact called an *emitter follower*. A closer look would show that the position of the resistance is different in common emitter circuits. Note the difference in where the output is collected. While in the common emitter mode, the output voltage is across the transistor's collector-emitter pin, in common collector mode, the output voltage (V_{out}) is across the resistance (R_E, acts as the load of the shown circuit).

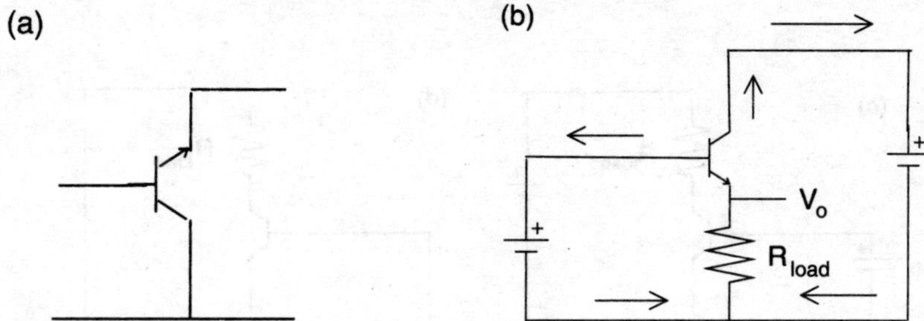

Figure 8.21: *The collector is common is both input and output in the common collector configuration (a). However, the common collector amplifier circuit appears as shown in (b), which looks very similar to the common emitter amplifier. However, there is a basic difference.*

How does the position of the resistance make a difference in the mode? To appreciate this, one has to go through Chapter 10. In short, one has to understand that a transistor amplifies input AC and produces an AC output with a help of the DC supply that a user provides. The two voltages, AC and DC act on the transistor linearly. That is super-position theorem holds valid on transistors.[10] To investigate whether the transistor circuit is in common emitter or common collector mode, the DC supplies are removed. Removing DC supply implies setting the voltage as zero volts, or more appropriate grounding those points where DC was being applied.

Figure(8.22a) shows an emitter follower circuit. The resulting circuit on removing DC sources is shown in fig(8.22b). Note, the grounded collector is common in both input and output (input is given between ground and base, not shown). For highlighting the method we have adopted for identifying the mode, fig(8.22c & d) is the common emitter mode circuit along with the DC sources grounded. Note, again input is fed to the base with respect to the ground and emitter is common both to the input and output.

As stated circuit shown in fig(8.21b) is called an emitter follower. When a small base current is fed to the transistor, an emitter current would flow in the resistance. Increasing base current leads to increase in emitter current, with output voltage $V_{out} = I_E R_E$ following the input voltage (i.e. output voltage is in phase with the input), hence the name "emitter follower".

Input Characteristics and Input Impedance

The input characteristics of a common collector amplifier would be identical to that of the common emitter (see fig 8.23), as a base current would flow as a result of applied base-collector

---

[10]The justifications of these statements made are in Chapter 10

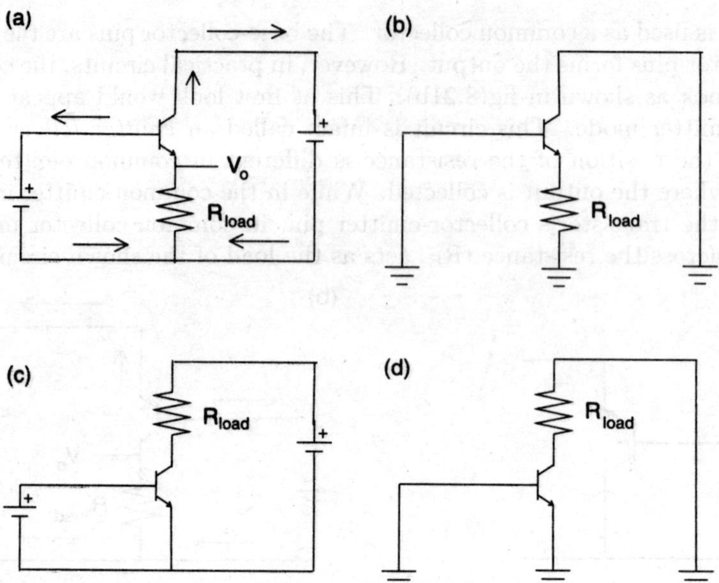

Figure 8.22: *The DC sources of the emitter follower (a) is grounded. As can now be seen in (b), the collector pin (grounded) is common in both the input and output side. Though the emitter follower looks very similar to the common emitter configuration (c), the difference can be appreciated from (d), that it is the positioning of the resistance which separates the two (in common emitter collector is not grounded).*

voltage. Essentially, the I-V characteristics of a pn diode would be obtained. Due to the similarity of the I-V characteristics of a common collector and common emitter mode transistor, one would expect the input impedance (h_{ic}) to be equal to h_{ie}, the input impedance of the common emitter mode transistor. Indeed it would have been so if the doping profile and size of the emitter and collector regions been identical. However as explained in the introductory note, the collector's charge carrier concentration is lower than that of the emitter. Thus, the collector has a lower conductivity or larger resistivity. A large resistivity implies the collector's region has higher resistance than the emitter region. The asymmetry between the emitter and collector region is compounded since the manufacturer gives the collector a larger dimension, thus further increasing the collector's resistance.

We revisit eq(8.6) to compare the magnitude of the common collector and common emitter's input impedance mathematically. The equation defines the common emitter's input impedance, h_{ie}, using the base current, I_B (here we shall call it I_{BE}), flowing from base to emitter. The input impedance of the common collector would be written as

$$h_{ic} = \frac{26mV}{I_{BC}}$$

Where, I_{BC} is the input current flowing between base and collector. Since, the path 'BC' (base-collector) is more resistive than 'BE' (base-emitter), the current drawn for the same applied

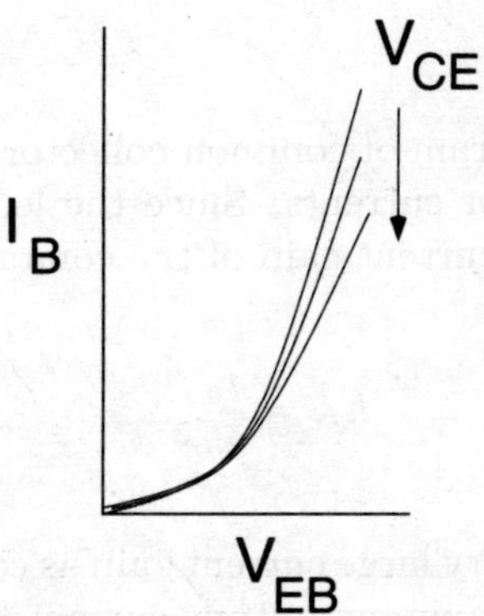

Figure 8.23: *The input IV characteristics of a transistor in common collector configuration. Arrow indicates direction of increase.*

voltage, would be less (Ohms law). Thus, $I_{BC} \ll I_{BE}$, resulting in the common collector having a larger input impedance than the common emitter mode.

Output Characteristics and Output Impedance

The output I-V characteristics of the common collector configuration is plot between the emitter current (I_E) and the emitter-collector voltage (V_{EC}). At first look one would think that the IV characteristics of this configuration would be identical to that of the common-emitter configuration. However, the saturation response (flat current response to voltage) seen for the CE configuration is absent for the common-collector configuration. Since here, the number of electrons crossing the reverse-biased collector-base junction into the collector (i.e. I_C) is not the question, but how much electrons (we are considering NPN transistors) are being generated in the emitter region and contributing to the current I_E. On increasing the emitter-collector voltage, the base-emitter gets more and more forward biased (remember fig 8.10). This increasing potential across the base-emitter increases the emitter current. The magnitude of the current depends on the charge carrier concentration of the emitter region. Since the emitter region is usually highly doped, the current increases for increasing voltage, explaining the low output impedance associated CC configuration.

The large input impedance of the CC configuration makes sure that this circuit would never load (draw too much current) any circuit to which it's input is applied. To prevent loading, remember, the input impedance of the second circuit (circuit whose input would be connected to the output of another circuit) should be greater than the output impedance of the first circuit (to which the second circuit is connected). This demands that the output impedance of a circuit should ideally be zero and it's input impedance should be infinity. This prevents loading. All circuits do not achieve this criteria. Since the common-collector circuit has these features, it can be inserted between two circuits to prevent loading. Such circuits are called **buffers**. Another requirement for making a good buffer is that it should have unity gain (i.e. the voltage gain should be equal to one). We now proceed to investigate the voltage gain of transistor in CC configuration.

Current Gain

It is apparent from the circuit diagram of common collector configuration that the load resistor receives both the base and collector currents. Since the load is in series with the emitter, the load current is $i_b + i_c$. Hence, the current gain of the common collector transistor is given as

$$h_{fc} = \frac{i_e}{i_b}$$

Thus, the CC configuration has a very large current gain as compared to the other configurations. Infact this configuration has the maximum output current for a given input current.

Voltage Gain

On examining fig(8.21b), it is clear that the output voltage of a transistor in common collector configuration is equal to $i_e R_E$. For an increasing input voltage (Δv_{in}) the base current increases. While the input voltage is increasing, the collector-base reverse biasing is decreasing, leading to an decrease in the emitter current. This implies that the input and output currents are out of phase (current gain would be represented by a negative number). The input loop equation (again we are assuming AC to be time varying DC voltages) is written as

$$V_{in1} = V_{BE} + I_{E1} R_E$$

when the input voltage increases (I_B increasing) the loop equation becomes

$$V_{in2} = V_{BE} + I_{E2} R_E$$

Thus, the change in output voltage with changing input voltage is given as

$$\Delta v_{in} = V_{in2} - V_{in1} = (I_{E2} - I_{E1}) R_E = -\Delta v_o$$

Hence, the voltage gain is

$$A_V = \frac{v_{in}}{v_o} = -1$$

The voltage gain of the transistor in common collector configuration is unity. Thus, as demanded, the CC amplifier will have unity gain and would make for an ideal buffer amplifier.

Table 8.1: Comparing the performance of the transistor in it's various modes.

	CE	CB	CC
Input Impedance	$\sim 1\text{K}\Omega$	$\sim 20\Omega$	very high ($\sim 750\text{K}\Omega$)
Output Impedance	$\sim 10\text{K}\Omega$	$\sim 1\text{M}\Omega$	very low ($\sim 50\Omega$)
Voltage Gain	~ 150	~ 500	≤ 1
Current Gain	high	1	high
Power Gain	best	worst	moderate
Application	Audio amplifier	Current source	buffer

8.5   Ebers-Moll Model

The Ebers-Moll model analysis the behaviour of the transistor similar to the piece-wise model discussed for the diode. However, the model essentially is used to describe the behaviour of the transistor when DC voltages are applied. AC model of the transistor is discussed in Chapter 10.

Consider the transistor to be symmetrical in nature. That is, the dimension of the collector and emitter region are identical with equal charge carrier concentration. While this is not the case, we make this the first assumption (we obviously refute this at the end of the model). The two regions (Base-Collector and Base-Emitter) of the NPN transistor can be viewed as two diodes kept back to back (see fig 8.24). The Base-Emitter being forward biased would have a current flowing through it, namely the diode current (eqn 8.5)

$$I_E = I_{ES}(e^{qV_{BE}/KT} - 1)$$

where I_{ES} is the reverse saturation current of the Base-Emitter region and the rest of the notations have their usual meanings. Of the electrons contributing to this current, some electrons would escape into the collector region, contributing a current $\alpha_E I_E$. The directions of both the currents are depicted in fig 8.24(a). Notice, the Base-Collector being depicted by an ideal diode can not explain the current flow and hence the current is shown from a dependent current source parallel to the ideal Base-Collector diode.

Now, if consider the Base-Collector to be forward biased (yes), then a Base-Collector current comes into existence (I_C) with electrons being injected into the Emitter being given as $\alpha_C I_C$. The senario being symmetrical is shown in fig 8.24(b). Considering superposition principle holds, the currents in the two regions can be written as sum of individual contributions, i.e.

$$I_E = I_{ES}(e^{qV_{BE}/KT} - 1) - \alpha_C I_{CS}(e^{qV_{BC}/KT} - 1) \tag{8.16}$$

(a)

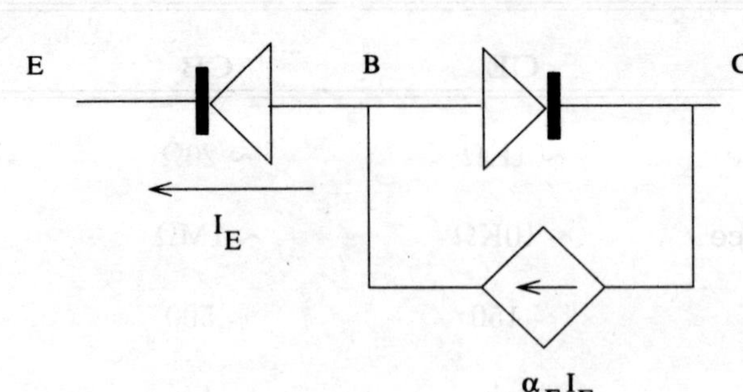

(b)

Figure 8.24: *The (a) forward biased Base-Emitter gives rise to a current I_E in it and a dependent current $\alpha_E I_E$ in the Base-Collector due to it. Similiar reasoning explains currents due to forward biasing of the Base-Collector (b).*

Similar arguments give the collector current as

$$I_C = \alpha_E I_{ES}(e^{qV_{BE}/KT} - 1) - I_{CS}(e^{qV_{BC}/KT} - 1) \tag{8.17}$$

The equations have been written with preferred direction of current from collector to the emitter. However, note the symmetry of the two equations (eqn 8.16 and 8.17). The complete model of the transistor can now be draw as shown in fig 8.25. Below we analyse the inferences that can be drawn from the Ebers-Moll model. From this figure it becomes clear why a transistor can not be made by connecting two diodes back to back. In such a case, the base would be very thick sending the collector-emitter current gain (α's=0). The lack of current sources in the model would mean one can not explain the existence of current in the circuit when the diode is reverse biased.

- **Normal Operation of transistor:** Under normal operation of the transistor, the Base-Emitter is forward biased (i.e. $V_{BE} \geq 0$) and the Base-Collector junction is reverse biased (i.e. $V_{CB} \leq 0$). Equation 8.16 and 8.17 would reduce to

$$\begin{aligned}
I_E &= I_{ES}(e^{qV_{BE}/KT} - 1) - \alpha_C I_{CS}(e^{-qV_{BC}/KT} - 1) \\
&= I_{ES}(e^{qV_{BE}/KT} - 1) - \alpha_C I_{CS}(-1) \\
&= I_{ES}(e^{qV_{BE}/KT} - 1) + \alpha_C I_{CS} \\
&\approx I_{ES}(e^{qV_{BE}/KT} - 1)
\end{aligned}$$
(8.18)

and

$$\begin{aligned}
I_C &= \alpha_E I_{ES}(e^{qV_{BE}/KT} - 1) + I_{CS} \\
&\approx \alpha_E I_{ES}(e^{qV_{BE}/KT} - 1)
\end{aligned}$$
(8.19)

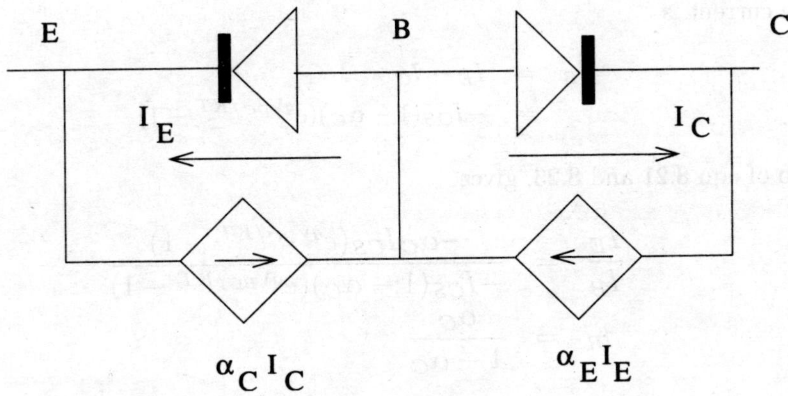

Figure 8.25: *The Ebers-Moll Model of the transistor.*

The base current can be calulated using eqn 8.18 and 8.19 using

$$\begin{aligned}
I_B &= I_E - I_C \\
&= I_{ES}(e^{qV_{BE}/KT} - 1) - \alpha_E I_{ES}(e^{qV_{BE}/KT} - 1) \\
&= I_{EN}(1 - \alpha_E)(e^{qV_{BE}/KT} - 1)
\end{aligned}$$
(8.20)

The ratio of eqn 8.19 and 8.20, gives

$$\begin{aligned}
\frac{I_C}{I_B} &= \frac{\alpha_E I_{ES}(e^{qV_{BE}/KT} - 1)}{I_{EN}(1 - \alpha_E)(e^{qV_{BE}/KT} - 1)} \\
\beta_N &= \frac{\alpha_E}{1 - \alpha_E}
\end{aligned}$$

where the subscript 'N' implies that the relation is valid for *normal* operation.

- **Inverted Operation of transistor:** Under the inverted operation the transistor's Base-Emitter is reversed biased (i.e. $V_{BE} \leq 0$) while the Base-Collector junction is forward biased (i.e. $V_{CB} \geq 0$). Equation 8.16 and 8.17 would reduce to

$$
\begin{aligned}
I_E &= I_{ES}(e^{-qV_{BE}/KT} - 1) - \alpha_C I_{CS}(e^{qV_{BC}/KT} - 1) \\
    &= -I_{ES} - \alpha_C I_{CS}(e^{qV_{BC}/KT} - 1) \\
    &\approx -\alpha_C I_{CS}(e^{qV_{BC}/KT} - 1)
\end{aligned}
\tag{8.21}
$$

and

$$
\begin{aligned}
I_C &= \alpha_E I_{ES}(e^{-qV_{BE}/KT} - 1) - I_{CS}(e^{qV_{BC}/KT} - 1) \\
    &\approx -I_{CS}(e^{qV_{BC}/KT} - 1)
\end{aligned}
\tag{8.22}
$$

The base current is

$$
\begin{aligned}
I_B &= I_E - I_C \\
    &= -I_{CS}(1 - \alpha_C)(e^{qV_{BC}/KT} - 1)
\end{aligned}
\tag{8.23}
$$

The ratio of eqn 8.21 and 8.23, gives

$$
\begin{aligned}
\frac{I_E}{I_B} &= \frac{-\alpha_C I_{CS}(e^{qV_{BC}/KT} - 1)}{-I_{CS}(1 - \alpha_C)(e^{qV_{BC}/KT} - 1)} \\
\beta_R &= \frac{\alpha_C}{1 - \alpha_C}
\end{aligned}
$$

where the subscript 'R' signifies that we are talking of the *reverse* operation. The transistor is never used in inverted mode and is fabricated such that the inverted current gain, β_R is far less than the forward/ normal current gain, β_N. This ofcourse is done by making the carrier concentration of the emitter region far greater than that of the collector.

- **Cut-off Region of the transistor:** The cut-off region of the transistor lies for region $I_B \leq 0$. When the base current is zero, set by leaving the base pin open, we have $I_E = I_C$. For biasing the transistor leaving the base open, we apply voltage V_{CE}, such that the Base-Emitter is forward biased and Base-Collector is reversed biased. From eqn 8.16 and 8.17, we have

$$
I_{ES}(e^{qV_{BE}/KT} - 1) + \alpha_C I_{CS} = \alpha_E I_{ES}(e^{qV_{BE}/KT} - 1) + I_{CS}
$$

which gives

$$
\begin{aligned}
(1 - \alpha_E)I_{ES}(e^{qV_{BE}/KT} - 1) &= I_{CS}(1 - \alpha_C) \\
\alpha_E I_{ES}(e^{qV_{BE}/KT} - 1) &= \alpha_E \frac{(1 - \alpha_C)}{(1 - \alpha_E)} I_{CS}
\end{aligned}
\tag{8.24}
$$

or (using eqn 8.17 and the above)

$$\alpha_E I_{ES}(e^{qV_{BE}/KT} - 1) + I_{CS} \;=\; \alpha_E \frac{(1 - \alpha_C)}{(1 - \alpha_E)} I_{CS} + I_{CS}$$

$$I_C|_{I_B=0} \;=\; \left[\alpha_E \frac{(1 - \alpha_C)}{(1 - \alpha_E)} + 1\right] I_{CS}$$

$$I_{CEO} = I_C|_{I_B=0} \;=\; \left[\frac{1 - \alpha_E \alpha_C}{1 - \alpha_E}\right] I_{CS}$$

where I_{CEO} is the collector leakage current when the base is left open circuited.

The model hence explains the behaviour of the transistor under various conditions of operation. An interesting exercise would be to see if the output characteristics of a transistor can be completely obtained by this model.

8.6   Transistor as a Switch

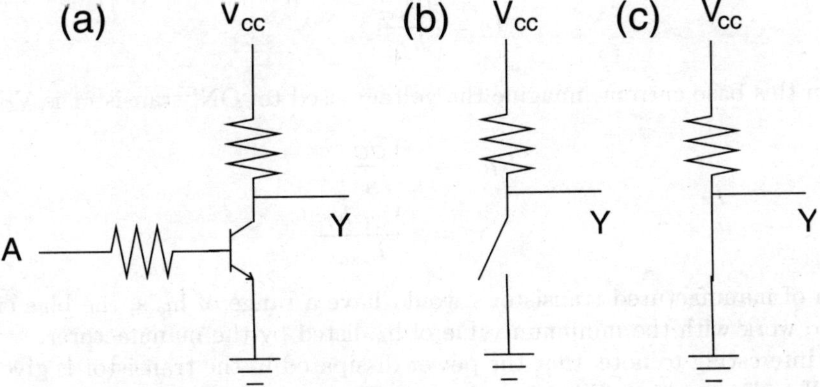

Figure 8.26: *A transistor in CE mode can be designed to work as a NOT gate (a). The switching behavior of the transistor give (b) high or (c) low output.*

One of the common uses of transistors is as a electronic switch, which goes 'ON' or 'OFF' allowing for the conduction or blocking of current. The switching of the transistor is suited for digital application. The switching action of the transistor for digital circuits is achieved by sending the transistor from its cut-off region to it's saturation region. In the "cut-off" region, the transistor (CE mode) has zero input base current and zero output collector current giving maximum (output) collector voltage. That is input is low and output is high. In "saturation", the BJT will have maximum amount of base current resulting in maximum collector current flow and minimum collector emitter voltage. Figure(8.26a) shows a NOT gate made using the transistor as a switch. As the name suggests the output of a NOT gate is "not" the same as the input. The truth table of a NOT gate is

Table 8.2: Truth Table of a NOT gate

A	Y
0	1
1	0

In fig(8.26b) the switch is open, i.e. zero volts (grounded[11]) is given at input/ base of transistor (point 'A' of fig 8.26 a). Since the transistor is off, i.e. it is not conducting and the collector current, I_C, will be zero. On the output side the load equation is

$$V_{CC} = V_{CE} + R_C I_C$$

Since no collector current flows, V_{CE} will rise to supply voltage (V_{CC}). When the transistor is 'ON', i.e. current is in base, I_C will flow and V_{CE} will be small (close to zero) if the base current is properly selected such that the transistor is forced into it's saturation region, i.e. $I_C = I_{Csat}$. That is, the required base current would be

$$I_B = \frac{I_{Csat}}{h_{fe}}$$

And to obtain this base current, imagine the voltage used to 'ON' transistor is V_{CC}, then

$$R_B = \frac{V_{CC}}{I_B}$$
$$= \frac{V_{CC} h_{fe}}{I_{Csat}}$$

Since a batch of manufactured transistor's would have a range of h_{fe}'s, the bias circuit should be designed to work with the minimum value of h_{fe} listed by the manufacturer.

Also, it is interesting to note, that the power dissipated in the transistor is given by $V_{CE} I_C$. In both cut-off and saturation, this is nearly zero. That is, minimum power is dissipated in the transistor, which obviously is unlike the case in analog circuits.

8.7 Differential Amplifier

A typical dual input balanced output differential amplifier circuit is shown in fig(8.27). As the name suggests, the circuit is given two inputs (V_{in1} and V_{in2}) while the output is the difference in the collector voltages of the two transistors. The differential amplifier amplifies the difference between the two input voltages. This holds true for both DC and AC inputs.

---

[11]Digital circuits made up by transistors are called TTL (Transistor-Transistor Logic) circuits. In TTL based circuits, leaving input hanging (i.e. not connected to ground) does not imply input is zero. On the contrary the circuit takes it as high state. Since base is not forced to ground, the collector supply, V_{CC}, forward biases the base-emitter (fig 8.10). With collector current flowing, V_{CE} is zero, and the NOT gate is function as if input was given while base is just left hanging.

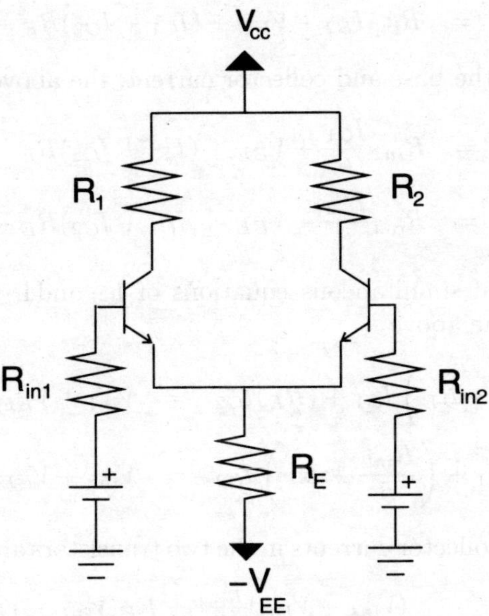

Figure 8.27: *Two transistors connected as shown forms a differential amplifier or more popularly operational amplifier.*

The DC analysis of the above circuit is simple and can be done with the knowledge of mathematics developed so far for the transistor. A similar exercise can be done for the AC input also, however, that requires the knowledge of AC models for transistors and hence would not be discussed here. As stated the output of a differential output is the difference in the collector voltages of the two transistors. The collector voltage can be found by analysing the output side of the common emitter configuration transistors, which give

$$
\begin{aligned}
V_o &= V_{C2} - V_{C1} \\
&= (V_{CC} - I_{C2}R_C) - (V_{CC} - I_{C1}R_C) \\
&= R_C(I_{C1} - I_{C2})
\end{aligned}
\tag{8.25}
$$

Thus, to find the output voltage in terms of the input voltages, all one has to do is write the expressions for the two collector currents. We know the collector current of transistors are related to it's base current (input as in the case of common emitter transistors). The base current and in turn the collector current can be found from the input equation of the circuit. For the two transistor's we have

$$
\begin{aligned}
V_{in1} &= R_{in1}I_{B1} + V_{BE1} + (I_{C1} + I_{C2})R_E - V_{EE} \\
V_{in2} &= R_{in2}I_{B2} + V_{BE2} + (I_{C1} + I_{C2})R_E - V_{EE}
\end{aligned}
$$

Usually, the transistors used in differential amplifiers are identical. The difference in currents is only due to the difference in the applied resistance R'_{in}s.

$$
V_{in1} = R_{in1}I_{B1} + V_{BE} + (I_{C1} + I_{C2})R_E - V_{EE}
$$

$$V_{in2} \;=\; R_{in2}I_{B2} + V_{BE} + (I_{C1} + I_{C2})R_E - V_{EE}$$

using the relation between the base and collector current, the above equation reduces to

$$V_{in1} \;=\; R_{in1}\frac{I_{C1}}{\beta} + V_{BE} + (I_{C1} + I_{C2})R_E - V_{EE}$$

$$V_{in2} \;=\; R_{in2}\frac{I_{C2}}{\beta} + V_{BE} + (I_{C1} + I_{C2})R_E - V_{EE}$$

The equations form a set of simultaneous equations of I_{C1} and I_{C2}, as can be understood by re-arranging the terms of the above

$$\left(\frac{R_{in1}}{\beta} + R_E\right)I_{C1} + (R_E)I_{C2} \;=\; V_{in1} + V_{EE} - V_{BE}$$

$$(R_E)I_{C1} + \left(\frac{R_{in2}}{\beta} + R_E\right)I_{C2} \;=\; V_{in2} + V_{EE} - V_{BE} \tag{8.26}$$

Solving for I_{C1} and I_{C2}, the collector currents in the two transistors are given as (assume $V_{BE} \to 0$)

$$I_{C1} \;=\; \frac{(V_{in1} + V_{EE})\frac{R_{in2}}{\beta} + R_E(V_{in1} - V_{in2})}{\left(\frac{R_{in1}}{\beta} + R_E\right)\left(\frac{R_{in2}}{\beta} + R_E\right) - R_E^2}$$

$$I_{C2} \;=\; \frac{(V_{in2} + V_{EE})\frac{R_{in1}}{\beta} + R_E(V_{in2} - V_{in1})}{\left(\frac{R_{in1}}{\beta} + R_E\right)\left(\frac{R_{in2}}{\beta} + R_E\right) - R_E^2} \tag{8.27}$$

Substituting the current expressions in eqn(8.25), we can compute the output voltage. The expression can be simplified by assuming $R_{in1}=R_{in2}=R_{in}$.

$$V_o \;=\; \frac{R_C}{\left(\frac{R_{in}}{\beta} + R_E\right)^2 - R_E^2}\left[(V_{in1} - V_{in2})\frac{R_{in}}{\beta} + 2R_E(V_{in1} - V_{in2})\right]$$

$$=\; \frac{R_C}{\left(\frac{R_{in}}{\beta} + 2R_E\right)\frac{R_{in}}{\beta}}\left[\frac{R_{in}}{\beta} + 2R_E\right](V_{in1} - V_{in2})$$

$$=\; \frac{\beta R_C}{R_{in}}(V_{in1} - V_{in2}) \tag{8.28}$$

It is intersting to note that the output voltage is proportional to the difference in the two input voltages and thus it's name 'differential amplifier'. This circuit and it's variants are at the heart of the popular electronic device called **operational amplifiers or simply op amps**. So vast is the scope of usage of this device that it is studied as a separate branch of electronics.

8.8   The Voltage Regulator Revisited

The modern regulators are more complex than the shunt and series regulators introduced in Chapter 7, where the regulation action was achieved by varying the resistance kept in parallel or series with the output load resistance. This variation is achieved by complex circuitry which come in IC forms. Fig (8.28) shows the internal circuit diagram of such a voltage regulator.

The output voltage V_{out} is sampled through the potential divider formed by R_1 and R_2. The voltage across R_2 is proportional to the output voltage and is compared with a standard voltage (if regulated output required is +5v, a zener of 5v can be used) by a comparator circuit, symbolized by the triangle in fig(8.28). The difference in the two inputs (the error) is amplified by the comparator, whose output is proportional to the difference in the input levels and is fed into the base of the transistor. This adjusts the current (I_c) and hence the resistance represented by the transistor.

Assume, initially that the output is open , i.e. no load is applied. This would make the output current, I_L, zero. The potential drop at the output would be

$$V_{out,\,nl} \quad = \quad (R_1 + R_2)(I_T - I_{gnd,\,nl})$$

where I_T is the input current that breaks into the transistor current I_{gnd}[12] and $I_T - I_{gnd}$ at the transistors collector node. The current $I_T - I_{gnd}$ further breaks into $I_T - I_{gnd} - I_L$ and I_L at the load's node, with $I_T - I_{gnd} - I_L$ being the current in the branch containing the resistances R_1 and R_2. The sub-

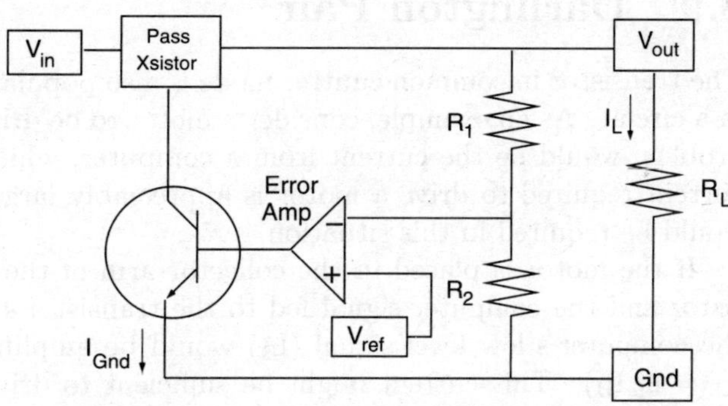

Figure 8.28: *The internal architecture of an shunt voltage regulator.*

script 'nl' is used to represent fact that *"no load"* is applied. If the output is open, it is as good as saying an infinitely large load has been applied. The output hence draws zero current and the voltage measured at the output would be the regulated voltage. Thus we can say the regulated voltage, V_r, should be

$$V_r = V_{out,\,nl} \quad = \quad (R_1 + R_2)(I_T - I_{gnd,\,nl}) \tag{8.29}$$

Now consider a resistance of relative low value is placed across the output. The choice of resistance R_L is such that it draws current or in other words loads the circuit. The output voltage would fall. The new output would be

$$V_{out} \quad = \quad (R_1 + R_2)(I_T - I_{gnd} - I_L) \tag{8.30}$$

Since, the output voltage has fallen, the sampled voltage by the comparator also falls. This would imply that the difference in the two input voltages decreases. Thus, the current fed into the transistor's base by the comparator would also fall. This brings down the transistor's current.

The change in the output voltage would be given as

$$\Delta V_{out} \quad = \quad V_r - V_{out}$$
$$= \quad (R_1 + R_2)(I_T - I_{gnd,\,nl} - I_T + I_{gnd} + I_L)$$

---

[12]The notation maintained in the datasheets of three pin voltage regulators have been maintained

$$= (R_1 + R_2)[(I_L - 0) - (I_{gnd,\,nl} - I_{gnd})]$$
$$= (R_1 + R_2)(|\Delta I_L| - |\Delta I_{gnd}|)$$

$$(8.31)$$

That is, with increasing load current, a corresponding change in the transistor current pulls down the variation in the output voltage from that of the regulated level. Obviously if the changing currents are balanced such that ΔV_{out} is zero, the regulator would be prefect.

8.9 Darlington Pair

The transistor in common emitter mode is also popularly used as a device to amplify the current in a circuit. As an example, consider a motor to be driven by instructions from a computer. The problem would be the current from a computer, which is very low (low milli-amps), while the current required to drive a motor is appreciably larger. For small motors, a current amplifier would be required in this situation.

If the motor is placed in the collector arm of the transistor and the computer signal fed to the transistor's base, the computer's low level signal (I_B) would be amplified to $I_C (= h_{fe}I_B)$. This current might be sufficient to drive the motor. But just what if the current I_C is not enough to drive the motor? One might look around for a transistor with a better h_{fe}. However, a more elegant solution would be to use Darlington pair of transistor as shown in fig(8.29). It is essentially two transistors cascaded. The effect of this cascading is a device with large input impedance (as compared to the individual transistors constituting the pair), decreased output impedance and large current gain. Trivial mathematics can show the current amplification capability of the Darlington pair. For the transistor Q_1, we have

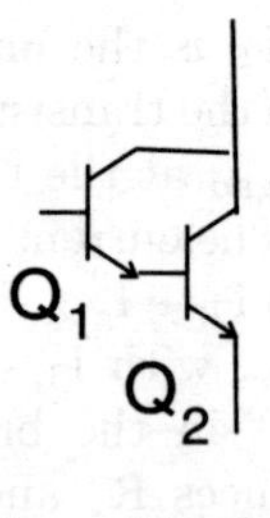

Figure 8.29: *Two transistors form a Darlington pair whose DC current gain β is the product of the two individual transistor's DC current gain.*

$$I_{E1} = (1 + h_{fe1})I_{b1}$$

Similarly, for the second transistor

$$I_{E2} = (1 + h_{fe2})I_{b2}$$

However, the base current of the second transistor is the emitter current of the first transistor, hence

$$\begin{aligned} I_{E2} &= (1 + h_{fe2})I_{E1} \\ &= (1 + h_{fe2}) \times (1 + h_{fe1})I_{b1} \\ &\approx h_{fe2} \times h_{fe1}I_{b1} \end{aligned}$$

The above equation relates the input and output current of the pair and in case of identical transistors, the current gain is given as

$$A_i = h_{fe}^2$$

Thus, the current gain is far greater than the gain of the individual transistors making up the pair. The detailed analysis of the Darlington pair's input impedance, current and voltage gain can be found in Chapter 10.

We have completed a brief overview of uses the transistor can be put to. As the book proceeds and as the student evolves, learning more and more from other sources, more applications of the transistor would be seen. However, hopefully, the basic ideas have been appreciated.

Exercise

Q1. Out of the emitter and collector region of a transistor, which has a higher doping concentration and why?

Q2. Why is the base of the transistor made thin with low doping concentration?

Q3. Explain whether the transistor is a current controlled or voltage controlled device.

Q4. The input signal $0.25\sin(\omega t)$ is fed to a CE amplifier of gain 200. What is the output signal?

Q5. In the above example if the input signal is 100mV (V_{rms}), the transistor is given a 10v DC supply and is positioned half way in it's active region (IV characteristics). Draw the expected output. Explain your answer.

Q6. For a basic CE transistor amplifier, given $V_{cc} = 15v$, $R_c = 1K\Omega$ and $R_E = 500\Omega$. Determine the operating point for maximum symmetrical swing.

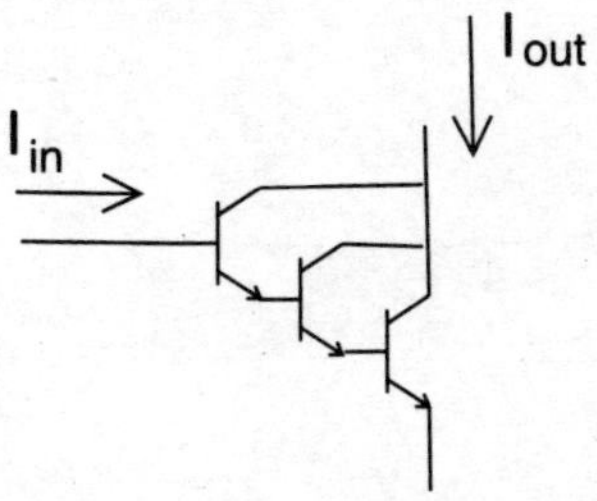

Figure 8.30:

Q7. Derive the current gain for the transistor combinations shown in fig(8.30).

Q8. Which of the transistor configurations (CE, CB or CC) has

   (i) highest input impedance, R_i

   (ii) highest output impedance, R_o

   (iii) lowest voltage gain, A_v

   (iv) lowest current gain, A_i

Q9. Which of the transistor configuration out of the CC, CB and CE acts as a buffer stage? Explain your answer.

Q10. What is a load line with reference to transistors.

Q11. Give two reasons why CE configuration is preferred over CB and CC configurations.

Q12. What are the main features of a buffer amplifiers. Which transistor configuration is used to design a buffer circuit?

Q13. Why can't a transistor be made by joining two diodes back to back?

Q14. Explain the output characteristics of the CB configuration qualitatively.

Q15. Draw the circuit of the transistor in CE configuration by which the IV characteristics of the transistor may be experimentally determined. Sketch the output characterisitcs and indicate the active, saturation and cut-off region.

Q16. What are the criteria for selecting the 'Q' point in an amplifier?

Q17. Describe the basic operation of a bipolar NPN transistor. Draw a diagram indicating the current flows in the transistor and state the relationship between the base current and collector current. Give a description between the base current and collector current.

Q18. Discuss the Ebers-Moll Model. Can a transistor behaviour be obtained by connecting two transistors back to back?

Q19. Can the output IV characteristics of a CE model transistor be obtained purely from the Eber-Moll's model?

Chapter 9

Transistor Biasing

One of the basic problems with transistor amplifiers is establishing and maintaining the transistor's operating point or the biasing point. The proper values of quiescent current I_{CQ} and voltage V_{CEQ} decides where in the transistor's active region would it operate. The processes of selecting the operating point (Q point) by proper selection of circuital elements is called biasing. Biasing also should ensure that the operating point is maintained despite variations in surrounding temperature, which cause changes in amplification and even distortion. Even if we are not concerning ourselves with the designing of an amplifier, for the proper functioning of the transistor, the base-emitter junction has to be forward biased while the collector-base junction should be reversed biased. In fig(8.21), the circuit shows two power supplies to properly bias the transistor (power supplies or batteries). Batteries are costly, hence methods to properly bias the junctions of the transistor with just one battery is necessary. That is exactly what we study here and more.

Various biasing methods can be used to accomplish these functions. Although there are numerous biasing methods, only four basic types will be discussed.

9.1 Base-Current Bias (Fixed Bias)

The first biasing method, called BASE CURRENT BIAS or sometimes FIXED BIAS, was used in fig (9.1). It consisted basically of a resistor (R_B) connected between the collector supply voltage and the base and R_C connected between the supply and the collector. The two resistances (R_B) and (R_C) selects the transistor's operating point. For the operating point to be half way in the active region of the transistor, the collector-emitter voltage should be $V_{cc}/2$ and the collector current should be $V_{cc}/2R_c$, i.e. half of the saturation current.[1] Thus, R_C selects the load line (V_{CEQ}), which is easily found by applying the voltage law on the output side,

$$V_{cc} = I_C R_C + V_{CE} \tag{9.1}$$

since the operating point lies on this line, we have

$$V_{cc} = I_{CQ} R_C + V_{CEQ}$$

---

[1]The saturation current is collector current when the collector-emitter voltage is zero, eq(9.1)

The base current is related to the collector current, hence for the required I_{CQ} we can find the required base current.

$$I_B \; = \; \frac{I_{CQ}}{\beta}$$

$$= \; \frac{V_{cc}}{2\beta R_C} \qquad (9.2)$$

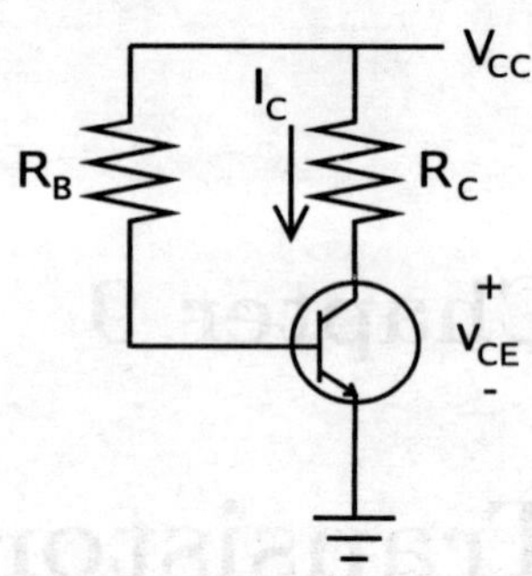

Figure 9.1: *The npn transistor is biased with a fixed bias circuit.*

On the input side applying Kirchoff's voltage law, we have

$$V_{cc} = I_B R_B + V_{BE} \qquad (9.3)$$

The required base current (eqn 13.3) is attained by the proper choice of the base resistance

$$R_B \; = \; \frac{V_{cc} - V_{BE}}{I_B}$$

$$R_B \; \sim \; \frac{V_{cc}}{I_B} \sim 2\beta R_C \qquad (9.4)$$

The expression is obtained by approximating $V_{BE} = 0$, hence we should choose R_B to be slightly smaller than $2\beta R_C$.

Example 9.1: A transistor is biased using a simple fixed bias circuit, with $V_{cc} = 15v$, $R_B = 200K\Omega$ and $R_C = 1K\Omega$. The transistor presently in circuit has $\beta=100$, and is damaged in use. If a new transistor with $\beta=300$ is used as a replacement, determine whether the new transistor would be properly biased. Determine the operating point in the two cases for the said resistances.

While for the first transistor eqn(9.4) is satisfied, the second transistor does not satisfy the equation. The base resistance would have to be replaced by a $600K\Omega$ resistor.

For calculating the Q point for the first case,

$$I_B \; = \; \frac{15 - 0.7}{200K}$$

$$= \; 71.5\mu A$$

The operating point I_{CQ} is thus selected by the base current

$$I_C \; = \; \beta I_B$$

$$I_{CQ} \; = \; 100 \times 71.5\mu A = 7.15 mA$$

The voltage drop across the transistor's collector-emitter is

$$V_{CEQ} \; = \; V_{cc} - I_{CQ}R_C$$

$$= \; 15 - 7.15 = 7.85v$$

The saturated collector current is

$$I_{sat} = \frac{V_{cc}}{R_C}$$
$$= 15mA$$

Since, for the transistor with $\beta = 100$, the two resistances give the 'Q' point nearly half way in the active region, the basing is proper.

For the second transistor, following the same steps as above we can find the Q point:

$$I_C = \beta I_B$$
$$I_{CQ} = 300 \times 71.5\mu A = 21.45mA$$

Hence, the voltage drop across the transistor's collector-emitter is

$$V_{CEQ} = V_{cc} - I_{CQ}R_C$$
$$= 15 - 21.45 = -6.45v$$

Thus the operating point is beyond the transistor's saturation region. Infact the collector-base is forward biased!!

As can be seen from the above example, the biasing circuit works only for a specific transistor given β.[2] Any replacement of transistor, the circuit is incapable of biasing properly. This is a problem, since with usage transistors do go bad and require replacement. Even with the present technological abilities, variation of β in the same batch of transistors is hard. Hence, ideally, biasing circuit's selection of the operating point should be independent of the transistor parameters.

Another objective that the biasing circuit should achieve is thermally stability. If the temperature of the transistor rises for any reason (due to a rise in ambient temperature or due to current flow through it), collector current will increase. This increase in current also causes the operating point, to move away from its desired position (level). This reaction to temperature is undesirable because it affects amplifier gain (the number of times of amplification) and could result in distortion. Infact the flow of current in the circuit is a prime reason of heating which leads to an increase in current (I_C) which pushes the temperature further. Thus, a cascading effect leads to what is called a **_Thermal run-away_**, destroying the transistor. The biasing circuit should be able to prevent any thermal run-away.

To understand the thermal run-away from mathematical equations established to explain the transistor, consider the equation

$$I_C = \alpha I_E + I_{CBO}$$

where I_{CBO} is the reverse saturation current across the reverse biased collector-base junction, caused by the minority charge carriers present in the semiconductor. Any increase in temperature increases this reverse saturation current. In operation condition, the collector current flows

---

[2] β acting as a transistor characterizing parameter

leading to heating. The increased temperature increases the trapped minority charge carriers, increasing I_{CBO}. This pushes the collector current higher, leading to more heating. Thus, a "chain reaction" or more appropriately a chain of events leads to a large collector current capable of destroying the transistor. It is not as if only the reverse biased junction is effected by temperature, even the forward biased base-emitter junction reacts to temperature. The functional response of this junction to temperature is given by the diode equation (eqn 5.2)

$$I_E = I_o(e^{qV_{BE}/kT} - 1)$$

The emitter current grows at 10% per °C, if V_{BE} is constant, the change obviously effects the collector current, in turn increasing heating.

Thus, the biasing circuit must make the collector current independent of I_{CBO} and β, to prevent thermal run-away and make circuit independent of the used transistor. Another transistor characterization parameter is V_{BE}. The biasing circuit should also make the collector current insensitive to variation in V_{BE}. Thus, the basic use of a biasing circuit is to stablise the transistor circuit. Three figures of merit are introduced, called the "*Stability Factor*", to quantify how well the biasing circuit has been able to achieve the said objective. They are,
(i) stability from thermal run-away caused to increase in I_{CBO} with temperature

$$S(I_{CBO}) \;=\; \frac{\partial I_C}{\partial I_{CBO}} \tag{9.5}$$

(ii) stability from transistor parameter, β

$$S(\beta) \;=\; \frac{\partial I_C}{\partial \beta} \tag{9.6}$$

(iii) stability from transistor parameter, V_{BE}

$$S(V_{BE}) \;=\; \frac{\partial I_C}{\partial V_{BE}} \tag{9.7}$$

For calculating the stability factor, the input equation of the transistor has to be used to set up an equation for the collector current which has all the three variable terms, i.e. I_{CBO}, β and V_{BE}. For the fixed bias circuit, the input and input equation is

$$V_{cc} = I_B R_B + V_{BE}$$

from the full relation between the collector current and base current we have (eqn 8.9)

$$I_C = \beta I_B + (1 + \beta)I_{CBO}$$

from which the base current can be written as

$$I_B = \frac{I_C - (1 + \beta)I_{CBO}}{\beta}$$

when we substitute the base current expression in the input equation, the resulting equation contains all the terms of interest (I_C, I_{CBO}, β and V_{BE}).

$$V_{cc} = \left[\frac{I_C - (1+\beta)I_{CBO}}{\beta}\right]R_B + V_{BE}$$

$$= \frac{I_C R_B}{\beta} - \frac{(1+\beta)I_{CBO}R_B}{\beta} + V_{BE} \qquad (9.8)$$

Differentiating the above equation w.r.t. I_{CBO}, we have

$$0 = \left(\frac{R_B}{\beta}\right)\frac{\partial I_C}{\partial I_{CBO}} - \frac{(1+\beta)R_B}{\beta}$$

$$= \left(\frac{R_B}{\beta}\right)S(I_{CBO}) - \frac{(1+\beta)R_B}{\beta}$$

On re-arranging, the stability factor is

$$S(I_{CBO}) = (1+\beta)$$

This means for a small change in the reverse leakage saturation current, there results a large change in collector current (β is large). Thus, the biasing circuit does nothing to prevent the transistor from running into thermal runaway condition. The smaller the stability factor, better is the circuit. Similarly working out $S(\beta)$ and $S(V_{BE})$ by differentiating eqn(9.8) w.r.t. β and V_{BE} respectively, we have

$$0 = -\left(\frac{R_B}{\beta^2}\right)I_C + \frac{R_B}{\beta}S(\beta) + \frac{1}{\beta^2}I_{CBO}R_B$$

from which we get

$$S(\beta) = \frac{1}{\beta}(I_C - I_{CBO})$$

and

$$0 = \left(\frac{R_B}{\beta}\right)S(V_{BE}) + 1$$

gives

$$S(V_{BE}) = -\frac{\beta}{R_B}$$

All stability factors are non-zero, thus, the fixed bias circuit fails to provide any stability to the transistors operation. However, the expression for $S(V_{BE})$ does given an idea. By making the base resistance large, we can make the circuit insensitive to variations in V_{BE}. This was the idea used when we reduced eqn(9.3) to eqn(9.4). We now proceed to investigate the performance of other biasing circuits.

9.2 Collector-Base Bias Circuit

A better method of biasing is obtained by inserting the bias resistor directly between the base and collector, as shown in figure. By introducing this resistance, a feedback voltage can be fed from the collector to the base. Let us first design the circuit, i.e. calculated the required to resistances. The input side voltage equation is

$$V_{cc} = (I_C + I_B)R_C + V_{CE} \qquad (9.9)$$

the saturation current would be

$$\begin{aligned}
V_{cc} &= (I_C + I_B)R_C \\
&= \left(1 + \frac{1}{\beta}\right) I_{C(sat)}R_C \\
I_{C(sat)} &= \frac{V_{cc}}{R_C}
\end{aligned}$$

$$(9.10)$$

for proper selection of the operating point, we demand the operating current to be half of the saturation current

$$I_{CQ} = \frac{V_{cc}}{2R_C} \qquad (9.11)$$

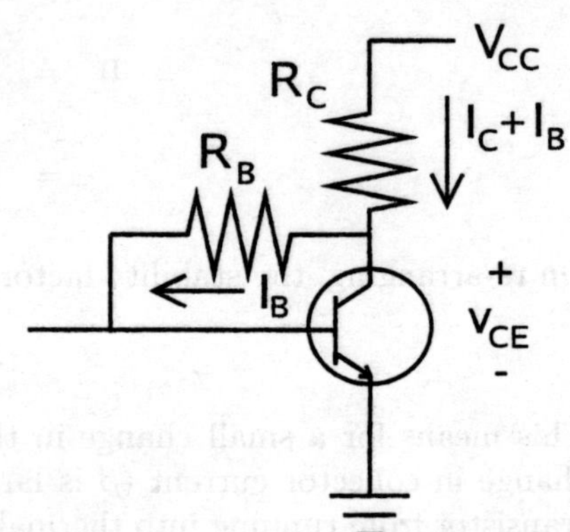

Figure 9.2: *A transistor with collector-base resistance biasing circuit.*

The required base current is obtained from

$$\begin{aligned}
I_B &= \frac{I_C}{\beta} \\
&= \frac{V_{cc}}{2\beta R_C}
\end{aligned} \qquad (9.12)$$

On the input side applying Kirchoff's voltage law, we have

$$V_{cc} = +(I_C + I_B)R_C + I_B R_B + V_{BE} \qquad (9.13)$$

The required base current (eqn 9.12) is attained by the proper choice of the base resistance

$$\begin{aligned}
R_B &= \frac{V_{cc} - V_{BE} - I_C R_C}{I_B} - R_C \\
R_B &\sim (\beta - 1)R_C
\end{aligned}$$

Now, on rearranging eqn(9.13), we have

$$I_B = \frac{V_{cc} - V_{BE} - I_C R_C}{(R_B + R_C)}$$

If an increase of temperature causes an increase in collector current, the base current (I_B) will fall. The decrease in base current will oppose the original increase in collector current and tend to stabilize it. The exact opposite effect is produced when the collector current decreases. This process of returning a part of the output back to its input is known as DEGENERATION or NEGATIVE FEEDBACK. Sometimes degeneration is desired to prevent amplitude distortion (as will be shown in chapter on negative feedback), however, it reduces amplification since the DC negative feedback also acts as a AC negative feedback (the collector signal feedback to the base cancels some of the input signal).

By arranging the input side voltage expression as

$$V_{cc} = R_c(I_B + I_C) + I_B R_B + V_{BE}$$

$$= I_B(R_B + R_C) + I_C R_C + V_{BE}$$

$$V_{cc} = \frac{(R_B + R_C)}{\beta} I_C - \frac{(1+\beta)(R_C + R_B)}{\beta} I_{CBO} + I_C R_C + V_{BE} \tag{9.14}$$

we can calculate the various stability factors:

$$S(I_{CBO}) = \frac{\partial I_C}{\partial I_{CBO}}$$

$$0 = \frac{(R_B + R_C)}{\beta} S - \frac{(1+\beta)(R_C + R_B)}{\beta} + R_C S$$

$$S(I_{CBO}) = \frac{(1+\beta)(R_c + R_B)}{R_B + (1+\beta)R_C} \tag{9.15}$$

Substituting our design requirement ($R_B = (\beta - 1)R_C$), the above expression reduces to

$$S(I_{CBO}) = \frac{(1+\beta)}{2}$$

Thus, the stability from thermal run-away has improved by introducing the negative feedback. Similarly,

$$S(V_{BE}) = \frac{\partial I_C}{\partial V_{BE}}$$

$$0 = \frac{(R_B + R_C)}{\beta} S + R_C S + 1$$

$$S(V_{BE}) = \frac{-\beta}{R_B + (1+\beta)R_C} \tag{9.16}$$

$$S(\beta) = \frac{\partial I_C}{\partial \beta}$$

$$0 = \frac{(R_B + R_C)}{\beta} S - \frac{(R_C + R_B)}{\beta^2} I_C + \frac{(R_C + R_B)}{\beta^2} I_{CBO} + R_C S$$

$$S(\beta) = \frac{(R_c + R_B)[I_C - I_{CBO}]}{\beta[R_B + (1+\beta)R_C]} \tag{9.17}$$

The appears of $(\beta + 1)R_C$ in the denominator makes the stability factor terms smaller than the counterpart of fixed bias circuits. Thus, indicating this circuits better stability.

9.3   Biasing with Emitter Resistor

This is similar to the fixed bias circuit studied earlier, except the emitter is not grounded directly but through an emitter resistance, R_E.

    Thus, the expression of the input and output side voltages would be similar to that expressed in eqn(9.1) and (9.3), except for an additional term, the voltage drop across the emitter resistance. Notice that this term appears both in the input and output side equations.

$$V_{cc} = I_C R_C + V_{CE} + I_E R_E \qquad (9.18)$$

and

$$V_{cc} = I_B R_B + V_{BE} + I_E R_E \qquad (9.19)$$

The emitter current is related to the collector current and base current by

$$
\begin{aligned}
I_E \;&=\; I_C + I_B \\
&=\; (1 + \beta)I_B \\
I_E \;&\sim\; I_C \sim \beta I_B
\end{aligned}
$$

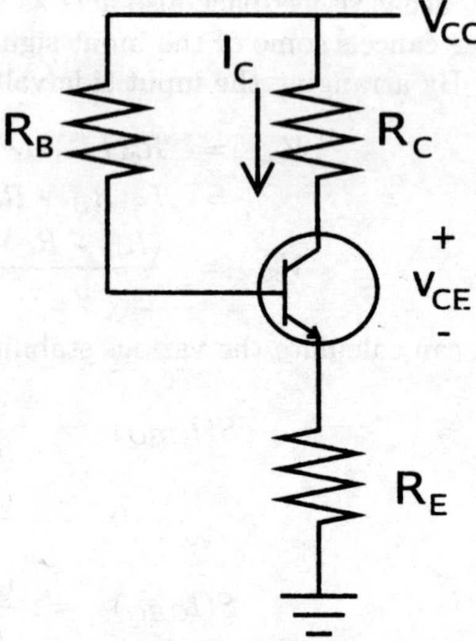

Figure 9.3: *A transistor with emitter resistance biasing circuit.*

For the operating current to be have way the saturation current for the approximation on emitter, collector and base currents made, we require (from the output side expression eqn 9.18)

$$I_{CQ} = \frac{V_{cc}}{2(R_C + R_E)}$$

Thus, the base current should be

$$I_B = \frac{V_{cc}}{2\beta(R_C + R_E)}$$

To attain this required base current the base resistance should be

$$
\begin{aligned}
I_B R_B \;&=\; V_{cc} - I_E R_E - V_{BE} \\
R_B \;&\sim\; \beta(2R_C + R_E) \qquad (9.20)
\end{aligned}
$$

So we have learnt how to design an emitter resistor biasing circuit, but how is it better? To understand this we re-write the input side voltage equation as

$$I_B = \frac{V_{cc} - V_{BE} - R_E I_E}{R_B}$$

As temperature of the device increases with current passing through it, the minority current of the reverse biased collector-base increases. Increasing current leads to increased voltage drop across the emitter resistance ($I_C \sim I_E$). This leads to a drop in the base current, I_B, which brings down the collector current ($I_C = \beta I_B$). Thus, the very cause of heating is brought down. Similar to the collector-base resistance bias circuit, this prevention from thermal run-away has been enabled due to the negative feedback in the circuit. The emitter resistance is present both on the input and output side. Any change in the output side is reflected in the input which modifies to correct operating points position. Of course the degree of stability obtained can be expressed by the various stability factors, which we proceed to calculate.

Again we start with the input side equation and express it in terms of the four currents of importance and differentiate it.

$$
\begin{aligned}
V_{cc} &= I_B R_B + V_{BE} + I_E R_E = I_B R_B + V_{BE} + (I_C + I_B)R_E \\
&= \left[\frac{I_C - (1+\beta)I_{CBO}}{\beta}\right](R_B + R_E) + V_{BE} + I_C R_E
\end{aligned}
$$

Differentiating the above equation w.r.t. I_{CBO}, we have

$$
\begin{aligned}
0 &= \left(\frac{R_B + R_E}{\beta}\right)\frac{\partial I_C}{\partial I_{CBO}} - \frac{(1+\beta)(R_B + R_E)}{\beta} + \left(\frac{\partial I_C}{\partial I_{CBO}}\right)R_E \\
&= \left(\frac{R_B + R_E}{\beta} + R_E\right)S(I_{CBO}) - \frac{(1+\beta)(R_B + R_E)}{\beta}
\end{aligned}
$$

On re-arranging, the stability factor is

$$
S(I_{CBO}) = \frac{(1+\beta)(R_B + R_E)}{R_B + (1+\beta)R_E}
$$

Since the resistive term is far less than unity, the stability factor of the emitter biased circuit is less than (better than) that of the fixed bias circuit, i.e.

$$
\begin{aligned}
[S(I_{CBO})]_{emitter} &< (1+\beta) \\
&< (S(I_{CBO}))_{fixed}
\end{aligned}
$$

Hence, the emitter biased circuit is better suited to prevent thermal run-away, where a small change in the reverse leakage saturation current produces in very small change in collector current. Now let us investigate whether the circuit is independent of the transistor used, differentiating w.r.t. β and V_{BE}

$$
0 = -\left(\frac{R_B + R_E}{\beta^2}\right)I_C + \frac{R_B + R_E}{\beta}S(\beta) + \frac{1}{\beta^2}I_{CBO}(R_B + R_E) + R_E S(\beta)
$$

from which we get

$$
S(\beta) = \frac{(R_B + R_E)(I_C - I_{CBO})}{\beta R_B + \beta(1+\beta)R_E}
$$

and

$$
0 = \left(\frac{R_B + R_E}{\beta}\right)S(V_{BE}) + 1 + R_E S(V_{BE})
$$

gives

$$S(V_{BE}) = -\frac{\beta}{R_B + (1 + \beta)R_E}$$

The appearance of the term $(1 + \beta)R_E$ in the denominator of all three stability expressions improves the stability of the circuit as compared to the fixed bias circuit. This circuit is also better of then the previous biasing circuit discussed, since the stability factors are independent of the collector resistance R_C. The gain of the amplifier that one designs depends on the collector resistance. As the stability is independent of R_C for the emitter bias circuit, there is no restriction of choice imposed on the designer. However, as discussed the emitter resistor brings stability to the circuit by acting as a negative feedback. In an amplifier, this negative feedback would try to bring down the gain! Thus failing the very propose that it is being designed for.

What is hence required is that R_E provides a DC negative feedback, but as far as an AC signal is concerned it should not appear in the circuit. Thus, we what the resistance to provide DC resistance and zero AC impedance. This is achieved by keeping a capacitor parallel to R_E whose AC impedance is far smaller than that of R_E. Usually

$$X_E = \frac{R_E}{10} \tag{9.21}$$

should do the trick. However, since this relation would stand only for a certain range of frequency, the amplifier's proper function (with no AC negative feedback and achieving of optimum gain) would be restricted to certain frequency, obtained from eqn(9.21). Finally, we now study the most popular transistor biasing circuit, the "voltage divider biasing circuit".

9.4   Voltage Divider Biasing Circuit

The forth biasing circuit is the voltage divider circuit shown in fig(9.4). One has to appreciate the development and superiority of each circuit. As shown, in all the biasing circuits, the gain and biasing base resistance (R_B) is related to R_C, the collector resistance. Thus, a designer always has to take care on selection of the collector resistance. Since selected R_B can influence the circuits input impedance (to be discussed in following chapter). Such restrictive selection on impedance is not approved off. We shall show that the voltage divider circuit gives a greater degree of freedom on selection of the base resistance.

To analyze the circuit shown in fig(9.4), imagine the input and output side of the transistor is being driven by two independent voltage sources, as shown in fig(9.5a). The Thevenin equivalent of the input side can be easily worked out with the Thenvenin resistance being

$$R_{TH} = \frac{R_1 R_2}{R_1 + R_2} \tag{9.22}$$

and the effective DC voltage works out as

$$V_{TH} = \left(\frac{R_2}{R_1 + R_2}\right) V_{CC} \tag{9.23}$$

Figure(9.5b) shows the equivalent circuit of the voltage divider biasing circuit, which makes our calculations for designing easier. The input and output side voltage equation of the equivalent circuit are given as

$$V_{TH} = I_B R_{TH} + V_{BE} + I_E R_E \qquad (9.24)$$

and

$$V_{cc} = I_C R_C + V_{CE} + I_E R_E \qquad (9.25)$$

The saturated collector current is given as

$$I_{Csat} = \frac{V_{cc}}{(R_C + R_E)} \qquad (9.26)$$

with

$$I_E \sim I_C \sim \beta I_B$$

Since the 'Q' point is half way in the transistor's active region (IV characteristics), the operating collector current is half the saturated collector current and is given as

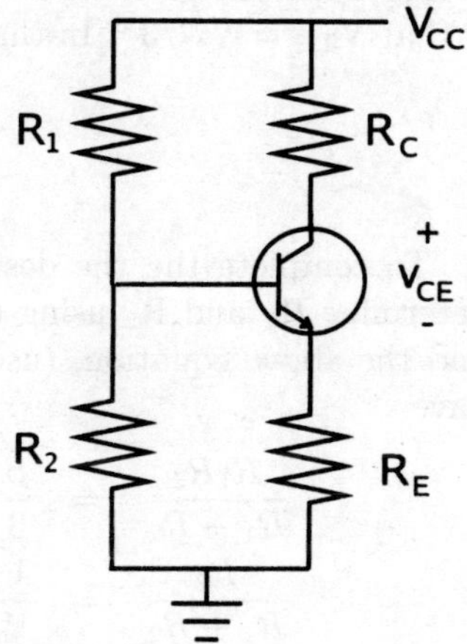

Figure 9.4: *The npn transistor is biased by a voltage divider circuit.*

$$I_{CQ} = \frac{V_{cc}}{2(R_C + R_E)} \qquad (9.27)$$

hence, the required base current is

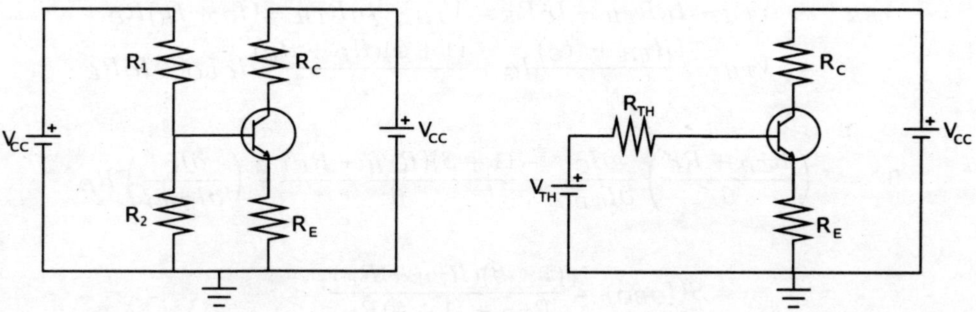

Figure 9.5: *The voltage divider can be reduced to it's Thevenin equivalent circuit for easier analyses.*

$$I_B = \frac{V_{cc}}{2\beta(R_C + R_E)}$$

From the input voltage equation one can work out R_{TH} as (assume $V_{BE} \sim 0$)

$$R_{TH} = 2\beta(R_E + R_C)\frac{V_{TH}}{V_{cc}} - (1 + \beta)R_E$$

The Thevenin's resistance is easily selected from the graph shown in fig(9.6). Obviously, only positive values of R_{TH} are of physical significance. As you can see, the circuit gives the designer the choice of vast range, for the biasing R_{TH} resistance. The thumb rule for selection is that $V_{TH} = V_{cc}/3$.[3] In that case

$$R_{TH} \sim \frac{\beta}{3}(2R_C - R_E)$$

To complete the the designing, we now have to determine R_1 and R_2 using the choice $V_{TH} = V_{cc}/3$ and the above equation, (use eqn 9.22 and 9.23) we have

$$\frac{R_1 R_2}{R_1 + R_2} = \frac{\beta}{3}(2R_C - R_E)$$

$$\frac{R_2}{R_1 + R_2} = \frac{V_{TH}}{V_{cc}} = \frac{1}{3}$$

gives

$$R_1 = \beta(2R_C - R_E)$$

$$R_2 = \frac{\beta}{2}(2R_C - R_E)$$

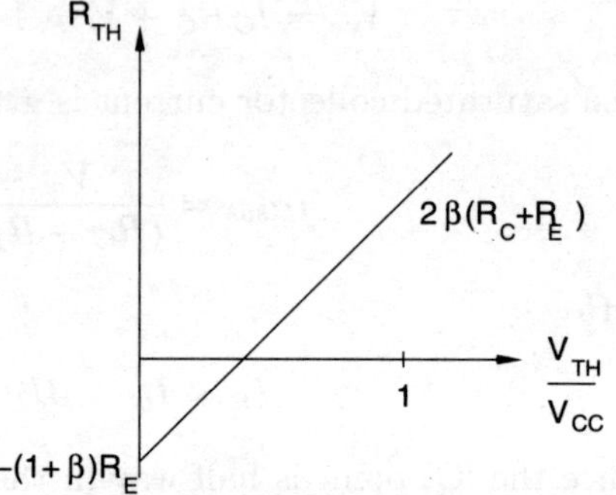

Figure 9.6: *The selection of* R_{TH} *with an voltage divider bias circuit is done by a proper selection of* V_{TH}.

Calculating the stability factor follows the same steps as for the preceding circuits, writing the input side voltage equation in terms of expressions of interest and differentiating.

$$V_{BE} = V_{TH} - I_B R_{TH} - I_E R_E = V_{TH} - I_B R_{TH} - (I_C + I_B)R_E$$

$$= V_{TH} - \frac{(R_{TH} + R_E)}{\beta}I_C + \frac{(1+\beta)(R_B + R_E)}{\beta}I_{CBO} - I_C R_E$$

$$0 = \left(\frac{R_{TH} + R_E}{\beta}\right)\frac{\partial I_C}{\partial I_{CBO}} - \frac{(1+\beta)(R_{TH} + R_E)}{\beta} + \left(\frac{\partial I_C}{\partial I_{CBO}}\right)R_E$$

$$S(I_{CBO}) = \frac{(1+\beta)(R_{TH} + R_E)}{R_{TH} + (1+\beta)R_E}$$

$$0 = +\left(\frac{R_{TH} + R_E}{\beta^2}\right)I_C - \frac{R_{TH} + R_E}{\beta}S(\beta) - \frac{1}{\beta^2}I_{CBO}(R_{TH} + R_E) - R_E S(\beta)$$

$$S(\beta) = \frac{(R_{TH} + R_E)(I_C - I_{CBO})}{\beta R_{TH} + \beta(1+\beta)R_E}$$

---

[3]The freedom of choice is important. The thumb rule is just a indicative value. User can obviously choose a value which give R_{TH} positive

Differentiating the modified input equation with respect to V_{BE} we get

$$1 = -\left(\frac{R_{TH} + R_E}{\beta}\right) S(V_{BE}) - R_E S(V_{BE})$$

$$S(V_{BE}) = -\frac{\beta}{R_{TH} + (1 + \beta)R_E}$$

The success of the voltage divider circuit is in removing the amplifier's dependence on it's transistor's parameters. This can be verified even without undertaking the exercise of calculating the stability factor 'S(β)'. We can show that the judicious choice of resistances can make the transistor amplifier circuit independent of the transistor parameter β. Repeating the calculation of the amplifer's operating point, from the input side equation

$$V_{TH} = I_B R_{TH} + V_{BE} + I_E R_E$$

we find the base current as

$$I_B = \frac{V_{TH} - V_{BE}}{R_{TH} + \beta R_E}$$

The collector current is related to the base current, hence the collector current can be written as

$$\begin{aligned}
I_C &= \frac{\beta(V_{TH} - V_{BE})}{R_{TH} + \beta R_E} \\
&= \frac{V_{TH} - V_{BE}}{R_E + R_{TH}/\beta}
\end{aligned}$$

For $R_E \gg R_{TH}/\beta$, the transistor's operating point collector current would be

$$\begin{aligned}
I_{CQ} &= \frac{V_{TH} - V_{BE}}{R_E} \\
&\approx \frac{V_{TH}}{R_E}
\end{aligned}$$

The collector-emitter voltage would also be found to be independent of β,

$$\begin{aligned}
V_{CEQ} &= V_{CC} - I_{CQ}(R_c + R_E) \\
&= V_{CC} - \frac{V_{TH}}{R_E}(R_C + R_E)
\end{aligned}$$

Thus, the voltage divider biasing circuit is a very useful circuit and one can now appreciate why most of the transistor circuits one will see henceforth is designed with this type of biasing.

Before proceeding to calculate the various resistances required to bias the transistor, one has to choose the collector resistance. Since this resistance influences the amplifier's gain, the value of the resistance is determined by the required voltage gain. In the following chapter we study the AC model of the transistor from which we relate the gain to the resistance. In the next chapter we develop the required theory, however, here without considering the voltage gain requirement we investigate the two methods used by circuit designers in designing the voltage divider biasing circuit. A final example is considered with gain requirements using the gain expression as a black box.

Circuit Designing

Method I

(i) Select an appropriate nominal operating point (I_{CQ}, V_{CEQ} and I_B). This selection would give R_E.

(ii) Arbitrarily select R_{TH} 10 to 15 times R_E. This selection makes sure of suitable stability for the circuit. As

$$S(I_{CBO}) = \frac{(1+\beta)(R_{TH}+R_E)}{R_{TH}+(1+\beta)R_E}$$

$$= \frac{(1+\beta)(1+\frac{R_{TH}}{R_E})}{\frac{R_{TH}}{R_E}+(1+\beta)}$$

As can be seen for a small value of R_{TH}/R_E, the stability factor is equal to 1. However if $R_{TH}/R_E \sim \infty$, the stability factor is large $(1+\beta)$. If R_{TH}/R_E is large and yet $(1+\beta) > R_{TH}/R_E$, we have

$$S(I_{CBO}) = \frac{(1+\beta)(1+\frac{R_{TH}}{R_E})}{\frac{R_{TH}}{R_E}+(1+\beta)}$$

$$= \frac{(1+\beta)(\frac{R_{TH}}{R_E})}{(1+\beta)}$$

$$= \frac{R_{TH}}{R_E}$$

Thus, for the made selection of R_{TH}, the stability factor would be 10 to 15.

(iii) Calculate V_{TH}, using eqn(9.24), followed by

(iv) evaluating required R_1 and R_2 using eqn(9.22) and eqn(9.23), which gives

$$R_1 = R_{TH}\frac{V_{CC}}{V_{TH}}$$

and

$$R_2 = \frac{R_{TH}R_1}{R_1 - R_{TH}}$$

Method II

Design of a transistor voltage amplifier with voltage divider biasing circuit, as shown in fig(9.4), is fairly simple by this method. Resistor R_E provides DC feedback to stabilize the emitter current and hence the operating point of the transistor. Hence, the first thing is to decide an appropriate emitter current. An emitter current of around 1 mA is usually quite satisfactory for audio amplification. Then select an emitter voltage (V_E). The higher this voltage, the greater is the emitter current stability in the presence of temperature variation and for variation in the value of h_{fe}. A value of V_E around 1 to 2 volts is quite satisfactory. If the supply voltage V_{cc} is large, V_E can be taken as 2volts and for small V_{cc}, a value of 1volts for V_E can be considered sufficient. The required R_E then would depend on the demanded voltage and current

$$R_E = \frac{V_E}{I_E}$$

Resistors R_1 and R_2 form a voltage divider which sets the base reference voltage. For proper biasing of the base-emitter, the voltage across R_2 should be greater than the voltage drop across the emitter resistance by 0.7volts in case of a silicon transistor. This is simply the forward voltage drop across the base to emitter diode junction. Hence, for a germanium transistor, this is close to 0.2volts higher than that at the voltage drop across the emitter resistance. Resistors R_1 and R_2 have to be selected such that it draws a current from the supply that is about 10 times the required base current. Let the sum of the voltages across the emitter resistance and forward biased base-emitter be designated V_B, we calculate R_1 and R_2 as follows:

$$\begin{aligned}
R_1 &= \frac{(V_{CC} - V_B)}{10 I_B} \\[2mm]
&= \frac{h_{fe}(V_{CC} - V_B)}{10 I_E}
\end{aligned} \qquad (9.28)$$

(The base current is equal to the collector current divided by h_{fe}, also note that collector current is nearly equal to emitter current.)

$$\begin{aligned}
R_2 &= \frac{V_B}{9 I_B} \\[2mm]
&= \frac{h_{fe} V_B}{9 I_E}
\end{aligned} \qquad (9.29)$$

R_2 calculation has divided by $9I_B$ and not as one tenth of the current (I_B) since I_B passes into the base itself.[4] We now work out a value for R_C so that the operating point is set correctly. As far as the signal is concerned, the available supply voltage is $(V_{CC} - V_E)$ and to make use of equal voltage swing either side of the operating point, the operating point is set half way between V_{CC} and V_E. For R_C, we calculate as follows:

$$\begin{aligned}
V_{CC} &= I_E R_c + V_{CE} + V_E \\[2mm]
I_E R_c &= V_{CC} - \frac{V_{CC}}{2} - V_E
\end{aligned}$$

$$R_C = \frac{(V_{CC} - 2V_E)}{2 I_E} \qquad (9.30)$$

---

[4]Applying KCL

$$10 I_B = I_B + 9 I_B$$

Arbitrarily selecting R_{TH} 10 to 15 times R_E in Method I achieves this. The selection makes sure the transistor's R_E does not load the Thevenin equivalent power supply.

Example 9.2: An transistor has a $\beta=100$. Design a common- emitter amplifier (as shown in fig 9.4). The collector current should be 1mA. Use a 12 volt DC supply. Calculate all of the components to properly bias the transistor and to get collector current as 1mA .

Since the operating current is 1mA, the saturation current for mid point operation would be twice of this. From

$$I_{Csat} = \frac{V_{CC}}{R_C + R_E}$$

$$2mA = \frac{12}{R_C + R_E}$$

the total resistance on the output side should be 6KΩ. Assuming for suitable feedback to take place the voltage across R_E is 2 volts. Then

$$R_E = \frac{V_E}{I_{CQ}}$$

$$= 2K\Omega$$

Hence, $R_C = 4K\Omega$. Now, take $R_{TH} = 10R_E$, we have $R_{TH} = 20K\Omega$. In

$$V_{TH} = I_B R_{TH} + V_{BE} + V_E$$

$$= \frac{I_{CQ}}{\beta} R_{TH} + V_{BE} + V_E$$

$$= \frac{2mA}{100} 20K + 0.7v + 2v$$

$$V_{TH} = 2.9 volts \tag{9.31}$$

Which gives $R_1 = 82.75K\Omega$ and $R_2 = 26.4K\Omega$.

Answer $R_1 = 82.75K\Omega$, $R_2 = 26.4K\Omega$, $R_C = 4K\Omega$ and $R_E = 2K\Omega$.

Example 9.3: Draw the whole circuit including generator and load. Why is it necessary to AC couple the generator and the load to the amplifier with capacitors? What value of capacitors would be suitable?

A amplifier circuit when practically used need additional capacitors, C_1 and C_2. The capacitors are isolation capacitors. The AC impedance offered by C_1 is negligible to the input frequency, however, it acts as an open circuit to any DC offset voltage that might be present in the audio source. This prevents any DC voltage from source entering the transistor base, thus leaving designed DC bias point unaffected. C_2, similarly isolate the circuit from the load resistor, i.e. while allowing the AC signal to reach the load it prevents any current division of DC collector current, I_C.

The impedance offered by the capacitors at AC frequency of interest should satisfy the condition,

$$X_{Ce} << R_E$$

and X_{C1} and X_{C2} should be far less than the transistor's input and output impedance respectively.

While Example 9.2 showed how a designer would calculate the operating point and select the various resistances for designing a transistor amplifier, the calculations would not be the same if you are investigating a circuit already designed. Infact the exercise is more that of back calculation. In the following example we study how this is done.

Example 9.4: Find the operating point of the transistor circuit shown in fig(9.7). Where in the active region is the 'Q' point placed? Given $\beta = 100$, $V_{cc} = 12v$, $R_1 = 60K\Omega$, $R_2 = 10K\Omega$, $R_c = 5K\Omega$ and $R_E = 1K\Omega$

The Thevenin's equivalent input circuit is considered and the input equation is written as

$$V_{TH} = I_B R_{TH} + V_{BE} + I_E R_E$$

With the given values of V_{CC}, R_1 and R_2 we have

$$R_{TH} = \frac{R_1 R_2}{R_1 + R_2} = 8.6 K\Omega$$

$$V_{TH} = \left(\frac{R_2}{R_1 + R_2}\right) V_{CC} = 1.714v$$

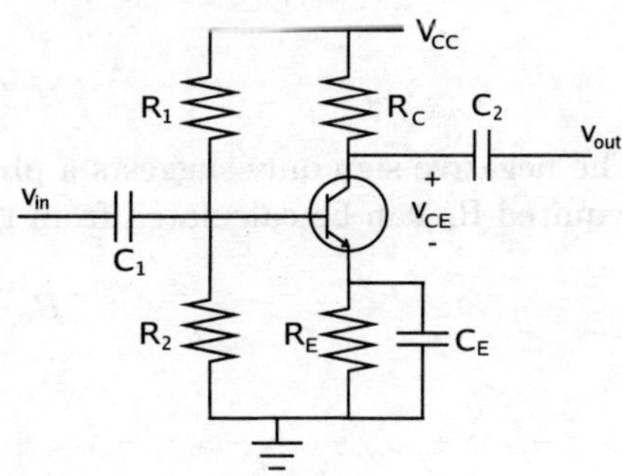

Figure 9.7: *The transistor biased with a voltage divider circuit with all the required capacitors as in a practical amplifier circuit.*

As we have no information as to the type of the transistor, i.e. whether it's Silicon or Germanium, we can consider $V_{BE}=0$ and also can safely say $I_E = I_C$. Hence, we now can find the base current as

$$\begin{aligned}
V_{TH} &= I_B R_{TH} + V_{BE} + I_E R_E \\
V_{TH} &= I_B R_{TH} + V_{BE} + \beta I_B R_E \\
1.714 &= (8.6 + 100) I_B
\end{aligned}$$

giving the base current to be $15.8\mu A$. Thus, the collector current is $1.58mA$ (βI_B). The collector-emitter voltage now is calculated from

$$\begin{aligned}
V_{CEQ} &= V_{CC} - I_C(R_C + R_E) \\
&= 12 - 1.58 \times 6 = 2.53v
\end{aligned}$$

The operating point of the transistor is circuit shown in fig(9.7) is

$$\boxed{\text{Answer } (V_{CEQ}, I_{CQ}) = (2.53v, 1.58mA)}$$

The point of intersection of the load line with the 'y' axis is the maximum current allowed in the designed circuit and found for $V_{CE} = 0$. We have

$$
\begin{aligned}
I_{Csat} &= \frac{V_{CC}}{R_C + R_E} \\
&= 2mA
\end{aligned}
$$

Comparing the co-ordinates of the operating point with the values of V_{CC} and I_{Csat}, it is clear that the transistor is biased between it's saturation region and it's active regions center.

$\boxed{\text{Example 9.5:}}$ Usually, a designer is interested not only to bias the transistor properly but also the design an amplifier. Thus, the value of R_c is decided by the required gain. Consider that a transistor with $h_{fe} = 50$ and $h_{ie} = 460\Omega$[5] has been provided and an amplifier of gain 400 is to be designed.

The task in hand would be to select proper resistances that would, not only bias the transistor but also achieve the objective of gain. In the following chapter it would be shown that the gain of the transistor amplifier is given as

$$A_v = -\left(\frac{h_{fe}}{h_{ie}}\right) R_c$$

The negative sign only suggests a phase difference between the input and output voltages. The required R_c can be calculated from this equation as

$$
\begin{aligned}
R_c &= \left(\frac{h_{ie}}{h_{fe}}\right) A_V \\
&= \frac{0.46}{50} \times 400 \approx 3.7k\Omega
\end{aligned}
$$

For having a suitable feedback, we demand $V_E = 0.6v$. Hence,

$$
\begin{aligned}
I_{CQ} R_E &= 0.6v \\
\frac{I_{Csat}}{2} R_E &= 0.6v
\end{aligned}
$$

As for the operating point to lie around the middle of the active region, $I_{CQ} = I_{Csat}/2$. Hence, in the output equation

$$V_{CC} = I_{Csat} R_c + I_{Csat} R_E$$

gives I_{Csat} as

$$
\begin{aligned}
I_{Csat} &= \frac{V_{CC} - I_{Csat} R_E}{R_c} \\
&= \frac{16 - 1.2}{3.7K} = 4mA
\end{aligned}
$$

---

[5]The meanings of these symbols would be explained in following chapters. They appear on the data-sheet shown in fig(8.9)

The operating DC current is $I_{CQ} = 2$mA, giving $R_E = 300\Omega$. The operating current I_{CQ} is determined by the base current I_B. To have $I_{CQ} = 2$mA, the base current has to be $I_B = I_{CQ}/\beta$ $=40\mu$A. This current is achieved by proper designing of the potential divider.

$$V_{TH} = I_B R_{TH} + V_{BE} + I_E R_E$$

Let us assign $V_{TH} = 4$v, then

$$4 = 0.04(R_{TH}) + 0.7 + 0.6$$
$$giving\ R_{TH} = 67.5 K\Omega$$

Using,

$$R_1 = R_{TH}\frac{V_{CC}}{V_{TH}}$$

and

$$R_2 = \frac{R_{TH}R_1}{R_1 - R_{TH}}$$

we select $R_1 = 270K\Omega$ and $R_2 = 90K\Omega$.

Consider you were analyzing the circuit of fig(9.8) and have to determine currents I_1 and I_2. Then firstly you would have to solve the voltage loop equation

$$V_{R_2} = V_{BE} + V_E$$
$$= 0.7 + 0.6 = 1.3v$$

The current I_2 is equal to $1.3v/90K\Omega \approx 14.4\mu$A. Usin KCL on the base node, we get $I_1 = 40+14.5 = 54.5\mu$A.

While the circuit works fine and gives the required gain, the stability factors would suggest circuit has left much more to desire. As an exercise calculate the various stability factor for the designed circuit.

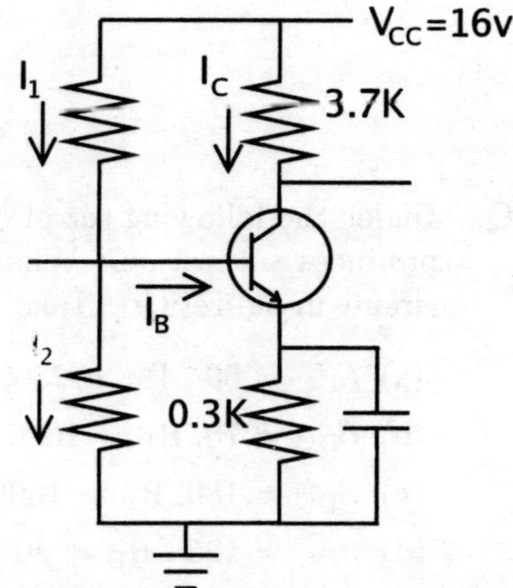

Figure 9.8: *A transistor amplifier circuit designed based on knowledge gained in this chapter.*

Exercise

Q1. For the silicon ($V_{BE} = 0.7$ V) transistor shown in the circuit in figure(9.9), determine whether transistor is in it's saturation region if

(a) $\beta_{DC} = 50$, $V_{BB} = 6.5$ V, $R_B = 2500\ \Omega$, $R_C = 500\ \Omega$, $V_{CC} = 6.0$ V, and $V_{CE(sat)} = 0.0$V;

(b) $\beta_{DC} = 75$, $V_{BB} = 4.0$ V, $R_B = 10$ kΩ, $R_C = 100\ \Omega$, $V_{CC} = 4.0$ V, and $V_{CE(sat)} = 0.3$V;

(c) $\beta_{DC} = 180$, $V_{BB} = 3.0$ V, $R_B = 16$ kΩ, $R_C = 400$ Ω, $V_{CC} = 12.0$, and $V_{CE(sat)} = 0.5$V;

(d) $\beta_{DC} = 125$, $V_{BB} = 3.9$ V, $R_B = 40$ kΩ, $R_C = 2.0$ kΩ, $V_{CC} = 9.5$, and $V_{CE(sat)} = 0.4$V.

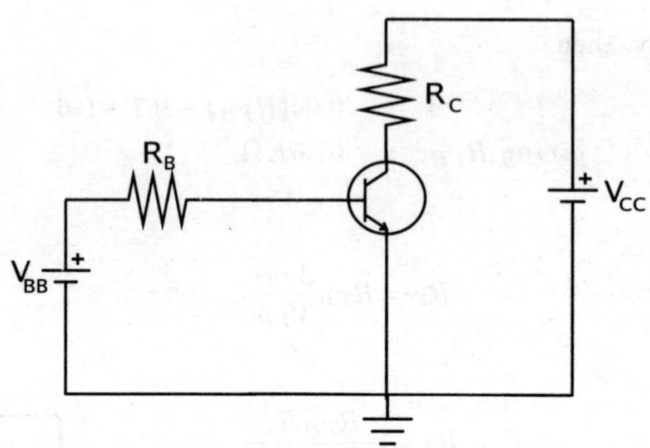

Figure 9.9:

Q2. Under the following set of conditions determine $I_{C(sat)}$. What is the minimum of I_B that produces saturation. What minimum value of V_{BB} produces saturation. Refer to the circuit in figure(9.9). Take $V_{BE} = 0.7$ V.

(a) $\beta_{DC} = 50$, $R_B = 2500$ Ω, $R_C = 500$ Ω, $V_{CC} = 6.0$ V, and $V_{CE(sat)} = 0.0$ V;

(b) $\beta_{DC} = 75$, $R_B = 10000$ Ω, $R_C = 100$ Ω, $V_{CC} = 4.0$ V, and $V_{CE(sat)} = 0.3$ V;

(c) $\beta_{DC} = 180$, $R_B = 16000$ Ω, $R_C = 400$ Ω, $V_{CC} = 12.0$ V, and $V_{CE(sat)} = 0.5$ V;

(d) $\beta_{DC} = 125$, $R_B = 40$ kΩ, $R_C = 2.0$ kΩ, $V_{CC} = 9.5$ V, and $V_{CE(sat)} = 0.4$ V.

Q3. For the following sets of conditions determine the Q-point. Determine the maximum peak value of the base current I_B . If the maximum value is exceeded, will the signal be driven into cutoff, or saturation, or both? Take $V_{BE} = 0.7$ V and $V_{CE(sat)} = 0.0$ V.

(a) $\beta_{DC} = 50$, $V_{BB} = 1.0$ V, $R_B = 2500$ Ω, $R_C = 500$ Ω, $V_{CC} = 6.0$ V;

(b) $\beta_{DC} = 25$, $V_{BB} = 1.0$ V, $R_B = 2500$ Ω, $R_C = 500$ Ω, $V_{CC} = 6.0$ V;

(c) $\beta_{DC} = 75$, $V_{BB} = 1.0$ V, $R_B = 2500$ Ω, $R_C = 500$ Ω, $V_{CC} = 6.0$ V.

Q4. The circuit (fig 9.10) shows a base-biased transistor. For the following sets of conditions determine the Q-point. Determine the maximum peak value of the base current. If the maximum value is exceeded, will the signal be driven into cutoff, or saturation, or both? Take $V_{BE} = 0.7$ V and $V_{CE(sat)} = 0.0$ V.

(a) $\beta_{DC} = 200$, $R_B = 75$ kΩ, $R_C = 200$ Ω, and $V_{CC} = 15.0$ V;

(b) $\beta_{DC} = 100$, $R_B = 75$ kΩ, $R_C = 200$ Ω, and $V_{CC} = 15.0$ V;

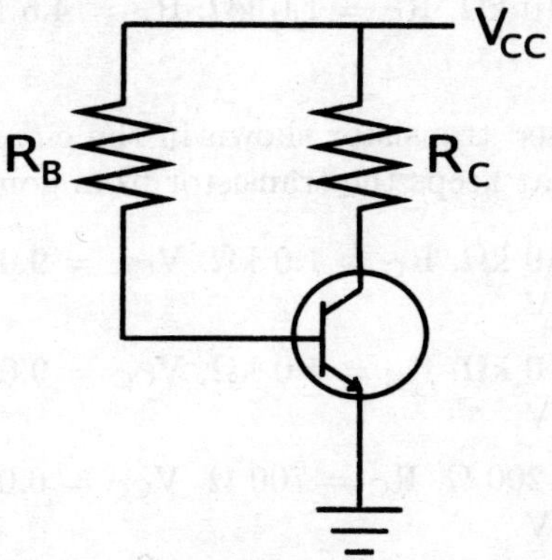

Figure 9.10:

(c) $\beta_{DC} = 300$, $R_B = 75$ kΩ, $R_C = 200$ Ω, and $V_{CC} = 15.0$ V.

Q5. The cicuit (fig 9.11) shows an npn emitter-biased transistor. For the following sets of conditions determine the Q-point. Determine the maximum peak value of the base current. If the maximum value is exceeded, will the signal be driven into cutoff, or saturation, or both? Take $V_{CE(sat)} = 0.0$ V.

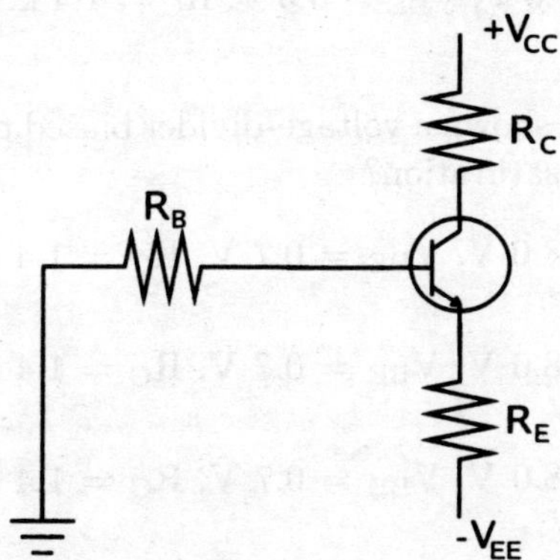

Figure 9.11:

(a) $\beta_{DC} = 150$, $R_B = 10$ kΩ, $R_C = 1.0$ kΩ, $R_E = 3.0$ kΩ, $V_{CC} = 9.0$ V, $V_{EE} = 9.0$ V, and $V_{BE} = 0.7$ V;

(b) $\beta_{DC} = 300$, $R_B = 10$ kΩ, $R_C = 1.0$ kΩ, $R_E = 3.0$ kΩ, $V_{CC} = 9.0$ V, $V_{EE} = 9.0$ V, and $V_{BE} = 0.6$ V;

(c) $\beta_{DC} = 150$, $R_B = 10$ kΩ, $R_C = 1.0$ kΩ, $R_E = 1.5$ kΩ, $V_{CC} = 9.0$ V, $V_{EE} = 9.0$ V, and $V_{BE} = 0.7$ V;

(d) $\beta_{DC} = 150$, $R_B = 10$ kΩ, $R_C = 1.0$ kΩ, $R_E = 4.5$ kΩ, $V_{CC} = 9.0$ V, $V_{EE} = 9.0$ V, and $V_{BE} = 0.7$ V.

Q6. For the npn emitter-biased transistor shown in the circuit (fig 9.11), what would be the minimum value of R_E that keeps the transistor from going into saturation?

(a) $\beta_{DC} = 150$, $R_B = 10$ kΩ, $R_C = 1.0$ kΩ, $V_{CC} = 9.0$ V, $V_{EE} = 9.0$ V, $V_{BE} = 0.7$ V, and $V_{CE(sat)} = 0.5$ V,

(b) $\beta_{DC} = 300$, $R_B = 10$ kΩ, $R_C = 1.0$ kΩ, $V_{CC} = 9.0$ V, $V_{EE} = 9.0$ V, $V_{BE} = 0.6$ V, and $V_{CE(sat)} = 0.4$ V,

(c) $\beta_{DC} = 220$, $R_B = 1200$ Ω, $R_C = 700$ Ω, $V_{CC} = 6.0$ V, $V_{EE} = 6.0$ V, $V_{BE} = 0.7$ V, and $V_{CE(sat)} = 0.7$ V.

Q7. The circuit in fig(9.4) shows a transistor with a voltage-divider bias. For the following sets of conditions determine the Q-point. Determine the maximum peak value of the base current. If the maximum value is exceeded, will the signal be driven into cutoff, or saturation, or both? Take $V_{CE(sat)} = 0.0$ Volts.

(a) $\beta_{DC} = 100$, $V_{CC} = 8.0$ V, $V_{BE} = 0.7$ V, $R_C = 1.4$ kΩ, $R_E = 600$ ΩW, $R_1 = 25$ kΩ, and $R_2 = 12$ $k\Omega$;

(b) $\beta_{DC} = 200$, $V_{CC} = 8.0$ V, $V_{BE} = 0.6$ V, $R_C = 1.4$ kΩ, $R_E = 600$ Ω, $R_1 = 25$ kΩ, and $R_2 = 12$ kΩ;

(c) $\beta_{DC} = 300$, $V_{CC} = 8.0$ V, $V_{BE} = 0.6$ V, $R_C = 1.4$ kΩ, $R_E = 600$ Ω, $R_1 = 25$ kΩ, and $R_2 = 12$ kΩ.

Q8. In the circuit of fig(9.4), is shown a voltage-divider biased transistor. What is the minimum value of R_2 which causes saturation?

(a) $\beta_{DC} = 100$, $V_{CC} = 8.0$ V, $V_{BE} = 0.7$ V, $R_C = 1.4$ kΩ, $R_E = 6.0$ kΩ, $R_1 = 25$ kΩ, and $V_{CE(sat)} = 0.0$ V;

(b) $\beta_{DC} = 200$, $V_{CC} = 8.0$ V, $V_{BE} = 0.7$ V, $R_C = 1.4$ kΩ, $R_E = 6.0$ kΩ, $R_1 = 25$ kΩ, and $V_{CE(sat)} = 0.0$ V;

(c) $\beta_{DC} = 300$, $V_{CC} = 8.0$ V, $V_{BE} = 0.7$ V, $R_C = 1.4$ kΩ, $R_E = 6.0$ kΩ, $R_1 = 25$ kΩ, and $V_{CE(sat)} = 0.5$ V;

Q9. An ideal npn bipolar junction transistor has a $\beta = 150$. Use it to design a common-emitter amplifier (fig 9.4) to work at a frequency of 10KHz. The collector current should be 2mA. Use a 10 volt DC supply. Show how to bias the transistor to get the 2mA current. Discuss the calculation of all components.

Q10. Calculate the stability factors, $S(I_{CBO})$, $S(\beta)$ and $S(V_{BE})$ for the circuit shown in fig(9.7). Also comment on the stability of the given circuit.

Q11. What meanings do the terms $S(I_{CBO})$, $S(\beta)$ and $S(V_{BE})$ hold?

Q12. What is meant by thermal run away?

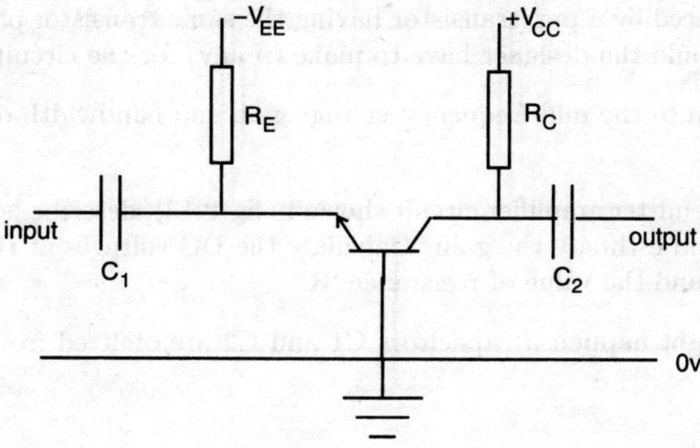

Figure 9.12:

Q13. Give three reasons for biasing a transistor.

Q14. What is the name given to the DC biasing scheme shown in figure(9.12).

Q15. Draw a small-signal equivalent circuit for the amplifier shown in figure(9.12).

Q16. Explain what is meant by the term "mid-point bias" or phrase "selecting Q point at the center" when one talks of amplifier circuits. Also explain why such biasing is usually adopted. Illustrate the bias point associated with midpoint bias on a sketch of current-voltage characteristics.

Q17. A npn transistor is used in a voltage divider circuit. Due to aging, the transistor burns out

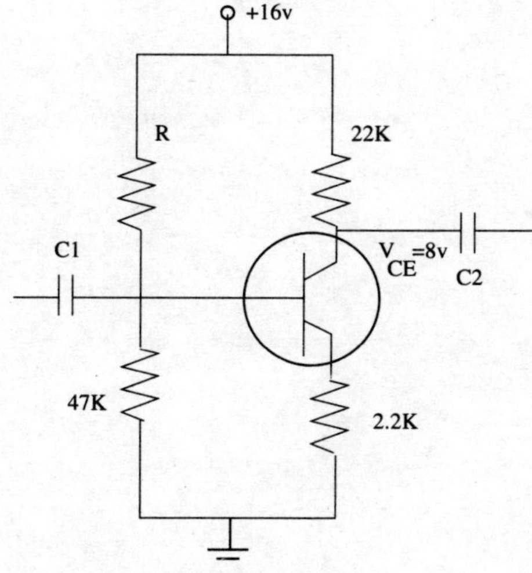

Figure 9.13:

and is to be replaced by a pnp transistor having the same transistor parameters (h_{fe} etc.). What changes would the designer have to make (if any) for the circuit to work properly?

Q18. What will happen to the mid-frequency voltage gain and bandwidth of an amplifier if C_E is removed?

Q19. For the common emitter amplifier circuit shown in fig(9.13), describe how the circuit works as an amplifier and estimate the gain. Calculate the DC voltages at the base and emitter of the transistor and the value of resistance 'R'.

Q20. Explain what might happen if capacitors C1 and C2 are omitted from the circuit shown in fig(9.13).

Chapter 10

Transistor Models

In the last chapter, we learnt how to bias the transistor and select a convenient operating point. Once, the point is selected, the transistor can be used as an amplifier by giving an AC input signal. The transistor would have both, a DC signal and an AC signal acting on it. The calculations for analysing the transistor circuit hence, would be involved. To analysis the transistor, it would be considerably simplified if we can treat the DC as a background and subtract it out (in our mathematical equations[1]). This would be possible using "***superposition theorem***". However, superposition theorem is only valid for linear circuital elements and our transistor is not a linear device. The way to overcome this problem is to approximate the transistor as a linear device. This is possible when we restrict the working of our transistor is a very small region of it's active region.[2] To achieve this we have to make sure that the input ac signal is small. This ensures that on applying the input AC signal the operating point (Q point) is only displaced slightly along the load line within the active region. This restriction we are imposing on the model, i.e. the amplitude of the input signals have to be small, gives our model the name "***small signal transistor models***".

To simplify the understanding of the transistor we are hence going to develop a model based on the black box concept (fig 10.1). We do the analysis by sending in a test voltage at the input and measuring the resulting input, output current and output voltage. Once we know this, we can create a circuit which mimics and behave identical to the transistor. These steps taken up to develop an equivalent circuit is nothing new and has been employed while converting 'T' section circuits to 'π' and visa versa (chapter 1). Basically, there are four parameters of interest v_1, i_1, v_2 and i_2 (input voltage and current along with output voltage and current). Typically, various models maybe developed by considering two of these parameters as depending on the values of the remaining two independent variables. Of these three of basic importance as far as

---

[1]suppose you are interested in measuring the intensity of light given out by a bulb. For this you would be required to use a photo-meter. If you do this measurement in daylight, the reading would be a result of the daylight and the light from the bulb. One possibility is you do the experiment in a dark room, i.e. you are forcing the background zero. The other possibility is you first measure the background light by switching off the bulb. After which, switch on the bulb and make the measurement. The intensity of the bulb would be obtained from substractng the two reading taken. This obviously is subjected to the condition that the reading of the photo-meter is directly proportional to the incident intensity, i.e. the photo-meter is a linear measuring device

[2]This is just like considering a small flat region on the surface of our curved earth

electronics industry is concerned is discussed here. Namely,

$$v_1 = f(i_1, v_2)$$
$$i_2 = f(i_1, v_2)$$

i.e. the input voltage and output current depends on the values of input current and output voltage. This gives a model of the transistor which is called the "**hybrid model**" of the transistor.

Similarly, the **admittance model** of a transistor is developed with

$$i_1 = f(v_1, v_2)$$
$$i_2 = f(v_1, v_2)$$

where the input and output currents depend on the input and output voltage. The **impedance model** of the transistor results from mathematical modelling that assumes the input and output voltages depend on the input and output currents, i.e.

$$v_1 = f(i_1, i_2)$$
$$v_2 = f(i_1, i_2)$$

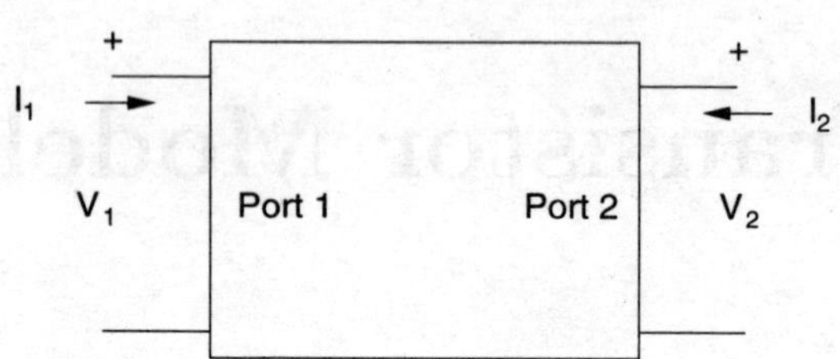

Figure 10.1: *Any two port network (two input pins and two output pins) can be treated as a black box, whose internal circuit is unknown. However, one can gain knowledge about the circuit by giving a test input voltage and measuring output voltage and input and output currents.*

Physical meaning can only be abscribed by the reader to these mathematical expressions as we develop the various models.

10.1   Hybrid Model and h Parameters

Consider, the input voltage and output current as dependent variables of the independent (variables) input current and output voltage. Mathematically, this is expressed as

$$v_1 = f(i_1, v_2)$$
$$i_2 = f(i_1, v_2) \tag{10.1}$$

Note, only small alphabets have been used. The current signals in the transistor are a mixture of AC and DC. The small AC signal combines with the DC biasing currents. It either adds or subtracts from the DC level. Using partial differential equation we can understand how the input voltage and output current varies with baring input current and output voltage.

$$dv_1 = \frac{\delta v_1}{\delta i_1} di_1 + \frac{\delta v_1}{\delta v_2} dv_2$$

$$di_2 = \frac{\delta i_2}{\delta i_1} di_1 + \frac{\delta i_2}{\delta v_2} dv_2 \tag{10.2}$$

The small variation represents the ac signal, with the dc background being filtered out. Under small signal approximation these variations (of above equation) can be considered to be constant.

That is, a linear relationship between the input voltage and output voltage as with the case of input voltage and input current etc. This linearity comes from the linearity of the transistor characters demanded by the small signal model. For picturizing the model, the small AC signal can be taken as slow time varying DC levels and hence voltages and currents have been represented in capitals. The above equations can be re-written as

$$V_1 = h_{11}I_1 + h_{12}V_2$$
$$I_2 = h_{21}I_1 + h_{22}V_2 \tag{10.3}$$

Based on our definition of eqn(10.3), practical measurements of h_{11}, h_{12}, h_{21} and h_{22} are done using DC sources. The above justifications are used to relate between experimentally measured entities and those mathematically used to describe/ model the transistors.

The above explanation were niceties required for mathematical completeness. Definitions that follow are important for circuit designers and students of electronics.

$$h_{11} = \left(\frac{V_1}{I_1}\right)_{V_2=0} = h_i$$

$$h_{12} = \left(\frac{V_1}{V_2}\right)_{I_1=0} = h_r$$

$$h_{21} = \left(\frac{I_2}{I_1}\right)_{V_2=0} = h_f$$

$$h_{22} = \left(\frac{I_2}{V_2}\right)_{I_1=0} = h_o \tag{10.4}$$

The term h_{11} has the dimensions of impedance. Since, the ratio is of the input voltage to the input current, it is the input impedance of the black box when the output voltage is zero ($V_2 = 0$). This measurement is made with the output shorted. Thus, h_{11} is the input impedance when the output is shorted. The transistor industry represents h_{11} by h_i,[3] where subscript 'i' represents impedance.

Similarly, h_{12} is the ratio is of the input voltage to the output voltage for open input (i.e. $I_1 = 0$). This might sound strange, as to how can you apply an input voltage, have input current zero and yet achieve an output voltage! Well, the measurement is done by giving input on the output side a voltage V_2 and then measuring the resulting voltage V_1 on the input side with a multimeter whose input impedance is infinite. Since, the multimeter's impedance is infinite, current drawn would be zero. As the measurement is made in the "reverse" (opposite) method and the expression is the "reverse" (inverse) of gain, the transistor industry represents h_{12} as h_r, where the subscript 'r' represents reverse.

The term h_{21} is the ratio of the output to the input current when output is shorted. Unlike the last term, this term is the current gain in the forward direction, hence, the transistor industry represents h_{12} as h_f, with the subscript 'f' obviously represents forward.

---

[3]The 11, 12, 21, and 22 subscripts are usually easier to remember for beginners as not only do they represent their position in a matrix array, the first subscript is same as that of the term of the numerator while the second subscript is the same as the denominators. However, one will obviously have to appreciate and follow industry conventions.

The last term, h_{22} is the ratio of the output current to output voltage with open input. The value is determined experimentally, similar to the method adopted for determining h_{12}. The dimensions suggests this term to be the output admittance. The subscript 'o' given implies it is the "output" admittance of the transistor. All the parameters have different dimensions and a model describing the transistor based on these parameters is called the hybrid model and these terms the hybrid parameters. See fig(8.9) that lists the values of 'h' parameters of transistor BC107.

Transistor Parameters

In the last chapter, we had approximated the values of the transistors input and output impedence, it's current and voltage gain from the input and output characteristics. We how calculate the same using the hybrid model developed. These parameters would be first calculated only for the transistor (with a load resistance R_L, across which the output voltage would appear), followed by doing the adding the various biasing circuits and calculating features for the amplifier circuit.

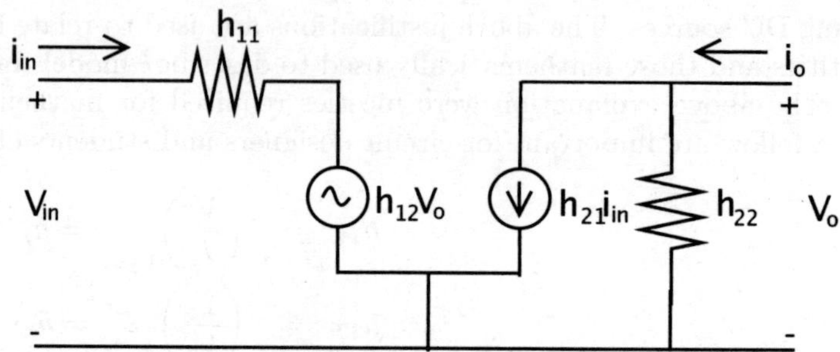

Figure 10.2: *The equation defined in eqn(10.3) describing the AC behavior of a transistor in small signal model is realised as a circuit.*

Figure (10.2) shows the hybrid equivalent circuit of a transistor. This is obtained by representing the input and output side mesh equation (eqn 10.4) as a circuit. The input voltage, (V_{in}), is a sum of the voltage appearing across the input impedance, h_{11}, due to the input current, I_{in} through it and the drop across the voltage source, $h_{12}V_o$. Similarly, The output current I_o breaks into two currents (a) current in the output admittance $h_{22}V_o$ which gives rise to a potential drop V_o at the output and (b) the current in the branch due to the current source $h_{21}I_{in}$.

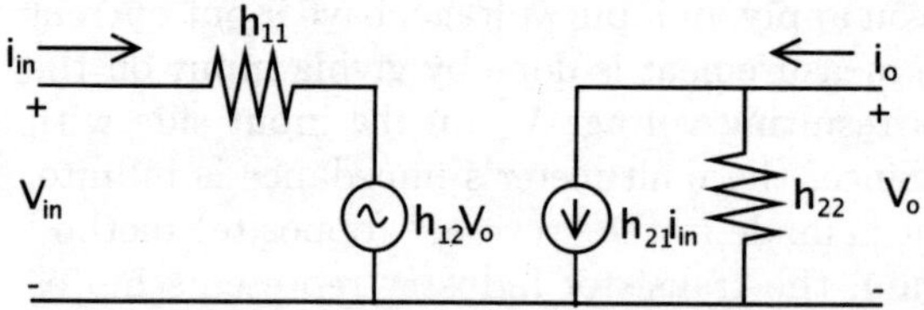

Figure 10.3: *Compact representation of a transistor's hybrid model shown in fig(10.2).*

Note that one of the pins of the transistor is common both to the input and output giving the two port like configuration. The short circuiting wire from the current and voltage sources to the common wire (ground) is redundant and can be done away with (see figure 10.3), however, it was included in fig(10.9) to emphasis that from the transistor's third pin to ground one might connect circuital elements. All three transistor configurations (CE, CB and CC) can also be represented by a hybrid model, however, the 'h' or hybrid parameters are vastly different in each mode. Thus, it is necessary to indicate which mode has been substituted by it's small signal hybrid equivalent model. This is done by including a second subscript, for example 'e' to imply the transistor being analyzed is in its CE mode. Similarly 'b'

and 'c' represent CB and CC respectively. Once you have identified the mode or configuration of the transistor, the subscript of the input and output currents would also have to be changed. For example the input and output currents for a CE mode transistor would be base current (i_b) and collector current (i_c) respectively. Also, since the emitter pin is common on both sides, i.e. is common to both input and output, the two loops representing the two equations are inter-connected (fig 10.3). As stated, there would be a modification in circuital representation (which we shall discuss later) when the biasing circuital elements would be added. Study the fig(10.4) to understand the difference in the hybrid models of the CE, CB and CC connected transistor.

Without using any extra circuital element, let us use fig(10.3) for estimating the output impedance of a transistor. The output current is given as

$$i_o = h_f i_{in} + h_o v_o \qquad (10.5)$$

The matrix subscript representation is left and the counter-part representation given in eqn(10.4) is being used. The input current can be written as (using eqn 10.3)

$$i_{in} = \frac{v_{in} - h_r v_o}{h_i} \qquad (10.6)$$

Using eqn(10.6) in eqn(10.5) we have

$$i_o = h_f \left(\frac{v_{in} - h_r v_o}{h_i} \right) + h_o v_o$$

$$= \frac{h_f}{h_i} \left(\frac{1}{A_v} - h_r \right) v_o + h_o v_o$$

where A_v is the voltage gain of the transistor. Re-arranging the above equation gives us the transistor's output impedence

$$R_o = \frac{1}{h_o + \frac{h_f}{h_i} \left(\frac{1}{A_v} - h_r \right)} \qquad (10.7)$$

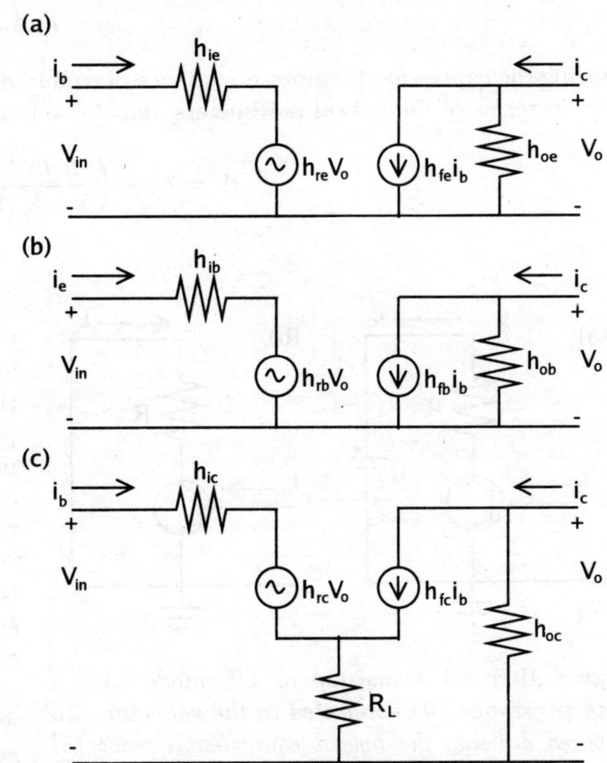

Figure 10.4: *The hybrid model of the transistor in (a) CE configuration, (b) CB configuration and (c) CC configuration.*

Similar calculation can be undertaken to estimate the input impedance of the transitor which would be on lines

$$R_{in} = \frac{v_{in}}{i_{in}} = \frac{h_i i_{in} + h_r v_o}{i_{in}}$$

$$= h_i + h_r \frac{v_o}{i_{in}} = h_i + h_r \frac{A_v v_{in}}{i_{in}}$$

$$= h_i + h_r A_v R_{in}$$

giving the transistor's input impedance as

$$R_{in} = \frac{h_i}{1 - h_r A_v} \tag{10.8}$$

The mathematics used would be used again in calculating the transistor's current gain, which is given as

$$A_i = \frac{i_o}{i_{in}} = h_f + h_o \frac{v_o}{i_{in}}$$
$$= h_f + h_o A_v R_{in}$$

Placing the expression for input impedance of eqn(10.8) we have the current gain of the transistor only in terms of the hybrid parameters and the voltage gain as

$$A_i = h_f + \left(\frac{h_o h_i A_v}{1 - h_r A_v} \right)$$

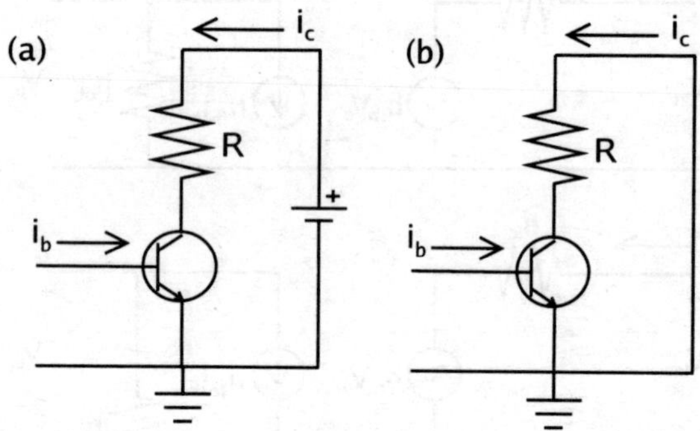

Figure 10.5: *A transistor in CE mode with a load resistance (R) connected to the collector. To proceed drawing the hybrid equivalent circuit (b) ground all DC sources.*

The expressions of input impedance, output impedance and current gain are all functions of the voltage gain. However, the voltage gain can not be written only interms of the hybrid parameters without making simplifing assumptions. Simplifications of the model is only discussed later. We now build on what we have learnt. The transistor is used for amplifying signals. In itself, the transistor is incapable of achieving this and is not useful. It is only along with circuital elements like resistances that the transistor can be used. Consider the simplest form of a transistor amplifier shown in fig(10.5a), where the transistor in CE mode is connected to a resistance. The output voltage is collected across this resistance.

To draw the hybrid equivalent circuit of the circuit shown in fig(10.5a), the following steps are undertaken. All the DC sources are grounded (shown in fig 10.5b). This is based on the assumption of the small signal model that the DC and AC signals are acting independently and hence superposition theorem can be used. Since we are involved in AC analysis of the transistor circuit, we ground DC sources. The transistor as a device is represented by the equivalent circuit of fig(10.3) after which the question is asked where in this circuit does the collector resistance (R) appears. From fig(10.5b), it is clear that this resistance is connected from the collector of the transistor and the ground. Hence, this load resistance would appear parallel to the transistor's output admittance (h_{22} or h_o), see fig(10.6).

Notice the path of the output current and output voltage as per the black box model is such that the current follows from lower potential to higher potential. Hence, as per the usual current-voltage convention, on applying the load resistance, the output voltage would be given as

$$v_o = -i_c R \qquad (10.9)$$

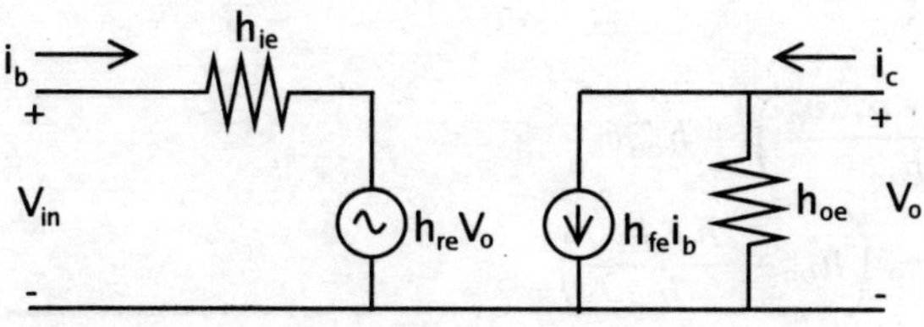

Figure 10.6: *The hybrid equivalent circuit of a transistor in CE mode with a load resistance (R) connected to the collector.*

In the CE mode, the input voltage is applied to the base w.r.t the emitter terminal. This results in a base current 'i_b', flowing into the base. The ratio of the input voltage to the base (input) current gives the input impedance of the transistor in CE mode, i.e.

$$\begin{aligned} R_{in} &= \frac{v_{in}}{i_b} = \frac{h_{ie}i_b + h_{re}v_o}{i_b} \\ &= h_{ie} + h_{re}\frac{v_o}{i_b} \end{aligned}$$

Substituting the output voltage (eq 10.9), we have

$$\begin{aligned} R_{in} &= h_{ie} + h_{re}\left(\frac{-i_c R}{i_b}\right) \\ R_{in} &= h_{ie} - h_{re}A_i R \qquad (10.10) \end{aligned}$$

Note, the input impedance R_{in} is different from the input impedance of the small signal model which is defined shorted output. No such restriction is imposed on R_{in}. Similarly, the ratio of output current to input current has not be substituted as h_{fe}, but A_i, since we have not shorted the output. h_{fe}, of course is defined as ratio of output to input current when output is shorted. Calculating A_i in terms of the newly defined terms, we have

$$\begin{aligned} A_i &= \frac{i_c}{i_b} = \frac{h_{fe}i_b + h_{oe}v_o}{i_b} \\ &= h_{fe} + h_{oe}\frac{v_o}{i_b} \end{aligned}$$

Again, substituting eq(10.9), the above equation is reduced to

$$\begin{aligned} A_i &= h_{fe} + h_{oe}\left(\frac{-i_c R}{i_b}\right) \\ &= h_{fe} - h_{oe}A_i R \end{aligned}$$

Collecting terms of A_i on the l.h.s we get

$$A_i = \frac{h_{fe}}{1 + h_{oe}R} \qquad (10.11)$$

The difference and the relation between A_i and h_{fe} can now be seen. For computing the output impedance of the transistor, we would require to replace the term i_b in the below equation with terms related to either i_c or v_o

$$i_c = h_{fe}i_b + h_{oe}v_o$$

For which, we can modify the input side equation and write

$$i_b = \frac{v_{in} - h_{re}v_o}{h_{ie}}$$

Hence, we write

$$
\begin{aligned}
i_c &= h_{fe}\left(\frac{v_{in} - h_{re}v_o}{h_{ie}}\right) + h_{oe}v_o \\
&= \frac{h_{fe}}{h_{ie}}v_{in} + v_o\left(h_{oe} - \frac{h_{fe}h_{re}}{h_{ie}}\right) \\
&= -\frac{h_{fe}v_o}{h_{ie}A_v} + v_o\left(h_{oe} - \frac{h_{fe}h_{re}}{h_{ie}}\right)
\end{aligned}
$$

Giving, the output impedance as

$$\frac{1}{R_o} = -\frac{h_{fe}}{h_{ie}}\frac{1}{A_v} + \left(h_{oe} - \frac{h_{fe}h_{re}}{h_{ie}}\right) \tag{10.12}$$

Equation(10.12) has a term we still have to compute, the voltage amplification of the transistor. For which we follow

$$
\begin{aligned}
A_v &= \frac{v_o}{v_{in}} = \frac{-i_c R}{v_{in}} \\
&= \frac{-i_c R}{h_{ie}i_b + h_{re}v_o} \\
&= \frac{-i_c R}{h_{ie}i_b - h_{re}i_c R}
\end{aligned}
$$

Dividing the numenator and denominator by i_c, we have

$$A_v = \frac{-A_i R}{h_{ie} - h_{re}A_i R}$$

Now to write the expression of gain in terms of experimentally determinable quantities we use eqn(10.11)

$$A_v = \frac{-\left(\frac{h_{fe}}{1+h_{oe}R}\right)R}{h_{ie} - h_{re}R\left(\frac{h_{fe}}{1+h_{oe}R}\right)} \tag{10.13}$$

Instead of simplifying this expression, the better option is to make realistic approximations and simplify our hybrid model of the transistor.

Simplification

The transistor is used as a device for amplification, both current and voltage is amplified in the case of transistor acting in common emitter mode. That would imply v_o/v_{in} would be an appreciably large quantity. In turn, h_{re} would be a small and negligible quantity. Similarly, we had established that the output impedance $(1/h_{oe})$ of the transistor in CE mode is very large. Thus, for a practical value of R_L applied to the transistor, which appears parallel to $1/h_{oe}$, we may neglect the influence of $1/h_{oe}$. Basically, the two assumptions

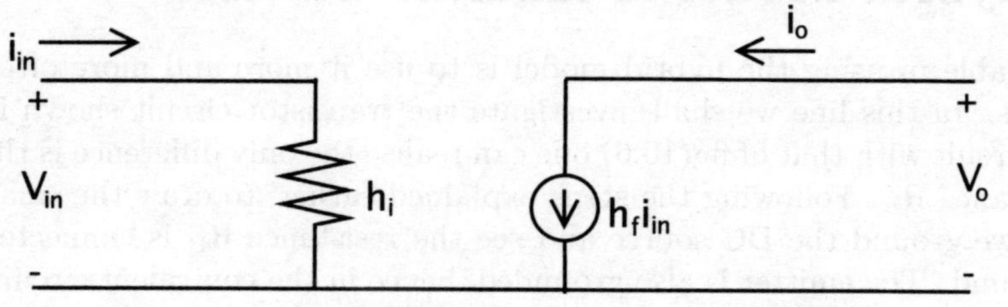

$$h_{re} \rightarrow 0$$
$$\frac{1}{h_{oe}} \gg R_L$$
$$h_{oe} \rightarrow 0$$

reduces the hybrid model of fig(10.2) to that shown in fig(10.7).

Figure 10.7: *The simplified hybrid model of a transistor.*

Using the simplified terms, the input-output impedance, as also current and voltage gain of the transistor expressed in eqn(10.10), eqn(10.11), eqn(10.12) and eqn(10.13) would reduce to

$$R_{in} = h_{ie}$$
$$A_i = h_{fe}$$

and

$$A_v = -\left(\frac{h_{fe}}{h_{ie}}\right) R_L \tag{10.14}$$

Using this we write the output impedance as

$$\frac{1}{R_o} = -\frac{h_{fe}}{h_{ie}}\frac{1}{A_v} + h_{oe}$$
$$= \frac{1}{R_L} + h_{oe}$$

i.e. the parallel combination of the load resistance and $1/h_{oe}$. Since, $1/h_{oe} \gg R_L$, the output impedance is equal to the applied load resistance, $R_o = R_L$.

Equation(10.14) is of interest. Note, the above equation obtained from the small signal model of the transistor, modifies the voltage gain relation obtained in the previous chapter (eqn 8.15). For designing a transistor amplifier to obtain maximum voltage gain, seeing eqn(10.14), one might be tempted to select a transistor with the highest current transfer ratio h_{fe}. In fact, this would be of no avail as voltage gain is essentially dependent on two factors, namely the collector current (I_c) and the output load resistance or collector resistance (R_L, whichever is lower), but not h_{fe}.

If you disagree with this statement, all you have to do is substitute value of h_{ie} of eqn(8.7) in eqn(10.14) and you have the voltage gain as

$$A_v = -h_{fe}\frac{R_L I_C}{h_{fe}(26mV)}$$
$$= -\frac{R_L I_C}{26mV}$$

As we had stated, the calculation shows that the voltage gain only involves the values of R_L and I_C and not h_{fe}.

10.2   Using the Hybrid Model to Analyse Circuits

The only way to get comfortable in using the hybrid model is to use it more and more often on different transistor circuits. In this line we shall investigate the transistor circuit shown in fig(9.1). On comparing this circuit with that of fig(10.6) one can realise the only difference is the inclusion of the biasing resistance R_B. Following the steps explained earlier, to draw the small signal model of this circuit, we ground the DC source and see the resistance R_B is connected between the base and the ground. The emitter is also grounded, hence in the equivalent circuit, R_B is connected between base and emitter. The equivalent model of our circuit is shown in fig(10.8).

The input impedance of the transistor circuit in fig(10.8) is given as

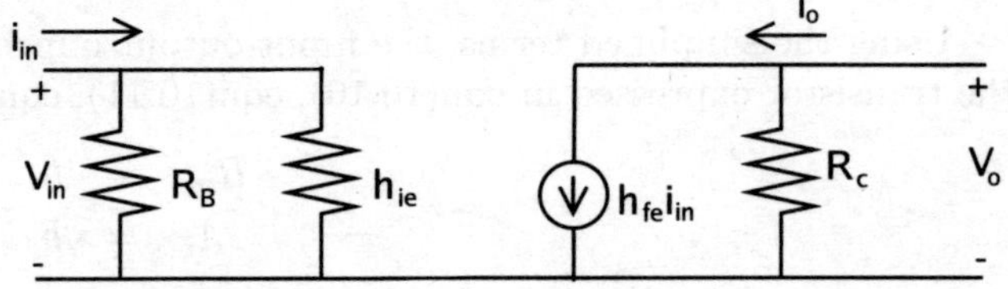

$$R_{in} = \frac{v_{in}}{i_b}$$

$$= \frac{(h_{ie} \| R_B) i_b}{i_b}$$

$$= h_{ie} \| R_B$$

Figure 10.8: *The simplified equivalent hybrid model of a circuit shown in fig(9.1).*

The current gain of the circuit is would not be modified (remains h_{fe}) as the output circuit and inturn output current is not influenced by R_B. However, the voltage gain of this circuit is modified and is given as

$$\begin{aligned}
A_v &= \frac{v_o}{v_{in}} = \frac{-i_c R_c}{v_{in}} \\
&= \frac{-i_c R_c}{(h_{ie} \| R_B) i_b} \\
&= \frac{-A_i R_c}{(h_{ie} \| R_B)} \\
&= \frac{-h_{fe} R_c}{(h_{ie} \| R_B)}
\end{aligned}$$

Finally we examine the output impedance of the circuit. The input voltage is

$$v_{in} = (h_{ie} \| R_B) i_b$$

The base or input current can be calculated from this equation which can then be substituted in the expression for the output current, giving

$$\begin{aligned}
i_o &= h_{fe} \left(\frac{v_{in}}{h_{ie} \| R_B} \right) + h_{oe} v_o \\
&= \left[-\frac{h_{fe}}{(h_{ie} \| R_B) A_v} + h_{oe} \right] v_o
\end{aligned}$$

$$\approx \frac{-h_{fe}}{(h_{ie}\|R_B)A_v}v_o$$

thus, the output impedance is written as

$$R_o = -\frac{(h_{ie}\|R_B)A_v}{h_{fe}}$$
$$= R_c$$

The expression for the circuit's input impedance and voltage gain are comparable with the circuit considered in the previous section (without biasing resistance R_B). The only difference is that h_{ie} is replaced by $h_{ie}\|R_B$. The transistor's input impedance is small (in CE and CC mode), hence $h_{ie}\|R_B \longrightarrow h_{ie}$. Both expressions reduces to those of the previous section. Therefore, $R_{in} = h_{ie}$ and $A_v = -h_{fe}R_c/h_{ie}$.

Example 10.1: Calculate the voltage gain of the circuit shown in fig(10.9). Given $V_{CC} = 20v$. Take values of h_{fe} and h_{ie} from fig(8.9).

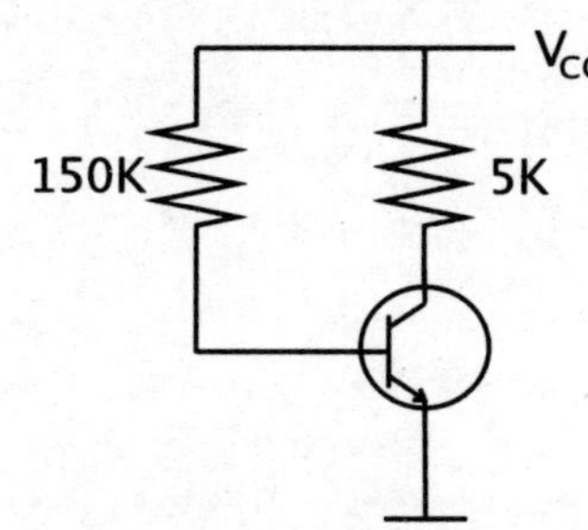

Figure 10.9: *The circuit for Example (10.1).*

The small signal AC equivalent circuit of this circuit is shown in fig 10.8. The input voltage is given as

$$v_{in} = i_b(R_B\|h_{ie}) \approx i_b h_{ie}$$

Also, the output voltage is given as

$$v_o = -i_c R_C$$

Thus, the gain of the amplifier is

$$A_v = -\frac{h_{fe}}{h_{ie}}R_C$$

To calculate the voltage gain we require the value of h_{ie} and h_{fe}. The datasheet in fig(8.9) specifies the same only for a given collector current. Hence, using DC loop equations, we have to find the collector current.

$$V_{CC} = V_{CE} + I_C R_C$$
$$= \frac{V_{CC}}{2} + I_C R_C$$

Hence, the collector current is 2mA. For this value of collector current the given value of h_{ie} is $\approx 1.6\text{K}\Omega$ and that of $h_{fe} \approx 125$. The gain can now be calculated and is found to be 391.

Answer $A_v \approx 391$

Consider the voltage divider biasing circuit shown in fig(9.4). Figure(10.10) shows the steps taken to reduce the amplifier circuit to it's hybrid equivalent circuit. The AC analysis of this circuit can now be done trivially. The two voltage dividing resistances, R_1 and R_2 come in parallel to the input circuit can can be considered as a single resistance $R_B = R_1 || R_2$. The input impedance may appear to be the parallel-series combination of three resistances at the input side. However, it is not so simple and evaluation is a bit more involved. This is because the resistance R_E is common in both the input and output circuits and hence two currents follow in it. The input voltage is given as

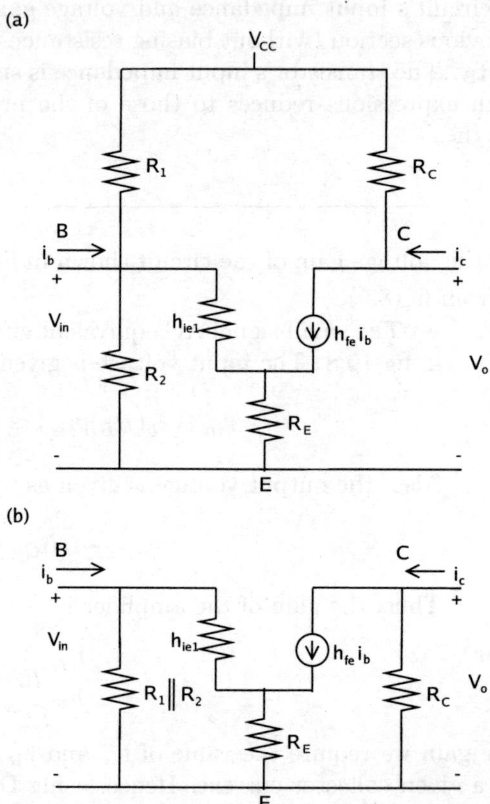

Figure 10.10: *First in the circuit of fig(9.4) (a) replace the transistor symbol by the simplified hybrid model of a transistor followed by grounding the DC source* V_{cc}. *The resistances* R_1, R_2 *and* R_c *turns out to be connected from base to ground while the emitter resistance is connected from emitter to ground, as shown in (b).*

$$v_{in} = R_B i_{b1} = h_{ie} i_{b2} + R_E (i_{b2} + i_c) \qquad (10.15)$$

where input current i_b breaks up into two

$$i_b = i_{b1} + i_{b2} \qquad (10.16)$$

From eqn(10.15), we have

$$i_{b1} = \frac{(h_{ie} + R_E)}{R_B} i_{b2} + \frac{R_E}{R_B} i_c$$

Therefore,

$$
\begin{aligned}
i_b &= \frac{(h_{ie} + R_E)}{R_B} i_{b2} + \frac{R_E}{R_B} i_c + i_{b2} \\
&= \frac{(h_{ie} + R_E)}{R_B} i_{b2} + \frac{R_E}{R_B} h_{fe} i_{b2} + i_{b2}
\end{aligned}
$$

where we have used the transistor base-collector current relation, $i_c = h_{fe} i_{b2}$, giving

$$i_{b2} = \frac{R_B i_b}{h_{ie} + (1 + h_{fe}) R_E + R_B} \tag{10.17}$$

hence, the input voltage (eqn 10.15) can be now rewritten as

$$
\begin{aligned}
v_{in} &= (h_{ie} + R_E) i_{b2} + R_E i_c \\
&= [h_{ie} + (1 + h_{fe}) R_E] i_{b2} \\
&= [h_{ie} + (1 + h_{fe}) R_E] \left[\frac{R_B i_b}{h_{ie} + (1 + h_{fe}) R_E + R_B} \right]
\end{aligned} \tag{10.18}
$$

giving the input resistance as

$$R_{in} = \frac{R_B [h_{ie} + (1 + h_{fe}) R_E]}{h_{ie} + (1 + h_{fe}) R_E + R_B} \tag{10.19}$$

Essentially, the input impedance is as a parallel connection of R_B is made to $h_{ie} + (1 + h_{fe}) R_E$. Thus, in this particular circuit the transistor's emitter resistor and it's current gain influences the input impedance. The current gain and the output impedance of the circuit remains h_{fe} and R_L respectively. However, the voltage gain would be effected as the resistor R_E acts as a negative feedback between the output and input. We now calculate the voltage gain using simplified eqn(10.18).

$$
\begin{aligned}
A_v &= \frac{v_o}{v_{in}} \\
&= \frac{-i_c R_L}{\left[\frac{R_B [h_{ie} + (1 + h_{fe}) R_E]}{h_{ie} + (1 + h_{fe}) R_E + R_B} \right] i_b} \\
&= \frac{-h_{fe} R_L i_{b2}}{\left[\frac{R_B [h_{ie} + (1 + h_{fe}) R_E]}{h_{ie} + (1 + h_{fe}) R_E + R_B} \right] i_b}
\end{aligned}
$$

Using eqn 10.17, we have

$$A_v = \frac{-h_{fe}R_L}{\left[\frac{R_B[h_{ie}+(1+h_{fe})R_E]}{h_{ie}+(1+h_{fe})R_E+R_B}\right]} \times \frac{R_B}{h_{ie}+(1+h_{fe})R_E+R_B}$$

$$= \frac{-h_{fe}R_L}{[h_{ie}+(1+h_{fe})R_E]} \qquad (10.20)$$

As is evident from the above equation, the larger the emitter resistance is, the larger is the negative feedback, which reduces the amplifier's gain. As discussed in the last chapter, this negative feedback is required to prevent any thermal runaway, caused by heating by the DC currents. The equation makes it evident this negative feedback also brings down the gain. Hence, what we require is a circuital element in R_E place, that presents a finite DC resistance however is an AC short. We achieve this by placing a capacitor in parallel to R_E. The DC resistance of this combination is R_E, while the AC impedance offered is zero.[4] Thus, the AC input impedance and voltage gain would not be the expression given by eqn(10.19) and eqn(10.20) respectively. They would be

$$R_{in} = \frac{R_B h_{ie}}{h_{ie}+R_B}$$

and

$$A_v = \frac{v_o}{v_{in}}$$

$$= \frac{-i_c R_L}{h_{ie}i_b} \approx \frac{-h_{fe}R_L}{h_{ie}}$$

$\boxed{\text{Example 10.2:}}$ Draw the hybrid model of the Darlington pair (see fig 8.29) and evaluate it's input impedence, current and voltage gain. Applying KVL on the input side we have

$$v_{in} = h_{ie1}i_{b1} + h_{ie2}i_{b2}$$

where i_{b2} is the current in impedance h_{ie2}. It is the sum of current from h_{ie1} and the current source $h_{fe1}i_{b1}$, hence the above equation reduces to

$$v_{in} = h_{ie1}i_{b1} + h_{ie2}(1+h_{fe1})i_{b1} \qquad (10.21)$$

Therefore, the input impedence is

$$R_{in} = h_{ie1} + h_{ie2}(1+h_{fe1})$$

for identical constituent transistors and assuming $h_{fe1} \gg 2$, the input impedance of the Darlington pair would be given as

$$R_{in} = h_{ie}(2+h_{fe})$$

$$= h_{ie}h_{fe}$$

---

[4]The task to draw the equivalent hybrid model for the circuit shown in fig(9.7) is left to the reader as an exercise

For calculating the current gain we consider the output current equation using KCL. The output current breaks into two currents

$$i_{out} = h_{fe1}i_{b1} + h_{fe2}i_{b2}$$

As can be understood from the derivation of the input impedance, the base current of the second transistor (i_{b2}) is the current flowing through h_{ie2}. This is given as

$$\begin{aligned} i_{b2} &= i_{b1} + h_{fe1}i_{b1} \\ &= i_{b1}(1 + h_{fe1}) \end{aligned}$$

Thus, the output current interms of the input current, i_{b1} is

$$i_{out} = h_{fe1}i_{b1} + h_{fe2}i_{b1}(1 + h_{fe1})$$

giving the current gain as

$$A_i = h_{fe1} + h_{fe2}(1 + h_{fe1})$$

again for identical transistors and $h_{fe1} \gg 2$, the current gain is given as

$$\begin{aligned} A_i &= h_{fe}(2 + h_{fe}) \\ &= h_{fe}^2 \end{aligned}$$

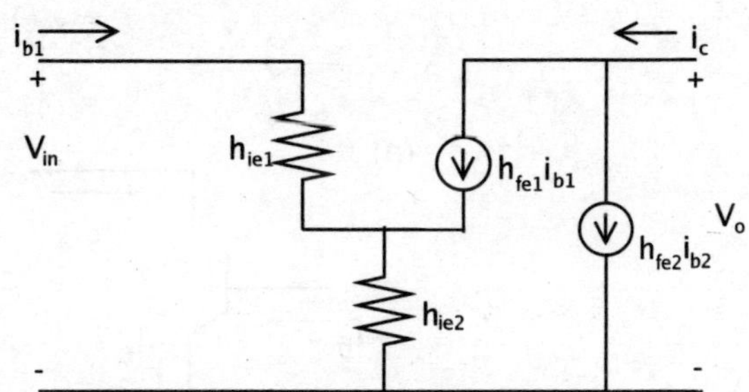

Figure 10.11: *The hybrid model of the Darlington pair of transistors shown in fig(8.29).*

The voltage gain is evaluated from the output voltage equation. Consider a load resistance R_L kept parallel to h_{oe}. Since, the output impedance of a transistor is large, the parallel combination of the load resistance and the transistor's output impedance would be equal to the smaller value resistance (R_L). The output voltage would be due to the flow of currents $h_{fe1}i_{b1}$ and $h_{fe2}i_{b2}$ through R_L. Therefore

$$\begin{aligned} v_o &= -h_{fe1}i_{b1}R_L - h_{fe2}i_{b2}R_L \\ &= -h_{fe1}i_{b1}R_L - h_{fe2}i_{b1}(1 + h_{fe1})R_L \\ &= -[h_{fe1} + h_{fe2}(1 + h_{fe1})]i_{b1}R_L \end{aligned}$$

Dividing the above equation with the expression for input voltage (eqn 10.21), we have

$$\begin{aligned} A_v &= -\frac{[h_{fe1} + h_{fe2}(1 + h_{fe1})]i_{b1}R_L}{h_{ie1}i_{b1} + h_{ie2}(1 + h_{fe1})i_{b1}} \\ &= -R_L\left[\frac{h_{fe1} + h_{fe2}(1 + h_{fe1})}{h_{ie1} + h_{ie2}(1 + h_{fe1})}\right] \end{aligned}$$

Following the assumptions made above, the voltage gain is

$$\begin{aligned} A_v &= -R_L\left[\frac{h_{fe}(2 + h_{fe})}{h_{ie}(2 + h_{fe})}\right] \\ &= -\frac{h_{fe}}{h_{ie}}R_L \end{aligned}$$

The derviations show that a Darlington pair has higher input impedance and current gain as compared to an individual transistor (increased by a factor of h_{fe}), while the voltage gain remains constant. These results complete the discussion on Darlington pair started in Chapter 8.

Uptil now, we have analysed the CE transistor amplifier using the AC small signal model. However, the model is equally applicable to CB and CC amplifiers. In fact the 'h' parameters of CB and CC configuration can be written in terms of the hybrid parameters of CE configuration. All that is required for conversion is the relation

$$\begin{aligned} i_e &= i_c + i_b \\ &= (1 + h_{fe})i_b \end{aligned} \tag{10.22}$$

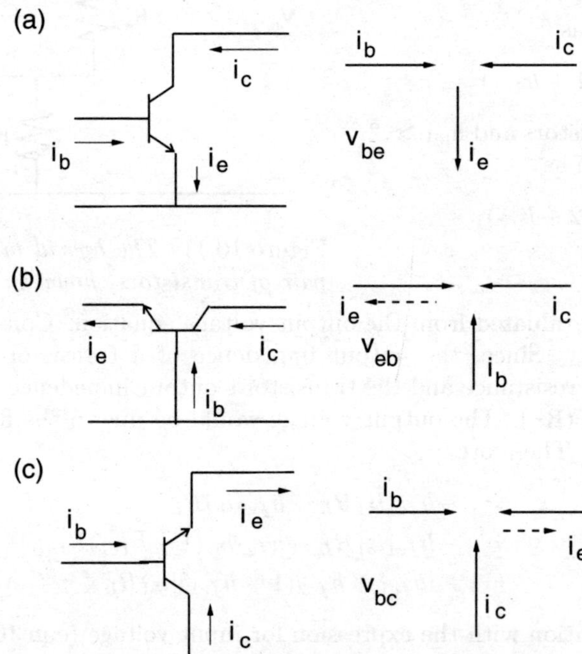

Figure 10.12: *The transistor in (a) CE, (b) CB and (c) CC configuration along with the KCL for the transistors. The dashed lines of the emitter current in (b) and (c) show the actual direction of current which however is opposite to the direction of input and output currents respectively of the two port model.*

and a basic understanding of which is the input voltage/ current and output voltage/ current in each configuration. This can be understood from fig(10.12).

$$\begin{aligned} v_{in} &= v_{be} = h_{ie}i_b \\ &= v_{eb} = h_{ib}i_{in} = -h_{ib}i_e \end{aligned}$$

From fig(10.12) it is clear that an AC input to a transistor in CE configuration is 'v_{be}' while in case of CB it is v_{eb}. Thus, the input of CE is -ive with respsect to CB's input. Also, while in CE mode the input current is $+i_b$, it is $-i_e$ in case of CB mode. Thus,

$$-(-h_{ib}i_e) = h_{ie}i_b$$

Using eqn(10.22), we have a relation between h_{ib} and h_{ie}.

$$h_{ib}(1 + h_{fe})i_b \;=\; h_{ie}i_b$$

$$\boxed{h_{ib} = \frac{h_{ie}}{1+h_{fe}}}$$

A relation between h_{fb} and h_{fe} can be evaluated from the current gain of the CB amplifier.

$$A_i \;=\; \frac{i_{out}}{i_{in}} = \frac{i_c}{i_e}$$

$$=\; \frac{h_{fe}i_b}{-(1+h_{fe})i_b}$$

the current gain is approximated as 'h_f', hence

$$\boxed{h_{fb} = \frac{-h_{fe}}{1+h_{fe}} \approx -1}$$

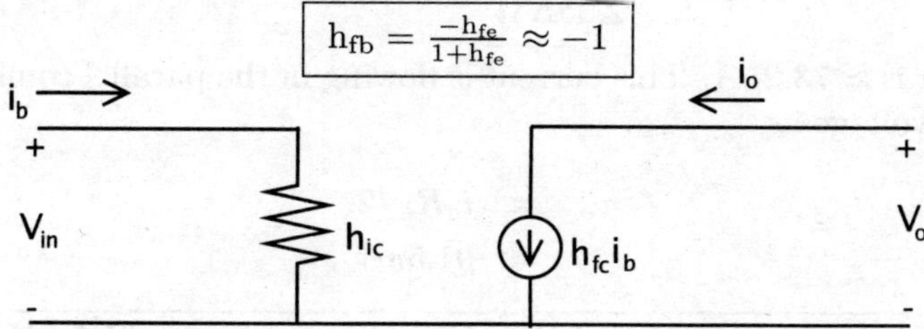

Figure 10.13: *The simplified hybrid model of the transistor in CC configuration.*

The -ive sign shows reversal in current at the output in common base. Since the small signal model on simplification only requires h_i and h_f, further derivations have been left out. Similar relationships between the 'h' parameters of a transistor in common collector configuration can be established with h_{ie} and h_{fe}. To the two port circuit (whether inside the black box you have a CE or a CC circuit) on giving identical input you have

$$h_{ic}i_b = h_{ie}i_b$$

or

$$h_{ic} = h_{ie}$$

The current gain of the CC in terms of it's emitter and base current (use fig 10.12) is given as

$$A_i \;=\; \frac{i_o}{i_{in}} = -\frac{i_e}{i_b}$$

hence,

$$h_{fc} = -(1 + h_{fe})$$

The performance of the transistor in various modes was compared in Table(8.1). The results obtained above are in agreement with those listed in the table.

Example 10.3: Assume the CE amplifier has an input impedance of 2KΩ and an output impedance far greater than the applied collector resistor of 2.5KΩ (this is the usual case). The output signal from the collector is delivered in to a load of 2500Ω. Is this a matched load? A signal generator at the input to the amplifier has an internal resistance of 50Ω and an emf of amplitude 1mV (i.e. the open circuit terminal voltage is 1mV) at a frequency of 10KHz. What is the AC voltage in the load? (take $\beta = 150$)

(a) This is a matched load because $R_L = R_C$.

The input voltage has been given as 1mV and the net resistance at the input is $R_{in} + R_s = 2000+50= 2050\Omega$, hence, the base current is given as

$$i_b = \frac{1mV}{2.05K\Omega} = 0.48\mu A$$

The collector current is $\approx 73.2\mu$A. This current is flowing in the parallel combination of R_C and R_L, giving the load voltage as

$$\begin{aligned}
v_o &= i_c R_L/2 \\
&= 91.5mV
\end{aligned}$$

The output voltage of the given CE amplifier is 91.5mV

10.3   Admittance Model and Y Parameters

This model assumes the output and input currents to be dependent on the input and output voltages. The functional form is represented as

$$\begin{aligned}
i_1 &= f(v_1, v_2) \\
i_2 &= f(v_1, v_2)
\end{aligned} \tag{10.23}$$

The variation of the currents with input and output voltages is given as

$$di_1 = \frac{\delta i_1}{\delta v_1}dv_1 + \frac{\delta i_1}{\delta v_2}dv_2$$

$$di_2 = \frac{\delta i_2}{\delta v_1}dv_1 + \frac{\delta i_2}{\delta v_2}dv_2 \tag{10.24}$$

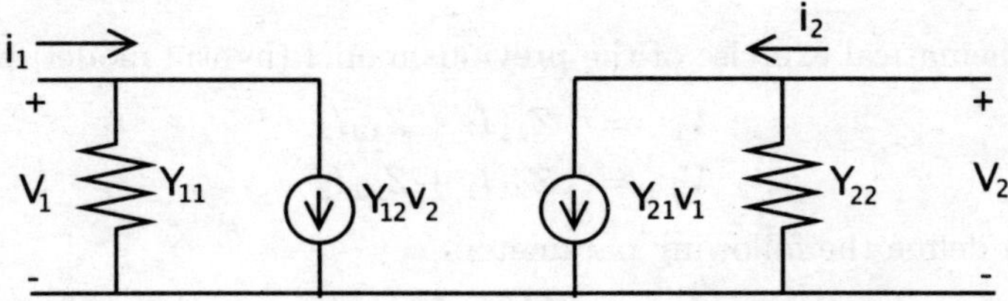

Figure 10.14: *The admittance model of a transistor.*

We follow the arguments used to develop the hybrid model and write

$$I_1 = Y_{11}V_1 + Y_{12}V_2$$
$$I_2 = Y_{21}V_1 + Y_{22}V_2 \tag{10.25}$$

defining,

$$Y_{11} = \left(\frac{I_1}{V_1}\right)_{V_2=0} = Y_i$$

$$Y_{12} = \left(\frac{I_1}{V_2}\right)_{V_1=0} = Y_r$$

$$Y_{21} = \left(\frac{I_2}{V_1}\right)_{V_2=0} = Y_f$$

$$Y_{22} = \left(\frac{I_2}{V_2}\right)_{V_1=0} = Y_o \tag{10.26}$$

Where Y_{11} (or Y_i) is the input admittance when the output is shorted and Y_{12} (Y_r) is the reverse admittance when the input is shorted. The reverse admittance is measured by giving a voltage V_2 at the output side and measuring the current (I_1) at the input side, while the input itself is shorted. The forward admittance (Y_{21} or Y_f) is similarly evaluated by measuring the output current I_2 while output is shorted and input is fed a voltage of V_1. Y_{22} (or Y_o) is the output admittance. The circuit modeled on eqn(10.25) is shown in fig(10.14).

10.4   Impedance Model and Z Parameters

In this model the input and output voltages are taken as dependent variables of the independent (variables) input and output currents. Mathematically, this is expressed as

$$v_1 = f(i_1, i_2)$$
$$v_2 = f(i_1, i_2) \tag{10.27}$$

The input and output voltage varies with varying input and output current as

$$dv_1 = \frac{\delta v_1}{\delta i_1}di_1 + \frac{\delta v_1}{\delta i_2}di_2$$

$$dv_2 = \frac{\delta v_2}{\delta i_1}di_1 + \frac{\delta v_2}{\delta i_2}di_2 \tag{10.28}$$

Repeating the mathematical exercise of the previous model (hybrid model) we have

$$V_1 = Z_{11}I_1 + Z_{12}I_2$$
$$V_2 = Z_{21}I_1 + Z_{22}I_2 \qquad (10.29)$$

From which we can define the following parameters

$$Z_{11} = \left(\frac{V_1}{I_1}\right)_{I_2=0} = Z_i$$

$$Z_{12} = \left(\frac{V_1}{I_2}\right)_{I_1=0} = Z_r$$

$$Z_{21} = \left(\frac{V_2}{I_1}\right)_{I_2=0} = Z_f$$

$$Z_{22} = \left(\frac{V_2}{I_2}\right)_{I_1=0} = Z_o \qquad (10.30)$$

Z_{11} has the dimensions of input impedance when the output is open, resulting in zero current flow in the output. The more popular representation is Z_i, where subscript 'i' represents input, while 'Z' obviously indicates we are talking of impedances.

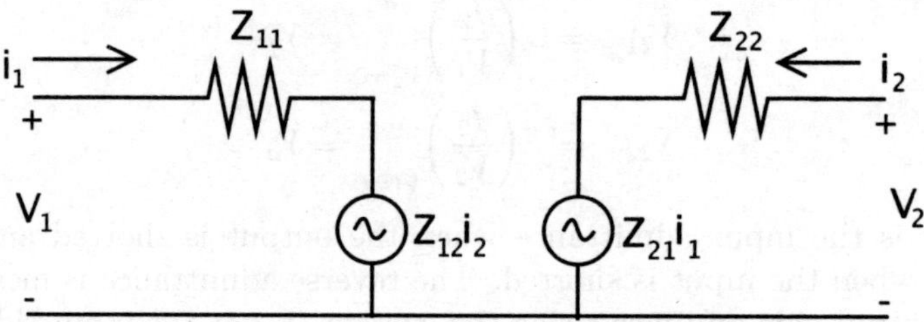

Figure 10.15: *The impedance model of a transistor.*

Similarly, Z_{12} or Z_r is the ratio is of the input voltage to the output current for open input (i.e. $I_1 = 0$). Again the dimensions are that of impedance, however, the voltage and current measurements are not of the same sides. Hence, the ratio is called the "reverse" impedance. The term Z_{21} or Z_f is the ratio of the output voltage to the input current when output is open. Unlike the last term, this term is the impedance in the forward direction, hence, the subscript 'f'. The last term, Z_{22} is the ratio of the output voltage to output current with open input. The dimensions suggests this term to be the output impedance. The subscript 'o' given implies it is the "output" impedance of the transistor.

Note, all four parameters of this model has the dimensions of impedances. Hence, the representation of the transistor (for that matter any circuit) by this model is called the "*impedance model*".

10.5   Relation Between Various Parameters

A examination of the transistor manufacturer's data-sheet shows that they only give the values of 'h' parameters and no mention is made of the 'Z' or 'Y' parameters. The reason for this is the popularity of the hybrid model as far as circuit analysist are concerned. Also, since these parameters are indicative of properties of the same transistor (same black box), there should obviously exist some relationship between these parameters. Hence, given the 'h' parameters, in principle one should be able to relate it to 'Z' and 'Y' parameter. We do exactly that in the following section.

10.5.1   Converting h to Z Parameters

A comparison of the definitions made in the two models show

$$
\begin{aligned}
Z_o &= \left(\frac{V_2}{I_2}\right)_{I_1=0} \\
&= \frac{1}{\left(\frac{I_2}{V_2}\right)_{I_1=0}} \\
&= \frac{1}{h_o}
\end{aligned}
\tag{10.31}
$$

Thus, the output impedance of the impedance model is just the inverse of the output admittance of the hybrid model

$$\boxed{Z_o = \frac{1}{h_o}}$$

The remaining three parameters can not be trivially found as not just do the ratios have to match but also the condition is which the measurement are made and hence, how the parameters are defined have to match. The hybrid model is obtained from the equations (10.3)

$$
\begin{aligned}
V_1 &= h_i I_1 + h_r V_2 \\
I_2 &= h_f I_1 + h_o V_2
\end{aligned}
$$

Writing the expression for V_2 from the second equation we have

$$
V_2 = \frac{-h_f I_1 + I_2}{h_o}
$$

The reader might have understood by now, since the impedance model parameters relate voltages with two different currents, the equation of the hybrid model has been re-written to reflect this. The forward imepdence Z_f can be obtained from the above equation by substituting $I_2 = 0$, we get

$$
\begin{aligned}
(V_2)_{I_2=0} &= -\frac{h_f}{h_o} I_1 \\
\left(\frac{V_2}{I_1}\right)_{I_2=0} &= -\frac{h_f}{h_o} \\
Z_f &= -\frac{h_f}{h_o}
\end{aligned}
\tag{10.32}
$$

Thus,

$$\boxed{Z_f = -\frac{h_f}{h_o}}$$

A substitution of the expression of V_2 in the first of the two expression involved in the hybrid model (10.3), we have

$$V_1 = h_i I_1 + \left(\frac{h_r}{h_o}\right)(I_2 - h_f I_1) \tag{10.33}$$

The equation relates input voltage to the input and output current and hence would allow us to find the relation of Z_i and Z_r with hybrid parameters. For the first relation, take output current zero, we have

$$(V_1)_{I_2=0} = h_i I_1 - \left(\frac{h_r}{h_o}\right)(h_f I_1)$$

$$\left(\frac{V_1}{I_1}\right)_{I_2=0} = h_i - \left(\frac{h_r h_f}{h_o}\right)$$

$$Z_i = h_i - \left(\frac{h_r h_f}{h_o}\right) \tag{10.34}$$

$$\boxed{Z_i = h_i - \left(\frac{h_r h_f}{h_o}\right)}$$

The forth relation is obtained by setting input current zero in eqn(10.33)

$$(V_1)_{I_1=0} = \left(\frac{h_r}{h_o}\right)(I_2)$$

$$\left(\frac{V_1}{I_2}\right)_{I_1=0} = \left(\frac{h_r}{h_o}\right)$$

$$Z_r = \left(\frac{h_r}{h_o}\right) \tag{10.35}$$

$$\boxed{Z_r = \left(\frac{h_r}{h_o}\right)}$$

Thus, if the hybrid model parameters of a transistor is given, a designer can convert it into an impedance model using the four relations obtained above.

10.5.2   Converting h to Y Parameters

As in the last case, we search for a 'Y' parameter, whose definition matches that of the 'h' parameter (this is best done by keeping both model's defined parameters along side each other). Close examination gives

$$Y_i = \left(\frac{I_1}{V_1}\right)_{V_2=0}$$

$$= \frac{1}{\left(\frac{I_1}{V_1}\right)_{V_2=0}}$$

$$= \frac{1}{h_i} \tag{10.36}$$

Thus, the input admittance of the circuit is just the inverse of the input impedance of the hybrid model

$$\boxed{Y_i = \tfrac{1}{h_i}}$$

h_f is defined for shorted output (i.e. $V_2 = 0$), thus to compare, put $V_2 = 0$ in eqn(10.25). Then,

$$I_2 = Y_{21}V_1$$
$$and, \ I_1 = Y_{11}V_1$$

dividing the above two equations, we get

$$h_{21} = \left(\frac{I_2}{I_1}\right)_{V_2=0} = \frac{Y_{21}}{Y_{11}}$$

using eqn(10.36), we have

$$Y_{21} = \frac{h_{21}}{h_{11}} \ or \ Y_f = \frac{h_f}{h_i} \tag{10.37}$$

$$\boxed{Y_f = \tfrac{h_f}{h_i}}$$

In eqn(10.25) substitute $I_1 = 0$. The equation reduces to

$$0 = Y_{11}V_1 + Y_{21}V_2 \tag{10.38}$$

which gives

$$\left(\frac{V_1}{V_2}\right)_{I_1=0} = h_r = -\frac{Y_{12}}{Y_{11}}$$

using eqn(10.36), we have

$$h_r = -Y_{12}h_i$$

$$\boxed{Y_r = -\tfrac{h_r}{h_i}}$$

For the last conversion, use eqn(10.38) and calculate V_1.

$$V_1 = -\frac{h_{12}}{h_{11}}V_2$$

This is voltage V_1 for current $I_1 = 0$. In expression $h_{21}V_1 + h_{22}V_2 = I_2$, if we substitute expression of V_1, we have

$$I_2 = \frac{h_{11}h_{22} - h_{12}h_{21}}{h_{11}}V_2$$

or

$$Y_o = \left(\frac{I_2}{V_2}\right)_{I_1=0} = \frac{h_{11}h_{22} - h_{12}h_{21}}{h_{11}}$$

In short

$$\boxed{Y_o = \frac{h_i h_o - h_r h_f}{h_i}}$$

Table 10.1: Lookup table to convert one parameter to the other.

to ↓ from ⟶	Z		Y		h	
Z			$\dfrac{Y_o}{\Delta Y}$	$-\dfrac{Y_r}{\Delta Y}$	$\dfrac{\Delta h}{h_o}$	$-\dfrac{h_r}{h_o}$
			$-\dfrac{Y_f}{\Delta Y}$	$\dfrac{Y_i}{\Delta Y}$	$-\dfrac{h_f}{h_o}$	$\dfrac{1}{h_o}$
Y	$\dfrac{Z_o}{\Delta Z}$	$-\dfrac{Z_r}{\Delta Z}$			$\dfrac{1}{h_i}$	$-\dfrac{h_r}{h_i}$
	$-\dfrac{Z_f}{\Delta Z}$	$\dfrac{Z_i}{\Delta Z}$			$\dfrac{h_f}{h_i}$	$\dfrac{\Delta h}{h_i}$
h	$\dfrac{\Delta Z}{Z_o}$	$\dfrac{Z_r}{Z_o}$	$\dfrac{1}{Y_i}$	$-\dfrac{Y_r}{Y_i}$		
	$-\dfrac{Z_f}{Z_o}$	$\dfrac{1}{Z_o}$	$\dfrac{Y_f}{Y_i}$	$\dfrac{\Delta Y}{Y_i}$		

10.6   Summary Table

The above two sections show how to convert 'h' parameters into 'Z' and 'Y' parameters. The reverse, i.e. 'Z' parameter into 'h' parameter and similarly 'Y' parameter into 'h' parameter, can be done. However, that mathematical exercise is best left as an assignment. To verify the results obtained compare with those listed in Table(10.1). For convenience of representation in Table(10.1), Δh represents $(h_i h_o - h_r h_f)$. Similarly, ΔY and ΔZ represents $(Y_i Y_o - Y_r Y_f)$ and $(Z_i Z_o - Z_r Z_f)$ respectively. Also, the relations are written in matrix representation, with element (1,1) representing the input parameter and so on.

Example 10.4: Find the h-parameters for the circuit shown in fig(1.29)

h_i is defined for output shorted, hence, Z_C is shorted. Thus, the impedance of the circuit is the parallel combination of Z_A and Z_B.

$$h_i = \frac{Z_A Z_B}{Z_A + Z_B}$$

Similarly for the shorted Z_C, we have to calculate h_f. If current I_1 is fed on port 1, the current in port 2, i.e. current in the short would be the current in Z_B. Thus, using current division equation

$$I_2 = \frac{Z_A}{Z_A + Z_B} \times I_1$$

$$h_f = \frac{Z_A}{Z_A + Z_B}$$

The remaining parameters are found by making measurements at port '2' keeping port '1' open (i.e. $I_1 = 0$). h_o is the output admittance measured by placing a multimeter at port '2' (port '1' open).

$$\frac{1}{h_o} = \frac{Z_C(Z_A + Z_B)}{Z_A + Z_B + Z_C}$$

$$h_o = \frac{Z_A + Z_B + Z_C}{Z_C(Z_A + Z_B)}$$

h_r is measured by calculating the voltage across Z_A (port '1') when a voltage is applied across Z_C. Using the potential divider equation

$$V_2 = \frac{Z_A}{Z_A + Z_B} V_1$$

$$h_r = \frac{V_1}{V_2} = \frac{Z_A + Z_B}{Z_A}$$

10.7 Transconductance Model

The transistor is mostly used in it's CE configuration. The analysis of this circuit is popularly done using the admittance model. This model is customized for the CE configuration and renamed as the ***Transconductance Model***. When the transistor in CE configuration is viewed as a two port device, the base current is the input current and the collector current is it's output current. Similarly, v_{be} and v_{ce} are the input voltage (v_1) and output voltage (v_2), respectively.

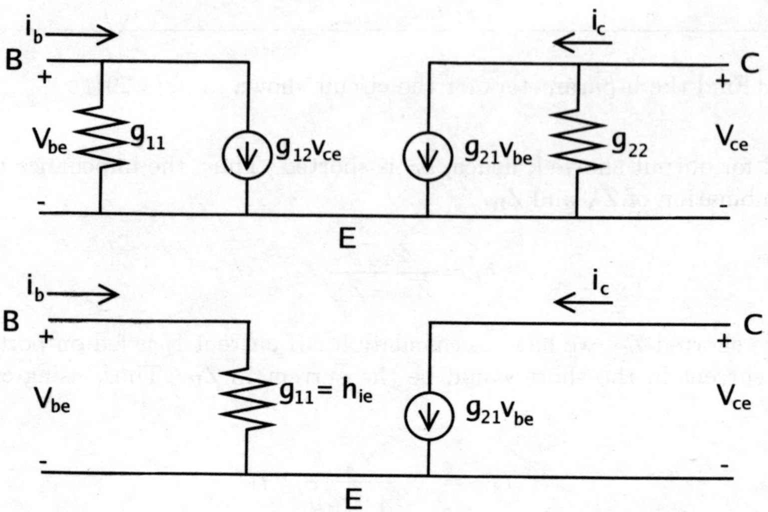

Figure 10.16: *The complete transconductance model of a transistor and it's simplified form.*

Hence, eqn(10.23) and the following definitions for the transistor in CE configuration are restated as

$$i_b = f(v_{be}, v_{ce})$$
$$i_c = f(v_{be}, v_{ce})$$

giving,

$$i_b = g_{11}v_{be} + g_{12}v_{ce}$$
$$i_c = g_{21}v_{be} + g_{22}v_{ce} \tag{10.39}$$

where

$$g_{11} = \left(\frac{i_b}{v_{be}}\right)_{v_{ce}=0}$$

$$g_{12} = \left(\frac{i_b}{v_{be}}\right)_{v_{be}=0}$$

$$g_{21} = \left(\frac{i_c}{v_{ce}}\right)_{v_{ce}=0} = g_m$$

$$g_{22} = \left(\frac{i_c}{v_{ce}}\right)_{v_{be}=0} \tag{10.40}$$

g_m is called the **transconductance**. Fig(10.16) shows the circuital realization of the equations (eqn 10.39). Simplifying based on results of the conversion table (Table 10.1), g_{11} and g_{22} are negligible. The simplified circuit is also shown in fig(10.16). This circuit highlights the importance of this model. *It relates the output current to the input voltage.* This model is of immense importance for studying FETs which is a voltage controlled device. However, even the transistor circuits at high frequencies are analyzed using a the transconductance model. The use of the model is discussed at length in the following chapter where the frequency response of the transistor is discussed.

10.8   T Model

The impedance model circuit can be re-arranged in the form of the alphabet 'T', to give what is called the T-model of the transistor. The T-model is preferred for simplifying and analyzing differential amplifiers (opamps) built with transistors. The first equation of eqn (10.29) can be written as

$$\begin{aligned} v_1 &= Z_{11}i_1 + Z_{12}i_2 \\ &= Z_{11}i_1 - Z_{12}i_1 + Z_{12}i_1 + Z_{12}i_2 \end{aligned}$$

i.e.

$$v_1 = (Z_{11} - Z_{12})i_1 + Z_{12}(i_1 + i_2) \tag{10.41}$$

Similarly, the second equation of eqn(10.29) is modified as

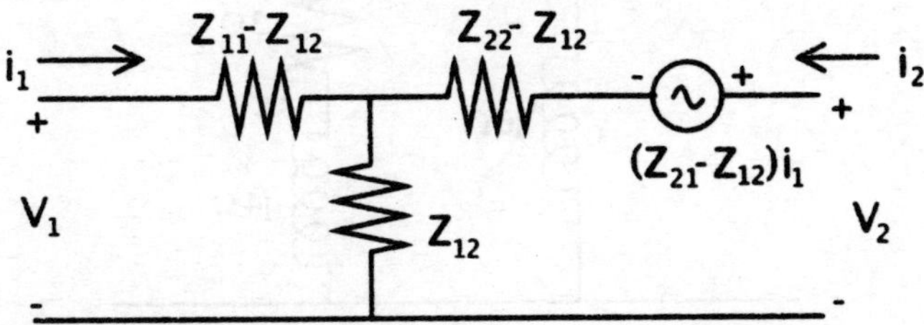

Figure 10.17: *The impedance model of a transistor can be converted to represent the transistor in a 'T' configuration, giving the name T Model.*

$$\begin{aligned} v_2 &= Z_{21}i_1 + Z_{22}i_2 \\ &= Z_{21}i_1 - Z_{12}i_1 + Z_{12}i_1 - Z_{12}i_2 + Z_{12}i_2 + Z_{22}i_2 \\ &= (Z_{21} - Z_{12})i_1 + Z_{12}(i_1 + i_2) + (Z_{22} - Z_{12})i_2 \end{aligned} \tag{10.42}$$

Eqn(10.41) and eqn(10.42) together give the input and output side loop equations, with $Z_{12}(i_1 + i_2)$ term common in both equations. This implies Z_{12} impedance is common a branch common to both the input and output loop. The 'T' circuit is realized by using the two equations (eqn 10.41 and eqn 10.42) is shown in fig(10.17).

Exercise

Q1. Find the h parameters for the circuit shown in fig(10.18).

Q2. If the circuit shown in fig(10.18) is converted to it's equivalent 'T' circuit, would the calculated 'h' parameters change? Verify.

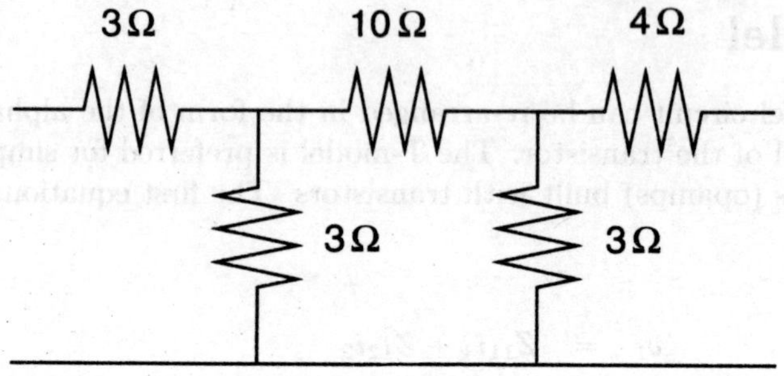

Figure 10.18:

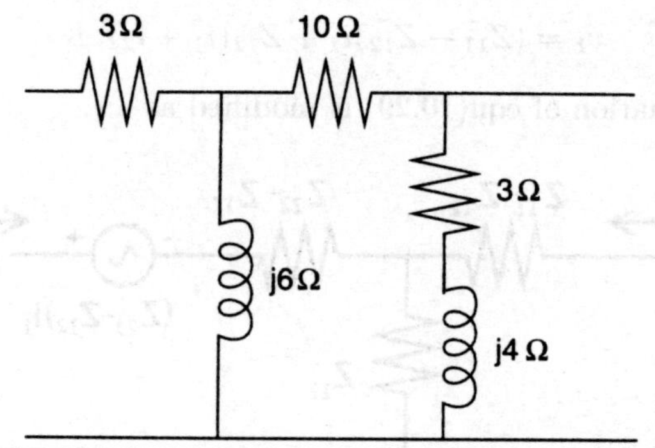

Figure 10.19:

Q3. Find the h parameters for the circuit shown in fig(10.19).

Q4. For the circuit used for Q1, determine the Z parameters.

Q5. Confirm the results of Q4, using the conversion relations of Table(10.5.2).

Q6. Show that the h-parameters do not exist for two port networks if $Z_{22}=0$.

Q7. What would be the value of output voltage for Example 10.3 if the load resistance is increased to 25KΩ?

Q8. Given the h-parameters for CB configuration as $h_{ib} = 30\Omega$ and $h_{fb} = -0.98$. Find the h-parameters for CE configuration for this very transistor.

Q9. Give the transconductance model of the transistor in CE configuration.

Q10. The following equation gives the currents at the two ports of a 2 port network:

$$I_1 = 0.3V_1 - 0.04V_2$$
$$I_2 = -0.04V_1 + 0.2V_2$$

Compute the Z-parameters for the network.

Q11. The following results were obtained as a result of measurements made on a 2-port network:

    i. with port 2 open circuited, a voltage of 20v applied to port 1 results in $I_1 = 1A$ and $V_1 = 3V$.

    ii. with port 1 open circuited, a voltage of 10v applied to port 2 results in $I_1 = 0.4A$ and $V_2 = 1.2v$.

Find the Z-parameter.

Q12. With a 5v ideal voltage source connected across port 1 of Q11, find the voltage across the resistor connected across port 2.

Q13. What determines whether a signal is "small" or "large"?

Q14. Why are no DC supplies shown in the linear models?

Q15. Convert circuit shown in fig. 10.20 into it's equivalent T-section circuit.

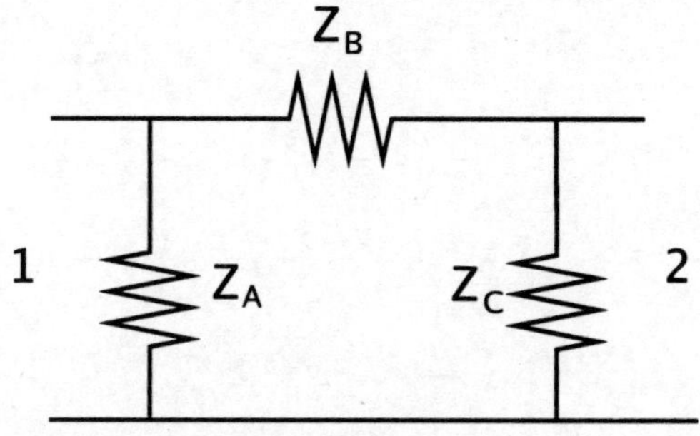

Figure 10.20:

    (i) Find the 'Z''-parameters of the above circuit.

    (ii) Verify whether the 'Z''-parameters of the above circuit and that of it's T-equivalent circuit are the same.

    (iii) Derive the relationship between the 'Z''-parameters and the hybrid parameters.

Q16. Draw the hybrid model of transistor is CE mode biased with a voltage-divider circuit. Calculate the AC gain of the amplifier designed in Q15.

Q17. What is the maximum AC input (peak to peak) you can give to the circuit designed in Q15? Explain with help of diagram what would happen if this precaution is not taken.

Q18. If the transistor you are using in Q15 is replaced with a new transistor which carries a precaution "Maximum Power=8mW", what changes in circuit would you have to make.

Q19. Given the h-parameters of a black box circuit (h_i, h_o, h_f and h_r). With notations having their usual meanings) draw the equivalent 'T' network for the given black box.

Q20. Given a transistor of h_{fe}=120. The maximum current that the transistor can handle is 2mA. The DC power supply available is of 20v. Complete the following exercise:

    a. Design a voltage-divider biasing circuit.

    b. Calculate the stability factor $S(I_{CBO})$, $S(V_{BE})$ and $S(\beta)$ for the circuit you designed.

    c. Explain the significance of these numbers.

    d. If R_E is doubled in value, repeat (a) and (b). Comment on the new values and discuss if such change of R_E is desirable.

Q21. The following equation gives the currents at the two ports of a 2 port network:

$$I_1 = 0.5V_1 - 0.2V_2$$
$$I_2 = -0.2V_1 + V_2$$

Compute the Y and Z-parameters for the network. Also find its equivalent π network.

Chapter 11

Frequency Response

In chapter one we showed circuital elements like inductors and capacitors respond or more precisely impede differently at different frequencies. This feature is with merits and has applications where circuits are required to select or reject certain frequencies. Let us now consider some of the simple frequency responding circuits and discuss their uses.

Consider the circuit shown in fig(11.1). The output voltage is taken across the resistance. The voltage divider equation can be used to find the output voltage.

$$v_o = \frac{R_i}{R_i + X_c} v_{in}$$

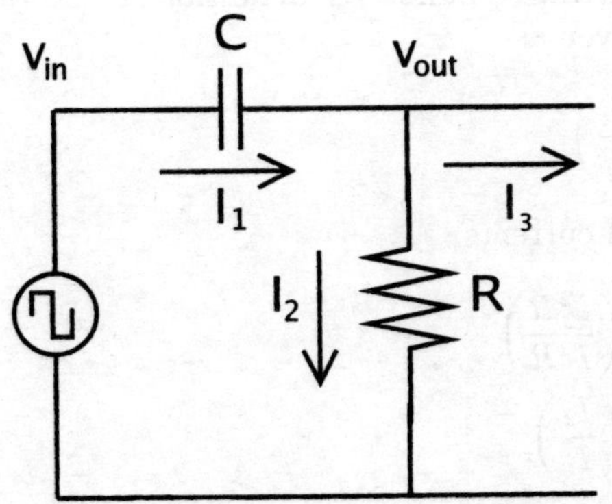

Figure 11.1: *The RC circuit for high pass filter.*

Since, the imepdance offered by the capacitor is frequency dependent and the input voltage's frequency can also be varied, the output voltage is also a function of the frequency. This would imply the gain of the circuit would also vary with the frequency

$$v_o(\omega) = \frac{R_i}{R_i - j/\omega C} v_{in}(\omega)$$

$$A_v(\omega) = \frac{1}{1 - j/\omega R_i C}$$

$$A_v(\omega) = \frac{1}{1 - j/2\pi f R_i C} \tag{11.1}$$

In our short introduction of complex numbers (chapter one) we had discussed the Euler's formula given as

$$ae^{j\Theta} = a\cos(\Theta) + ja\sin(\Theta)$$

Comparing the eqn(11.1) with the above equation, it is clear that the second term of the denominator has to be dimensionless. The 'f' in the term $1/2\pi f R_i C$ is frequency of the input supply.

Since the dimension of frequency is 1/T, to make the term dimensionless we have to demand

$$f_c = \frac{1}{2\pi R_i C} \qquad (11.2)$$

that is the dimensions of RC is T.[1] Thus eq(11.1) can be written as

$$A_v(\omega) = \frac{1}{1 - jf_c/f} \qquad (11.3)$$

The gain magnitude's variation with frequency is given as

$$|A_v(\omega)| = \frac{1}{\sqrt{1 + (f_c/f)^2}} \qquad (11.4)$$

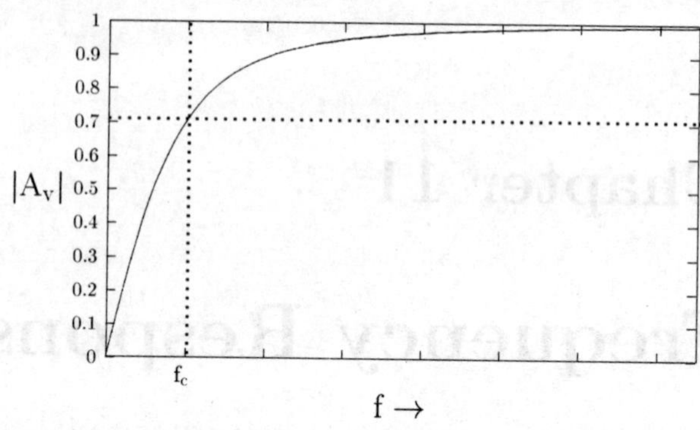

Figure 11.2: *The gain versus frequency of a high pass filter as described by eq(11.4).*

Figure(11.2) is a graphical representation of eq(11.4). Engineers and scientist however, represent the above equation on the logarithmic scale. Let us first understand the logarithmic scale.

11.1 Logarithmic Scale

The ratio of similar quantities like power, voltages etc. when are measured as a function of some variable (like frequency) are often expressed in logarithmic units. By definition, if the ratio is of two powers P_2 and P_1, the logarithmic ratio is given as

$$X = 10 log_{10} \left(\frac{P_2}{P_1} \right)$$

Logarithmic units may also be used for voltages and currents

$$X = 10 log_{10} \left(\frac{I_2^2 R}{I_1^2 R} \right)$$

$$X = 20 log_{10} \left(\frac{I_2}{I_1} \right)$$

or

$$X = 20 log_{10} \left(\frac{V_2}{V_1} \right)$$

The units of gain in logarithmic units is decibel (dB). The wisdom behind the definition of such ratio is clear from an example.

---

[1]The explanation from Euler's formula is given as an alternative. RC has a dimension of time is evident from the solution of differential equations used to explain RC networks as filters in chapter 6.

Example 11.1: Consider an amplifier which when fed an input voltage of 1v gives 10Kv (just an example!) in the output. Calculate the voltage gain.

The voltage gain is a ratio of the output voltage to the input voltage.

$$A_v = \frac{10Kv}{1v} = 10,000$$

This is a very large number and difficult to communicate. Another convenient way to communicate this number would be to represent this number in scientific notations, i.e. 10^4. This is still not an elegant number to communicate. However, in terms of logarithmic ratio, the gain would be

$$A_v = 20log_{10}(10^4) = 80dB$$

The gain of this amplifier circuit is 80dB. A far more elegant and easier value to communicate.

The magnitude of the above example may sound unrealistic. However, in communication circuits such as receivers circuits, we require the circuit should have very small noise/ disturbance as compared to the audio signal received by the antenna. Since the ratio of Signal to Noise (SNR) for a good receiver would be very large in magnitude (the noise in μV and the signal in 'V'), expressing it in dB makes more sense.

Going back to eqn(11.4) and expressing it in terms of decibels, we have

$$\begin{aligned}
20log(|A_v(\omega)|) &= 20log\left[\frac{1}{\sqrt{1+(f_c/f)^2}}\right] \\
&= 20log(1) - 10log\left(1+\frac{f_c^2}{f^2}\right) \\
&= -10log\left(1+\frac{f_c^2}{f^2}\right)
\end{aligned}$$

Figure(11.3) shows the variation of gain (dB) w.r.t. frequency. The three important regions of the graph are

$$20log(|A_v|) = -10log(\infty) = -\infty dB \quad (f = 0)$$

$$20log(|A_v|) = -10log(1) = 0dB \quad (f = \infty)$$

After the two extreme possibility, consider case f = f_c

$$20log(|A_v|) = -10log(2) = -3dB \quad (f = f_c)$$

The frequency (f_c) is called the cut-off frequency. While in the logarithmic scale, the gain at the cut-off frequency is -3dB, in eq(11.4) the gain at the cut-off frequency is $1/\sqrt{2}$. This is why popularly the cut-off point is called the $1/\sqrt{2}$ point.

One might ask *"what is so sacramental about the cut-off point?"* Well truly speaking there is nothing. Consider eqn(11.4) again, had the equation been derived for power gain rather then voltage gain, the equation would be

$$|A_P| = \frac{1}{1 + \frac{f_c^2}{f^2}}$$

Hence, the power gain at the cut-off would be 0.5 or 50%. This level is by convention adopted as acceptable level of performance as far as the amplifier or circuit is concerned. The 50% point works out to be the $1/\sqrt{2}$ point on the voltage/ current gain graph. As from fig(11.2), the voltage signals at frequencies below f_c have been attenuated, i.e. magnitude of output has diminished as compared to the input, and as per convention since is below 50% of the initial value isn't strong enough to contribute in the output. Signals at frequencies above f_c, the strength of the signal is appreciable. Thus, the RC circuit under discussion (fig 11.1), rejects signals of frequencies below f_c while allowing higher frequencies to pass through. Hence, the circuit is called a high pass filter.

To best appreciate the filtering ability of this circuit, input a square wave to the circuit and study the output. A square wave is periodic and repeats itself. As explained in chapter 5, any periodic wave can be written as a sum of cosine and sine terms (Fourier series). The Fourier series of the square wave

$$v(t) = 1, \quad 0 \geq t \geq \pi/2$$
$$v(t) = -1, \quad \pi/2 \geq t \geq \pi$$

is given as

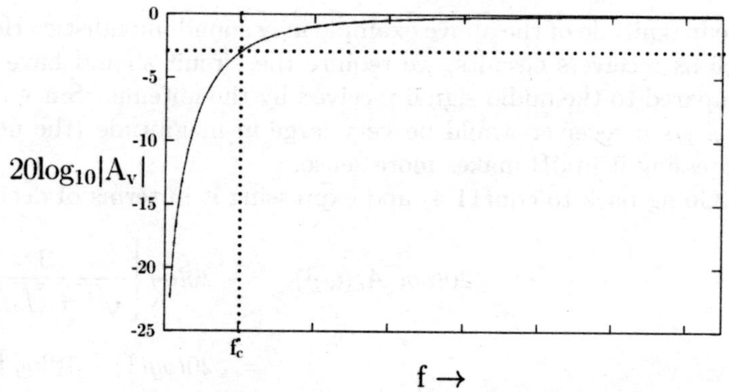

Figure 11.3: *The frequency response of a high pass filter, where the gain is shown in decibels.*

$$v(t) = \left(\frac{4}{\pi}\right) \sum_{n=0}^{\infty} \frac{\sin(2\pi(2n+1)ft)}{2n+1}$$

$$= \left(\frac{4}{\pi}\right) \left[\sin(x) + \frac{\sin(3x)}{3} + \frac{\sin(5x)}{5} + \ldots\ldots\ldots \right] \tag{11.5}$$

where x = 2πft. The n=1 frequency correspondence to the fundamental frequency while n 1 are the first, second and higher harmonics. The low pass filter depending on the cut-off frequencies remove some of the initial terms of the series given by eqn(11.5). Figure(11.4) compares similar two series. Notice that the flat region of the square wave diminishes as lower frequency terms are removed.

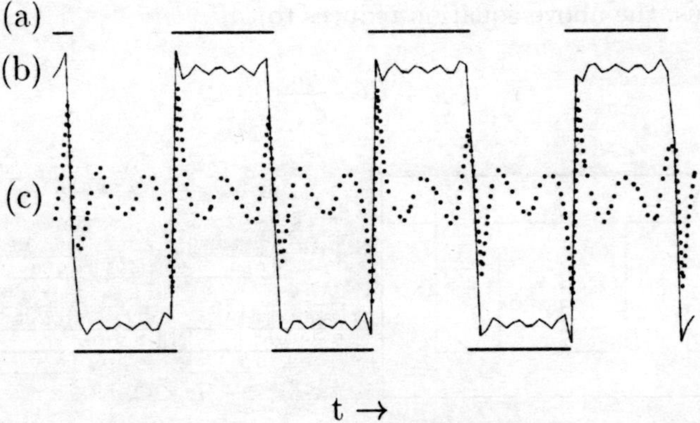

Figure 11.4: *The (a) ideal square wave is compared to limited series of eqn(11.5) where (b) is series $\sum_{n=0}^{10} \frac{\sin[2\pi(2n+1)ft]}{2n+1}$ and (c) is series $\sum_{n=2}^{10} \frac{\sin[2\pi(2n+1)ft]}{2n+1}$.*

Example 11.2: A square wave of frequency 1Hz is fed into a high-pass filter. If the components used in making the filter is R = 1KΩ and 25μF, find the frequencies that are rejected. Also, upto what harmonics are removed from the output?

The cut-off frequency is

$$f_c = \frac{1}{2\pi \times 1 \times 10^3 \times 25 \times 10^{-6}}$$
$$= 6.4Hz$$

Hence, frequencies below 6Hz would be rejected. To find the harmonics being removed we have to solve the inequality

$$(2n + 1)f = 6$$

where f is the fundamental frequency, in this example 1Hz. Thus, the filter removes the fundamental frequency along with the first and second harmonics (remember n is an integer).

We can predict another ability of this RC circuit using simple circuit analysis. If the current drawn by output circuit (since must likely the output is being viewed by an oscilloscope, the input impedance of the oscilloscope has to be large), then $i_1 = i_2$.

$$i_2 = i_1$$
$$\frac{v_{out}}{R} = C\frac{d(v_{in} - v_{out})}{dt}$$

While selecting the components for designing the filter, we make sure the product of RC is a specified (f_c), however, we are at liberty of choosing R to be small. This ensures v_{out} is small compared to v_{in}. Thus, the above equation reduces to

$$v_{out} = RC \frac{dv_{in}}{dt}$$

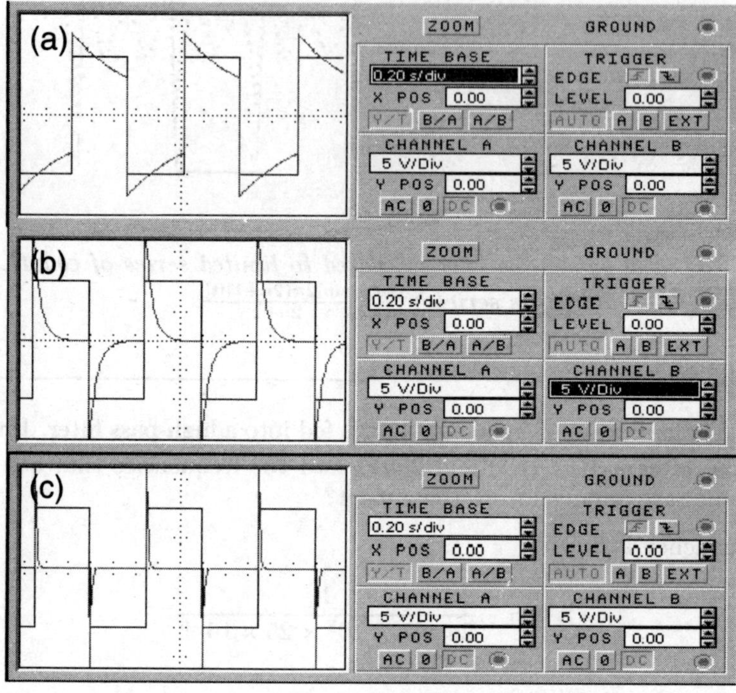

Figure 11.5: *The output of an differentiating/ high-pass filtering RC circuit for resistance of* 1.3KΩ *and (a) 500μF, (b) 50μF and (c) 10μF.*

i.e. the mathematical equivalent of differentiation of the input is done. Figure(11.5) shows the differentiated output of the high-pass RC filter. The resistance for all three cases is 1.3KΩ with different capacitance (a) 500μF, (b) 50μF and (c) 10μF. The frequency of the input square wave was 1Hz. Hence, the cut-off frequencies for the three cases are 0.25Hz, 2.45Hz and 122.5Hz respectively. Since more and more lower frequencies are being removed by decreasing the capacitance, the output waveform is becoming sharper and sharper, resembling a spike at the input square waveforms positive and negative going edge. Thus, a high pass filter would also function as a differentiator.

Example 11.3: Consider a triangular waveform is given as an input to a high-pass filter. Predict the nature of the output waveform?

Without undertaking the tedious exercise of finding the Fourier series of the triangular waveform we may consider the triangular wave to be made up of two different straight lines. One with a positive slope for half the cycle and for the remaining half cycle with negative slope. Put mathematically,

$$\begin{aligned}
v(t) &= mt, \quad 0 \geq t \geq T/2 \\
v(t) &= -mt, \quad T/2 \geq t \geq T
\end{aligned}$$

On differentiating we have

$$\begin{aligned}
v(t) &= m, \quad 0 \geq t \geq T/2 \\
v(t) &= -m, \quad T/2 \geq t \geq T
\end{aligned}$$

This is a square wave. Thus, a triangular waveform when given as an input to a high-pass filter gives a square wave in the output.

Consider the circuit shown in fig(11.6). This is the second possible configuration of a voltage divider that you can get using a RC combination. The output voltage is taken across the capacitor. The output voltage is given as

$$v_o = \frac{X_c}{R_i + X_c} v_{in}$$

Repeating the mathematics of the previous circuit, we have the gain as

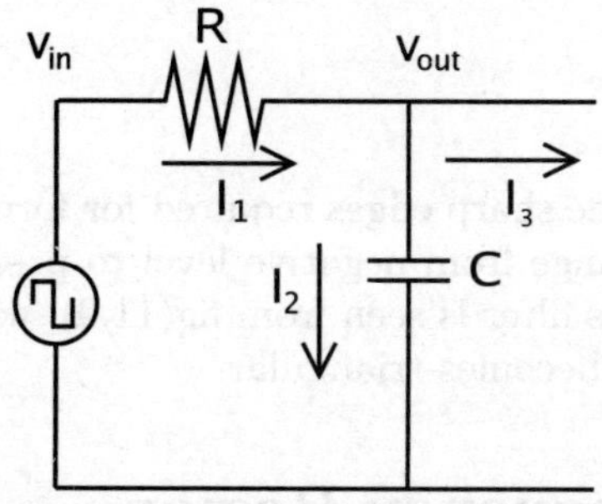

Figure 11.6: *The RC circuit for low pass filter.*

$$v_o(\omega) = \frac{1}{1 + \frac{R_i}{X_c}} v_{in}(\omega)$$

$$A_v(\omega) = \frac{1}{1 - j\omega R_i C}$$

$$A_v(\omega) = \frac{1}{1 - j\left(\frac{f}{f_c}\right)} \tag{11.6}$$

where, f_c is same as eqn(11.2). The magnitude of the gain is given as

$$|A_v(\omega)| = \frac{1}{\sqrt{1 + (f/f_c)^2}} \tag{11.7}$$

or for the purpose of comparing with eqn(11.2) can be written as

$$|A_v(\omega)| = \frac{(f_c/f)}{\sqrt{1 + (f_c/f)^2}} \tag{11.8}$$

Figure(11.7) shows how the circuit rejects frequencies greater than the cut-off frequencies, while low frequencies are allowed to pass. Thus, the circuit of fig(11.6) acts as a low-pass filter. Using the same methodology we used to show the differentiating ability of this high-pass RC filter, we can show that the loss-pass filter acts as an integrating circuit.

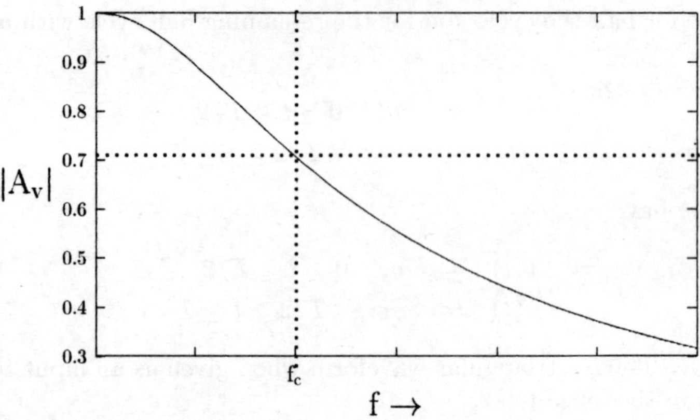

Figure 11.7: *The frequency response of a low pass filter is shown (as given by eq 11.8).*

Again assuming the current drawn by output circuit is small and choosing R to be large so that v_{out} is small compared to v_{in}, then first, $i_1 = i_2$.

$$i_2 = i_1$$
$$C\frac{dv_{out}}{dt} = \frac{(v_{in} - v_{out})}{R}$$

and secondly

$$v_{out} = \frac{1}{RC} \int v_{in} dt$$

Figureure(11.8) shows how on removing higher frequencies the sharp edges required for forming a square wave is removed and the output shows gradual change from negative level to positive level and visa versa. The integrating behavior of the low-pass filter is seen from fig(11.9), where for an appropriate value of RC the output of a square wave becomes triangular.

11.2   Transistor Amplifier in Low-frequency Region

Any response or put simpler, any change in behavior of a circuit with frequency is due to the capacitor or inductor present in the circuit (remember fig 1.22 of chapter 1). Since, the transistor amplifier circuit (fig 9.7) contains many capacitors it is but natural to expect the circuit to behave differently at different frequency. Basically, the transistors amplification factor (A_v) would depend on the frequency. However, from eq(10.14)

$$A_v = -\left(\frac{h_{fe}}{h_{ie}}\right) R_L$$

the expression for the voltage gain we had derived using the hybrid model of the CE transistor amplifier does not suggest any frequency dependency. Well that is because we had made many simplifications while deriving the expression and the most important of that was that

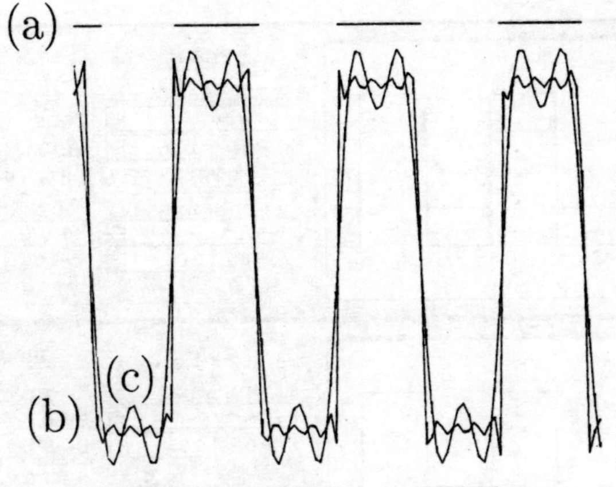

Figure 11.8: *The (a) ideal square wave is compared to limited series of eqn(11.5) where (b) is series* $\sum_{n=0}^{10} \frac{\sin[2\pi(2n+1)\mathrm{ft}]}{2n+1}$ *and (c) is series* $\sum_{n=0}^{2} \frac{\sin[2\pi(2n+1)\mathrm{ft}]}{2n+1}$.

the capacitor C_e, is a short in the frequency range of interest. This is usually true for high frequencies and frequency regions where this approximation holds true (the amplification factor is independent of the frequency) is called the **mid-frequency** region/range of the amplifier. It is hence better to write the above expression of voltage gain with an subscript 'm', suggesting that it is restricted to mid-frequencies

$$A_m = -\left(\frac{h_{fe}}{h_{ie}}\right) R_L \tag{11.9}$$

If we now consider the frequency to be appreciably low, by which the capacitor C_e can not be considered a short and offering impedance comparable to that of the emitter resistance (R_e), then the simplifications of the previous hybrid model would have to be corrected. The emitter to ground arm would be the basic change as shown in fig(11.10). The transistor amplifier circuits gain would now obviously be effected by the effective impedance offered by the **parallel** combination of emitter resistance and capacitor, which is given as

$$R'_e = R_e \| X_{CE} = \frac{R_e}{1 + j\omega R_e C_e}$$

We repeat the exercise of calculating the voltage gain of the given amplifier circuit of fig(11.10). We assume $R_1 \| R_2 \gg (h_{ie} + R'_e)$, this allows us to estimate the input current (i_b) trivially by applying KVL on the input side

$$\begin{aligned} v_{in} &= h_{ie}i_b + R'_e(i_b + i_c) \\ &= h_{ie}i_b + R'_e(i_b + h_{fe}i_b) \\ &= (h_{ie} + R'_e + h_{fe}R'_e)i_b \end{aligned}$$

The input current is

$$i_b = \frac{v_{in}}{h_{ie} + (1 + h_{fe})R'_e} \tag{11.10}$$

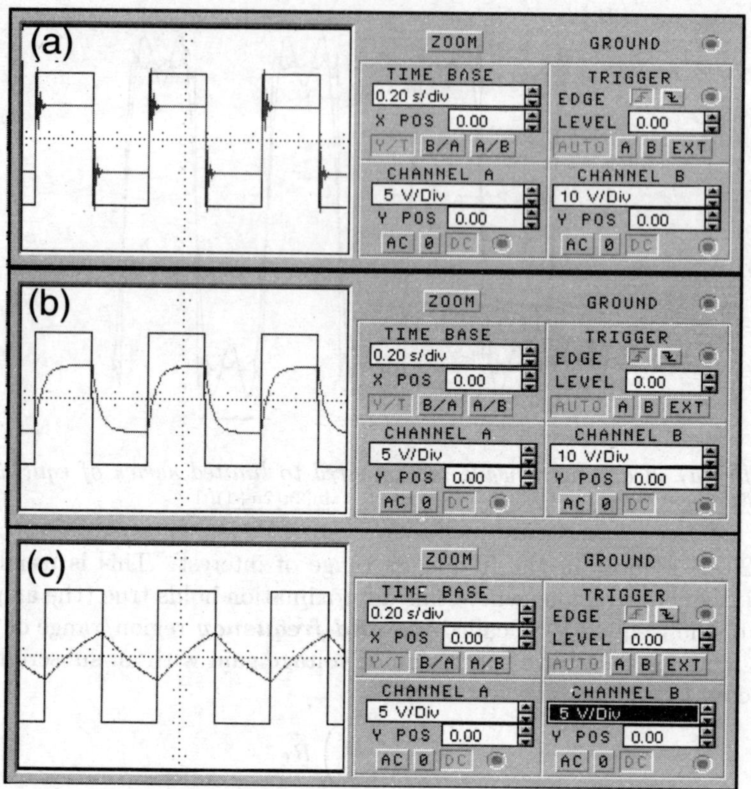

Figure 11.9: *The output of an integrating/ low pass filtering RC circuit for resistance of* $1.3K\Omega$ *and (a)* $1\mu F$, *(b)* $50\mu F$ *and (c)* $500\mu F$.

The output voltage is taken across R_c, however, due to the assumed output voltage polarity and the assumed direction of the current (assumptions made during the development of the two port models), the output voltage is given as

$$v_o = -i_c R_c = -h_{fe}i_b R_c$$

Substituting the expression for input current, we have

$$v_o = \frac{-h_{fe}R_c v_{in}}{h_{ie} + (1 + h_{fe})R_e'}$$

From which the amplifier's voltage gain can be written as

$$\begin{aligned} A_v &= \frac{-h_{fe}R_c}{h_{ie} + (1 + h_{fe})R_e'} \\ &= \frac{-h_{fe}R_c}{h_{ie} + (1 + h_{fe})\frac{R_e}{1 + j\omega R_e C_e}} \end{aligned}$$

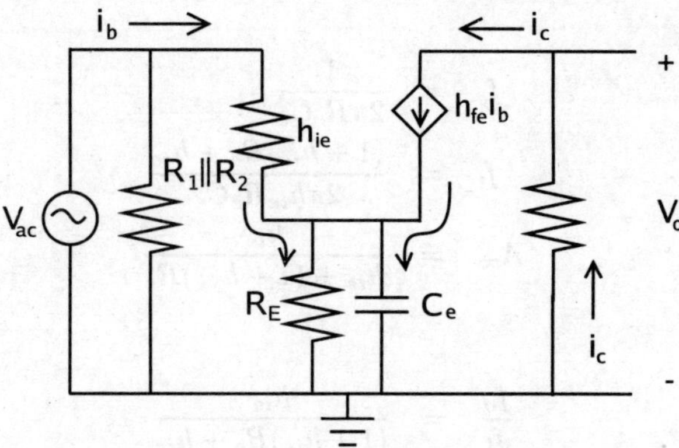

Figure 11.10: *The hybrid model of a transistor amplifier in CE model at low frequencies where the impedance offered by the emitter capacitor can not be neglected. The usual assumption that the output impedance of the CE transistor amplifier is infinite is maintained.*

$$= \frac{-h_{fe}R_c(1 + j\omega R_e C_e)}{[h_{ie} + (1 + h_{fe})R_e] + j\omega h_{ie}R_e C_e}$$

The above expression is quite large and difficult to analyze. To get some meaningful information, we now take up some mathematics and re-write the above expression as

$$A_v = \frac{-h_{fe}R_c}{h_{ie} + (1 + h_{fe})R_e}\left[\frac{1 + j\omega R_e C_e}{1 + j\left(\frac{\omega h_{ie}R_e C_e}{h_{ie}+h_{fe}R_e+R_e}\right)}\right]$$

$$= \left[\left(\frac{-h_{fe}R_c}{h_{ie}}\right)\frac{h_{ie}}{h_{ie} + (1 + h_{fe})R_e}\right]\left[\frac{1 + j\omega R_e C_e}{1 + j\left(\frac{\omega h_{ie}R_e C_e}{h_{ie}+h_{fe}R_e+R_e}\right)}\right]$$

What we have achieved is an expression relating the low frequency voltage gain to the mid frequency gain. Note, the imaginary terms of the above expression have to be dimensionless. In that case the RC terms would be indicative of certain important frequencies. Let us investigate this

$$A_v = A_m A_\infty \left(1 + j\frac{\omega R_e C_e}{1 + j\frac{\omega h_{ie}R_e C_e}{h_{ie}+h_{fe}R_e+R_e}}\right)$$

$$\frac{A_v}{A_m} = A_\infty \frac{1 + j\frac{f}{f_o}}{1 + j\frac{f}{f_l}} \tag{11.11}$$

where

$$f_o = \frac{1}{2\pi R_e C_e}$$

$$f_l = \frac{(1 + h_{ie})R_e + h_{ie}}{2\pi h_{ie} R_e C_e}$$

$$A_\infty = \frac{h_{ie}}{h_{ie} + (1 + h_{fe})R_e} \tag{11.12}$$

Also,

$$\frac{f_o}{f_l} = \frac{h_{ie}}{(1 + h_{ie})R_e + h_{ie}}$$

$$= A_\infty \tag{11.13}$$

We are now ready to investigate the variation of voltage gain of an transistor amplifier with frequency. All we have to do is find the magnitude of eq(11.11)

$$\frac{|A_v|}{|A_m|} = |A_\infty|\sqrt{\frac{1 + \frac{f^2}{f_o^2}}{1 + \frac{f^2}{f_l^2}}} \tag{11.14}$$

Figure(11.11) is a plot of the expression given by eqn(11.14) which explains how the voltage

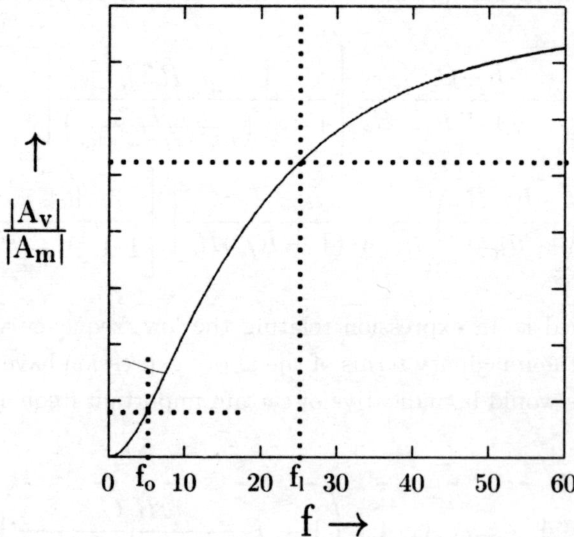

Figure 11.11: *The variation of gain with frequency for a transistor in low frequency region, as given by eq(11.14) is shown.*

gain varies with increasing frequency. The gain increases till of course at a very high frequency where the amplifier's gain becomes constant (the mid-frequency region). Thus, here we have plotted the low-frequency response of the transistor amplifier. Though fig(11.11) shows the plot of eqn(11.14), it was done using a computer and is not an easy task. There is a better and simpler way which the engineers use to understand eqn(11.14). As explained earlier, they take the logarithm of the equation[2] which gives

$$20log\left(\frac{|A_v|}{|A_m|}\right) = 20log|A_\infty| + 10log\left(1 + \frac{f^2}{f_o^2}\right) - 10log\left(1 + \frac{f^2}{f_l^2}\right) \tag{11.15}$$

For low frequencies, i.e $f \to 0$, we have

$$20log\left(\frac{|A_v|}{|A_m|}\right)_{f\sim 0} = 20log|A_\infty|, \quad f \to 0$$

This is shown by the dotted line (i) of fig(11.12). At $f = f_o$, the gain in decibel is given as

$$20log\left(\frac{|A_v|}{|A_m|}\right)_{f=f_o} = 20log|A_\infty| + 3 - 10log\left(1 + \frac{f_o^2}{f_l^2}\right), \quad f = f_o$$

$$= 20log|A_\infty| + 3 - 10log\left(1 + A_\infty^2\right) \tag{11.16}$$

With $A_\infty \ll 1$, the above equation reduces to

$$20log\left(\frac{|A_v|}{|A_m|}\right)_{f=f_o} \sim 20log|A_\infty| + 3$$

(see dotted line (ii) of fig 11.12). The -3dB point is at f_l and is calculated by

$$20log\left(\frac{|A_v|}{|A_m|}\right)_{f=f_l} = 20log|A_\infty| + 10log\left(1 + \frac{f_l^2}{f_o^2}\right) - 3, \quad f = f_l$$

$$20log\left(\frac{|A_v|}{|A_m|}\right)_{f=f_l} = 20log|A_\infty| + 10log\left(1 + \frac{1}{|A_\infty|^2}\right) - 3$$

$$= 20log|A_\infty| + -20log|A_\infty| - 3$$

$$= -3dB$$

As discussed earlier, the cut-off point (f_l) which represents the $1/\sqrt{2}$ point is called the -3dB point on the logarithmic scale. For $f_o \ll f \ll f_l$, eq(11.15) can be written as

$$20log\left(\frac{|A_v|}{|A_m|}\right) = 20log|A_\infty| + 20log\left(\frac{f}{f_o}\right) - 10log(1)$$

$$= 20log|A_\infty| - 20log(f_o) + 20log(f)$$

---

[2]Another advantage of plotting on the log scale is that a normal transistor amplifier would have a response of interest over a large frequency range, physically plotting the same would be difficult. Plotting on a log scale ('y axis' representing $log_{10}f$) is easier. Also the plot representing eqn(11.14) can easily be shown as three straight line in different frequency ranges, thus simplifying the problem of plotting. For more detailed understanding of eqn(11.14) one uses the Bode Plot. A Bode Plot is a pair of curves plotted on the same frequency axis, one curve showing the gain of the circuit and the other showing the phase shift, i.e. simultaneously the phase and amplitude of the circuits response is exhibited.

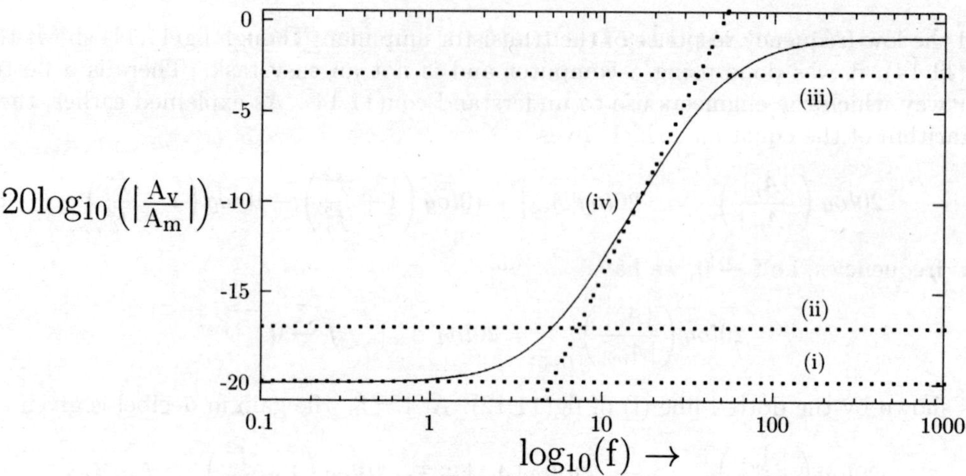

Figure 11.12: *The variation of gain in decibels with the logarithmic of frequency. Dotted line (iii) shows the -3dB point.*

Thus, in the frequency region of interest, the gain variation (dB) with frequency (log) is linear (dotted line iv).

$\boxed{\text{Example 11.4:}}$ Show that the variation of transistor gain in its low frequency region is identical to that of a low pass filter.

While this is evident by comparing the two figures (11.1) and (11.11), eq(11.14) can be shown to reduce to eq(11.4) for the appropriate condition ($f_o \ll f$).

$$\frac{|A_v|}{|A_m|} = |A_\infty|\sqrt{\frac{1+\frac{f^2}{f_o^2}}{1+\frac{f^2}{f_l^2}}}$$

$$\frac{|A_v|}{|A_m|} = \frac{|A_\infty|}{f_o}\frac{f}{\sqrt{1+\frac{f^2}{f_l^2}}}$$

$$\frac{|A_v|}{|A_m|} = |A_\infty|\left(\frac{f_l}{f_o}\right)\frac{1}{\sqrt{1+\frac{f_l^2}{f^2}}}$$

Thus the two equations (11.4 and 11.14) are identical.

11.3   Amplifier in High-frequency Region

The transistor in high frequency can not be analyzed with the hybrid model discussed until now. To understand why, one has to appreciate the physical attributes of the transistor, especially those that make an appearance only at high frequencies. The most important feature arising due to the reverse biased base-collector junction. The depletion width of the base-collector junction acts as a capacitor. The magnitude of the capacitance offered is of the order of pico-farad (pF). As can be understood from fig(11.13), the smaller the capacitance, the larger is the magnitude of impedance offered by it at low frequencies. This explains why until now, the base-collector capacitor was not considered in the model circuits. The impedance offered is so large that the path of conduction was considered open.

This base-collector capacitor, $C_{b'c}$, appears as a connection between the input and output of the hybrid model for the transistor in common emitter configuration (see fig 11.14a). Parallel to this capacitor, the resistance of the base-collector appears ($r_{b'c}$). The magnitude of this resistance is high and usually is considered open. However, one has to take into consideration the contribution of this resistance once one takes into account the parallel impedance offered by the capacitor.

Even though the base-emitter is forward biased there is a phase lag as charge carriers switch positions with switching field at high frequencies. This gives rise to a very-very small storage capacitance at the base-emitter junction, $C_{b'e}$. Analogous to the previous case a resistance $r_{b'e}$ appears across this capacitor. For obvious reasons related to the physical dimension of the transistor,[3] $r_{b'c} > r_{b'e}$. Typically, $r_{b'c}$ =1MΩ and $r_{b'e}$ =1KΩ. Notice, the subscripts of the parameters used to describe the parameters. The base is not represented by '(b)' but '(b')'. This is done to indicate that the tran-

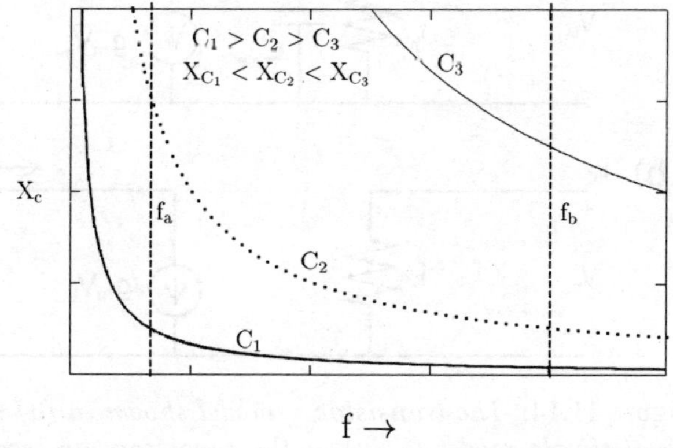

Figure 11.13: *The frequency response of three different capacitors.*

sistor currents flow through an imaginary node (B', see fig 11.14a) within the base. Thus, from the point of metal contact of the base to this imaginary node, the physical dimension of the base contributes a resistance, $r_{bb'}$, which typically is of the order of few hundred ohms.

The transconductance model is used to analyze the circuit. Hence, the current source of the output side is in terms of 'g_m', the transistor's transconductance. The hybrid model would demand a knowledge of current (input) through resistance $r_{b'e}$ (see fig 11.14a) which would be difficult to establish. However, the voltage $v_{b'e}$ can be approximated to be equal to the input voltage (provided $r_{b'b} \sim 0$), hence, the output current can be written interms of the input voltage (as per the transconductance model). The current of the diode is related to the voltage applied across it and is given by eqn (5.2). This equation can be used to the forward biased

---

[3]the collector is always physically larger than the emitter. However, the carrier concentration of the emitter is more than that in the collector region.

base-emitter junction as

$$I_E = I_{sat} exp\left(\frac{V_{BE}}{V_T}\right)$$

Since the collector current can be approximated as equal to the emitter current, we have

$$I_C = I_{sat} exp\left(\frac{V_{BE}}{V_T}\right)$$

We continue modeling the 'AC' signal as a slowly varying 'DC' signal, hence on differentiating the above equation we can write

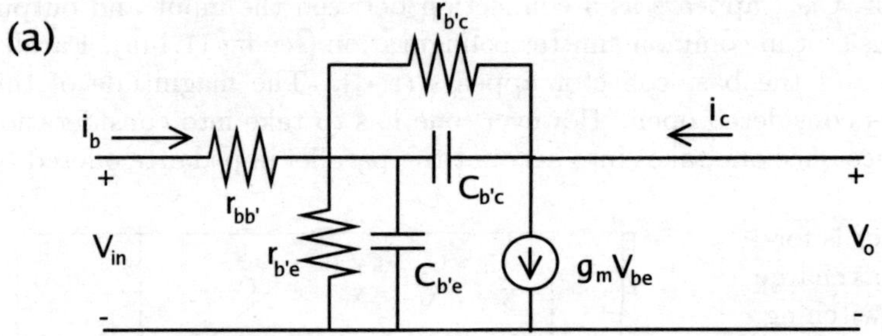

$$\frac{I_C}{V_{BE}} = \frac{I_{sat}}{V_T} exp(\frac{V_{BE}}{V_T})$$

$$\frac{di_c}{dv_{be}} = \frac{i_c}{V_T}$$

$$g_m = \frac{i_c(mA)}{26mV}$$

'g_m' is the transconductance of the transistor and relates the output current (collector current) to the input voltage of the transistor in CE configuration.

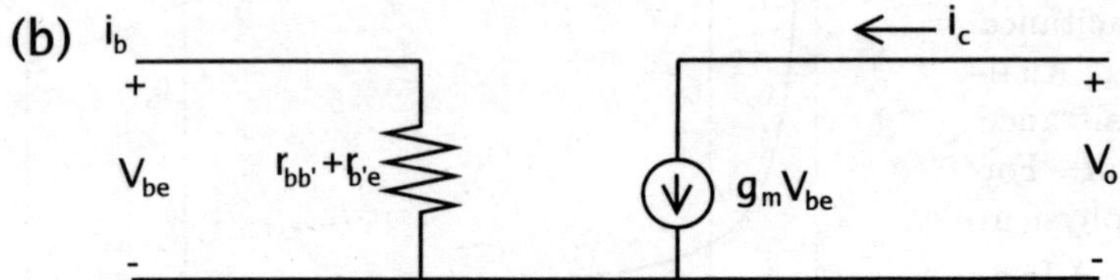

If we reduce the π hybrid model of the transistor used for analyzing high frequency to the mid frequency regime, the model would look like that in fig(11.14b). The

Figure 11.14: *The transistor π model shown in (a) reduces to the usual simple model (b) when the capacitors are considered open (in mid-frequency) range.*

gain of the transistor circuit in term of transconductance can now be calculated for the mid-frequency region.

$$\begin{aligned} v_o &= -i_c R_L \\ &= -g_m v_{be} R_L \end{aligned}$$

The gain is the ratio of the output voltage to the input voltage, giving

$$A_v = -g_m R_L \tag{11.17}$$

Similarly, the current gain is given as

$$\begin{aligned} i_c &= g_m v_{be} \\ &= g_m(r_{bb'} + r_{b'e})i_b \end{aligned}$$

giving

$$A_i = g_m(r_{bb'} + r_{b'e}) \tag{11.18}$$

Comparing the above result with that obtained in terms of hybrid parameters

$$g_m = \frac{h_{fe}}{h_{ie}}$$

as also

$$h_{ie} = r_{bb'} + r_{b'e}$$

11.3.1 Miller Effect

Miller effect is a way of dealing with a situation where the voltages at both terminals of a capacitor changes at the same time. Consider, the case shown in fig(11.15), where the potential at the first pin is described as v_{in} while that at pin 2 is given as v_o. The two voltages are varying with time and may or may not be dependent on each other (if v_o is dependent on v_{in}, one may write $v_o = A_v v_{in}$). The potential difference present across the capacitor is given as $v_{in} - v_o$, hence the current through the capacitor can be calculated

$$i \;=\; C\frac{d}{dt}\left(v_{in} - v_o\right)$$

It would be easier to understand the situation by considering the output voltage to be dependent on the input voltage, i.e.

$$i \;=\; (1 - A_v)C\frac{dv_{in}}{dt}$$

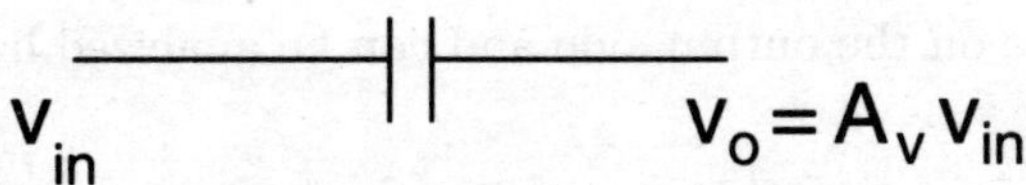

Figure 11.15: *Two varying potentials present at each terminal of a capacitor.*

That is the capacitor 'C' present in the circuit, effectively appears as $(1 - A_v)C$ in the circuit. Thus, due to two varying voltages appearing at the pins of the capacitor, an effective capacitance rather than the true capacitance appears in the circuit (Miller Effect).

Miller effect plays an important role in CE mode transistor amplifiers where the junction capacitance appearing due to the reverse biased base-collector comes into the picture. A more detailed consideration would be required to understand the Miller Effect's role on the input and output side of transistor amplifier. For this we consider the collector-base capacitance appearing in the two port model as shown in fig(11.16). The input current divides into two current, I_1 into the black box and I_2 into the feedback capacitor (called feedback since it is connected to both the input and output sides).

$$
\begin{aligned}
I_i &= I_1 + I_2 \\
\frac{V_i}{Z_i} &= \frac{V_i}{R_i} + \frac{V_i - V_o}{X_{C_f}}
\end{aligned}
\tag{11.19}
$$

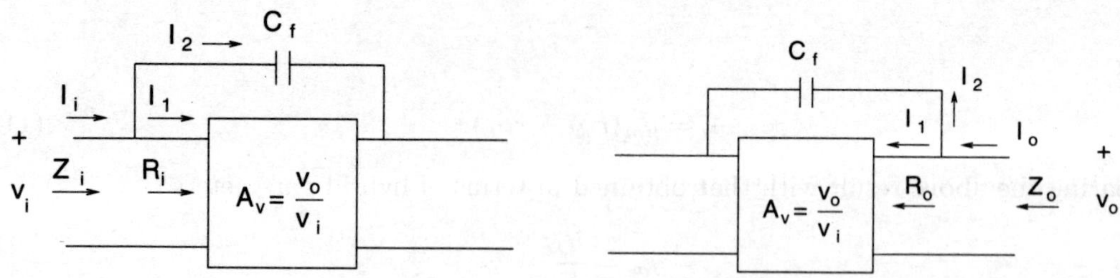

Figure 11.16: *The capacitor connecting the input and output side of the two port effects the circuit's input and output impedance.*

$$\frac{V_i}{Z_i} = \frac{V_i}{R_i} + \frac{(1 - A_v)V_i}{X_{C_f}}$$

$$\frac{1}{Z_i} = \frac{1}{R_i} + \frac{1 - A_v}{X_{C_f}}$$

The above equation indicates that the impedance $(1 - A_v)/X_{C_f}$ appears parallel to the input impedance (R_i). Thus the capacitance of the apparent capacitor at the input is

$$X_{C_i} = \frac{X_{C_f}}{1 - A_v}$$

$$C_i = C_f(1 - A_v) \tag{11.20}$$

As expected the Miller effect has bought in an apparent capacitance $(1 - A_v)C_f$ which offers an impedance that appears parallel across the input impedance of the transistor. Also, since the CE transistor's gain is negative (indicating the phase reversal of the output), the Miller effect introduces a large capacitance on the input.

The feedback capacitance would also play a role on the output side and can be analyzed by

$$I_o = I_1 + I_2$$

$$\frac{V_o}{Z_o} = \frac{V_o}{R_o} + \frac{V_o - V_i}{X_{C_f}} \tag{11.21}$$

Since, the output resistance (R_o) offered by a transistor in CE mode is appreciable, we have

$$\frac{V_o}{Z_o} = \frac{(1 - V_i/V_o)V_o}{X_{C_f}}$$

$$\frac{1}{Z_o} = \frac{A_v - 1}{A_v X_{C_f}}$$

Hence, the output impedance is capacitive in nature and the output capacitance is given as

$$X_{C_o} = \frac{A_v X_{C_f}}{A_v - 1}$$

$$C_o = \frac{C_f(1 - A_v)}{A_v}$$

$$C_o \sim C_f \tag{11.22}$$

The net impedance offered by the capacitors on the input side is (see eq 11.20)

$$C_d = C_{be} + C_{bc}(1 - A_v)$$

$$C_d = C_{be} + C_{bc}(1 + g_m R_L)$$

The input loop equation after using the Miller capacitance is

$$v_{be} = (r_{bb'} + r_{b'e}\|X_{C_d})i_b$$

$$v_{be} = \left(r_{bb'} + \frac{r_{b'e}}{1 + j\omega r_{b'e}C_d}\right)i_b$$

Assuming $r_{b'b} \sim 0$, we have

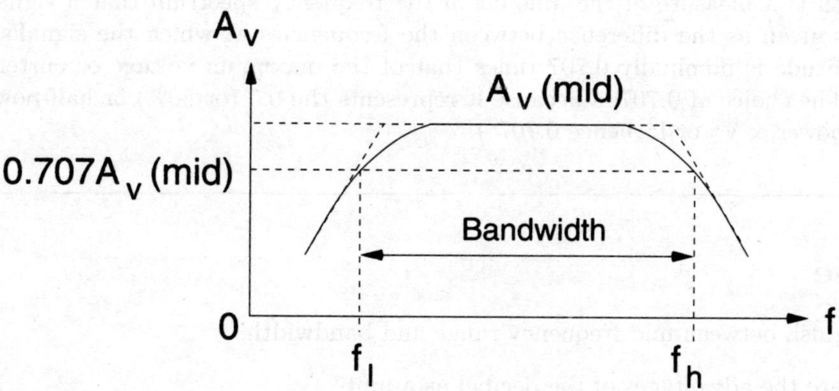

Figure 11.17: *The frequency response of a transistor amplifier in normal scale. The bandwidth has been indicated.*

$$v'_{be} = \left(\frac{r_{b'e}}{1 + j\omega r_{b'e}C_d}\right)i_b$$

$$g_m v'_{be} = g_m\left(\frac{r_{b'e}}{1 + j\omega r_{b'e}C_d}\right)i_b$$

$$i_c = g_m\left(\frac{r_{b'e}}{1 + j\omega r_{b'e}C_d}\right)i_b$$

Re-arranging, we get the transistor's current gain in high frequency regime as

$$A_i = g_m\left(\frac{r_{b'e}}{1 + j\omega r_{b'e}C_d}\right)$$

$$A_i = g_m\left(\frac{r_{b'e}}{1 + j\frac{f}{f_h}}\right)$$

where,

$$f_h = \frac{1}{2\pi r_{b'e} C_d}$$

$$= \frac{1}{2\pi R_e\left[C_{b'e} + C_{b'c}(1 - A_v)\right]} \tag{11.23}$$

f_h is the cut-off point in the high frequency response. The magnitude of current gain is

$$|A_i| = \frac{g_m r_{b'e}}{\sqrt{1 + \frac{f^2}{f_h^2}}} \tag{11.24}$$

The expression indicates that the gain of the transistor falls with increasing frequency. The fall of course is from the constant gain in the mid-frequency region. Combining, and plotting all the three regions response on one graph, we get the frequency response shown in fig(11.17). The response gives as important information of the circuit's performance, called the **bandwidth**.

Bandwidth is a measure of the amount of the frequency spectrum that a signal occupies. Usually, it is given as the difference between the frequencies at which the signal's voltage or current amplitude is nominally 0.707 times that of the maximum voltage or current given by the circuit. The choice of 0.707 is because it represents the 0.5 (or 50%) or half-power point of the circuit (power $\propto V^2$ or I^2, hence 0.707^2).

Exercise

Q1. Distinguish between mid-frequency range and bandwidth.

Q2. What are the advantages of the decibel as a unit?

Q3. In the voltage divider biased circuit (fig 9.7) assume $V_{cc} = 15v$, $\beta = 50$, $R_1 = 100K\Omega$, $R_2 = 20K\Omega$, $R_c = 4K\Omega$ and $R_E = 500\Omega$. At the signal frequency, the reactance of all capacitors are negligable.

    a. Using necessary assumptions determine bias collector current.

    b. Draw the small signal model for the amplifier.

    c. Calculate the voltage gain.

Q4. Consider the fixed bias circuit of fig(9.1) designed using a silicon BJT, whose $\beta = 50$. The collector resistance (R_c) is 5KΩ. Stating any assumption, predict I_c and specify R_B to establish V_{CE} at 2*volts*.

Q5. In the voltage divider biased circuit (fig 9.7), $V_{cc} = 10v$, $\beta = 100$, $R_1 = 58K\Omega$, $R_2 = 142K\Omega$, $R_c = 8K\Omega$ and $R_E = 3.9K\Omega$. At the signal frequency, the reactance of all capacitors are negligable. Using necessary assumptions, determine the DC current I_c and voltage V_{CE}.

Q6. What is Miller Effect?

Q7. Explain why Miller Effect does not come into play in Common-Base and Common-Collector transistor amplifiers.

Q8. What do you mean by the band-width of an amplifer circuit?

Q9. How do you descide the cut-off points. Explain significance of selection.

Q10. Derive an expression for the Gain-Bandwidth product of a transistor amplifer, assuming the high cut-off frequency is far greater than the low cut-off point.

Q11. Derive an expression for the Gain-Bandwidth product of a transistor amplifer without following the above questions assumption.

Q12. Define the gain-bandwidth product. Why is it useful?

Chapter 12

Multistage Transistor Amplifiers

The single transistor amplifier typically has a voltage gain of the order of 100 to 250. This is not sufficient for most practical purposes. Especially in telephone communication where you might have audio signals in micro-volts (μV), thus even on amplifying the output voltage is in milli-volts. Since this is not enough, then how to achieve higher amplification? A simple way out is to feed the output of the first transistor amplifier to the input of the second amplifier, the output of the second amplifier would hence be sufficiently large. Examining the voltage gain of the two transistor acting together, we have

$$A_v = \frac{v_{out2}}{v_{in1}}$$
$$= \frac{v_{out2}}{v_{in2}} \times \frac{v_{in2}}{v_{in1}}$$

The numbers 1 and 2 in the subscript indicates the contribution to/ by the first and second transistor amplifier. Since the output of the first transistor is the input of the second transistor amplifier,

$$A_v = \frac{v_{out2}}{v_{in2}} \times \frac{v_{out1}}{v_{in1}}$$
$$= A_{v1} A_{v2}$$

Had identical transistor amplifiers been used, each with gain 100 ($A_{v1} = A_{v2} = 100$), the net gain of the two stage transistor amplifier would be

$$A_v = 10,000$$
$$or \quad 80dB$$

This is quite appreciable. Extending this idea from two stage cascaded amplifier to 'n' stage cascaded transistor amplifier, we have the net amplification factor as

$$A_{vn} = (A_v)^n \tag{12.1}$$

Interesting details of the cascaded transistor amplifier can be found by trivial mathematics on eqn(12.1). The individual amplifiers (identical) constituting the cascaded amplifier would have the frequency response (gain's variation with low frequency, A_{vl} and gain's variation with high frequency, A_{vh}) and cut-off points (f_l in the low frequency and f_h in the high frequency) given in Chapter 11, i.e.

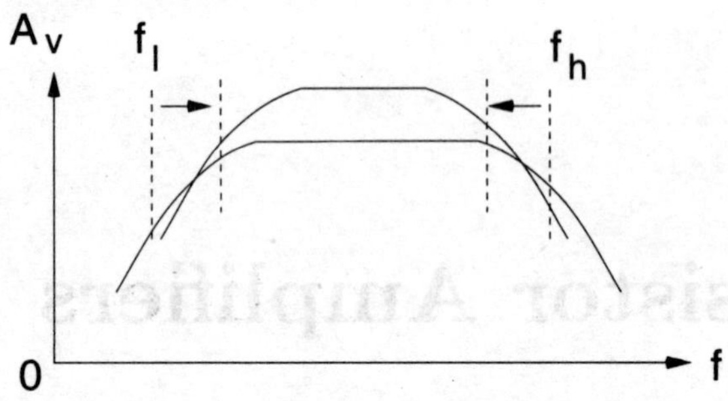

$$A_{vl} = \frac{A_{vm}}{\sqrt{1 + \frac{f_l^2}{f^2}}}$$

$$A_{vh} = \frac{A_{vm}}{\sqrt{1 + \frac{f^2}{f_h^2}}}$$

where A_{vm} is the mid frequency gain of the individual transistor amplifier. The nature of the frequency response of the cascaded amplifier would remain the same, however, one would expect that on cascading the cut-off points should vary. This is obvious since f_l and f_h depends on the amplifiers input impedance and Miller capacitance respectively. These parameters of the cascaded amplifier can not be expected to be the same as that of the single stage transistor amplifier.

Figure 12.1: *The bandwidth decreases with increasing gain.*

Thus, on considering the low frequency response of the cascaded amplifier, eqn(12.1) can be modified as

$$\frac{(A_{vn})_m}{\sqrt{1 + \frac{f_{ln}^2}{f^2}}} = \left[\frac{(A_{vm})}{\sqrt{1 + \frac{f_l^2}{f^2}}}\right]^n \tag{12.2}$$

where f_{ln} is the low cut-off frequency of the cascaded amplifier. Since, the mid-frequency amplification is also related by eqn(12.1), the above reduces to

$$1 + \frac{f_{ln}^2}{f^2} = \left(1 + \frac{f_l^2}{f^2}\right)^n \tag{12.3}$$

To find where the cut-off point is for the cascaded amplifier with respect to 'f_l', the cut-off point of individual amplifier, subsititute $f = f_{ln}$ in eqn(12.3). At the frequency $f = f_{ln}$[1]

$$1 + \frac{f_l^2}{f_{ln}^2} = 2^{1/n}$$

which gives the low frequency cut-off point of the cascaded amplifier as

$$f_{ln} = \frac{f_l}{\sqrt{2^{1/n} - 1}}$$

---

[1]If calculation is done for the frequency $f = f_l$, we get the result

$$f_{ln} = f_l \sqrt{2^n - 1}$$

Thus, on cascading the 0.707 or half-power point occurs at a higher frequency as compared to that of it's constituting single transistor amplifier (see fig 12.1). Similar calculations for the high frequency response gives

$$1 + \frac{f^2}{f_{hn}^2} = \left(1 + \frac{f^2}{f_h^2}\right)^n$$

at the frequency $f = f_{hn}$[2]

$$f_{hn} = f_h \sqrt{2^{1/n} - 1}$$

The high frequency cut-off point thus comes down on cascading (see fig 12.1). It thus follows, while the amplification factor on cascading increases, the band width comes down.

12.1   Direct Coupled Amplifiers

The problem of direct coupled transistor amplifier is that the DC current of the first stage is not isolated from that of the second stage. Thus, the DC biasing of both transistors are effected. This leads to the operating points and load lines of both transistors to be displaced from the center of the active region. Consider the example in Chapter IX (Example 9.5) where a single stage transistor amplifier was designed. The voltage V_C of the circuit is 8.6v ($V_C = V_{CC} - I_C R_C = 16 - 3.7K\Omega \times 2mA$) and V_B is 1.3v (V_{R_2}).

Now if two identical transistors of Example 9.5 are coupled the collector volatge V_C comes parallel to V_B. With no resistance/impedances connected between the two, the lower voltage V_B of the second transistor tries to pull down the collector voltage, V_C of the first. This results in a large current flowing into the base of the second terminal. This poses a difficult task to a circuit designer to match the two amplifiers.

12.2   RC Coupled Amplifiers

The biasing problem of direct coupled amplifier is overcome by DC isolating the two transistor amplifiers giving rise to what is called the RC coupled transistor amplifier. The combined amplification factor of course is greater than that of each of the individual amplifier.

Voltage Gain

To calculate the gain of an RC coupled transistor amplifier, we consider the mid-frequency small signal AC model of the circuit shown in fig(12.4). For the condition $h_{ie} \ll R_B$ (where $R_B = R_1 \| R_2$), the input voltage is given as

---

[2]If calculation is done for the frequency $f = f_h$, we get the result

$$f_{hn} = \frac{f_h}{\sqrt{2^n - 1}}$$

The trend of cut-off variation however, remains the same.

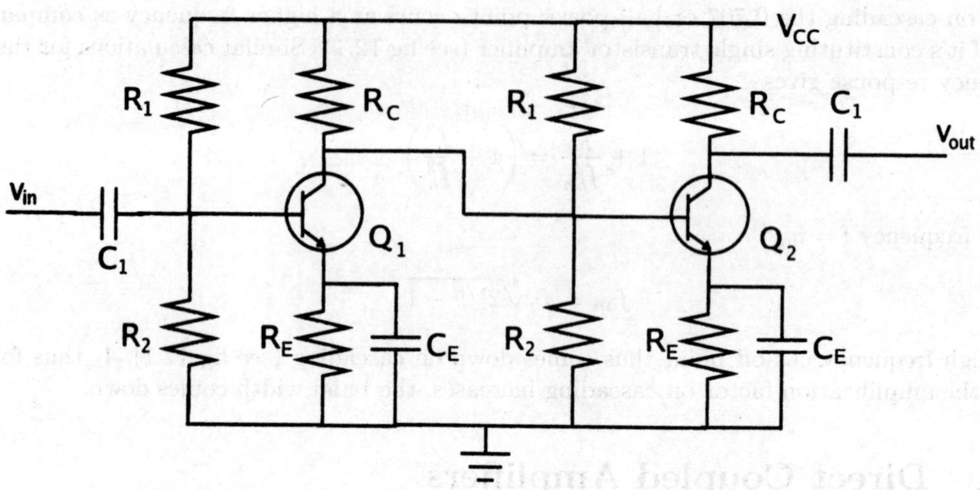

Figure 12.2: *Two amplifiers are cascaded by direct coupling.*

$$v_{in} = h_{ie}i_b \tag{12.4}$$

The output voltage at the collector of the first transistor would be different from a single stage transistor amplifier as R_c (collector resistance of the first transistor) comes parallel to R_B and h_{ie}[3] of the second transistor. Thus, the output voltage is

$$v_{o1} = -i_c Z$$

$$= -i_c \frac{h_{ie}\left(\frac{R_b R_c}{R_b + R_c}\right)}{h_{ie} + \left(\frac{R_b R_c}{R_b + R_c}\right)}$$

The current flowing into the input impedance (h_{ie}) of the second transistor is given as

$$i_{o1} = -\frac{\left(\frac{R_b R_c}{R_b + R_c}\right)}{h_{ie} + \left(\frac{R_b R_c}{R_b + R_c}\right)} h_{fe}i_b$$

This is the input current of the second transistor. The negative sign implies this current is in the opposite direction of the input current of the two port model. Since this is the input current for the second transistor, the output current of the second transistor would be $i_{o2} = -h_{fe}i_{o1}$, or

$$i_{o2} = -\frac{\left(\frac{R_b R_c}{R_b + R_c}\right)}{h_{ie} + \left(\frac{R_b R_c}{R_b + R_c}\right)} h_{fe}^2 i_b$$

The output voltage is a result of this current flowing through R_C. Hence,

$$v_o = -i_{o2}R_C$$

---

[3]we assume the two stages are identical

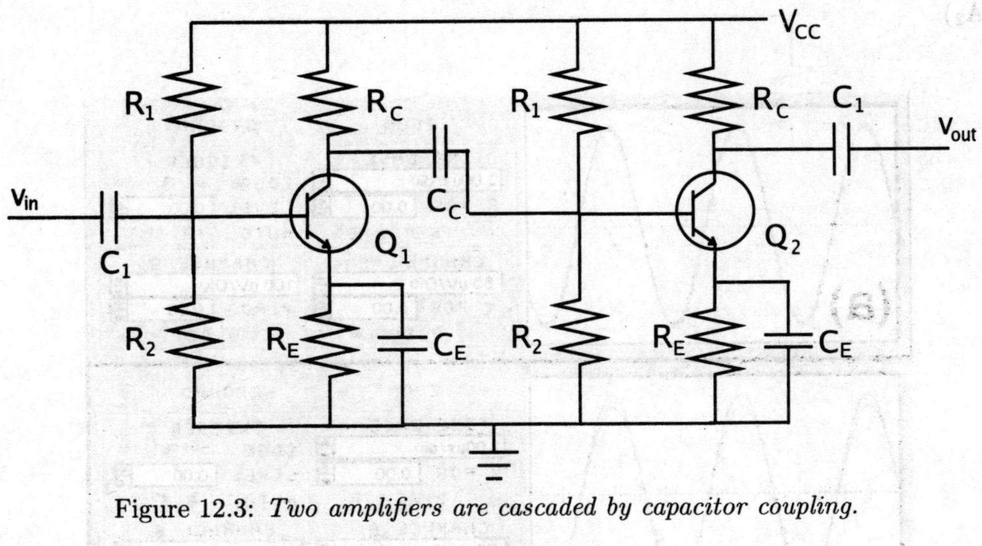

Figure 12.3: *Two amplifiers are cascaded by capacitor coupling.*

$$= \frac{\left(\frac{R_b R_c}{R_b + R_c}\right) R_c}{h_{ie} + \left(\frac{R_b R_c}{R_b + R_c}\right)} h_{fe}^2 i_b$$

The net gain of the two stage RC coupled transistor amplifier is given as

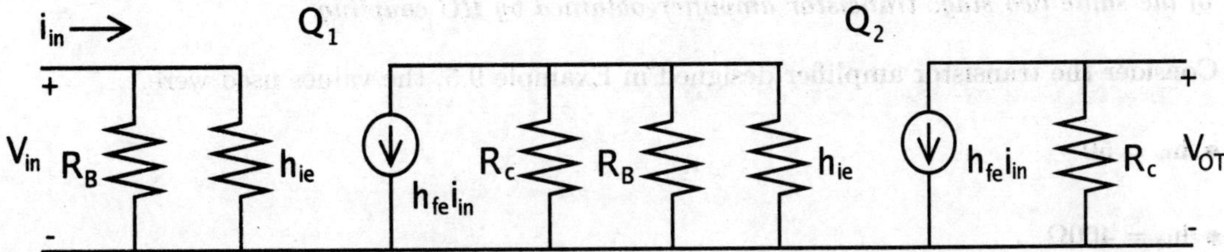

Figure 12.4: *The small signal hybrid model of an RC coupled transistor amplifier in mid-frequency region.*

$$
\begin{aligned}
A_{vm} &= h_{fe}^2 i_b \frac{\left(\frac{R_b R_c}{R_b + R_c}\right) R_c}{h_{ie} + \left(\frac{R_b R_c}{R_b + R_c}\right)} \times \frac{1}{h_{ie} i_b} \\[2mm]
&= \frac{h_{fe}^2 \left(\frac{R_b R_c}{R_b + R_c}\right) R_c}{h_{ie} \left[h_{ie} + \left(\frac{R_b R_c}{R_b + R_c}\right) \right]} \\[2mm]
&= \frac{h_{fe}^2 R_c^2 R_b}{h_{ie}[h_{ie}(R_b + R_c) + R_b R_c]}
\end{aligned}
\tag{12.5}
$$

Equation(12.5) is the net voltage amplification achieved by a two stage RC coupled transistor amplifier. This is less than the gain one expects on cascading ($A_V = A_1 \times A_2 = h_{fe}^2 R_c^2 / h_{ie}^2$ if

$A_1 = A_2$).

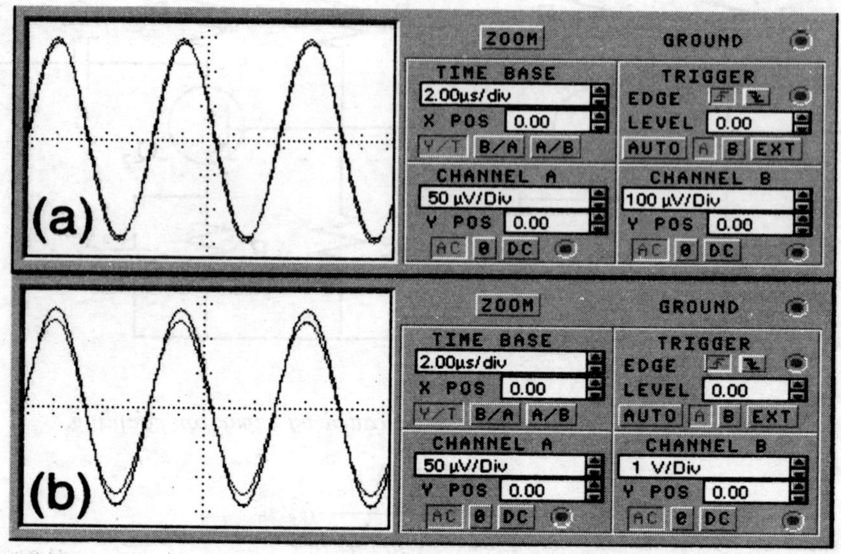

Figure 12.5: *The oscilloscope trace of the output of (a) a two stage direct coupled amplifier and (b) of the same two stage transistor amplifier obtained by RC coupling.*

Consider the transistor amplifier designed in Example 9.5, the values used were,

- $h_{fe} = 50$

- $h_{ie} = 460\Omega$

- $R_c = 3.7K\Omega$

- $R_1 = 270K\Omega$, and

- $R_2 = 90K\Omega$.

The gain of an individual transistor was found to be $A_v = 394$. The theoretical gain as a result of cascading should have been 155.2×10^3. However, since the resistances R_1, R_2 and h_{ie} of the second transistor comes in parallel of R_c of the first transistor, the net volatge gain is given by eqn(12.5). Consider two identical transistor amplifiers with of Example 9.5 are cascaded using a capacitor, using eqn(12.5) we have net voltage gain as 17.8×10^3. Figure(12.5) compares the output voltage of a direct coupled two stage amplifier with that of a RC coupled amplifier. The two amplifiers used are of Example 9.5.

Current Gain

The net current in the output loop of the first transistor is given by the current source on the output side:

$$i_c = h_{fe}i_b$$

The current in the input impedance of the second transistor, h_{ie}, is obtained by the current division equation.

$$i_{o1} = \frac{\left(\frac{R_b R_c}{R_b + R_c}\right)}{h_{ie} + \left(\frac{R_b R_c}{R_b + R_c}\right)} h_{fe}i_b$$

This is the input current of the second transistor which then appears amplified at the output of the second stage transitor as $i_{o2} = -h_{fe}i_{o1}$. As explained above the negative signal is based on the two port model of the transistor. Hence, the net current gain is

$$A_{im} = \frac{i_{o2}}{i_b} = -\frac{\left(\frac{R_b R_c}{R_b + R_c}\right)}{h_{ie} + \left(\frac{R_b R_c}{R_b + R_c}\right)} h_{fe}^2$$

$$= -\frac{h_{fe}^2 R_b R_c}{h_{ie}(R_b + R_c) + R_b R_c}$$

The negative sign indicates the phase reversal of the output current with respect to the input current.

Low Frequency Response

At low frequencies the capacitors with low impedances (X_C) contributes to the amplifier's frequency response (see fig 11.13). Two possible capacitors that can contribute to the frequency response are C_c, the coupling capacitor and C_E, the emitter feedback capacitor. If we select $C_c \ll C_E$ (or $X_{C_c} \gg X_{C_E}$), the half-power point of the amplifier would be decided by the value of capacitor C_c. Since the role of the coupling capacitor is just to block DC signals, a capacitor with very low capacitance would do (i.e. value of C_c can be very small). Thus, unlike the case of the single stage transistor amplifier where feedback capacitor C_E contributes, in RC coupled transistor amplifiers, the coupling capacitor contributes to the frequency response.

The small signal AC equivalent circuit is shown in fig(12.6a). It is easier to analyse the output side if the current source is converted to a voltage source (see fig 12.6b). The voltage loop equation at the output side gives

$$h_{fe}i_b R_c = \left[R_c + X_{C_c} + \frac{R_b h_{ie}}{R_b + h_{ie}}\right] i$$

giving the net current in the circuit as

$$i = \frac{h_{fe}i_b R_c}{\left[R_c + X_{C_c} + \frac{R_b h_{ie}}{R_b + h_{ie}}\right]}$$

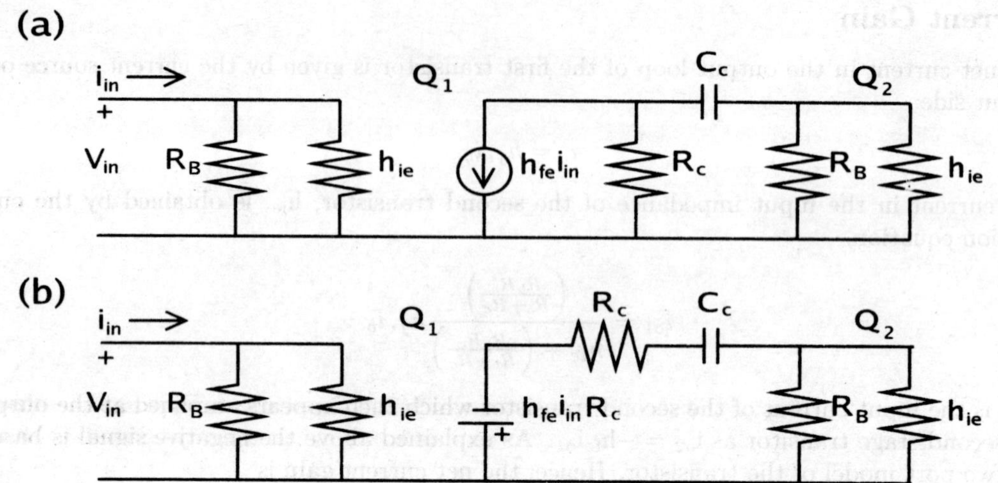

Figure 12.6: *The coupling capacitor that is used to couple the two stage amplifier introduces a frequency response in low frequencies. The small signal hybrid of the same is shown in (a). The current source is converted to a voltage source (b) for analyzes.*

This current splits into R_b and h_{ie} (of the second transistor). The current in h_{ie} is given as

$$i_1 = \left(\frac{R_b}{R_b + h_{ie}}\right) i$$

$$= \left(\frac{R_b}{R_b + h_{ie}}\right) \left[\frac{h_{fe} i_b R_c}{R_c + X_{C_c} + \frac{R_b h_{ie}}{R_b + h_{ie}}}\right]$$

Again this is the input current for the second stage transistor whose output current is $i_2 = -h_{fe} i_1$. Hence, the current gain of the transistor amplifier of an RC coupled transistor amplifier is

$$A_{il} = -\left(\frac{R_b}{R_b + h_{ie}}\right) \left[\frac{h_{fe}^2 R_c}{R_c + X_{C_c} + \frac{R_b h_{ie}}{R_b + h_{ie}}}\right]$$

$$= \frac{-h_{fe}^2 R_c R_b}{R_c R_b + R_c h_{ie} + R_b h_{ie} + X_{C_c}(R_b + h_{ie})}$$

Further simplification gives

$$= \frac{-\left[\frac{h_{fe}^2 R_c R_b}{R_c R_b + R_c h_{ie} + R_b h_{ie}}\right]}{1 + \left[\frac{X_{C_c}(R_b + h_{ie})}{R_c R_b + R_c h_{ie} + R_b h_{ie}}\right]}$$

$$= \frac{A_{im}}{1 + \left[\frac{X_{C_c}(R_b + h_{ie})}{R_c R_b + R_c h_{ie} + R_b h_{ie}}\right]}$$

$$= \frac{A_{im}}{1 - j\left[\dfrac{(R_b + h_{ie})}{2\pi f(R_c R_b + R_c h_{ie} + R_b h_{ie})C_c}\right]}$$

The numerator of the above expression is the current gain in the mid-frequency regime. Substituting f_l,

$$f_l = \frac{(R_b + h_{ie})}{2\pi(R_c R_b + R_c h_{ie} + R_b h_{ie})C_c}$$

we have

$$A_{il} = \frac{A_{im}}{1 - j\frac{f_l}{f}}$$

or

$$|A_{il}| = \frac{|A_{im}|}{\sqrt{1 + \frac{f_l^2}{f^2}}}$$

This equation is same as eqn(11.4) and represents variation of gain at low frequencies. The response would be same as that shown in fig(11.11) with f_l being the $1/\sqrt{2}$ (cut-off) point.

High Frequency Response

The reversed biased junction of the collector-base contributes a capacitance (junction capacitance) whose effect is only observed at very high frequencies. The contribution of this capacitance was on both the input and output of the transistor amplifier, all this was discussed in Chapter 11. Consider the Miller capacitance on the output side is C_o and on the input side it is C_d. On cascading, C_o of the first stage transistor comes in parallel to C_d of the second stage. Since Miller effect makes $C_d \gg C_o$, the cascaded transistor amplifier's frequency response is essentially due to C_d.

Figure(12.7) shows the small signal AC equivalent circuit for the two stage RC coupled amplifier at high frequency. Following the derivation of the last section, the current source is converted to a voltage source, where $R'_c = \frac{R_B R_c}{R_B + R_c}$. The net current in the output circuit is

$$i = \frac{h_{fe} i_b R'_c}{\left[R'_c + \frac{X_C h_{ie}}{X_C + h_{ie}}\right]}$$

Once the net current is known, the current in the second transistor's input impedance (h_{ie}) is obtained by the current division equation. That is

$$i_{o1} = \left(\frac{X_C}{X_C + h_{ie}}\right)\left[\frac{h_{fe} i_b R'_c}{R'_c + \frac{X_C h_{ie}}{X_C + h_{ie}}}\right]$$

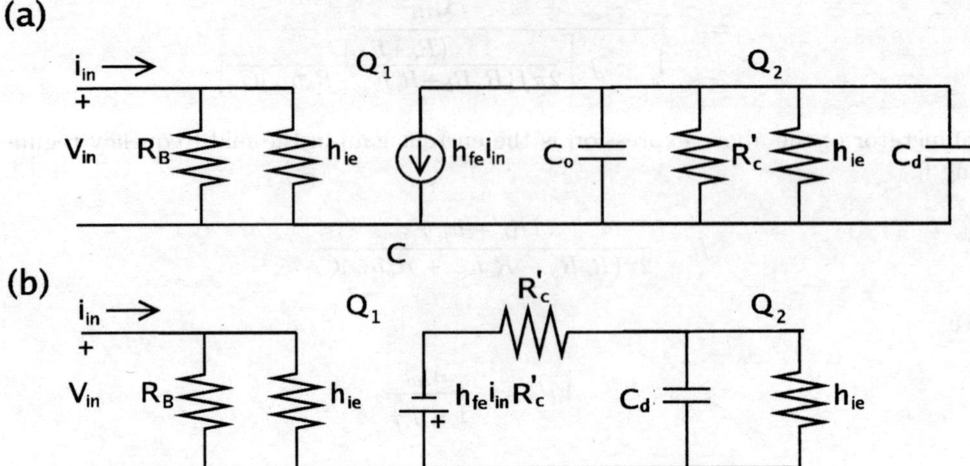

Figure 12.7: *The Miller capacitors introduces a frequency response in high frequencies. The small signal hybrid of the same is shown in (a). The current source is converted to a voltage source (b) for analyzes.*

The current is the output of the second transistor is ($i_{o2} = -h_{fe}i_{o1}$)

$$i_{o2} = -\left(\frac{X_C}{X_C + h_{ie}}\right)\left[\frac{h_{fe}^2 i_b R_c'}{R_C' + \frac{X_C h_{ie}}{X_C + h_{ie}}}\right]$$

Hence, the net current gain is given as

$$A_{ih} = -\frac{h_{fe}^2 R_c' X_C}{R_c' h_{ie} + X_C(h_{ie} + R_c')}$$

$$= -\frac{h_{fe}^2 R_c'}{j\omega R_c' h_{ie} C + (h_{ie} + R_c')}$$

(12.6)

Substituting, the expression for R_c' and A_{im} and simplifying, we get

$$A_{ih} = -\frac{h_{fe}^2 R_b R_c}{j\omega R_b R_c h_{ie} C + h_{ie}(R_b + R_c) + R_b R_c}$$

which on expanding results in

$$= -\frac{\left[\dfrac{h_{fe}^2 R_b R_c}{h_{ie}(R_b + R_c) + R_b R_c}\right]}{1 + j\left[\dfrac{\omega R_b R_c h_{ie} C}{h_{ie}(R_b + R_c) + R_b R_c}\right]}$$

$$= \frac{A_{im}}{1 + j\left[\dfrac{\omega R_b R_c h_{ie} C}{h_{ie}(R_b + R_c) + R_b R_c}\right]}$$

substitute

$$f_h = \frac{h_{ie}(R_b + R_c) + R_b R_c}{2\pi R_b R_c h_{ie} C}$$

we have

$$A_{ih} = \frac{A_{im}}{1 + j\frac{f}{f_h}} \tag{12.7}$$

or

$$|A_{ih}| = \frac{|A_{im}|}{\sqrt{1 + \frac{f^2}{f_h^2}}}$$

The above expression is the same as eqn(11.24), the equation which showed the frequency response of a single stage transistor amplifier at high frequencies. The $\frac{1}{\sqrt{2}}$ cut-off point is given by f_h. The value of the cut-off frequency of the two stage transistor amplifier is less than that of a single stage (verify). This is expected as, we had on the onset itself stated, as the gain increases the bandwidth comes down. This gives an idea of new constant, we discuss in the next section.

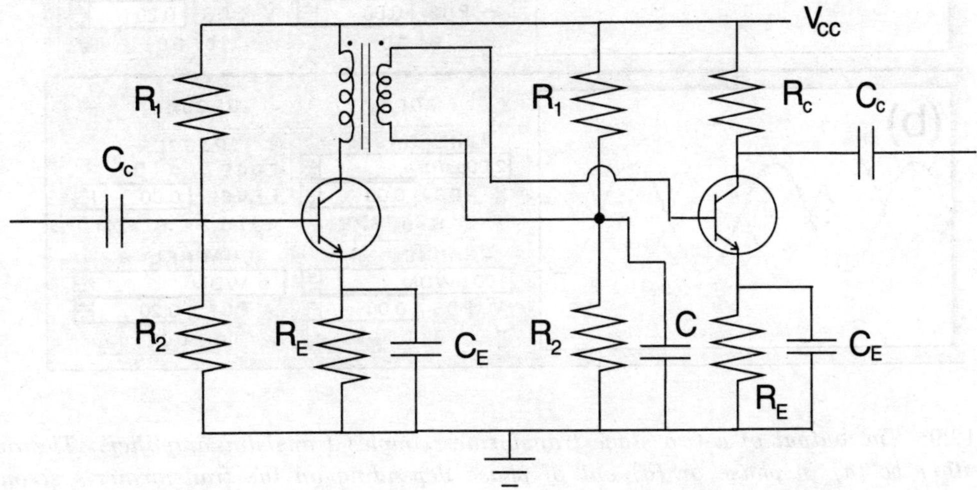

Figure 12.8: *Two transistors coupled using a transformer.*

Another Constant, Gain-Bandwidth Product

The bandwidth is found by calculating the difference between the two cut-off points, i.e. f_h-f_l. However, since under normal case of circuit design, $f_h \gg f_l$, the bandwidth is approximated as f_h. This makes it easier to calculate a new constant, the gain-bandwidth product (G.BW). It is essentially a product of the amplifier's gain in the mid-frequency region with the high frequency cut-off point.

$$G.BW \;=\; \left[\frac{h_{fe}^2 R_b R_c}{h_{ie}(R_b + R_c) + R_b R_c}\right] \times \left[\frac{h_{ie}(R_b + R_c) + R_b R_c}{2\pi C_o R_b R_c h_{ie}}\right]$$

$$=\; \frac{h_{fe}^2}{2\pi C_o h_{ie}} \tag{12.8}$$

This constant highlights the variation (decrease) of the amplifier's gain with increasing gain.

12.3 Transformer Coupled Amplifiers

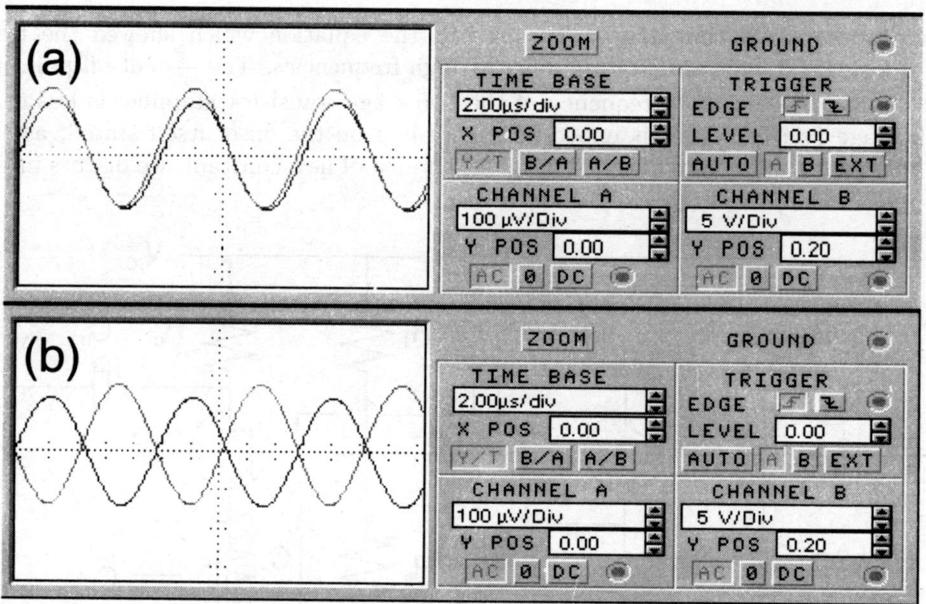

Figure 12.9: *The output of a two stage transformer coupled transistor amplifier. The output would either be (a) in phase or (b) out of phase depending on the transformer's secondary connection.*

In transformer coupled amplifiers a transformer is used to pass the signal from the collector of one transistor to the base of the next. The transformer coupling provides DC isolation, i.e. the DC voltages at the collector of the first stage can not effect the bias of the next, just as the coupling capacitor did in the case of a RC coupled multistage amplifier. However, the transformer allows for impedance matching and hence is generally done when the input impedance (load of the first circuit) of the second stage is very small. Remember, the gain of the two stage RC coupled amplifier was found to be less than the theoretical gain (A_v^2). This

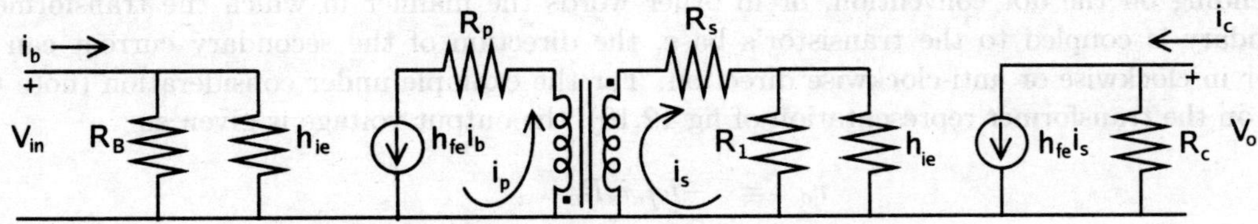

Figure 12.10: *The small signal AC model of the transformer coupled amplifier in mid-frequency range.*

was due to the fact that the second stage was loading the preceeding stage resulting from it's input impedance being small.

Figure 12.8 is the circuit diagram of two transistor amplifiers coupled using a transformer. The important point to note is that the biasing resistance (R_2) is AC shorted using a capacitor kept in parallel. This parallel capacitor allows for the AC input to be fed into the the base of the second transistor with respect to ground. The mid-frequency response of the circuit can be evaluated using the small signal equivalent circuit shown in fig(12.10). In the mid-frequency and low frequency range we assume the impedance (ωL) offered by the coils of the transformer are negligible. The input voltage is due to the input current i_b flowing through h_{ie}, i.e.

$$v_{in} = h_{ie} i_b$$

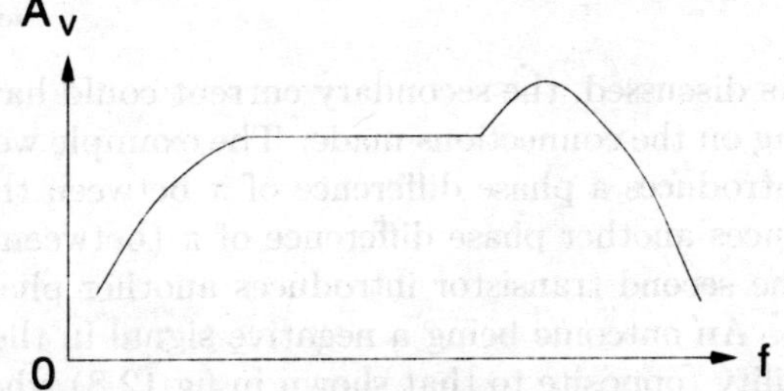

Figure 12.11: *The frequency response of the transformer coupled amplifier.*

This expression is obtained on using the usual assumption $R_B \gg h_{ie}$, where R_B is the parallel combination of R_1 and R_2. The collector current of the first transistor is related to the input current via relation

$$i_c = h_{fe} i_b$$

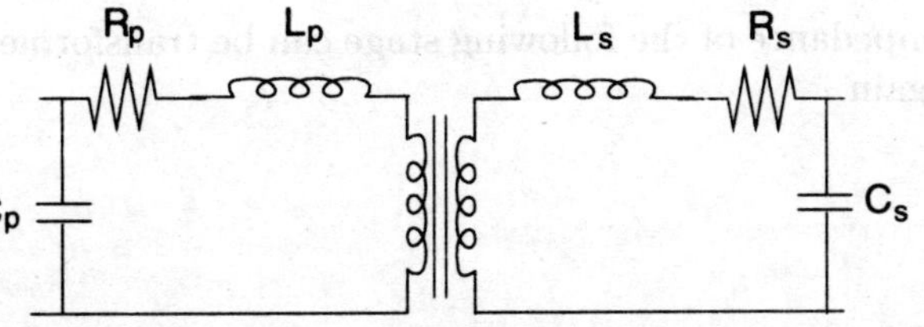

Figure 12.12: *A section of the small signal AC model of the transformer coupled amplifier at high frequency.*

since this current is fed into the coupling transformer's secondary having N_P coils which present a stray resistance R_P, the current can also be named as i_p. The subscript implying that it is associated with the primary of the transformer. Using eqn(4.6), the secondary current is given as

$$i_s = \left(\frac{N_p}{N_s}\right) i_p$$
$$= \left(\frac{N_p}{N_s}\right) h_{fe} i_b$$

Depending on the dot convention, or in other words the manner in which the transformer's secondary is coupled to the transistor's base, the direction of the secondary current can be either in clockwise or anti-clockwise direction. For the example under consideration (note the dots on the transformer representation of fig 12.10), the output voltage is given as

$$v_o \;=\; -h_{fe}i_s R_c$$

$$\;=\; -h_{fe}^2 \left(\frac{N_p}{N_s}\right) R_c i_b$$

Thus, the mid-frequency gain of the transformer couled amplifier is given as

$$A_v = -\frac{h_{fe}^2}{h_{ie}} \left(\frac{N_p}{N_s}\right) R_c \qquad (12.9)$$

As discussed, the secondary current could have been either clockwise or anti-clockwise depending on the connections made. The example we have discussed has the senario, the first transistor introduces a phase difference of π between the input current i_b and i_p. The transformer introduces another phase difference of π (between i_p and i_s). Thus, i_s is in phase with i_b. However, the second transistor introduces another phase differenc of π between output current (i_o) and i_s. An outcome being a negative signal in the result shown in eqn(12.9). For the second possibility (opposite to that shown in fig 12.8) where the transformer introduces no phase difference between the primary and secondary current, the output signal would be in phase with the input current. The gain of the circuit would be same as that given in eqn(12.9) except for the -ive sign. This difference is highlighted in the oscilloscope traces shown in fig(12.9).

The transformer coupling is not used in modern audio amplifiers due to it's disadvantages, however it does have the following advantages:

Advantage:

The output impedance and the input impedance of the following stage can be transformed for optimum matching and maximum gain.

Disadvantages:

It is expensive.

The transformer causes sound quality losses.

Consider cascading two amplifiers that was designed in Example 9.5 (Chapter 9). The gain of the amplifier was 400, with $R_c = 3.7K\Omega$, $R_E = 300\Omega$, $R_1 = 270K\Omega$ and $R_2 = 90K\Omega$. The transistor parameters were $h_{ie} = 460\Omega$ and $h_{fe} = 50$. The first transistor's collector resistance is replaced with a transformer with $N_p/N_s = 4$. The voltage gain using eqn(12.9) works out to be ≈ 80434 or 98dB. The output of this circuit is shown in fig(12.9). The value of gain is better than that of a similar exercise we did for RC coupled amplifier, however, the poor frequency response of this cascading method makes it unpopular.

The frequency response of transformer coupled amplifier is shown in fig(12.11). The constant region is usually over a small range, with a hump appearing in the high frequency range. This comes into picture due to the tank circuits forming on the primary and secondary side of the transformer, as seen in the small signal equivalent model of the amplifier (fig 12.12). At high frequencies, the impedance offered by the transformer coils (ωL) can not be neglected. Along with this, the Miller capacitors of the transistors form a tank circuit. At the resonant frequency (for details see Chapter 17) a peak forms which looks like a hump on the transformer coupled amplifier's frequency response.

Exercise

Q1. What is meant by "cascading"? Where is it useful?

Q2. Why is it common for amplifier circuits to use multiple stages of transistors, rather than just one transistor?

Q3. Describe the function of each component shown in fig(12.3), the circuit of a two-stage RC coupled amplifier circuit.

Q4. Label the transformer's polarity using "dot" notation in order to achieve (a) no inversion and (b) inversion of signal from input to output (see fig 12.8).

Q5. What advantage(s) does a transformer-coupled amplifier have over circuits using other methods of coupling? Are there any disadvantages to using a transformer for signal coupling between transistor stages? Explain in detail.

Q6. A 3-stage RC coupled amplifier uses identical stages of gain 100 with cut-off frequencies of $f_l=50Hz$ and $f_h=500KHz$. Find the overall voltage gain and bandwidth.

Q7. What would happen to the mid-frequency voltage gain and bandwidth of an RC coupled amplifier if the coupling capacitor is removed?

Chapter 13

Feedback Circuits

Feedback circuits are circuits that sample the output and place it back at the input. This action is done to improve the stability of the circuit. Any change in the output is negated by the adjustment made at the input. How much adjustment is to be made is decided by the magnitude of variation produced at the output which is monitored (sampled). This has already be seen in the emitter biasing circuit (see Chapter 9) which senses the increase in collector current and brings the input base current down. The decrease in base current brings the output current down, stabilizing the output and preventing possible damage to the transistor by thermal run away.

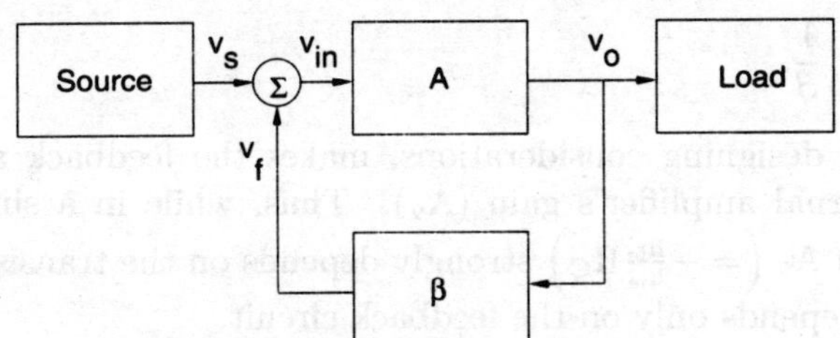

Figure 13.1: *The block diagram of feedback circuits. The input to the internal amplifier (v_{in}) is a sum of the signal from the source (v_s) and the feedback (β) circuit.*

A simple feedback circuit is shown in figure(13.1). The first block is the internal amplifier whose gain is given as 'A_v', where

$$A_v = \frac{v_o}{v_{in}} \qquad (13.1)$$

The β block of fig(13.1) is the feedback circuitry which senses the output voltage v_o. Without modifying this output, a fraction of it is feedback or added with the input (v_s). It should be appreciated that the sampling of the output can be done either by sensing the output voltage or output current. In the example of interest (fig 13.1) the feedback circuit senses the output voltage. Thus, the net input, the input given to the amplifier with feedback is given as

$$v_s = v_{in} + \beta v_o \qquad (13.2)$$

where β is the fraction of the sensed output that is being fed back to the input. As implied by *"fraction"*, β is a number between zero and one. Using eqn(9.1), the above equation reduces to

$$v_s = \frac{v_o}{A_v} + \beta v_o$$

$$v_s = v_o \left(\frac{1 + A_v\beta}{A_v} \right)$$

$$\frac{v_s}{v_o} = \left(\frac{1 + A_v\beta}{A_v} \right)$$

or

$$\frac{v_o}{v_s} = \left(\frac{A_v}{1 + A_v\beta} \right)$$

$$A_f = \frac{A_v}{1 + A_v\beta} \tag{13.3}$$

The gain of the amplifier with feedback circuit comes down and hence we say that the circuit has negative feedback.

13.1   Advantage of Negative Feedback

Gain Stability

Even though this might appear as counter productive, negative feedback circuits do have some advantages. Consider the situation where the designer has achieved the condition $A_v\beta \gg 1$, then eqn(13.3) reduces to

$$A_f = \frac{A_v}{1 + A_v\beta} = \frac{A_v}{A_v\beta}$$

$$= \frac{1}{\beta}$$

i.e. the feedback circuit under proper designing considerations, makes the feedback amplifier's gain (A_f) independent of the internal amplifier's gain (A_v). Thus, while in a single stage common emitter amplifier whose gain $A_v \left(= -\frac{h_{fe}}{h_{ie}} R_C \right)$ strongly depends on the transistor parameters, the feedback amplifier's gain depends only on the feedback circuit.

Reduced Sensitivity

Even if the condition of $\beta A_v \gg 1$ is not achieved, by using a negative feedback, the sensitivity of the amplifier can be bought down. To appreciate this different eqn(13.3) with respect to A_v, we have

$$\frac{dA_f}{dA_v} = \frac{(1 + \beta A_v) - A_v\beta}{(1 + \beta A_v)^2}$$

$$= \frac{1}{(1 + \beta A_v)^2}$$

Modifying the above equation we write

$$\frac{dA_f}{dA_v} = \frac{A_v}{(1 + \beta A_v)} \times \frac{1}{(1 + \beta A_v)} \times \frac{1}{A_v}$$

$$dA_f = A_f \times \frac{1}{(1 + \beta A_v)} \times \frac{dA_v}{A_v}$$

$$\frac{dA_f}{A_f} = \frac{1}{(1 + \beta A_v)} \times \frac{dA_v}{A_v}$$

This shows even if there is a change in the gain of the internal amplifier, the change in the gain of the amplifier with feedback is not much (reduced by a factor $1 + \beta A_v$). Thus, adding a feedback reduces the sensitivity of the amplifier. The term $1 + \beta A_v$ is also addressed as the **desensitivity factor**.

Reduction of Noise

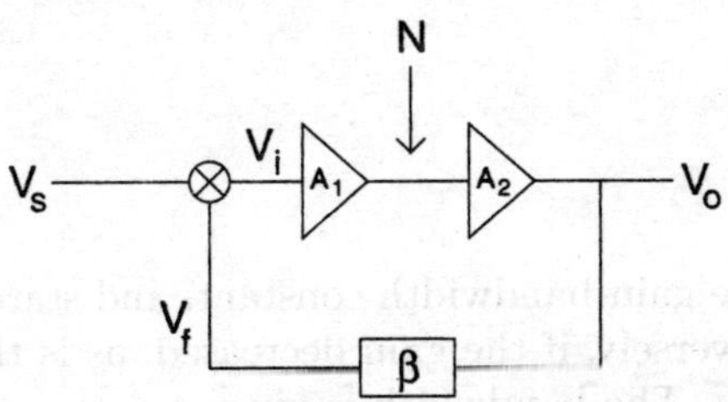

Figure 13.2: *The circuit with feedback to understand how negative feedback reduces noise. To develop the mathematics it is assumed that noise appears in the circuit in the output of amplifier A_1.*

The negative feedback reduces the effect of noise in the circuit. To understand how this is achieved, we consider the block diagram of fig(13.2). To make the case general, the block diagram shows two internal amplifiers being cascaded and the noise is introduced at the output of the first amplifier. The output of the second amplifier is sampled and feed back at the input of the first internal amplifier. The output would be given as

$$V_o = V_i A_1 A_2 + N A_2$$

with the voltage (V_i) being the sum of the input signal voltage and the feedback signal, i.e.

$$V_i = V_s + V_f$$

Hence

$$\begin{aligned} V_o &= A_1 A_2 (V_s + V_f) + N A_2 \\ &= A_1 A_2 (V_s + \beta V_o) + N A_2 \end{aligned}$$

as the amount of feedback is proportional to the output voltage ($V_f = \beta V_o$). Thus, the output voltage interms of the input voltage (V_s) is

$$\begin{aligned} V_o(1 - A_1 A_2 \beta) &= A_1 A_2 V_s + N A_2 \\ V_o &= \frac{A_1 A_2}{1 - A_1 A_2 \beta} \left[V_s + \frac{N}{A_1} \right] \end{aligned} \qquad (13.4)$$

As can be seen from the above equation, the contribution of the noise is significantly reduced, especially if gain A_1 is large.

Even if only a single amplifier is being used and noise is being introduced at the input itself, the noise level in the output is reduced by $1 - A_2 \beta$ (take $A_1 = 1$). In practical circuits of measuring devices however, two stage amplification is preferred since it makes sure that the signal to noise ratio (SNR) is appreciable. i.e.

$$\frac{V_o}{N} = \frac{A_1 A_2}{1 - A_1 A_2 \beta} \left[\frac{V_s}{N} + \frac{1}{A_1} \right] \to \infty$$

Reduction of Non-linear Distortion

The non-linear distortion in an amplifier is a result of the non-linear I-V characteristics associated with the transistor. This distortion is seen as a flattening of the peaks of the sine wave. The negative feedback brings down the gain and hence also the distortion. Similar to that of gain levels, the distortion level after feedback is given as

$$D' = \frac{D}{1 + \beta A_v}$$

Increase in Bandwidth

We had in the last chapter (Chapter 12) talked of the gain-bandwidth constant, and stated that if the gain increased the bandwidth decreased. Conversely, if the gain decreased, as is the case of giving negative feedback, the bandwidth increases. The bandwidth is given as

$$B.W = f_h - f_l$$

with the higher cut-off point (f_h) usually being far greater than the lower cut-off point (f_l), hence it is appropriate to approximate

$$B.W \approx f_h$$

Eqn(13.3) holds over all frequency range, even high frequencies, hence to show that we are restricting our discussion to high frequencies a subscript h is included in eqn(13.3)

$$(A_h)_f = \frac{(A_h)_v}{1 + \beta (A_h)_v}$$

where the expression relating variation of the gain (in high frequency regime) with frequency of the internal amplifier and amplifier's gain in the mid-frequency regime is given by eqn(12.7) of Chapter 12.

$$A_h = \frac{A_m}{1 + j\frac{f}{f_h}}$$

Using this equation in eqn(13.3), we have

$$(A_h)_f = \frac{\frac{(A_m)_v}{1 + j\frac{f}{f_h}}}{1 + \beta \frac{(A_m)_v}{1 + j\frac{f}{f_h}}}$$

or

$$(A_h)_f = \frac{(A_m)_v}{1 + \beta(A_m)_v + j\frac{f}{f_h}}$$

dividing the numerator and denominator by $1 + \beta(A_m)_v$, we have

$$(A_h)_f = \frac{\frac{(A_m)_v}{1+\beta(A_m)_v}}{1 + j\frac{f}{f_h[1+\beta(A_m)_v]}}$$

The expression

$$\frac{(A_m)_v}{1 + \beta(A_m)_v}$$

relates to the voltage gain of the amplifier with feedback in the mid-frequency regime, i.e. it is $(A_m)_f$. Hence the above equation can be written as

$$(A_h)_f \;=\; \frac{(A_m)_f}{1 + j\frac{f}{f_h[1+\beta(A_m)_v]}}$$
$$=\; \frac{(A_m)_f}{1 + j\frac{f}{(f_h)_f}}$$

Comparing the above equation with eqn(12.7) the term $f_h[1 + \beta(A_m)_v]$ shows the new cut-off point is the high frequency regime of the amplifier with feedback. As can be seen, the cut-off point of the amplifier increases with negative feedback (i.e. $f_h \to f_h[1 + \beta(A_m)_v]$). Subject to the condition discussed at the onset, this indicates an increase in the amplifier's bandwidth on giving a negative feedback.

| Example 13.1: | What happens to the lower cut-off point on using negative feedback? |

The expression relating how the gain of an amplifier changes with frequency and the amplifier's gain in the mid-frequency for very low frequencies is given in Chapter 12.

$$A_l \;=\; \frac{A_m}{1 + j\frac{f_l}{f}} \tag{13.5}$$

Using this equation in eqn(13.3), we have

$$(A_l)_f = \frac{\frac{(A_m)_v}{1+j\frac{f_l}{f}}}{1 + \beta\frac{(A_m)_v}{1+j\frac{f_l}{f}}}$$

or

$$(A_l)_f = \frac{(A_m)_v}{1 + \beta(A_m)_v + j\frac{f_l}{f}}$$

dividing the numerator and denominator by $1 + \beta(A_m)_v$, we have

$$(A_l)_f = \frac{\frac{(A_m)_v}{1+\beta(A_m)_v}}{1 + j\frac{f_l}{f[1+\beta(A_m)_v]}}$$

The expression

$$\frac{(A_m)_v}{1 + \beta(A_m)_v}$$

relates to the voltage gain of the amplifier with feedback in the mid-frequency regime, i.e. it is $(A_m)_f$. Hence the above equation can be written as

$$(A_l)_f = \frac{(A_m)_f}{1 + j\frac{f_l}{f[1+\beta(A_m)_v]}}$$

$$= \frac{(A_m)_f}{1 + j\frac{(f_l)_f}{f}}$$

Comparing this equation with eqn(13.5) the term $\frac{f_l}{[1+\beta(A_m)_v]}$ is the new cut-off point in the low frequency regime of the amplifier with feedback. As can be seen, the cut-off point of the amplifier decreases with negative feedback (i.e. $f_l \gg f_h[1 + \beta(A_m)_v]$).

Positive Feedback

If β is negative, eqn(13.3) shows an increase in the feedback amplifier's gain, i.e.

$$A_f = \frac{A_v}{1 - A_v\beta}$$

such feedback circuits are said to have positive feedback (see Chapter 15). Note if you have positive feedback the gain increases while it decreases (eqn 13.3) in case of negative feedback. The question of how to make β positive or negative requires us to understand the topology of feedback circuits which we introduce in the following sections.

13.2   Voltage-Series Feedback

Consider the circuit shown in fig(13.3). The output "*voltage*" is sampled by a potential divider circuit in the feedback block (represented as β in the block diagram). If the potential divider does not load the output ($R_1 + R_2$ is large as compared to R_L), the output voltage remains unaffected by the sampling. The voltage drop across R_2 is then added in "*series*" to the input voltage (v_s). Thus, the shown circuit is an example of voltage-series feedback. The Kirchoff's voltage equation for the input side loop can be written as

$$v_s = v_{in} + v_{R_2} \tag{13.6}$$

where v_{R_2} is the voltage fed-back to the input (v_f). This equation can be re-written as

$$v_{in} = v_s - v_{R_2}$$

Note, the equation has been re-written to show that the feedback is done such that it subtracts from the input applied (v_s). This implies that this circuit is of a negative feedback. Thus, without calculating the gain expression one can make out if the circuit has negative or positive feedback, from the manner in which the feedback is fed at the input (whether it brings it down or increases it). We now calculate the gain of this circuit.

13.2.1 Gain

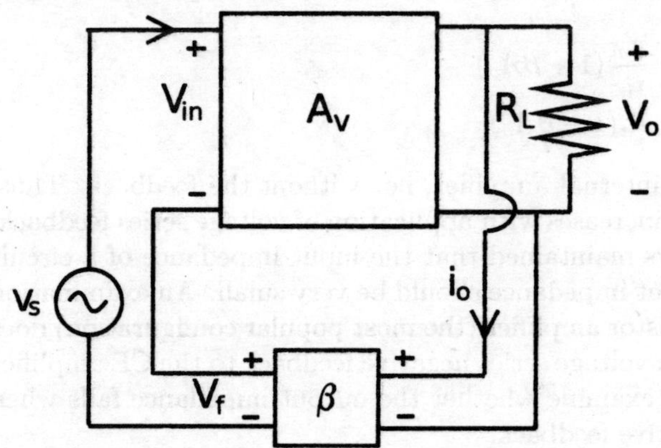

Figure 13.3: *The voltage series feedback circuit.*

Using the knowledge of the internal amplifier's gain, A_v $(=v_o/v_{in})$ and eqn(9.3), we have

$$v_s = \frac{v_o}{A_v} + \left(\frac{R_2}{R_1 + R_2}\right) v_o$$

$$= \left(\frac{1}{A_v} + \frac{R_2}{R_1 + R_2}\right) v_o$$

where $\frac{R_2}{R_1 + R_2}$ is the fraction of the output voltage that is sampled and fed back to the input, say

$$\beta = \frac{R_2}{R_1 + R_2} = \frac{v_f}{v_o}$$

we have

$$\frac{v_s}{v_o} = \left(\frac{1}{A_v} + \beta\right)$$

$$\frac{1}{A_f} = \left(\frac{1 + \beta A_v}{A_v}\right)$$

The net gain of the amplifier with feedback is related to the internal amplifier by

$$A_f = \frac{A_v}{1 + \beta A_v}$$

Thus, the voltage gain of an amplifier with voltage series feedback comes down. The nomenclature of the feedback should be appreciated. If the output voltage is being sampled and is given back to the input as a voltage, the feedback is called voltage series feedback. The term "series" arises since, at the input, voltages are added in series. Had current been sampled at the output, it would be added in parallel to the input current source, i.e it would have been added in shunt. The feedback in that case would have been *"current shunt"*.

13.2.2   Input Impedance

The input impedance of the amplifier circuit changes with application of feedback. This is computed using eqn(9.3)

$$v_s = v_{in} + v_f \tag{13.7}$$

where the fed-back voltage is a fraction of the output voltage. Hence, the circuit's input impedance with feedback is

$$
\begin{aligned}
R_{if} &= \frac{v_s}{i_{in}} = \frac{v_{in} + \beta v_o}{i_{in}} \\
&= \frac{v_{in} + \beta A_v v_{in}}{i_{in}} \\
&= \frac{v_{in}}{i_{in}}(1 + \beta A_v) \\
R_{if} &= R_i(1 + \beta A_v)
\end{aligned}
$$

where R_i is the input impedance of the internal amplifier, i.e. without the feedback. Thus, the input impedance of the amplifier circuit increases with application of voltage series feedback. In our discussion on loading, we have always maintained that the input impedance of a circuit should be as large as possible while the output impedance should be very small. An examination of Table(8.4) shows that the CE mode transistor amplifier (the most popular configuration) does not satisfy this criteria. Hence, after giving a voltage series negative feedback to the CE amplifier the input impedance increases. Let us now examine whether the output impedance falls when the amplifier is given as voltage series negative feedback.

13.2.3   Output Impedance

For evaluating the output impedance of the circuits with voltage series feedback or for that matter for any feedback based on sensing the output voltage, the current source of the hybrid model's output is converted to a voltage source. This was also done while analyzing the multi-stage transistor amplifier at high frequencies (see Chapter 12, fig 12.6 and fig 12.7). The magnitude of the voltage source is given as

$$v_o = h_{fe} i_{in} R_c$$

with R_c coming in series to the voltage source and the direction of the voltage source as shown in fig(13.4a). R_c is the output impedance of the amplifier without any feedback. The voltage gain of the internal amplifier is given as (the input voltage is obtained by converting the current source to a voltage source)

$$A_v = -\left(\frac{h_{fe} i_{in} R_c}{v_{in}}\right)$$

or

$$h_{fe} i_{in} R_c = -A_v v_{in} \tag{13.8}$$

Thus, the voltage source of fig(13.4a) can be written and represented interms of the internal amplifier's gain (fig 13.4b). The negative sign (is due to the phase difference introduced between the output and input signal) implies the polarity of the voltage source of fig(13.4b) has to be inverted as shown in fig(13.4c).

(a)

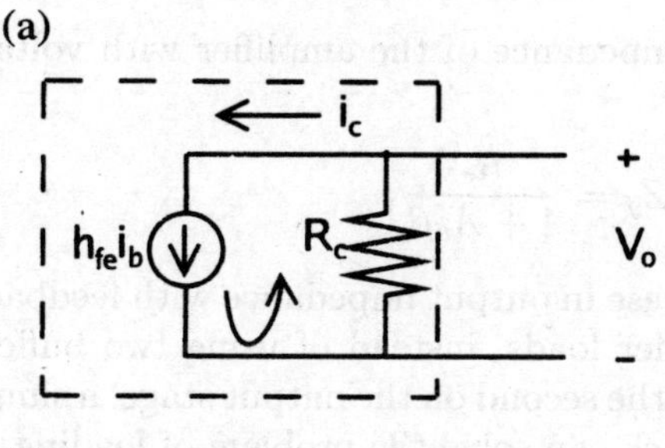

(b)

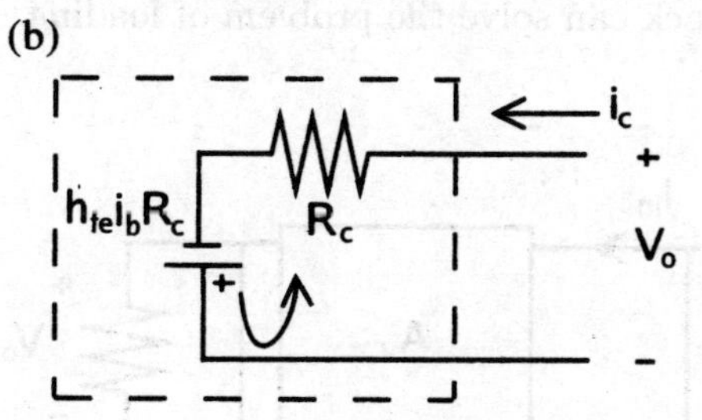

(c)

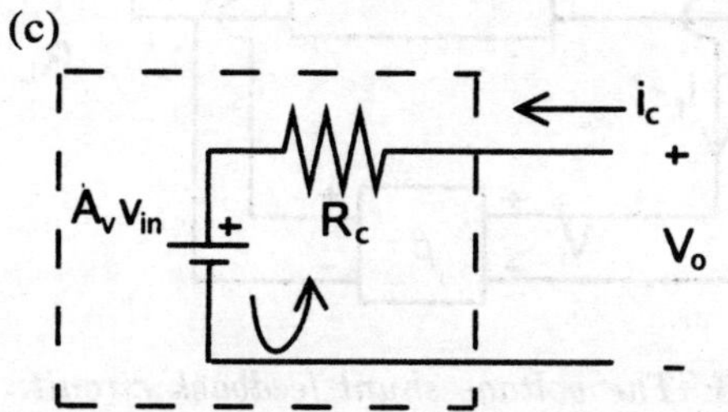

Figure 13.4: *The output side of the voltage series feedback circuit. Steps for converting the hybrid model's current source to voltage source is shown through (a) to (c).*

Now to evaluate the output impedance imagine a voltage source v_o is applied to the output terminals and that the input to the amplifier with feedback is zero (i.e. $v_s = 0$, see fig 13.5). This implies, from eqn(13.2), that $v_{in} = -\beta v_o$. Applying voltage source v_o results in current i_o being fed into the circuit. This splits into two currents, i_1 into resistance R_c and i_2 into the feedback circuitry. That is, the output impedance of the circuit with feedback is given as

$$Z_f = \frac{v_o}{i_o}$$

and

$$i_o = i_1 + i_2 \qquad (13.9)$$

On applying KVL on the output loop, we have

$$\begin{aligned} v_o &= i_1 R_c + A_v v_{in} \\ &= i_1 R_c - A_v \beta v_o \end{aligned}$$

hence, the current in the loop is

$$i_1 = \left(\frac{1 + A_v \beta}{R_c} \right) v_o$$

Also, the loop equation of the feedback circuitry gives

$$i_2 = \frac{v_o}{Z_\beta}$$

where Z_β is the impedance offered by the feedback circuit (we have already demanded that for loading to be prevented, $Z_\beta = R_1 + R_2$ has to be very large). Substituting the two currents in eqn(13.9) we have

$$i_o = \left(\frac{1 + A_v \beta}{R_c} \right) v_o + \frac{v_o}{Z_\beta}$$

or

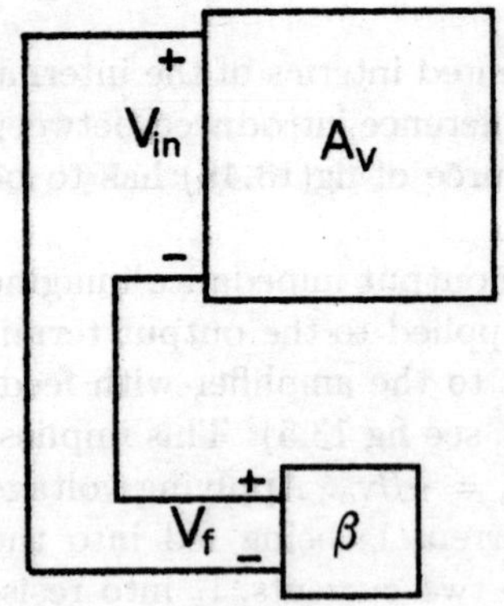

Figure 13.5: *The input side when input* $V_s = 0$.

$$\frac{i_o}{v_o} = \frac{1}{Z_f} = \frac{1 + A_v\beta}{R_c} + \frac{1}{Z_\beta}$$

$$\approx \frac{1 + A_v\beta}{R_c}$$

Hence, the output impedance of the amplifier with voltage series feedback is

$$Z_f = \frac{R_c}{1 + A_v\beta}$$

This shows the decrease in output impedance with feedback. Hence, if the amplifier loads, instead of using two buffers, one at the input and the second on the output stage, a simple voltage series feedback can solve the problem of loading.

13.3   Voltage-Shunt Feedback

As the name suggests, in circuits with voltage shunt feedback, the output voltage is sampled and is fed to the input in parallel (shunted). Since two signals are in parallel they have to be current sources. Thus, the feedback circuit has to convert the sampled voltage into current. Thus, the feedback faction (β) is given as

$$I_f = \beta V_o \qquad (13.10)$$

β has the units of mho. On the input side view how the feed-back signal is fed to the input (see fig 13.6). The Kirchoff's current law applied on the input mode gives

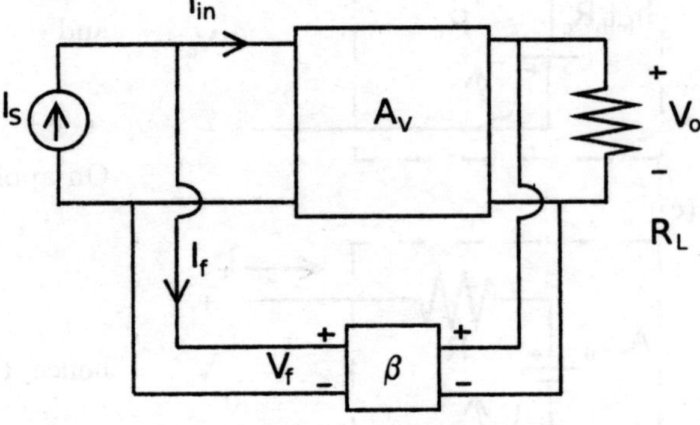

Figure 13.6: *The voltage shunt feedback circuit.*

$$
\begin{aligned}
I_s &= I_f + I_{in} \\
&= \beta V_o + I_{in} \qquad (13.11)
\end{aligned}
$$

The direction of I_f is such that it decreases I_s, hence the arrangement is of a negative feedback. The input impedance of the amplifier with voltage shunt feedback is evaluated below.

13.3.1   Input Impedance

The input impedance of the circuit with feedback is calculated using the following relation

$$R_{if} = \frac{V_{in}}{I_s}$$

Using eqn(13.11), the above equation may be written as

$$R_{if} = \frac{V_{in}}{I_{in} + \beta V_o} = \frac{V_{in}/I_{in}}{1 + \beta V_o/I_{in}}$$

$$= \frac{R_{in}}{1 + \beta R_m}$$

where R_{in} is the input impedance of the internal amplifier (circuit without feedback) and R_m is given as

$$R_m = \frac{V_o}{I_{in}} \tag{13.12}$$

R_m[1] may be defined as a gain parameter of our two port (internal amplifier) circuit, where on giving an input current I_{in}, the output obtained is in terms of voltage, V_o.

13.3.2   Gain

Using eqn(13.11) in eqn(13.12), we have

$$\begin{aligned} V_o &= R_m I_{in} \\ &= R_m(I_s - \beta V_o) \end{aligned}$$

or

$$V_o(1 + \beta R_m) = I_s R_m$$

Thus,

$$R_{mf} = \frac{V_o}{I_s} = \frac{R_m}{1 + \beta R_m}$$

The circuit's gain parameter decreases on applying a voltage shunt negative feedback.

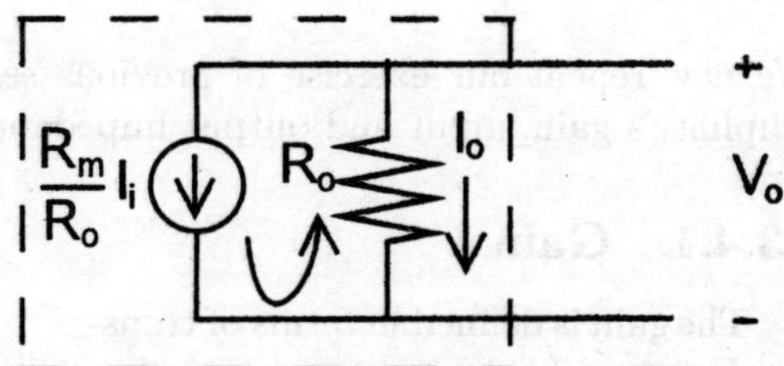

Figure 13.7: *The hybrid model's output side for the amplifier with voltage shunt feedback circuit.*

13.3.3   Output Impedance

To evaluate the output impedance, we again replace the output load (R_L) with a voltage source V_o (see fig 13.7). Also, the input current, I_s is removed. Hence eqn(13.11) gives

$$I_i = -\beta V_o$$

Applying KVL on the output loop, we have

$$\begin{aligned} V_o &= R_m I_i + R_o I_o \\ &= -\beta R_m V_o + R_o I_o \end{aligned}$$

---

[1]The current gain may be written as

$$h_{fe} = \frac{I_o}{I_i} = \frac{I_o}{V_o} \times \frac{V_o}{I_i}$$

or

$$h_{fe} = \frac{R_m}{R_o}$$

Hence, the current source on the output side of the small signal equivalent model of the transistor would be given as $h_{fe}i_b = R_m I_i/R_o$. When this current source is converted to a voltage source we have the magnitude of voltage as $R_m I_i$.

Re-arranging the above equation give us

$$\frac{V_o}{I_o} = \frac{R_o}{1 + \beta R_m}$$

$$Z_{of} = \frac{R_o}{1 + \beta R_m}$$

Thus, the output impedance and the input impedance of the amplifier with voltage shunt feedback decreases.

13.4 Current-Series Feedback

In an amplifier with current series feedback (fig 13.8), the output current is sampled and a proportional voltage is fed back to the input. That is, the feedback factor would be

$$\beta = \frac{V_f}{I_o}$$

We now repeat our exercise of previous sections and examine how this feed back effects the amplifier's gain, input and output impedance.

13.4.1 Gain

The gain is defined in terms of trans-conductance (g_m) since we are interested in the output current (we are sampling this) with respect to the input voltage. Hence, the gain of the internal amplifier is given as

$$g_m = \frac{I_o}{V_{in}}$$

The gain of the amplifier with feedback is given as

$$g_{mf} = \frac{I_o}{V_s}$$

The input sides loop equation is given by eqn(13.7) from which we can substitute V_s. Thus,

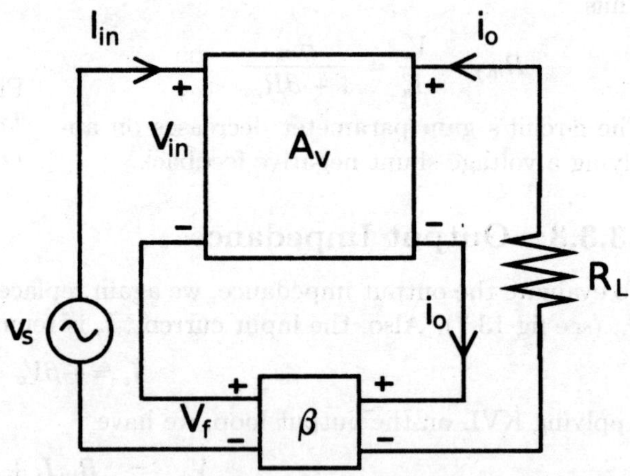

Figure 13.8: *The Current series feedback circuit.*

$$g_{mf} = \frac{I_o}{V_{in} + V_f}$$

$$= \frac{I_o}{V_{in} + \beta I_o}$$

Dividing the numerator and denominator with V_{in}, we have

$$g_{mf} = \frac{\frac{I_o}{V_{in}}}{1 + \beta \frac{I_o}{V_{in}}}$$

$$= \frac{g_m}{1 + \beta g_m}$$

As expected the gain of the negative feedback circuit decreases.

13.4.2   Input Impedance

The input impedance of the amplifier circuit with feedback is computed
using the following equation

$$R_{inf} = \frac{V_s}{I_i}$$

Using eqn(13.7), the above equation modifies to

$$R_{inf} = \frac{V_i + \beta I_o}{I_i}$$

$$= \frac{V_i + \beta g_m V_i}{I_i}$$

This gives the input impedance of the
circuit with feedback in terms of the
internal amplifier's input impedance.

$$R_{inf} = R_{in}(1 + \beta g_m)$$

The input impedance of the circuit with current series feedback increases.

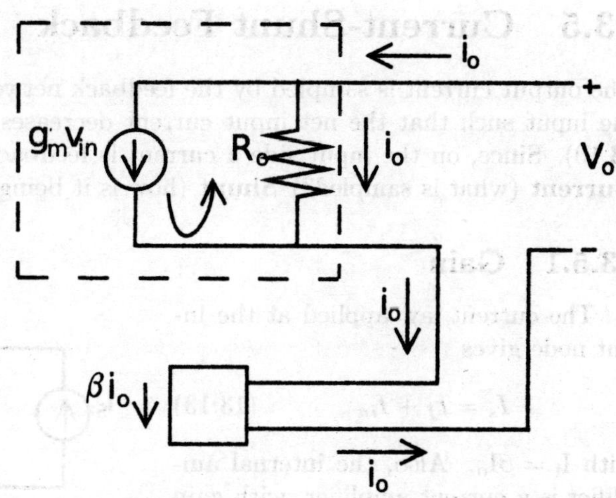

Figure 13.9: *The output side of an amplifier with current series feedback.*

13.4.3   Output Impedance

Following the procedure adopted in the previous sections, to measure the circuit's output
impedance, the load resistance is replaced by a voltage source of V_o and the applied voltage V_s is shorted. The output sides circuit is shown in fig(13.9). The voltage loop equation can
be written using this figure.

$$V_o = R_o(I_o - g_m V_{in}) + Z_\beta(I_o + \beta I_o)$$

$$= R_o(I_o + g_m \beta I_o) + Z_\beta(I_o + \beta I_o)$$

where Z_β is the impedance associated with the feedback network. Also, remember

$$V_s = V_{in} + V_f = V_{in} + \beta I_o$$

$$0 = V_{in} + \beta I_o$$

giving, $V_{in} = -\beta I_o$. Thus, the output impedance is

$$Z_{of} = R_o(1 + g_m\beta) + Z_\beta(1 + \beta)$$

As the feedback circuit is required to sense/ sample the output current without changing the current, the impedance associated with the feedback circuit has to be small. Thus, $Z_\beta(1 + \beta)$ is negligible. The circuit's output impedance increases with current series feedback.

$$Z_{of} = R_o(1 + g_m\beta)$$

13.5   Current-Shunt Feedback

The output current is sampled by the feedback network and a proportional current is added to the input such that the net input current decreases, i.e. a negative feedback is obtained (fig 13.10). Since, on the input side a current is feedback, it adds in parallel and hence the name **Current** (what is sampled?) **Shunt** (how is it being added with the input?) **Feedback**.

13.5.1   Gain

The current law applied at the input node gives

$$I_s = I_f + I_{in} \qquad (13.13)$$

with $I_f = \beta I_o$. Also, the internal amplifier is a current amplifier, with gain $A_i = I_o/I_{in}$. Hence, eqn(13.13) is written as

$$I_s = \beta I_o + \frac{I_o}{A_i}$$

which gives the gain with the feedback network as

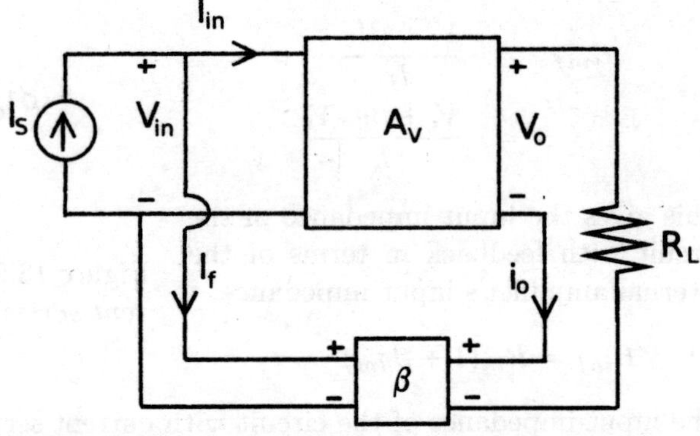

Figure 13.10: *The current shunt feedback circuit.*

$$A_f = \frac{I_o}{I_s} = \frac{A_i}{1 + \beta A_i}$$

Again, a negative feedback brings down the gain.

13.5.2   Input Impedance

The voltage drop across the input impedance with and without feedback is the same, i.e. $V_f = V_{in}$. That is,

$$\begin{aligned} Z_f I_s &= Z_i I_{in} \\ Z_f &= \frac{Z_i I_{in}}{I_s} \end{aligned}$$

Using eqn(13.13), the above equation reduces to

$$Z_f = \frac{Z_i I_{in}}{I_{in} + \beta A_i I_{in}}$$

$$= \frac{Z_i}{1 + \beta A_i}$$

The input impedance of an amplifier with current shunt feedback decreases. This is natural, as parallel (shunted) networks always have lower impedances.

13.5.3 Output Impedance

The procedure of evaluating the output impedance of a circuit has been standard throughout. The output load is removed and a voltage source V_o is applied (fig 13.11). Along with the load resistance, the input to the circuit with feedback is also removed. Hence, eqn (13.13) reduces to

$$0 = I_{in} + \beta I_o$$
$$I_{in} = -\beta I_o$$

The voltage loop equation of the output side (use fig 13.11) is

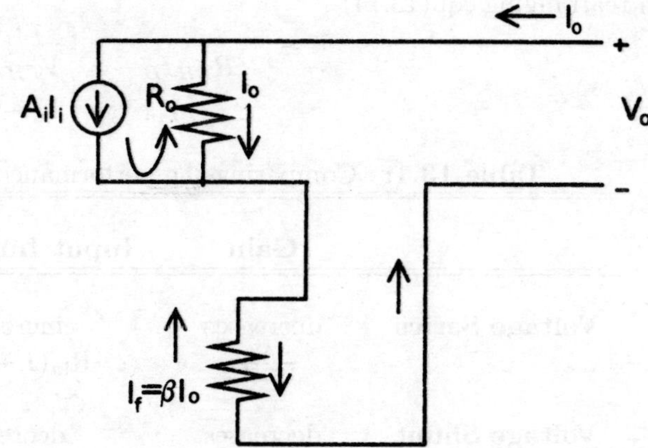

Figure 13.11: *The output side of an amplifier's small signal model with current shunt feedback.*

$$V_o = R_o I_o(1 + \beta A_i) + Z_\beta I_o(1 + \beta)$$

The feedback is sampling the output current and has to be done with a very small impedance. Hence $Z_\beta \approx 0$. Thus, the output impedance of the circuit with current shunt feedback is

$$Z_{of} = \frac{V_o}{I_o} = R_o(1 + \beta A_i)$$

The output impedance shows an increase with application of negative feedback. Table (13.5.3) summarizes and compares the performance of the amplifiers for various feedback modes.

Note, the voltage source should have infinite input impedance and zero output impedance. Though very hard to achieve, a voltage amplifier with voltage series feedback helps to attain this objective (nearly). However, the requirement for an ideal current source is different. It requires near zero input impedance and infinite output impedance. The current shunt feedback helps in designing a good current source.

The voltage-divider transistor biasing as also the emitter resistor biasing are practical examples of current series negative feedback[2]. The input and output loop equations for the circuit

---

[2]The inverting and non-inverting op-amp amplifiers are circuits which have voltage series and voltage shunt feedbacks. The study of op amp is a course in itself and hence have not been discussed here

shown in fig(9.5b) are

$$V_{TH} = V_{BE} + R_{TH}I_B + R_E I_E \tag{13.14}$$

and

$$V_{CC} = V_{CEQ} + R_C I_C + R_E I_E$$

respectively. The output current $I_C = \alpha I_E$ is sensed and is fed-back to the input interms of a voltage $R_E I_E$, which brings done the input voltage (i.e. is a case of negative feedback) as is clear on rearranging eqn(13.14)

$$R_{TH}I_B = V_{TH} - R_E I_E$$
$$V_{in} = V_s - V_f$$

Table 13.1: Comparing the performance of the various feedback circuits.

	Gain	**Input Impedance**	**Output Impedance**
Voltage Series	decreases $\frac{A_v}{1+\beta A_v}$	increases $R_{in}(1 + \beta A_v)$	decreases $\frac{R_o}{1+\beta A_v}$
Voltage Shunt	decreases $\frac{R_m}{1+\beta R_m}$	decreases $\frac{Z_i}{1+\beta R_m}$	decreases $\frac{R_o}{1+\beta R_m}$
Current Series	decreases $\frac{g_m}{1+\beta g_m}$	increases $R_{in}(1 + \beta g_m)$	increases $R_o(1 + \beta g_m)$
Current Shunt	decreases $\frac{A_i}{1+\beta A_i}$	decreases $\frac{Z_i}{1+\beta A_i}$	increases $R_o(1 + \beta A_i)$

As a closing remark it is important to pay tribute to the inventor of negative feedback (concept of), Harold Black. Harold Black graduated from Worcester Polytechnic Institute in 1921 and went on to become a research scientist at Bell Labs. He was assigned to work on the problem of telephony distortion, a problem that hampered long-distance telephony. The harmonic and frequency distortion added to the signal by the line and by the necessary chain of repeater amplifiers made voices unrecognizable and unintelligible. Black conceived the concept of negative feedback in 1927 where the idea came to him as a flash while commuting to work. He published his work, "Stabilized Feed-Back Amplifiers" in the January 1934 issue of Electrical Engineering, published by the American Institute of Electrical Engineering.

Exercise

Q1. What is negative feedback?

Q2. Show that the output impedance decreases with negative feedback if the sampled signal is fed back as a voltage (shunt and series both cases).

Q3. What are the characteristics of a good current amplifier? Which type of feedback is used in it?

Q4. An amplifier in the absense of any feedback has the following specifications:

Input resistance $R_i = 10K\Omega$

Output resistance $R_o = 1K\Omega$

Gain $A_v = 150$

What would be the gain of the amplifier if a 10% negative feedback is given in voltage series configuration?

Q5. Also calculate the change in input and output resistances for Q4.

Q6. Define Desensitivity. What would be the gain of an amplifier with feed back if the desensitivity factor is large.

Q7. Show that if the sampled output is fed back in series to the input, the circuit's input impedance increases.

Q8. An amplifier has $A_v = 500$ and input voltage of 0.01volts. A feedback is added until 1v is required as input to give the same output. What are the values of gain and designed β.

Q9. An amplifier with gain 75dB is connected to a negative feedback network. The feedback factor β is 0.02. If now, due to replacement of the transistor the internal gain is changed by 50%, what would be the percentage change in the gain of the amplifier with the feedback?

Q10. What change should be done in the circuit discussed in Q9 so that even with the change of the transistor, the gain of the circuit remains same as it was initially.

Chapter 14

Field Effect and Uni-Junction Transistors

One might be wondering as to why anymore devices are required, since it seems much of the desired control of electric signals can be achieved with the diode and transistor that we have studied in the previous chapters. However, inspite of the revolution the bipolar junction transistor (BJT) bought about, it suffers from the fact that it's input impedance is low (espeacially the CE and CB configuration which have useful application). This causes problems in matching impedances between interstage amplifiers calling for the use of buffers inbetween.

Research for overcoming led to the development of the **FIELD-EFFECT TRANSISTOR** (*FET*). Infact in 1960, twelve years after the coming of the BJT, Bell scientist John Atalla developed the JFET based on Shockley's original idea presented in 1940's. In contrast to the bipolar transistor, which uses bias current between base and emitter to control conductivity[1], the FET uses voltage to control an electrostatic field within the device, this inturn controls the conductivity of the device (see fig 14.1). This gives the device it's name. Infact the control of conductivity is achieved by an inbuilt diode. With the diode appearing at the input and the device functioning when it is reverse biased, the device has a very high input impedance. Having a high input impedance minimizes the interference with or "loading" of the signal source when a FET is put to use. Because the FET is voltage-controlled, just like a vacuum triode tube, it is sometimes called the "solid-state vacuum tube", (fig 14.2 shows what a vacuum tube looked like, nearly 40-50 times larger then todays transistor or FET).

The elements of one type of FET, the junction type (JFET), are compared with the gate value like behaviour of the bipolar transistor in fig 14.1. As fig(14.3) shows, the JFET is a three pin device. The "source" and "drain" pins of the JFET is connected to the ends of a solid bar, made either of N-type or P-type material, forming the channel through which charge carriers conduct. Between the Source and the Drain, the material acts as a resistor. The "gate" pin of the JFET is connected to a layer around the periphery of this channel and is of opposite doping type of the channel itself. That is, if the channel is N-type, the gate pin is connected to a P-type layer which encompasses the N-type channel. The "gate" corresponds very closely in

---

[1]as explained by the gate value example of the transistor

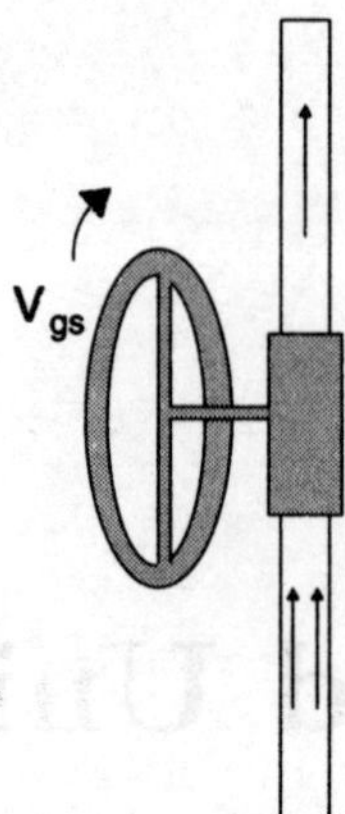

Figure 14.1: *The field effect transistor is again like a gate value where the flow of charge carriers is controlled by a voltage. While in transistors, the current in the base pin contols the flow of current, here in an JFET the voltage* V$_{GS}$ *applied to the gate pin does the controlling.*

Figure 14.2: *The vacuum tubes of yester years.*

operation to the base of the transistor and the grid of the vacuum triode tube[2] and as the name suggests, if the "gate" is "open" there is conduction in the channel from "source" to "drain", else on closing the "gate" no conduction takes place.

---

[2]A vacuum tube is just a glass tube evaluated and trapping a vacuum (an volume from which most gases have been removed). What makes it interesting is that when electrical contacts are put on the ends, you can get a current to flow though that vacuum. Thomas Edison noticed this first in 1883. While fiddling with lightbulbs that he had invented, he saw that he could get current to jump from the hot filament to a metal plate at the bottom. What Edison discovered (and it was promptly dubbed the "Edison effect") was that electrical current doesn't need a wire to move through. It could travel through gases at low pressure. Edison's discovery became useful only in 1904. Using this idea the British scientist, John A. Fleming made a vacuum tube known today as a diode. The diode was popularly known as a "valve," because it allowed current in the tube to travel only in one direction (the primitive PN diode). A decade later (around 1914) Lee De Forest invention (what he called the "audion") the triode value started replacing the diode valve. De Forest put a metal grid in the middle of the vacuum tube. By using a small on the grid, a stronger current could flowing between the two electrodes could be controlled. Turning weak currents into strong currents was crucial at that time. The transistor and FET are semiconductor devices which went on to replace the triode in it's application as an amplifier. It would be worth mentioning that De Forest had proposed ideas to include sound in motion movies.

14.1 FET: Similarity with BJT?

If the channel is made of N-type material, the device is called an N-channel JFET. In a P-channel JFET, the channel is made of P-type material and the gate of N-type material. In fig(14.3), schematic symbols for the two types of JFET are shown. Like the bipolar transistor types, the two types of JFET differ only in the configuration of bias voltages required and in the direction of the arrow within the symbol. Just as it does in transistor symbols, the arrow in a JFET symbol always points towards the N-type material. Thus the symbol of the N-channel JFET shows the arrow pointing toward the drain/source channel, whereas the P-channel symbol shows the arrow pointing away from the drain/source channel toward the gate. However, unlike the BJT that has three configurations, thus requiring a detailed study of each mode, the FET is a simple device to understand (as far as the theory of the device goes). The channel (source-drain pin) is the region of conduction with source grounded and common in the input and output circuits.

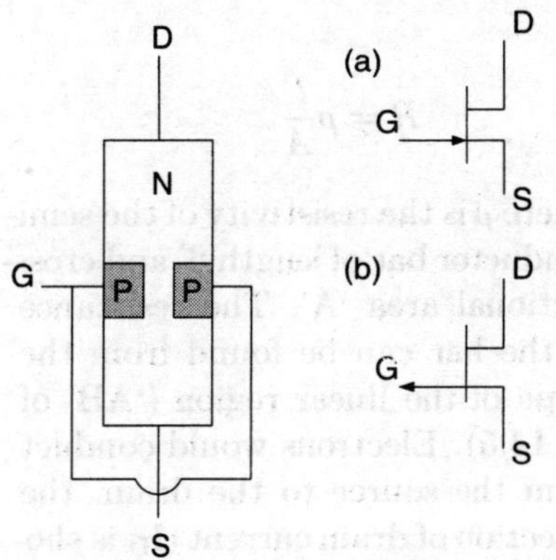

Figure 14.3: *The internal doping profile of a junction type field effect transistor (JFET) is shown here. The voltage V_{GS} applied to the gate w.r.t. source pin does the controlling. Also, shown along side is the circuital representation of the (a) N channel and (b) P channel FET.*

14.2 Channel Pinch-off

The key to FET operation is the effective cross-sectional area of the channel which is available for the conducting charge carriers. As stated, in the N-channel FET the majority charge carriers are the electrons. Since the industry prefers N-channel FETs over P-channel,[3] we restrict our explanation of the FETs I-V characteristics only to that of N-channels. However, it is trivial and easy to extend the explanation to that of P-channels.

Figure(14.4) shows the two voltages that are applied to the FET. Basically, the drain-source voltage (V_{DS}) and the gate control voltage (V_{GS}), between the gate and source. Consider the gate to be grounded, i.e. $V_{GS}=0v$. This is achieved by shorting the gate-source supply by a wire. In this condition, the channel is typically a bar of resistance, whose resistance is given

---

[3]The N-channel FET is prefered since the *"effective mass"* of the electrons are lower than that of the holes which constitutes the majority charge carriers of a P-channel FET. This gives the electrons a better mobility which inturn makes the device suitable for fast circuits. If electrons have lower effective mass compared to holes, it implies that the holes are heavy. But what mass can empty space (void) have? Note the difference. Holes do not have mass but are abscribed "effective mass". *What is "effective mass"?* Indeed it is difficult to explain what effective mass is in the footnote for a subject of semiconductor physics for which chapters are dedicated. However, for now it is enough to understand that though holes and electrons are quantum mechanical particles, macroscopically their response to an electric field can be written using Newton's equation

$$F = m^* \frac{dv}{dt} = qE$$

where m* is the effective mass showing the particles inertia for motion.

as

$$R = \rho \frac{l}{A}$$

where ρ is the resistivity of the semi-conductor bar of length 'l' and cross-sectional area 'A'. The resistance of the bar can be found from the slope of the linear region ('AB' of fig 14.5). Electrons would conduct from the source to the drain, the direction of drain current (I_D is shown in fig(14.4). The channel can now be modeled along the lines of fig(8.10) and can be considered to be a series combination of many resistances, i.e.

$$R = \rho \frac{l_1}{A} + \rho \frac{l_2}{A} + \rho \frac{l_3}{A} \quad (14.1)$$

where

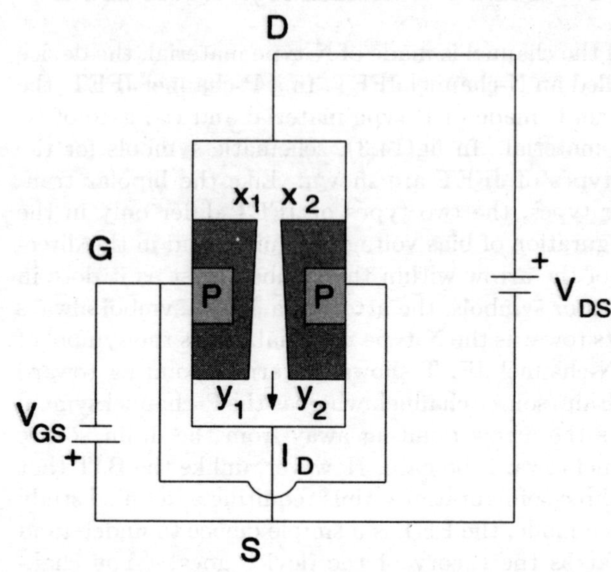

Figure 14.4: *The profile of the depletion width extending into the thickness of the JFET when properly baised with* V_{GS} *and* V_{DS}.

$$l_1 + l_2 + l_3 = l$$

Thus, there exists a potential gradient along the length of the channel. The highest potential being at the drain contact and the lowest at the source. Consider, a positive potential like +5v is applied to the channel and point 'x'_1 (or 'x'_2 at same distance from drain) is at +4v. Then the potential at 'y'_1 (or 'y'_2) would be less than +4v, say +1v. The reverse biasing of the PN diode at 'x'_1 would be greater than the reverse biasing at 'y'_1, Thus, the depletion width is more at 'x'_1 as compared to 'y'_1 (shaded region of fig 14.4). A gradient exists in the size of the depletion width along the channel's length similar to the potential gradient. A depleted region as the name suggests is "stripped off" or depleted of any free charge carriers, hence the region does not allow for conduction. Thus, the cross-sectional area available for conduction is less near the drain as compared to the source. Eqn(14.1) now can be written as

$$R = \rho \frac{l_1}{A(l_1)} + \rho \frac{l_2}{A(l_2)} + \rho \frac{l_3}{A(l_3)} \quad (14.2)$$

The dependence of the cross-sectional area on the potential gradient explains the deviation from linearity in the I-V characteristics (region 'BC'). If V_{DS} is increased further and further, the constraint '$x_1 x_2$' closes. That is the depletion regions grow so much that the entire cross-section of the channel becomes depleted. This is like pinching a rubber water pipe (say with a pinch cork) to stop the flow of water. However, here we are talking of the drain current and the pinching is done by the junction formed between gate and channel. Point 'C' on the

I-V characteristics represents the **pinch-off** point of the FET. You might ask how conduction (or flow of current) is possible beyond the pinch-off point represented by the saturation region (region 'CD' of fig 14.5). Remember, pinch-off occured due to the voltage gradient appearing along the length of the channel when the drain current flows. Due to the pinch-off the current flow stops removing the very cause of pinch-off. This leads to the depletion width to receed, allowing the drain current to flow again. As soon as the drain current flows, pinch-off reappears. The rapid opening and closing of the constraint gives the saturated current.

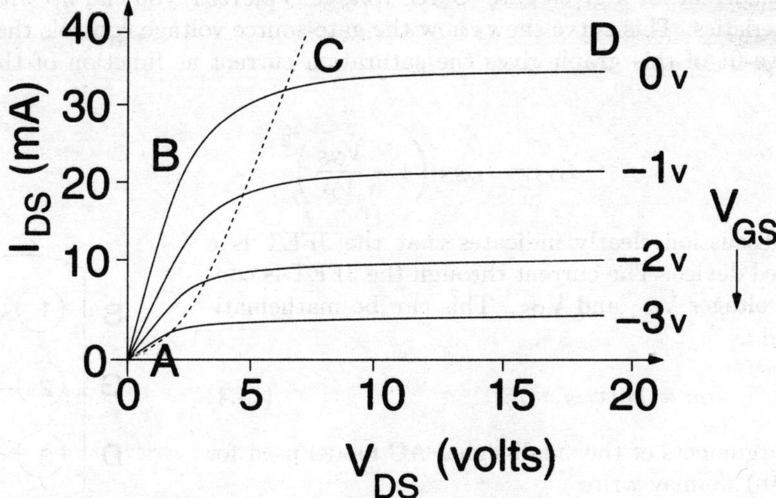

Figure 14.5: *The IV characteristics of the field effect transistor. The dashed line shows the pinch-off voltage for various gate-source voltages (*V$_{GS}$*).*

The application of a -ive voltage, i.e. keeping the gate at a lower potential then the source, provides an uniform depletion width along the channel length. This pre-existing depletion width (due to V$_{GS}$) makes it easier to achieve pinch-off, as seen can be achieved at a lower V$_{DS}$ in the family of curves. Infact the pinch-off point, or voltage V$_{DS}$ required to achieve pinch-off falls on a parabola (dotted lines on fig 14.5). Thus at a high reverse bias of gate-source, pinch-off is achieved at small drain-source voltage. Since the gate junction acts as the device's input, since it is reverse biased and because there is no minority carrier contribution to the flow through the device, the input impedance of the device is extremely high.

In summary and by looking at these curves we can see that the JFET has two areas of operation. At low

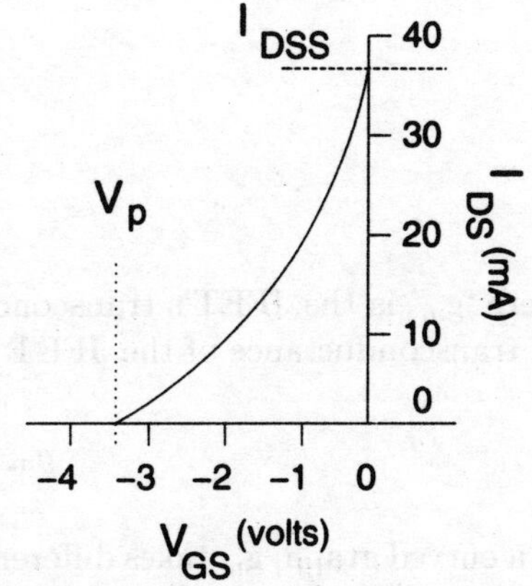

Figure 14.6: *The transfer characteristics of JFET.*

drain-source voltages it behaves like a variable resistance whose value is controlled by the applied gate-source voltage. At higher drain- source voltages it passes a current whose value depends on the applied gate-source voltage. In this 'high voltage' region the FET acts as a voltage controlled current source. The transfer characteristics (fig 14.6) for the JFET is useful for visualizing the gain from the device and identifying the region of linearity. The gain is proportional to the slope of the transfer curve. The current value I_{DSS} represents the maximum current for the device when the Gate is shorted to ground. This value will be part of the data supplied by the manufacturer. The Gate voltage at which the current reaches zero is called the "pinch voltage", V_P. The significance of the parabola in fig(14.5) is difficult to understand, however, if the saturation current for a given gate-source voltage is plotted, you end up with the JFET's transfer characteristics. This curve shows how the gate-source voltage controls the drain-source current. A curve fit of this graph gives the saturation current as function of the gate-source voltage

$$I_{DS} = I_{DSS}\left(1 - \frac{V_{GS}}{V_P}\right)^2 \qquad (14.3)$$

The above discussion clearly indicates that the JFET is a voltage controlled device. The current through the JFET is controlled by two voltages V_{GS} and V_{DS}. This can be mathematically represented as

$$i_{DS} = fn(v_{GS}, v_{DS}) \qquad (14.4)$$

Extending the arguments of the small signal AC model used for BJT (Chapter 10) we may write

$$I_{DS} = g_m V_{GS} + \frac{1}{r_d} \times V_{DS}$$

where

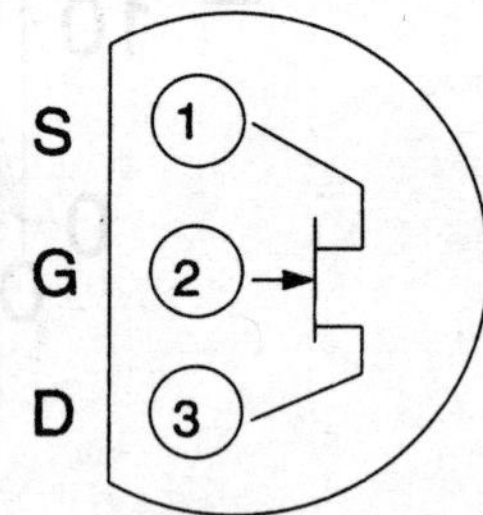

Figure 14.7: *The pin layout of the 2N3819 junction field effect transistor.*

$$g_m = \frac{di_{DS}}{dv_{GS}}\bigg|_{V_{DS}=0}$$

$$r_d = \frac{dv_{DS}}{di_{DS}}\bigg|_{V_{GS}=0}$$

where 'g_m' is the JFET's transconductance and 'r_d' is it's output impedance. Using eqn(14.3), the transconductance of the JFET is given as

$$g_m = -\frac{2I_{DSS}}{V_P}\left(1 - \frac{V_{GS}}{V_P}\right)$$

On a curved graph, g_m takes different values at different points on the curve. Hence, for designing purposes the value is calculated for the operating point or 'Q' point (as in the case of BJT), in other words for V_{GSQ}. It should be noted here, that just as β and V_{BE} are transistor parameters, varying from one make to another, I_{DSS} and V_P are JFET specific parameters. Combining the above information, with the fact that the JFET's input impedance is very high (ideally tending to infinity), the JFET can be modeled as shown in fig(14.8).

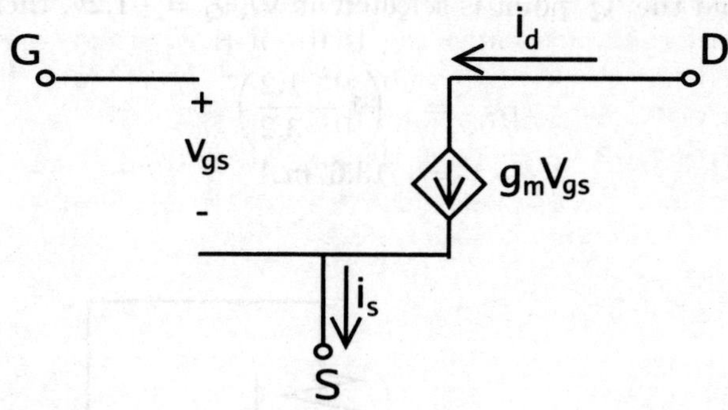

Figure 14.8: *The small signal AC model of the JFET.*

JFET as Voltage Controlled Resistance

The curves of the JFET's IV characteristics shown in fig (14.5) can be approximated by the mathematical relationship

$$I_{DS} = I_{Dsat}(1 - e^{-\lambda V_{DS}}) \tag{14.5}$$

where the functional relationship between I_{Dsat} and the gate-source voltage (V_{GS}) is given by eqn(14.3). For small values of Drain-Source voltage (V_{DS}), we can expand the exponential term in eqn(14.5) as

$$\begin{aligned} I_{DS} &= I_{Dsat}(V_{GS})\left[1 - (1 - \lambda V_{DS})\right] \\ &= \lambda I_{Dsat}(V_{GS})V_{DS} \end{aligned}$$

This shows the linear relationship between the applied drain-source voltage and current I_{DS} flowing from drain to source as a result of it. As in Ohms law, the slope gives the channel resistance. That is,

$$\frac{dV_{DS}}{dI_{DS}} = \frac{1}{\lambda I_{Dsat}(V_{GS})}$$

As can be seen, the resistance depends on gate-source voltage, V_{GS}. Decreasing V_{GS} decreases I_{Dsat} that leads to a large channel resistance. Thus, the JFET's channel acts as a voltage controlled resistance for low drain-source voltages.[4]

14.3   JFET Biasing

Consider that an amplifier is to be designed using a JFET. The basic circuit would be as shown in fig(14.9). We shall now put input practical use, the information of the transfer characteristics

---

[4]This feature was used by the author in a project, Ashima Katyal, Parul Gupta and P.Arun in Physics Education (India), **24** (2007) 49-52.

(fig 14.6). Let us demand the 'Q' point is selected at $V_{GSQ} = -1.2$v, then using eqn(14.3)

$$I_{DS} = \left(1 - \frac{1.2}{3.2}\right)^2$$

$$= 13.67 mA$$

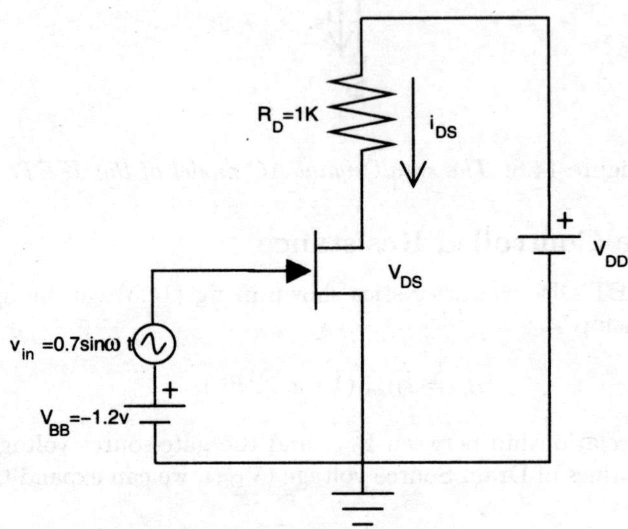

Figure 14.9: *Basic FET amplifier circuit.*

where the data $I_{DSS} = 35$mA and $V_P = -3.2$v is collected from the transfer characteristics (fig 14.6). If a 1.4v peak-to-peak sine wave is fed at the input the drain-source current varies between 5.7-24.9mA. The corresponding output voltage would be obtained using the voltage loop equation on the output side, i.e.

$$V_{DD} = i_{DS}R_D + v_{DS}$$

$$or$$

$$\Delta V_{DS} = R_D \Delta I_{DS}$$

(Again AC signal has been trivialised as a slowly varying DC signal.)

For the input variation 0.5v($I_{DS} = 24.91$mA) to 1.9v($I_{DS} = 5.77$mA), the output voltage is given as

$$\Delta V_{DS} = (24.91 - 5.77) \times 1K\Omega = 19.14v$$

Thus, the voltage gain of the circuit in question (fig 14.9) is $-19.14/1.4 \approx -13.6$.

The above amplifier example called for selecting a operating point on the transfer characteristics just as one was selected on the output characteristics of a BJT. However, in this case, the load line locating 'Q' point is called the **bias line**. Figure(14.10) is an example of a bias line used to select an operating point.

For biasing a JFET the voltage divider circuit used to bias a BJT can be put into use. Figure 14.11 shows the same circuit. The voltage divider circuit can be replaced by the Thevenin equivalent circuit. The voltage loop equation at the input can be written as

$$V_G = V_{GS} + I_{DS}R_S$$

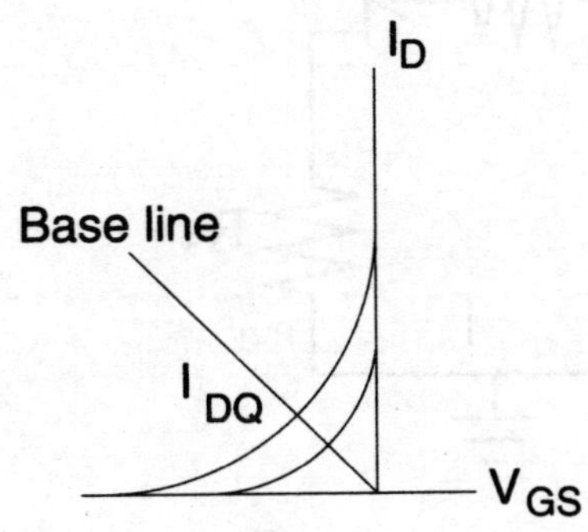

Figure 14.10: *Using a bias line to select the operating point.*

Since the input impedance of the JFET is infinte (very large) the term $I_S R_G$ does not exist. As usual the JFET is used in the high current region (i.e. saturation region or the linear region) hence using eqn(14.3) we have

$$V_G = V_{GS} + I_{DSS}\left(1 - \frac{V_{GS}}{V_P}\right)^2 R_S$$

Thus, selecting an operating point involves solving the above simultaneous equation

$$\left(\frac{I_{DSS}R_S}{V_P^2}\right)V_{GSQ}^2 + \left(1 - \frac{2I_{DSS}R_S}{V_P}\right)V_{GSQ} + (I_{DSS}R_S - V_G) = 0 \tag{14.6}$$

for V_{GSQ} such that it is negative in case of a 'N' channel JFET. The resistance R_D is selected on the thumb rule $V_{R_D} = V_{DS} = V_{R_S} = \frac{1}{3}V_{DD}$. Before designing a voltage amplifier to test our knowledge let us use the small signal AC model of the JFET to compute the voltage gain of the circuit.

Example 14.1: Select a bias line for a JFET assuming $I_{DSS} = 12mA$ and $V_P = -2v$.

For selecting a bias line, a judicious selection of the resistance R_S has to be done (see eqn 14.6). Demanding the operating point should be half of V_P, ($V_{GSQ} = V_P/2$), we write

$$\left(\frac{12R_S}{4}\right) + \left(\frac{V_P}{2} - \frac{2 \times 12R_S}{2}\right) + (12R_S - V_G) = 0$$

$$3R_S - 1 - 12R_S + 12R_S - V_G = 0$$

$$1 + V_G = 3R_S$$

V_G is the Thevenin voltage source that replaces the voltage divider circuit and is given as

$$V_G = \frac{R_2 V_{DD}}{R_1 + R_2}$$

The magnitude of V_G lies in the designer's choice, and we select $V_G = 5v$, hence $R_S = 2K\Omega$.

$$\boxed{R_S = 2K\Omega \text{ and } V_G = 5v}$$

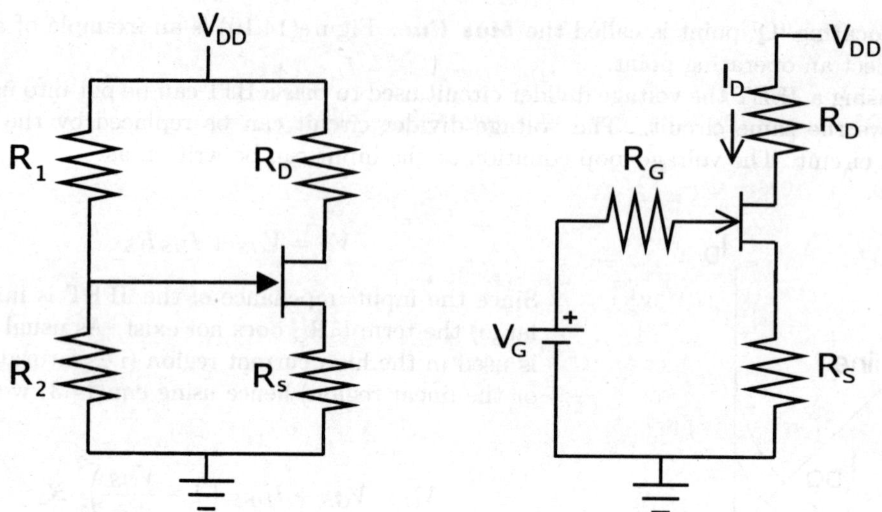

Figure 14.11: *A voltage divider circuit used for selecting the operating point. The second circuit shows the equivalent circuit where the voltage divider is replaced with the Thevenin equivalent voltage source and resistance.*

The selection of this bias line results in a drain-source current (use eqn 14.3)

$$I_{DS} = 12\left[1 - \left(\frac{-1}{-2}\right)\right]^2 = 3mA$$

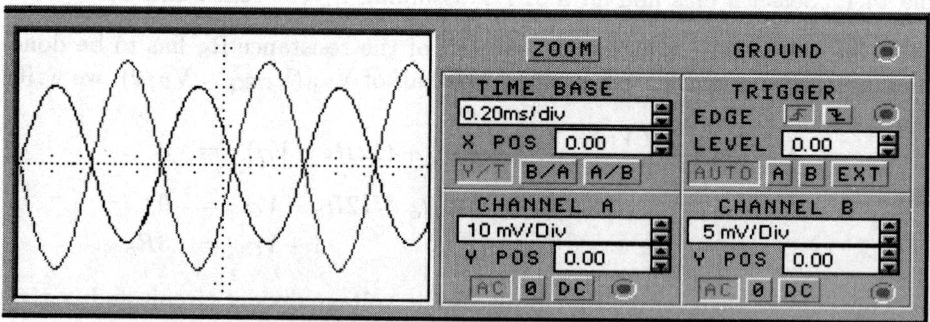

Figure 14.12: *The oscilloscope trace of the output signal with respect to the input signal for a JFET amplifier.*

The JFET amplifier shown in fig(14.11) can now be designed. If the DC source put into use is 15v(V_{DD}), to obtain $V_G = 5$v select $R_1 = 200K\Omega$ and $R_2 = 100K\Omega$). The output side loop equation is written as

$$V_{DD} = I_{DS}R_D + V_{DS} + I_{DS}R_S$$

The values of R_s and I_{DS} have been determined in Example(14.1). Also, the thumb rule demands $V_{DS} = V_{DD}/3$. Using these informations we can calculate the required R_D as

$$15 = 3R_D + 5 + 6$$
$$R_D = \frac{15 - 11}{3} K\Omega$$

Thus, the required R_D is 1.33KΩ. Figure 14.12 shows the oscilloscope output of the designed common-source JFET amplifier (source of JFET is common on both input and output side). The gain can be measured from the scope and is found to be -2.64, where the negative sign implies the phase difference of π between the output and input signal.

For designing an amplifier of a given gain, one would have to analyse the small signal AC model (see fig 14.13) of the voltage divider circuit shown in fig(14.11). The input voltage is given as

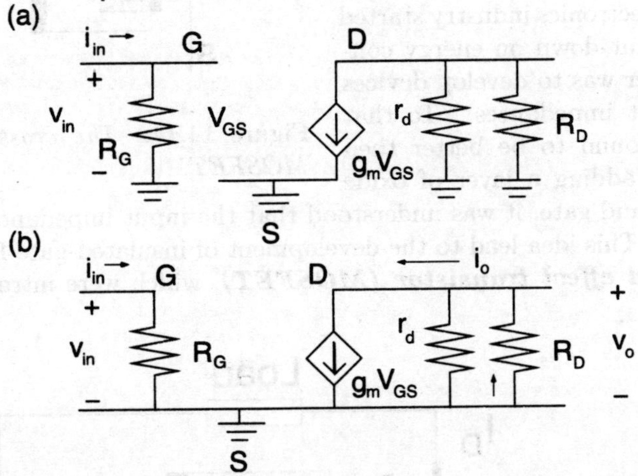

Figure 14.13: *The (a) small signal AC model of the JFET amplifier circuit shown in fig(14.11). The circuit has been drawn in a more compact manner in (b).*

$$v_{in} = (R_1 \| R_2)i_{in} = v_{gs} \tag{14.7}$$

while the output voltage is a result of the current $g_m v_{gs}$ through the parallel combination of r_d and R_D and is written as

$$v_o = -g_m v_{gs}(r_d \| R_D) \tag{14.8}$$

Using eqn(14.8) and eqn(14.7), the voltage gain is given as

$$A_v = \frac{-g_m v_{gs}(r_d \| R_D)}{v_g s}$$
$$= -g_m(r_d \| R_D)$$

The value of g_m for 2N3819 is typically $2000\mu s$ (micro-siemen or micro-mho). Siemen is the new SI unit for conductance. For $r_d \gg R_D$, we can confirm the result of our small signal analysis with that of our designed JFET amplifier. The voltage gain would be $-2000 \times 10^{-6} \times 1.33 \times 10^3 = -2.66$ which is approximately equal to that observed from the oscilloscope output (≈ -2.64). The input and output impedance of this amplifier is trivially shown to be $R_1\|R_2$ and $r_d\|R_D$, respectively.

14.4   MOSFET

After the advent of ICs, the rush for minia-turizing increased many fold. Followed by this came the requirement of mobility and hence freedom from the AC mains. This required for good battery technology, where the battery could supply energy for a long time. While work on that front was being picked up by chemical engineers, electronics industry started looking for ways to cut down on energy consumption. The answer was to develop devices with very large input impedances. To that effect, JFETs were found to be better then BJTs. However, by adding a layer of oxide

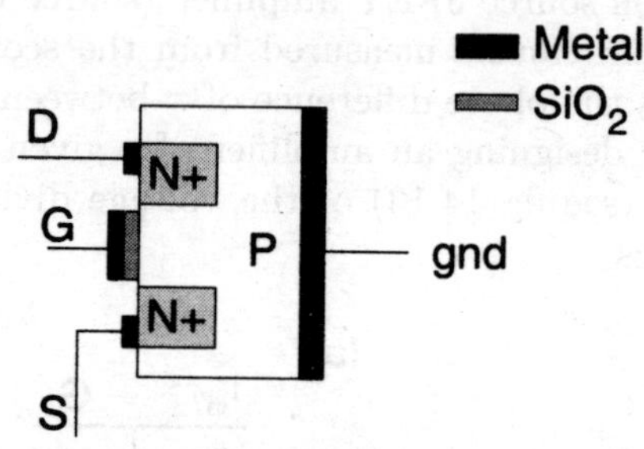

Figure 14.14: *The cross-section of a typical MOSFET.*

between the channel and gate, it was understood that the input impedance of the device could be increased further. This idea lead to the development of insulated-gate FET or ***metal oxide semiconductor field effect transistor (MOSFET)***, which were introduced in the market in 1970s.

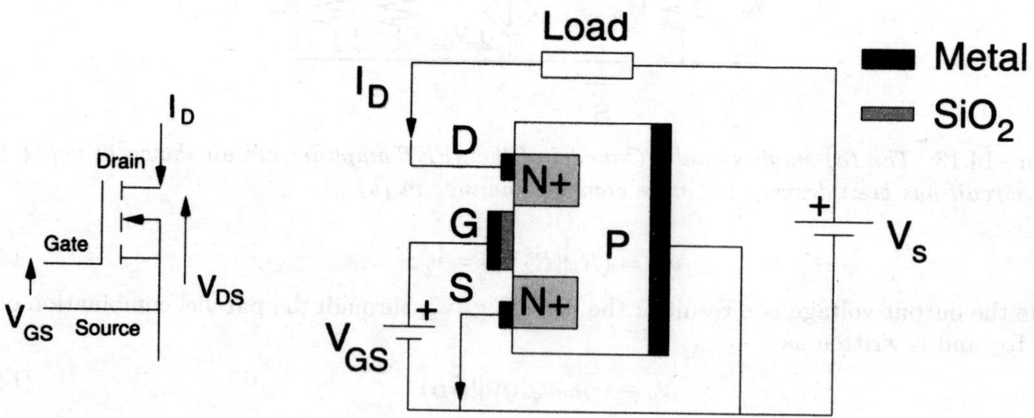

Figure 14.15: *The N-channel enhanced MOSFET in ON state with the appropriate applied voltages. Also, shown along with it is the circuit representation of the device.*

As suggested, the MOSFET is similar to the JFET but exhibits an larger input impedance due to a thin layer of silicon dioxide that is used to insulate the gate from the semiconductor channel. The magnitude of input impedance of a typical MOSFET is of the order 10^{10} to $10^{14}\Omega$. The insulating oxide layer forms a capacitive coupling between the gate and the body of the transistor. This geometry comes with a disadvantage, the insulating material acts as a the capacitor that can be easily damaged by the internal discharge of static charge developed during normal handling. This makes the MOSFET a bit fragile compared to the ragged BJT. The MOSFET is not only used in large-scale digital integrated circuits because of its high input impedance which result in very low power consumption per component but also because it's planar structure is best suited for IC fabrication techniques. Figure(14.14) shows the cross-section of a typical MOSFET.

Note, in fig(14.14), the source and drain is connected to heavily doped 'N' region which are separated by a 'P' type semiconductor in between. The scenario can be reversed with a 'N' type semiconductor separating the source and drain of heavily doped 'P' region. The two possibilities are the 'N' channel MOSFET and the 'P' channel MOSFET respectively. Each of the MOSFET can further be produced either as depletion and enhancement mode MOSFETs.

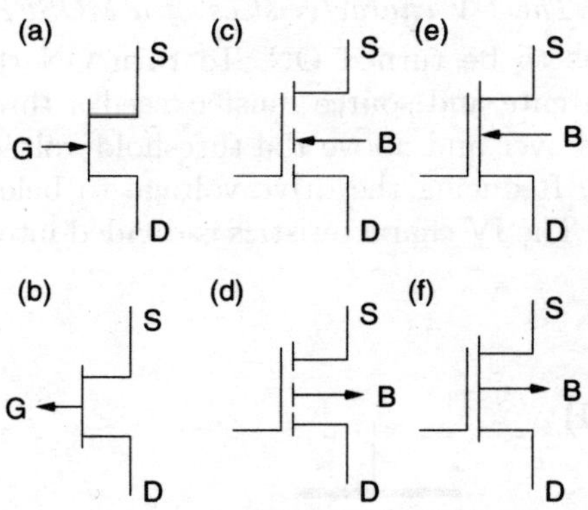

Figure 14.16: *Circuital schematics of (a) n channel JFET, (b) p channel JFET, (c) n channel enhancement type MOSFET, (d) p channel enhancement type MOSFET, (e) n channel depletion type MOSFET and (f) p channel depletion type MOSFET.*

To understand the difference between an enhancement mode MOSFET and a depletion mode MOSFET, let us understand the working of the N-channel enhancement mode MOSFET shown in fig (14.14). When a positive voltage is applied to the gate with respect to the source, the electric field will cause positive charges in the P-region to move away from the base inducing or ***enhancing*** a N-region in its place (details are discussed in the next section). Conduction can then take place between the N^+(drain)-N (enhanced region)-N^+ (source). Increasing or decreasing the gate voltage will cause the induced (enhanced) N channel to grow or decrease in size thus controlling conduction. Thus, the enhancement type is normally off, which means that the drain to source current increases as the voltage at the gate increases. Also, no current flows when no voltage is supplied at the gate.

By converse, the depletion mode MOSFET is normally "ON" and operates just like an JFET. In a ***depletion mode N-channel MOSFET*** a n-region exists connecting the highly doped N^+ drain and source. On applying V_{DS} between the drain and source, electrons flow, giving rise to current I_D. Now to stop the flow of this current, the gate has to be reverse biased with respect to the source. Holes from the P-region moves into the N-region due to the negative field around the gate. Recombination of holes with electrons in the N-region removes charge carriers for conduction which discourages conduction between drain and source, however, does not make it zero. Further increase in reverse bias of gate, inverts the channel to P-type layer. This sends the current to zero and thus, the MOSFET is switched "OFF".

14.5   I-V Characteristics of MOSFET

Fig(14.15) shows an 'N' channel enhancement type MOSFET with appropriate voltages applied. Also notice that the channel has come into existence, i.e. the MOSFET is in it's ON state. The channel is not uniform along the length, this can be understood after understanding the MOSFET's IV characteristics (see fig 14.17). Before going into the details of the IV characteristics it is worthwhile to know that 'N' channel enhancement type MOSFETs are the most popular as it remains

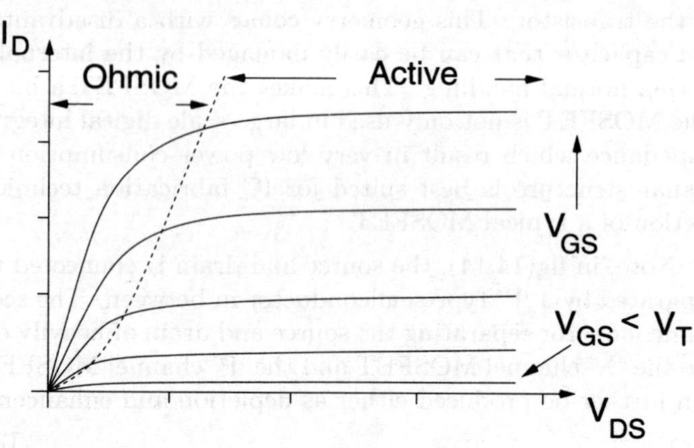

Figure 14.17: *The I-V characteristics of a MOSFET.*

OFF all the time (no power consumption) and has to be turned ON. To turn ON the device a drive voltage or the voltage applied between gate and source must exceed a threshold value (V_T) although in practical applications values over and above the threshold voltage are used to ensure the MOSFET is fully switched ON. Reducing the drive voltage to below the threshold voltage causes the MOSFET to turn OFF. The IV characteristics is divided into three regions (see fig 14.17).

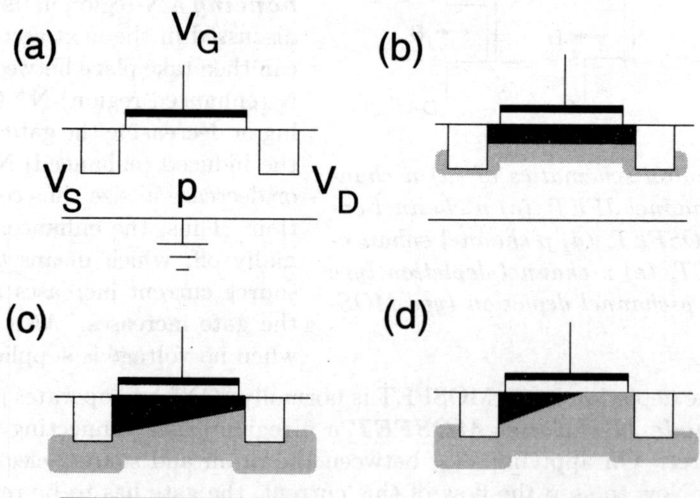

Figure 14.18: *The potential at important points of the n-channel enhancement type MOSFET. The channel cross-section appears (b) uniformly when the gate-source voltage is greater than the threshold voltage and the drain-source voltage is small. As the drain-source voltage increases (c) the channel cross-section becomes non-uniform. Finally, for a larger drain-source voltage (d) pinch-off occurs.*

A) **Cut-off Region:** If $V_{GS} < V_T$, then the n-channel is not induced and no current flows (fig 14.18a).

B) **Ohmic Region:** If $V_{GS} > V_T$ and V_{DS} is small, then the n-channel is continuous (connects the Source and Drain, fig 14.18b). The current flowing between the Source and Drain is controlled by the external resistance connected to the Source and Drain pin as also the resistance of the channel. The drain current increases if the voltage drop between S and D (V_{DS}) increases. This explains the Ohmic nature of the IV characteristics. The channel resistance depends on how much charge is injected at the Source, which in turn is controlled by the Gate-Source voltage (V_{GS}). The Drain current (I_D) depends on both V_{GS} and V_{DS}, this is functionally same as that expressed in eqn(14.4) for JFET.

C) **Saturation Region:** As V_{DS} increases the effective voltage near the drain is $V_{GS} - V_{DS}$. If this voltage ($V_{GS} - V_{DS}$) is greater than V_T, the inversion and formation of channel takes place. Since on applying a drain voltage, the potential drops as one moves from the drain to the source. Moving along the same direction, $V_{GS} - V_{DS}$ increases, which results in the channel cross-section (inversion layer) increasing as one moves from source to drain (see fig 14.18c). However, if $V_{GS} - V_{DS} < V_T$, inversion would not occur and the channel disappears. Any increase in V_{DS} from that applied for case fig(14.18c) leads to the n-channel disappearing near the drain. This is pinch-off of the channel. The condition shown in fig(14.18d) represents the condition of pinch-off. After pinch-off, the current I_{DS} saturates. It should be noted that the carriers (electrons coming from the source in n-channel enhancement type MOSFET) moving into the pinched off region are no longer confined to the inversion layer near the surface, but begin to move away from the surface into the bulk.

Some important MOSFET Parameters that would be available on the manufacture datasheet are

- **Maximum Drain-Source Voltage, (V_{DS}):** V_{DS} is the maximum instantaneous operating voltage.

- **Drain Current, I_D:** I_D is the maximum current the MOSFET can continuously carry. It is a function of temperature.

- **Maximum Pulsed Drain Current, I_{DM}** is greater than I_D and specified for a particular pulse width and duty cycle.

- **Maximum Gate-Source Voltage, V_{GS}:** V_{GS} is the maximum voltage that can be applied between gate and source without damaging the gate insulation.

- **Gate Threshold Voltage, V_T or V_{TH} or $V_{GS(th)}$:** V_T is the minimum gate voltage at which the MOSFET will turn ON.

The MOSFET is now used intensively in ICs because of it's high input impedance. It can also be used as a variable resistance, whose resistance depends on the gate-source voltage. To maintain the "ON" state in a MOSFET (a voltage controlled device) a very small current is required. Infact, only $1/5^{th}$ or $1/10^{th}$ of current that is required for a BJT (a current controlled device). Hence, power consumption of MOSFETs are very low. Also, the MOSFET's are unipolar devices and by using N channel (i.e. electrons are conducting charge carriers), their switching speed is faster than the BJT by orders of magnitude. These features along with it's planar geometry (ideal for IC fabrication) makes it score over JFETs also.

14.6   Uni-Junction Transistor

The basic structure of a unijunction transistor (UJT) is shown in Figure(14.19). It is essentially a bar of N type semiconductor material into which P type material has been diffused somewhere along its length, usually away from the half way point along the length of the N type semiconductor. An UJT, whose body is made up of N type semiconductor is called N channel UJT. Similarly, one can also have P channel UJTs. However, by industry standards N channel UJTs are more prelevent. The speed of the UJT as a device depends on the speed with which charge carriers travel along the channel. Since electrons have better mobility then holes due to their lower effective mass, N channel UJTs are more faster then the P type UJTs. The three contacts of the UJT are referred to as the emitter, Base 1 and Base 2 respectively. Figure(14.19) also shows the schematic symbol used to denote a UJT in circuit diagrams.

The equivalent circuit shown in fig (14.20) can be used to explain how the device works. When emitter is left open, we define R_{BB} is known as the inter-base resistance or the resistance of the channel, and is the sum of R_{B1} and R_{B2}:

$$R_{BB} = R_{B1} + R_{B2} \tag{14.9}$$

V_{RB1} is the voltage developed across R_{B1}, since the channel is a combination of resistances in series, this voltage is given by the voltage divider rule, i.e.

$$V_{RB1} = \left(\frac{R_{B1}}{R_{B1} + R_{B2}}\right) V_{BB}$$

or, using equation (14.9)

$$V_{RB1} = \left(\frac{R_{B1}}{R_{BB}}\right) \times V_{BB} \tag{14.10}$$

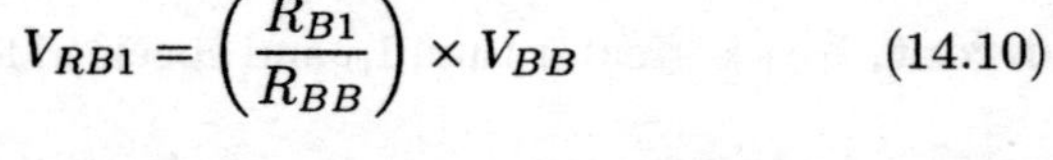

Figure 14.19: *The internal structure of an UJT with it's circuital representation.*

The ratio R_{B1}/R_{BB} is referred to as the intrinsic standoff ratio and is denoted by the Greek letter eta (η). If an external voltage V_E is connected to the emitter, the equivalent circuit along with a series combination of resistances would include a PN diode as shown in fig(14.20).

If the applied DC voltage V_E is less than the voltage drop across the UJT's equivalent circuit's R_{B1}, i.e. $V_E \leq V_{RB1}$, the diode is reverse biased and the circuit behaves as though the emitter was open circuit. However, if V_E is increased so that it exceeds V_{RB1} by at least 0.7V, the diode becomes forward biased and emitter current (I_E) flows into the base region represented by R_{B1}. With the doping concentration of the P region of the UJT being very large, on conduction large number of holes are injected into the base region (R_{B1}). Due to this, the resistivity of this region and in turn the value of R_{B1} decreases. Further increase in V_E causes the emitter current to increase which in turn reduces R_{B1} and this causes a further increase in current. This repeative/ iterative effect is termed as regeneration. The value of emitter voltage at which this occurs is known as the peak voltage V_P, and is given as

$$V_P = \eta V_{BB} + V_D \tag{14.11}$$

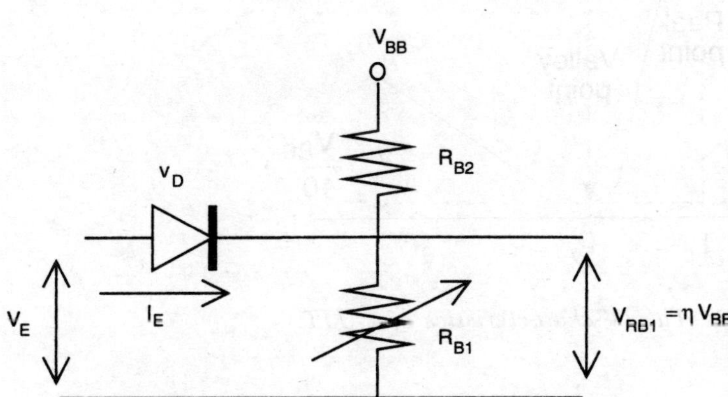

Figure 14.20: *The equivalent circuit of an UJT.*

The characteristics of the UJT are illustrated in it's IV characteristics (emitter voltage versus emitter current) as shown in fig (14.21). As the emitter voltage is increased, the current is very small (in microamps). When the peak point is reached, the current rises rapidly, until at the valley point the device runs into saturation. At this point R_{B1} is at it's lowest value, which is known as the saturation resistance.

The simplest application of a UJT is as a relaxation oscillator, which is defined as one in which a capacitor is charged gradually and then discharged rapidly. The basic circuit is shown in fig(14.22a). However, external resistances are essential in the practical circuit as shown in fig(14.22b). The external resistance, R_3, limits the emitter current and provides an output voltage pulse. Figure(14.22c) shows the waveforms occurring across the capacitor (voltage drop across the emitter) and base 1 region. The first is typically that of a capacitor charging and discharging and can be approximated as a sawtooth waveform. Once the voltage across the capacitor exceeds the peak voltage, the UJT is send from it's "relaxing position" of OFF state to ON state and the capacitor discharges across the low resistance base 1 region. Hence, the second waveform which is the capacitor discharging through a zero resistance giving a pulse/ spike like waveform of short duration.

To evaluate the time between two pulses, one has to find how the voltage across the capacitor develops. When the UJT is in OFF state and the circuit is put into operation, we have a RC circuit connected to the source. Hence the current in the capacitor is

$$i_c(t) = \frac{V_{BB}}{R_1} e^{-t/R_1 C}$$

Using this, the charging/ rising voltage across the capacitor is evaluated

$$\begin{aligned} v_c(t) &= \frac{1}{C} \int i_c(t) dt \\ &= -V_{BB} e^{-t/R_1 C} + constt \end{aligned}$$

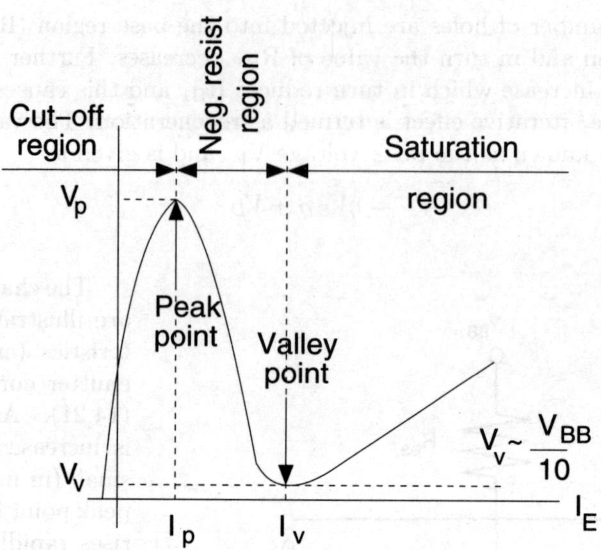

Figure 14.21: *The IV characteristics of a UJT.*

Since, on the start of the charging process the capacitor had no charge across it, $v_c(t) = 0$. Even in steady state condition, the initial condition holds since when the UJT is sent to it's conducting state, the capacitor discharges and losses all charges present. Hence,

$$v_c(t) \;=\; V_{BB}(1 - e^{-t/R_1 C})$$

In the relaxation oscillator, the UJT is sent from it's OFF state to ON state, when the capacitor voltage crosses V_P. The time instant this voltage is attained is easily computed from the above equation. And is given as

$$t_1 \;=\; R_1 C ln\left(\frac{V_{BB}}{V_{BB} - V_P}\right)$$
$$\;=\; R_1 C ln\left(\frac{1}{1 - \eta}\right) \tag{14.12}$$

the above result is obtained by the realistic assumption $V_D \to 0$. The intrinsic stand-off ratio or η is typically between 0.5-0.8 as can be seen in fig(**??**) which shows various important parameters of UJT 2N2646.

The relaxation oscillator sends pulses of large amplitude for very short duration. The gate triggering of SCRs which would be introduced in Chapter 18 requires such pulses. Thus, the relaxation oscillator is an ideal candidate and finds use for periodic gate triggering of an SCR.

Designing a Relaxation Oscillator

To appreciate what we have learnt it would be interesting to design a relaxation oscillator using an UJT. The most common UJT available is the 2N2646 UJT, whose relevant specifications are

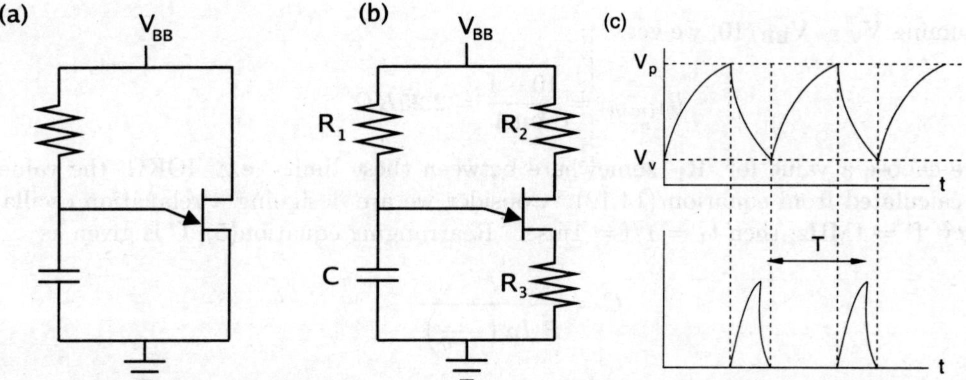

Figure 14.22: *The (a) basic circuit of a relaxation oscillator made using an UJT with (b) the practical circuit and (c) the voltage waveform of the relaxation oscillator's output.*

- V_{EB2O} =30V

- I_E(peak)=2A

- P_{TOT}(max), i.e. maximum power that can be dissipated in the UJT=300mW

- I_P(max), current flowing into emitter when V_p is attained=5μA

- I_V(max), is the current flowing into emitter when saturation of valley point (V_V) is attained=4mA

It is important that the value of 'R_1' is small enough to allow the emitter current to reach I_P when the capacitor voltage reaches V_P and large enough so that the emitter current is less than I_V when the capacitor discharges to V_V. The limiting values for 'R_1' are given by:

$$R_{1(max)} = \frac{V_{BB} - V_P}{I_P}$$

and

$$R_{1(min)} = \frac{V_{BB} - V_V}{I_V}$$

Thus, for calculating the value of resistance required, we first should know the peak voltage (V_P) and the valley point potential (V_V). Both values are not listed in the datasheet. However, V_P can be calculated. From the specifications for the 2N2646 the average value of η is (0.56 + 0.75)/2= 0.655. Substituting this value in eqn(14.11) and assuming V_D=0.7V and supply voltage of 10v, we have

$$\begin{aligned} V_P &= 0.655 \times 10 + 0.7 \\ &= 7.25V \end{aligned}$$

So

$$R_{1(max)} = \frac{10 - 7.25}{5\mu A} = 550K\Omega$$

and assuming $V_V \sim V_{BB}/10$, we get

$$R_{1(min)} = \frac{10 - 1}{4mA} = 2.25 K\Omega$$

If we choose a value for 'R_1' somewhere between these limits, e.g. 10KΩ, the value of C can be calculated from equation (14.12). Consider we are designing a relaxation oscillator of frequency 'f' = 1MHz, then t_1 = 1/f = 1msec. Rearranging equation(5), C is given as

$$C = \frac{t}{R_1 ln\left(\frac{1}{1-\eta}\right)}$$

Substituting the values, C works out approx 84nF. In practice one would use a variable resistance (or a variable resistance in series with a fixed resistance) for R_1 so that the frequency of oscillation could be adjusted to give the required value. R_2 is not essential; if it is included, a value of 470Ω is appropriate for the 2N2646. The value of R_3 should be small in comparison with R_{BB}, with which it is in series, so as to prevent it from affecting the value of the peak voltage. A value of 47Ω or thereabouts is satisfactory.

Exercise

Q1. Explain how a depletion region may act like an insulator in a JFET yet conduct current in a PN junction?

Q2. Draw the structure of an ideal P Channel JFET, indicating appropriate voltages on the terminals. Indicate direction of expected current flow.

Q3. A JFET having I_{DSS}=16mA and V_P=-4v is to be used in a common-source amplifier with a mid-band, ac voltage gain ≥ 4 when connected between a source with output resistance of 1kΩ and a 10kΩ load resistance. Assume that $r_d = 40k\Omega$ [eqn(14.3) holds]

    (a) Sketch a circuit diagram, using preferred biasing, for this amplifier and provide values for all the resistors used with a 15volt power supply.

    (b) Determine the quiescent values of all node voltages and all branch currents in your circuit.

    (c) Calculate the actual value of the overall voltage gain of your amplifier. Justify all values.

Q4. Define the pinch off voltage, V_P.

Q5. How does the FET behave for

    (i) small values of V_{DS}.

    (ii) large values of V_{DS}.

Q6. Define (a) transconductance (g_m) and (b) drain resistance (r_d).

Q7. How is a FET used as a voltage variable resistance? Explain.

Q8. Explain how a bias line is selected.

Q9. Drwa a circuit for a common-source JFET amplifier. What is it's voltage gain?

Q10. How can an oxide layer in an electronic device be of any help?

Q11. Discuss the working of an N channel enhancement type MOSFET.

Q12. Discuss the channel formation in the above question based on modeling the metal- oxide-p type bulk semiconductor as a capacitor.

Q13. Draw a diagram, explicitly showing the charge accumulation in a p-channel enhancement type MOSFET.

Q14. If a circuit demands a resistance whose value can be controlled by a user just by varying a voltage, explain how to realize the same using a MOSFET. Draw a suggestive circuit diagram with working condition.

Q15. Explain the IV characteristics of the MOSFET.

Q16. Explain the process of pinch-off within a MOSFET device?

Q17. Explain why an oscillator designed using an UJT is called a "relaxation oscillator"?

Chapter 15

Oscillators

Oscillators are circuits that produce sinusoidal, square, sawtooth, triangle, or any complex waveforms in the output. The output is essentially AC, hence the circuit converts DC power to AC power. This hence, is the opposite of rectifiers that converts AC power to DC power. Oscillators that produce sine waves are commonly used in radio/ TV transmitter circuits or in other words communication circuits. Oscillator circuit was first invented by Armstrong, who notices that if a portion of the output is fed back to the input, sustained oscillations were obtained. This obviously is the feedback concept, where the theoretically ideal of feedback was developed in Chapter 13.

To re-cap let us revisit eqn(13.3) of Chapter 13. If β is negative the equation shows an increase in the feedback amplifier's gain, i.e.

$$A_f \;=\; \frac{A_v}{1 - A_v\beta}$$

Such feedback circuits, where β is negative, are said to have positive feedback. Note if you have positive feedback the gain increases. The question is what would happen if $A_v\beta = 1$? Mathematically we would have $A_f = \infty$. Physically, this interpret as an output voltage obtained for zero input. Thus, power is absorbed from the DC source and the amplifier circuit produces AC on the output without receiving any input!

The equality $A_v\beta = 1$, is referred to as **Barkhausen's criteria** and gives the condition needed to achieve stable and sustained oscillations. In the following sections we discuss various oscillations circuits that have been developed down the ages.

15.1 Phase Shift Oscillators

The phase shift oscillator (fig 15.1), is a sine-wave generator that uses a resistive-capacitive (RC) network as its frequency dependent feedback device. The common-emitter transistor amplifier forms the internal amplifier of this circuit. The resistances R_1, R_2, R_E and and R_C provide base and collector bias. Capacitor parallel to R_E bypasses it and prevents any AC feedback. The transistor introduces a $180°$ phase difference between the base and the collector voltage.

To obtain positive feedback a circuit is required which introduces another phase shift of 180°
between the output and input signal.

An RC network consisting of minimum
three RC sections can provide this positive
feedback. Each section introduces a phase
shift of 60°. Since the impedance of an RC
network is capacitive, the current flowing
through it leads the applied voltage by a
specific phase angle. The phase angle is
determined by the value of resistance and
capacitance of the RC section.

When power is applied to the circuit,
noise (these are random electrical variations
generated internally in any electronic components essentially due to the motion of the
electrons). These noise signals, which form
a change in the flow of base current results
in an amplified change in collector current
which is phase-shifted by 180°. When the
signal is returned to the base, it is further
shifted by 180° by the RC network. However, this positive feedback is only experienced by one frequency, and only that particular frequency is amplified. All other frequencies die out. Thus, the output is purely
sine and of only one frequency.

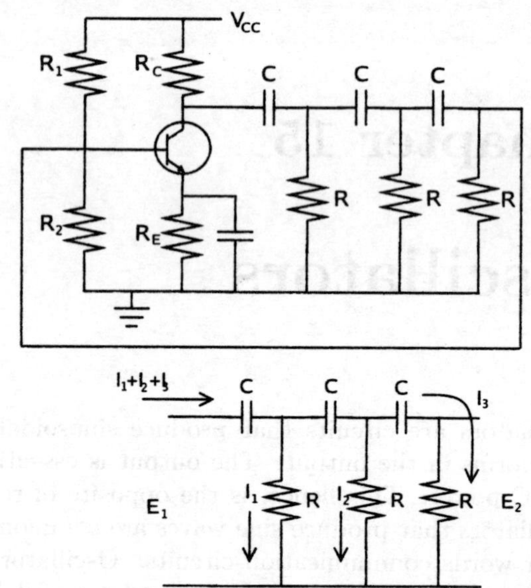

Figure 15.1: *The RC phase shift oscillator circuit.
Positive feedback is obtained with the help of the
transistor and the three sections of resistance and
capacitor. The various currents in the feedback
circuit has been indicated.*

15.1.1   With FET

The circuit shown in fig(15.1) though shows a BJT is fit for generating oscillations only if an
FET is used. However, before going into the reasons, we can develop the mathematics required
to design a RC phase shift oscillator using an FET. The voltage E_2, across the third resistance
is fed back to the circuit's input.

$$I_3 R = E_2 \tag{15.1}$$

Using KVL on the third loop we have (see fig 15.1)

$$I_2 R = I_3 (R + X_c) \tag{15.2}$$

Similarly, KVL on the second loop

$$I_1 R = I_2 (R + X_c) + I_3 X_c \tag{15.3}$$

Applying KVL on the first loop, we have

$$E_1 = (I_1 + I_2 + I_3) X_c + I_1 R \tag{15.4}$$

We now re-arrange the above equations to obtain expression for the currents, I_1, I_2 and I_3

$$I_3 = \frac{E_2}{R} \qquad (15.5)$$

Using eqn(15.5) in eqn(15.2), we have

$$I_2 = \frac{E_2}{R^2}(R + X_c) \qquad (15.6)$$

Similarly, using eqn(15.5) and eqn(15.6) in eqn(15.3), we have

$$I_1 = \frac{E_2}{R^3}(R + X_c)(R + X_c) + \frac{E_2 X_c}{R^2}$$
$$= \frac{E_2}{R^3}(R^2 + X_c^2) + \frac{3E_2 X_c}{R^2} \qquad (15.7)$$

Finally, substitute eqn(15.5), eqn(15.6) and eqn(15.7) in eqn(15.4), we have

$$E_1 = \left[\frac{E_2 X_c}{R^3}(R^2 + X_c^2) + \frac{3E_2 X_c^2}{R^2}\right] + \left[\frac{E_2 X_c}{R^2}(R + X_c)\right] + \left[\frac{E_2 X_c}{R}\right] + \left[\frac{E_2}{R^2}(R^2 + X_c^2) + \frac{3E_2 X_c}{R}\right]$$

Simplifying, we have

$$E_1 = \frac{6E_2 X_c}{R} + \frac{5E_2 X_c^2}{R^2} + E_2 + \frac{E_2 X_c^3}{R^3}$$

Thus, the amount of feedback provided by the circuit is

$$\frac{1}{\beta} = \frac{E_1}{E_2} = \frac{6X_c}{R} + \frac{5X_c^2}{R^2} + 1 + \frac{X_c^3}{R^3} \qquad (15.8)$$

To solve the above equations, we separate the real and imaginary terms. Collecting the **imaginary** terms we have

$$0 = \frac{6X_c}{R} + \frac{X_c^3}{R^3}$$
$$0 = 6 + \frac{X_c^2}{R^2} \qquad (15.9)$$

Solving, we get the frequency where phase shift introduced by this circuit is π[1]

$$(\omega RC)^2 = \frac{1}{6}$$
$$f = \frac{1}{2\pi\sqrt{6}RC} \qquad (15.10)$$

---

[1]When we collect the real and imaginary part what we had in our above example was

$$\frac{1}{\beta}cos\pi + \frac{j}{\beta}sin\pi = \frac{E_1}{E_2} = \frac{6X_c}{R} + \frac{5X_c^2}{R^2} + 1 + \frac{X_c^3}{R^3}$$

Showing the phase shift involved is π.

Collecting the real parts of eqn(15.8), we have

$$\frac{1}{\beta} = \frac{5X_c^2}{R^2} + 1$$

Substituting the result of eqn(15.9) we have

$$\frac{E_1}{E_2} = \frac{1}{\beta} = -29 \tag{15.11}$$

The negative sign associated with β indicates the positive feedback being rendered by the feedback circuit. Also, by Barkhausen's criteria, for stable and sustained oscillations $A_v\beta$ should be equal to unity. Using eqn(15.11) we understand that the internal amplifier should have a gain of atleast 29 for having sustained and stable oscillations.

Example 15.1: Find the phase shift introduced by each RC section in phase shift oscillator.

The calculation of the phase difference can be done by extending the calculation done in Chapter XI, where the frequency response of the RC network was done. The output voltage E_2 (of network shown in fig 15.1), can be calculated by finding the current in resistance across which the output voltage is collected, i.e.

$$E_2 = RI_3$$

for evaluating I_3, we find the net impedance offered by the network to E_1, from which the net current is given as

$$I_1 + I_2 + I_3 = \frac{E_1}{Z_T}$$

where

$$Z_T = X_C + \frac{R(X_C + R')}{R + R' + X_C}$$

with

$$R' = \frac{R(X_C + R)}{2R + X_C}$$

using the current division equation and some *"hammering mathematics"*, we have

$$I_3 = \frac{R}{(2R + X_C)} \times \frac{RE_1}{Z_T(R + R' + X_C)}$$

$$E_2 = RI_3 = \frac{R^3 E_1}{Z_T(2R + X_C)(R + R' + X_C)}$$

Substituting values of Z_T and R', we have

$$\frac{E_2}{E_1} = \frac{1}{1 + 6\frac{X_C}{R} + 5\frac{X_C^2}{R^2} + \frac{X_C^3}{R^3}}$$

$$\beta = \frac{1}{\left(1 - \frac{5}{\omega^2 R^2 C^2}\right) - j\left(\frac{6}{\omega RC} - \frac{1}{\omega^3 R^3 C^3}\right)}$$

representing this in polar co-ordinate system we have

$$\beta = \frac{1}{\sqrt{\left(1 - \frac{5}{\omega^2 R^2 C^2}\right)^2 + \left(\frac{6}{\omega RC} - \frac{1}{\omega^3 R^3 C^3}\right)^2}} \angle tan^{-1}\left[\frac{\frac{1}{\omega RC}\left(6 - \frac{1}{\omega^2 R^2 C^2}\right)}{\left(1 - \frac{5}{\omega^2 R^2 C^2}\right)}\right]$$

For oscillaions, the feedback should be positive and should introduce a phase difference of π between E_1 and E_2. For this the term within the square bracket, should fall equal to zero. This gives the frequency at which the output oscillates as

$$6 - \frac{1}{\omega^2 R^2 C^2} = 0$$

$$f = \frac{1}{2\pi\sqrt{6}RC}$$

The above result is in agreement with our earlier result, abid different derivation. By current division we have currents in the second and first section as

$$I_2 = \frac{(R + X_C)}{(2R + X_C)} \times \frac{RE_1}{Z_T(R + R' + X_C)}$$

and

$$I_1 = \frac{(R' + X_C)E_1}{Z_T(R + R' + X_C)}$$

respectively, resulting in voltage across the resistance

$$V_{I_2} = \frac{(R + X_C)}{(2R + X_C)} \times \frac{R^2 E_1}{Z_T(R + R' + X_C)}$$

and

$$V_{I_1} = \frac{(R' + X_C)RE_1}{Z_T(R + R' + X_C)}$$

which when represented in their polar form would give the phase difference introduced by each RC section.

$$\frac{V_{I_2}}{E_1} = \frac{\left(1 - \frac{j}{\omega RC}\right)}{\left(1 + 5\frac{X_C^2}{R^2}\right) + 6\frac{X_C}{R} + \frac{X_C^3}{R^3}}$$

$$= \frac{\sqrt{1 + \frac{1}{\omega^2 R^2 C^2}}}{\sqrt{\left(1 - \frac{5}{\omega^2 R^2 C^2}\right)^2 + \left(\frac{6}{\omega RC} - \frac{1}{\omega^3 R^3 C^3}\right)^2}} \angle tan^{-1}\left[\frac{\frac{1}{\omega RC}\left(6 - \frac{1}{\omega^2 R^2 C^2}\right)}{\left(1 - \frac{5}{\omega^2 R^2 C^2}\right)}\right] - \angle tan^{-1}\left(\frac{1}{\omega RC}\right)$$

For the given condition of oscillation (frequency), we have $\omega RC = 1/6$, the phase difference at the end of the second RC section is

$$= \angle \pi - tan^{-1}(\sqrt{6})$$
$$= \angle \pi - 67.8^o = 112.2^o$$

Finally, for the first RC section, we have

$$\frac{V_{I_1}}{E_1} = \frac{\left(1 - \frac{1}{\omega^2 R^2 C^2}\right) - j\frac{3}{\omega RC}}{\left(1 - \frac{5}{\omega^2 R^2 C^2}\right) - j\left(\frac{6}{\omega RC} - \frac{1}{\omega^3 R^3 C^3}\right)}$$

$$= \frac{\sqrt{\left(1 - \frac{1}{\omega^2 R^2 C^2}\right)^2 + \frac{9}{\omega^2 R^2 C^2}}}{\sqrt{\left(1 - \frac{5}{\omega^2 R^2 C^2}\right)^2 + \left(\frac{6}{\omega RC} - \frac{1}{\omega^3 R^3 C^3}\right)^2}} \angle tan^{-1}\left[\frac{\frac{1}{\omega RC}\left(6 - \frac{1}{\omega^2 R^2 C^2}\right)}{\left(1 - \frac{5}{\omega^2 R^2 C^2}\right)}\right]$$

$$- \angle tan^{-1}\left(\frac{\frac{3}{\omega RC}}{1 - \frac{1}{\omega^2 R^2 C^2}}\right)$$

Again the phase difference at the oscillating frequency can be calculated and we find it to be

$$= \angle \pi - \angle tan^{-1}\left[\frac{3\sqrt{6}}{1 - 6}\right]$$

$$= 55.77^o$$

The three RC sections of the phase shift oscillators introduce phase differences in and around 60°, such that the net phase difference introced by the three together is 180° or π. The phase difference introduced after each stage is

$$After\ first\ RC\ section \ = \ 55.77^o$$
$$After\ second\ RC\ section \ = \ 112.2^o,\ and$$
$$After\ third\ RC\ section \ = \ 180^o.$$

15.1.2   With BJT

If you recollect the small signal AC model of the transistor amplifier, you would recollect that the collector resistance R_c comes in parallel to the load resistance, thus reduces the effective load. Since, the output load in the case of this circuit is the feedback circuit, it is liable to effect the above calculations (see fig 15.2). To investigate this we proceed to convert the current source to a voltage source and analysis the three loops after incorporating the necessary changes.

As in the previous mathematical analysis, we proceed to write the KVL for the last two loops

$$I_3 R = E_2 \tag{15.12}$$

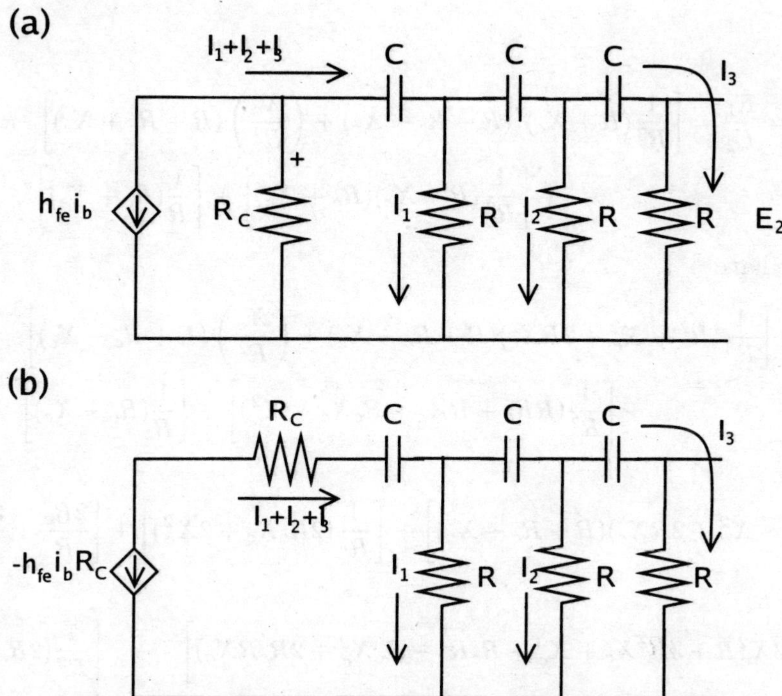

Figure 15.2: *The output side of the transistor amplifier in it's small signal AC model has been shown with the feedback circuit. The current source of (a) has been converted to a voltage source in (b). Note this consideration is being made because of the non-zero/ finite output impedance of the transistor circuit that can result in the feedback circuit loading the amplifier.*

and

$$I_2R = I_3\left(R + X_c\right) \tag{15.13}$$

and

$$I_1R = I_2(R + X_c) + I_3X_c \tag{15.14}$$

However, the third loop will differ, since the internal amplifier's output is being collected across the amplifier's collector resistance, R_c. The output voltage is i_cR_c, however, on converting the current source to voltage source, we have

$$E_1 = -h_{fe}i_bR_c = (I_1 + I_2 + I_3)(R_c + X_c) + I_1R \tag{15.15}$$

Individual currents are calculated from the above expression and substituted in eqn(15.15), giving

$$E_1 = \left[\frac{E_2}{R^3}(R + X_c)^2(R + R_c + X_c) + \left(\frac{E_2X_c}{R^2}\right)(R + R_c + X_c)\right]$$
$$+ \left[\frac{E_2}{R^2}(R + X_c)(R_c + X_c)\right] + \left[\frac{E_2}{R}(R_c + X_c)\right] \tag{15.16}$$

Resulting in

$$\frac{1}{\beta} = \frac{E_1}{E_2} = \left[\frac{1}{R^3}(R+X_c)^2(R+R_c+X_c) + \left(\frac{X_c}{R^2}\right)(R+R_c+X_c)\right]$$

$$+ \left[\frac{1}{R^2}(R+X_c)(R_c+X_c)\right] + \left[\frac{1}{R}(R_c+X_c)\right] \qquad (15.17)$$

Simplifying, we have

$$\frac{1}{\beta} = \left[\frac{1}{R^3}(R^2+X_c^2+2RX_c)(R+R_c+X_c) + \left(\frac{X_c}{R^2}\right)(R+R_c+X_c)\right]$$

$$+ \left[\frac{1}{R^2}(RR_c+RX_c+R_cX_c+X_c^2)\right] + \left[\frac{1}{R}(R_c+X_c)\right]$$

$$= \left[\frac{1}{R^3}(R^2+X_c^2+2RX_c)(R+R_c+X_c)\right] + \left[\frac{1}{R^2}(2R_cX_c+2X_c^2)\right] + \left[\frac{2R_c}{R}+\frac{3X_c}{R}\right]$$

$$= \left[\frac{1}{R^3}(R^3+3X_c^2R+3R^2X_c+X_c^3+R_cR^2+R_cX_c^2+2R_cRX_c)\right] + \left[\frac{1}{R^2}(2R_cX_c+2X_c^2)\right]$$

$$+ \left[\frac{2R_c}{R}+\frac{3X_c}{R}\right]$$

$$= \left[\frac{1}{R^3}(R^3+X_c^3+R_cX_c^2)\right] + \left[\frac{1}{R^2}(4R_cX_c+5X_c^2)\right] + \left[\frac{3R_c}{R}+\frac{6X_c}{R}\right]$$

Collecting the real and imaginary parts of eqn(15.16), we have

$$0 = \frac{X_c^3}{R^3} + \frac{4R_cX_c}{R^2} + \frac{6X_c}{R}$$

$$0 = \frac{X_c^2}{R^2} + \frac{4R_c}{R} + 6$$

$$0 = \frac{X_c^2}{R^2} + 4k + 6 \qquad (15.18)$$

where

$$k = \frac{R_C}{R} \qquad (15.19)$$

Solving, the frequency of the sustained oscillations obtained from this circuit is given as

$$0 = -\frac{1}{\omega^2 R^2 C^2} + 4k + 6$$

$$f = \frac{1}{2\pi RC\sqrt{6+4k}} \qquad (15.20)$$

Collecting the real parts of eqn(15.16), we have

$$\frac{1}{\beta} = 1 + \frac{R_c X_c^2}{R^3} + \frac{5X_c^2}{R^2} + \frac{3R_c}{R}$$

$$= 1 + \frac{kX_c^2}{R^2} + \frac{5X_c^2}{R^2} + 3k$$

Substituting the value of X_c/R_c from eqn(15.18), we have

$$\frac{1}{\beta} = 1 - (k+5)(4k+6) + 3k$$

$$= -29 - 23k - 4k^2 \tag{15.21}$$

or

$$\frac{1}{\beta} = -29 - 23k - 4k^2 = \frac{E_1}{E_2} = \frac{-h_{fe}i_b R_c}{E_2}$$

$$\frac{-h_{fe}i_b R_c}{I_3 R} = -29 - 23k - 4k^2$$

$$\frac{h_{fe}i_b}{I_3} = \frac{29}{k} + 23 + 4k$$

Thus, design consideration demands,

$$h_{fe} > \frac{29}{k} + 23 + 4k$$

The suitable value of k is decided by finding the extrema

$$-\frac{29}{k^2} + 4 = 0$$

$$k = 2.7$$

Thus, for this value of k, eqn(15.21) demands $\frac{1}{\beta} = -120$. The negative sign indicates the nature
of feedback to be positive. In Barkhausen's criteria

$$A\beta = 1$$

$$A = \frac{1}{\beta} \sim 120 \tag{15.22}$$

The gain requirement of the internal amplifier is more in this case than that of the FET
example. This implies, due to the loading taking place on connecting the feedback circuit to
the transistor, a high-gain transistor must be used. If the internal amplifier's gain remains low
(as in the case of an FET circuit, $A_v = 29$) and loading takes place, sustained oscillations would
not be possible. Thus, requiring high gain internal amplifier's for transistor circuits.

Also, one might be wondering at this point if a phase shift oscillator can only be made with
three-section RC network? Infact, it is better to have more than three RC sections, since the
overall signal loss within the network decreases with more sections. Consider using a four RC

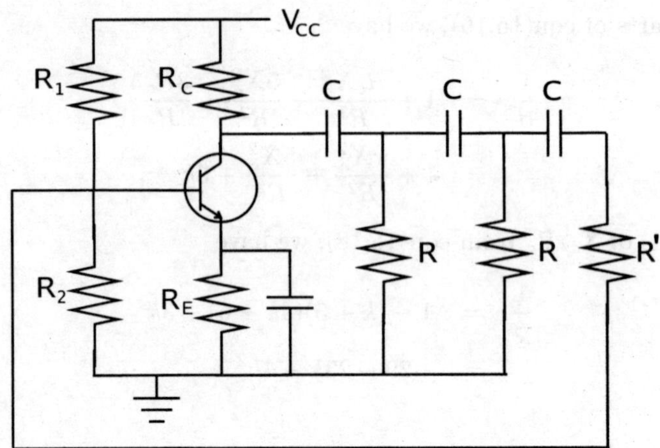

Figure 15.3: *The small signal AC model of the transistor amplifier suggests that the bias resistance* R_1 *and* R_2 *come parallel to the feedback circuits third resistance R, leading to a non-π phase shift introduced by the feedback circuit. This destroys any case of oscillations. The third resistance R is replaced as shown by R' such that the combination of R' and* R_1, R_2 *is equal to R.*

section feedback circuit, each section would introduce a phase shift of in and around 45°, thus the loss for each section is lowered as compared to the three section circuit. As for whether a phase shift oscillator can be made with only one or two RC sections, it is only a trivial exercise to follow the mathematics of Example 15.1 and comment on the feasability. The polar representation of the output voltage to the input voltage of a two section RC network is given as

$$\frac{E_2}{E_1} = \frac{1}{\sqrt{\left(1 - \frac{1}{\omega^2 R^2 C^2}\right)^2 + \frac{9}{\omega^2 R^2 C^2}}} \angle tan^{-1}\left[\frac{\frac{3}{\omega RC}}{1 - \frac{1}{\omega^2 R^2 C^2}}\right] \tag{15.23}$$

This shows that the circuit with two RC sections would only be set into oscillation if the value of R or C takes up an impractical value of infinity (remember the phase difference should be π, resulting in $tan^{-1}[\pi] = 0$.) The analysis of a single section RC also results in such a conclusion. This shows that a practical phase shift oscillator can not be fabricated using a single or two RC sections and require a minimum of three sections.

In an oscillator you have the transistor that introduces a phase shift of π and also the phase shift network that introduces a phase shift. The transistor introduces this phase shift of π at all frequencies, however, only for the frequency for which the network introduces shift π does stable and sustained oscillations take place. The difficulty in terms of fabrication makes a three section RC network more popular than the four section counterpart. However, it is seen that the circuit giving larger value of $d\phi/d\omega$ has the more stable oscillator frequency (ϕ being the phase shift). As can be seen from fig(15.4), four sections of RC has a larger value of $d\phi/d\omega$ then three section RC circuits, making it a better phase-shift network then for oscillator.

A major problem with these phase shift oscillators were tunability, i.e. if you wanted to change the frequency of oscillation, you needed to change the whole circuit. New circuits thus were designed that not only improved on stability but also on tunability.

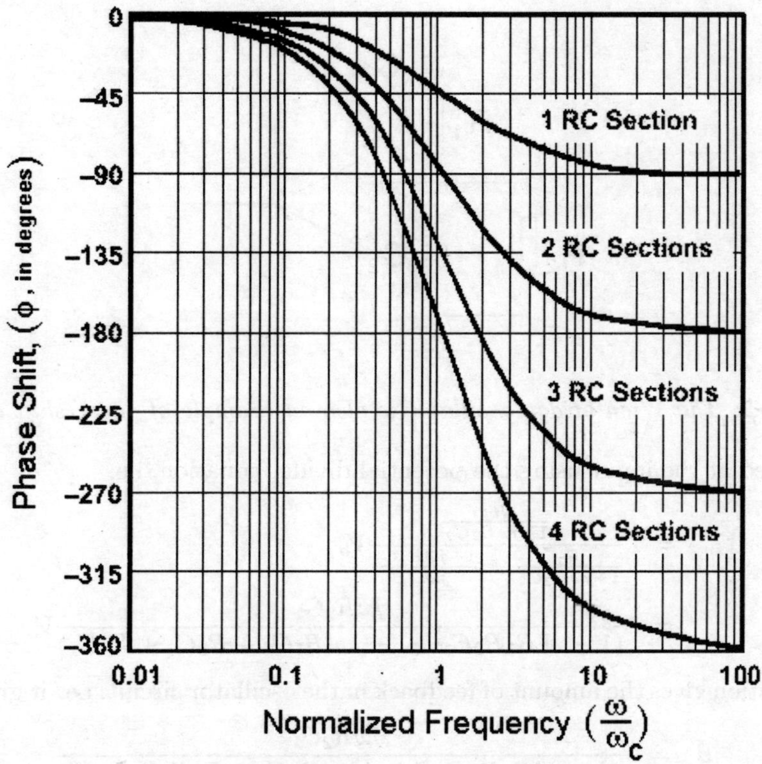

Figure 15.4: *The phase shift introduced by 'n' number of RC combinations as a function of frequency (courtesy Texas Instrumentation).*

15.2 Wien Bridge Oscillators

The Wien-bridge oscillator was designed and developed in 1940[2] by William Hewlett and David Packard. The basic arrangement of the Wien Bridge circuit is shown in fig(15.5). The circuit became quite popular because of it's stability, relatively low distortion and ease of tuning. The bridge is a phase shift network. However, to understand the points highlighted above, one has to understand the working of this circuit.

The resistance R_1 is in series with capacitor C_1 which is connected in series to the parallel combination of R_2 and C_2. The triangle is a common representation of an amplifier. The output voltage of the amplifier is fed across the resistance-capacitor network. The voltage across the parallel combination of R_2 and C_2 is given as an input to the amplifier (see fig 15.5). The network hence forms the phase-shift network. To compute the amount of shift we compute the

---

[2]You might wonder how this circuit was designed in 1940, a good seven years before the invention of the transistor? The circuit was designed with triode values. When the transistor came along the triode was replaced by the transistor

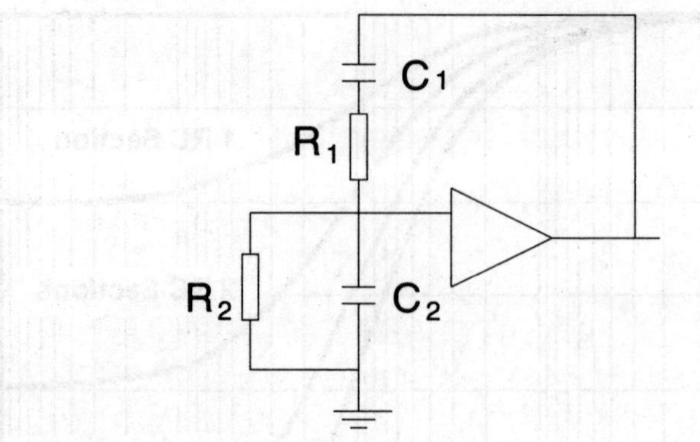

Figure 15.5: *The Wien bridge oscillator is also an example of phase shift oscillator.*

voltage being fed at the input using the potential divider equation, i.e.

$$V_{in} = \frac{\frac{R_2}{1+j\omega R_2 C_2}}{\frac{R_2}{1+j\omega R_2 C_2} + \frac{1+j\omega R_1 C_1}{j\omega C_1}} V_o$$

$$= \frac{j\omega R_2 C_1}{(1 - \omega^2 R_1 R_2 C_1 C_2) + j\omega(R_1 C_1 + R_2 C_2 + R_2 C_1)} V_o$$

The above equation gives the amount of feedback in the oscillator circuit, i.e. it gives β. Re-write

$$\beta = \frac{j\omega R_2 C_1}{(1 - \omega^2 R_1 R_2 C_1 C_2) + j\omega(R_1 C_1 + R_2 C_2 + R_2 C_1)}$$

$$= \frac{-\omega R_2 C_1}{j(1 - \omega^2 R_1 R_2 C_1 C_2) - \omega(R_1 C_1 + R_2 C_2 + R_2 C_1)} \quad (15.24)$$

To successfully attain oscillation (Barkhausen's criteria), the phase shift network should introduce a phase difference of π (the remaining phase difference of π required for positive feedback comes from the amplifier). That is feedback network should achieve, $\beta\pi$, for which the above equation's imaginary terms should fall equal to zero. That gives the frequency of oscillation as

$$f = \frac{1}{2\pi\sqrt{R_1 R_2 C_1 C_2}}$$

In the above circuit, if $R_1 = R_2 = R$ and $C_1 = C_2 = C$, the frequency of oscillation becomes

$$f = \frac{1}{2\pi RC} \quad (15.25)$$

With this condition, collecting the real terms of eqn(15.24), we have

$$\beta = \frac{1}{3}$$

or

$$\beta = \frac{R_2 C_1}{(R_1 C_1 + R_2 C_2 + R_2 C_1)}$$

The gain of the internal amplifier should be $A_v \geq 3$, for getting sustained oscillations. The Tunability of this circuit is usually achieved by varying capacitances. However, it should be noted that by tuning the maximum frequency practically obtained is with audio range.

15.3 Oscillators for Radio Frequencies

At high frequencies the capacitors contribute zero impedance and hence are shorts, bringing the three resistances parallel to each other and the feedback contributed would not be as required. Thus, the phase shift oscillator is only useful for generating sine waves in audio frequency range. However, in communication applications one requires higher frequency output (radio frequency range). These circuits are fabricated using resonant or tank circuit made with an inductor and a capacitor. The tank circuit consists of a capacitor and a coil connected in parallel. To understand how the LC circuit works, consider the capacitor to be charged. The capacitor's plate stores electrons, and is said to be charged. Now, when a resistance is connected to the capacitor (forming a closed circuit/ loop) the capacitor discharges, thus making the capacitor's plates neutral. However, instead of a resistance if a coil is connected across the charged capacitor, the discharging capacitor's current generates a magnetic field around the coil. This magnetic field generates a voltage across the coil that opposes the direction of electron flow. Because of this, the capacitor does not discharge right away. Now once the capacitor is fully discharged through the coil, the magnetic field starts to collapse around the coil. The voltage induced from the collapsing magnetic field recharges the capacitor with the opposite polarity. Then the capacitor begins to discharge through the coil again, generating a magnetic field. This process ideally continuous indefinitely. The differential equation (KVL) for the tank circuit would be

$$L\frac{di}{dt} + \frac{1}{C}\int i dt + v_c(t=0) \;=\; 0$$

$$L\frac{d^2i}{dt^2} + \frac{i}{C} \;=\; 0$$

$$\frac{d^2i}{dt^2} \;=\; -\frac{i}{LC} = -\omega^2 i$$

This equation is the famous S.H.M (simple harmonic motion) equation describing the oscillatory motion of a pendulum. The standard solution of which is of the form

$$i(t) = A sin\omega t$$

The frequency of oscillation is given by ω, i.e.

$$f = \frac{1}{2\pi\sqrt{LC}}$$

The frequency produced by the oscillator circuit having tank circuits would be the same as the resonant frequency (given by the above equation) of the tank circuit put into use. Practically, the basic LC circuit loses voltage with every oscillation. To overcome this, the oscillator circuit (amplifier circuit) provides additional voltage to replinish the lost voltage and sustaining the oscillation.

Various circuits are in use and they differ in the method which the feedback signal is coupled with the internal amplifier. While one method is to take some of the energy from the inductor and feed it back to the input, the second method is to pinch energy from the capacitor. Taking energy from the inductor can be done in two different ways, (a) using a **TICKLER COIL** (fig 15.6a, Oscillators using tickler coils are called Armstrong oscillator) and (b) using a split coil (fig 15.6b). Circuits using split coils are called Hartley Oscillator. Similarly, you have Colpitt oscillator where feedback energy is obtained from the split capacitor. Note the oscillators are named after their designers.

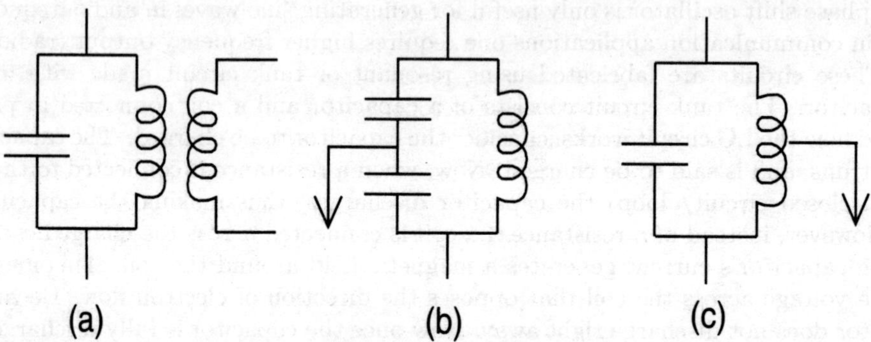

Figure 15.6: *The three different methods by which a tank circuit can be used in an oscillator circuit. Feedback voltage can be given by (arrow indicates end to be connected to the input) (a) using a tickler circuit, where tickler is the small coil used to input feedback voltage, (b) using a split capacitor or (c) using a split inductor.*

15.3.1   Armstrong Oscillators

The Armstrong oscillator, named after it's inventor Edwin Howard Armstrong,[3] uses a tickler circuit to produce a sine-wave output of constant amplitude and of fairly constant frequency in the radio frequency range. Consider circuit shown in fig(15.7). The tickler coil forms the collector impedance of the internal amplifier circuit, while the feedback is achieved by coupling (mutual inductance) between the tickler coil and the LC tank circuit. Basically, as in the case of the phase shift oscillators discussed above, the internal amplifier's transistor contributes a phase difference of π between the input and output voltages. Another phase difference of π is introduced in the feedback path by the transformer like coupling between the tickler and tank coils.

---

[3] Edwin Howard Armstrong (1890-1954) worked in New York City. Around 1930 Lee De Forest's had developed his triode tube however, much applications of the device was not developed since many did not understand how it worked. Armstrong discovered that the gain of a triode amplifier could be enormously increased by feeding some of the amplifier output back into the input, i.e. by using positive feedback. Given enough feedback, the amplifier became a stable and powerful oscillator, perfect for driving radio transmitters. Based on this invention Armstrong went on to give his greatest invention, frequency modulation (FM). Armstrong's life was a tradedy since many of his inventions were ultimately claimed by others in patent lawsuits. He was a wonderful experimentalist as his invention and statement suggests:
"Anyone who has had actual contact with the making of the inventions that built the radio art knows that these inventions have been the product of experiment and work based on physical reasoning, rather than on the mathematicians' calculations and formulae."

Let us examine the Armstrong oscillator further. Initially, consider the transistor's base to be forward biased. To picturise this, consider V_{cc} is switched on at time t=0. A base base current flows through the biasing resistance R_2 which sets the transistor into conduction. A collector current is established through the tickler coil L_1. The rising collector current causes the voltage across the coil to increase ($v_{L_1} = L_1 di_c/dt$). The mutual inductance leads to an inverted voltage setting up in the tank circuit's coil L_2. This circuit is such that the induced voltage in L_2 ensures that the base is made positive w.r.t to ground and is in phase with the base voltage (phase difference of 2π). The positive signal is now coupled to the base of the transistor. Two important points are to be noted, as positive potential is building across L_2, the same potential is appearing across the parallel capacitor and this potential is forward biasing the transistor's base more and more. The collector current increases, since the transistor conducts better and better. With the transistor conducting harder and more collector current flowing through the tickler coil, a larger voltage is induced into L_2, sending the transistor into further conduction until it saturates. On saturation, the collector current is at a maximum value (I_{Csat}) and cannot increase any further. With a steady current through the tickler coil, there is no changing magnetic flux lines cutting L_2 and hence no voltage is induced in it.

With no external voltage applied, capacitor across L_2 acts as a voltage source and discharges. As the voltage across the capacitor decreases, its energy is transferred to L_2 (i.e. electrostatic energy is converted to magnetic energy). As the capacitor discharges, the forward bias on the transistor decreases. This results in decrease of the collector current. A decrease in collector current allows the magnetic field of L_1 to fall. This, action continues till the transistor is sent to it's cutoff (attained when the potential across the capacitor is zero). The capacitor continues to discharge and attains potential $-V_c$, the negative sign indicating that the polarity is opposite to that when the capacitor was charging. After the complete discharging action of the capacitor, the capacitor now discharges in the opposite direction, i.e. tries to go from $-V_c$ to $+V_c$. In the process when the voltage across it is greater than zero, the base is sent forward and the process described above repeats itself.

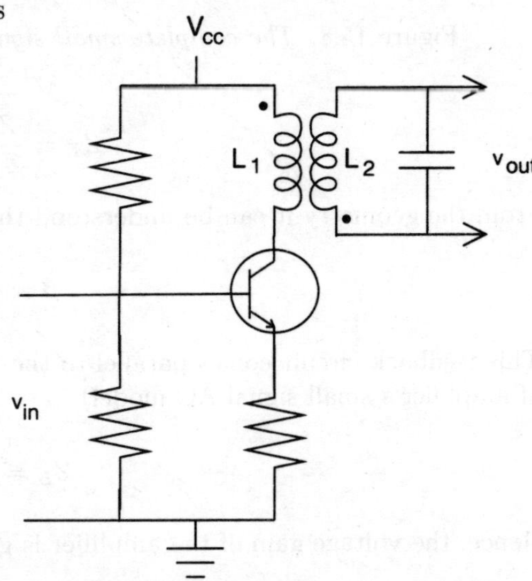

Figure 15.7: *The Armstrong oscillator using a tickler coil.*

During the oscillatory charging and discharging behavior taking place in the tank circuit, damping does occur. The lost energy is fed back during the period collector current grows from zero to it's saturation value. An important point to note is that the transistor base is forward biased only for half the oscillation, i.e. it is sent into conduction only for 50% of the time. Thus, the transistor has to be suitably biased, in what is called a Class-C configuration. Details of biasing of Class C would be discussed in Chapter 16.

The oscillator in fig(15.7) is a tuned-base oscillator, because the tank circuit is connected to the base of the transistor amplifier circuit. If it is in the collector circuit, it is called a tuned-collector oscillator.

15.3.2   Oscillators with Split Tank Circuits

While the Armstrong oscillator is simple to fabricate, it's limitation is in it's tunability. A stable and tunable oscillator can be made by splitting one of the tank circuit's elements. Consider the general feedback circuit depicted in fig(15.8), the total impedance offered by this circuit is

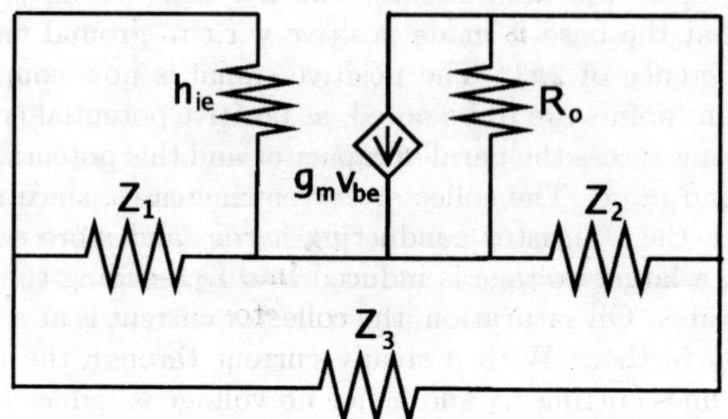

Figure 15.8: *The complete small signal circuit using the feedback circuit is shown.*

$$Z_T = \frac{Z_3(Z_1 + Z_2)}{Z_1 + Z_2 + Z_3} \tag{15.26}$$

From the geometry it can be understood that the sampled output voltage used for feedback is

$$\beta = \frac{Z_1}{Z_1 + Z_2}$$

This feedback circuit comes parallel to the internal amplifier's output impedance (see fig 15.8b of amplifier's small signal AC model)

$$Z_L = \frac{R_o Z_T}{R_o + Z_T} \tag{15.27}$$

Hence, the voltage gain of the amplifier is given as

$$\begin{aligned}
A_v &= -g_m Z_L \tag{15.28} \\
&= -\frac{g_m R_o Z_3(Z_1 + Z_2)}{R_o(Z_1 + Z_2 + Z_3) + Z_3(Z_1 + Z_2)}
\end{aligned}$$

Applying Barkhausen's criteria, we have

$$A_v \beta = 1$$

$$-\frac{g_m R_o Z_1 Z_3}{R_o(Z_1 + Z_2 + Z_3) + Z_3(Z_1 + Z_2)} = 1$$

To understand the import of this expression we have to investigate the nature of Z_1, Z_2 and Z_3 in actual circuits. There are essentially two circuits of importance. Z_1 and Z_2 are inductors while Z_3 is an capacitor or Z_1 and Z_2 are capacitor while Z_3 is an inductor. That is, either you have

$$-\frac{g_m R_o X_{L1} X_C}{jR_o(X_{L1} - X_C + X_{L2}) + X_C(X_{L1} + X_{L2})} = 1 \qquad (15.29)$$

or

$$-\frac{g_m R_o X_{C1} X_L}{jR_o(X_L - X_{C1} - X_{C2}) + X_L(X_{C1} + X_{C2})} = 1 \qquad (15.30)$$

Consider the two cases individually,

1) Collecting the imaginary parts of eqn(15.29), we have

$$\begin{aligned} X_{L1} - X_C + X_{L2} &= 0 \\ X_{L1} + X_{L2} &= X_C \end{aligned} \qquad (15.31)$$

The frequency for which sustained oscillations are obtained is given by this equation as

$$f = \frac{1}{2\pi\sqrt{(L_1 + L_2)C}} \qquad (15.32)$$

While equating the real parts of eqn(15.29) we get

$$\begin{aligned} -\frac{g_m R_o X_{L1}}{(X_{L1} + X_{L2})} &= 1 \\ -g_m R_o &= \frac{X_{L1} + X_{L2}}{X_{L1}} \end{aligned}$$

The term $-g_m R_o$ is related to the gain (see eqn 15.28) of the voltage gain of the internal amplifier which as per requirement has to be large. Hence, the above equation reduces to

$$-g_m R_o \geq \frac{L_2}{L_1} \qquad (15.33)$$

The usage of Barkhausen's criteria gives both the frequency of oscillation (eqn 15.32) and the required internal amplifier's gain (eqn 15.33) for designing a Hartley oscillator (we shall learn more about this oscillator in the next section).

2) Collecting the imaginary parts of eqn(15.30), we have

$$\begin{aligned} X_L - X_{C1} - X_{C2} &= 0 \\ X_{C1} + X_{C2} &= X_L \end{aligned} \qquad (15.34)$$

The frequency for which sustained oscillations are obtained is given by this equation as

$$f = \frac{1}{2\pi}\sqrt{\frac{1}{L}\left(\frac{1}{C_1} + \frac{1}{C_2}\right)} \qquad (15.35)$$

While equating the real parts of eqn(15.30) we get

$$-\frac{g_m R_o X_{C1}}{(X_{C1} + X_{C2})} = 1$$

$$-g_m R_o = \frac{X_{C1} + X_{C2}}{X_{C1}}$$

Again, using the knowledge that the gain of the internal amplifier is large

$$g_m R_o = \frac{C_1}{C_2} \tag{15.36}$$

Equation(15.35) and eqn(15.36) give the frequency of oscillation and the required internal amplifier's gain respectively, for designing a Colpitt oscillator (see details in the next section).

The derivations above show the general design requirements of oscillators with split tank circuit. As stated there are two possiblities that you end up with, the Hartley oscillator and the Colpitt oscillator. We know investigate the two circuits in more detail.

Hartley's Oscillator

Hartley oscillator oscillators are similar to the Armstrong oscillator, the difference between the two is that the tickler coil is part of the LC tank circuit. Hartley oscillators have the advantage of having one centre tapped inductor and one tuning capacitor. This arrangement simplifies the construction of a Hartley oscillator circuit and allows for easy tunability, i.e. one can easily obtain variable frequency.

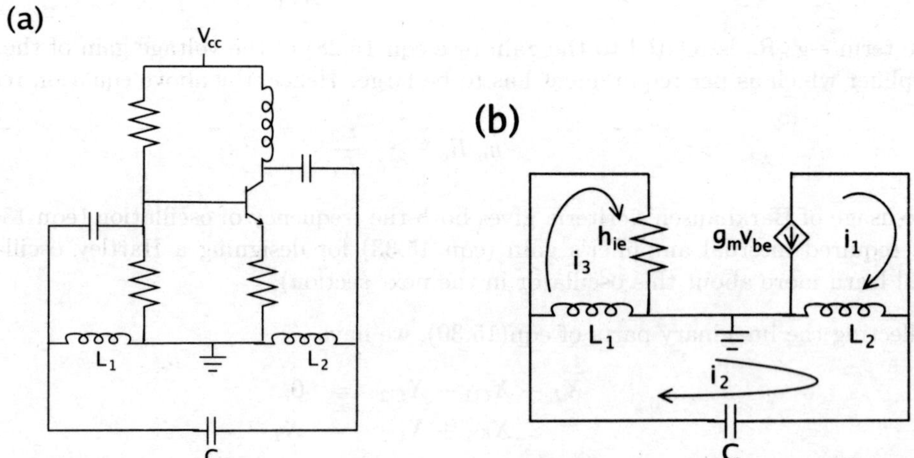

Figure 15.9: *The (a) Hartley oscillator circuit with (b) the small signal equivalent circuit.*

One version of Hartley oscillator is shown in figure 15.9. The tank circuit consists of the tapped coil (L_1 and L_2) and capacitor C. The feedback circuit is from the tank circuit to the base of transistor through the coupling capacitor C. Coupling capacitors prevents the low DC resistance of L_1 from placing a short across the emitter-base junction and resistor R_2. The coils L_1 and L_2 are coupled by mutual inductance and is also coupled capacitively by the capacitor C. Current flow through L_2 induces a voltage in the coil L_1. The phase difference introduced by the mutual coupling is π and along with the π phase difference introduced by the transistor amplifier, the net feedback develop is regenerative (positive feedback).

To have a complete and exhaustive characterisation of the Hartley oscillator, we apply KVL on the three loops of the small signal model (see fig 15.9b). The output voltage develops across L_2 and is given as

$$\begin{aligned} v_o &= j\omega(L_2 + M)(i_1 - i_2) \\ &= j\omega(L_2 + M)i_1\left(1 - \frac{i_2}{i_1}\right) \end{aligned}$$

where M is the mutual inductance[4] existing between the two coils L_1 and L_2. The current in the loop is decided by the current source.

$$i_1 = -h_{fe}i_b = -g_m v_{be}$$

Substituting this in the loop equation, we put the voltage gain as

$$\begin{aligned} v_o &= -j\omega g_m(L_2 + M)v_{be}\left(1 - \frac{i_2}{i_1}\right) \\ A_v &= \frac{v_o}{v_{be}} = -j\omega g_m(L_2 + M)\left(1 - \frac{i_2}{i_1}\right) \end{aligned}$$

Using the information that the input voltage develops across h_{ie}, we have the amount of feedback as

$$\beta = \frac{v_{be}}{v_o} = \frac{h_{ie}i_3}{j\omega(L_2 + M)i_1\left(1 - \frac{i_2}{i_1}\right)}$$

Using the gain and feedback expression in Barkhausen's criteria ($A_v\beta = 1$), we have

$$-j\omega g_m(L_2 + M)\left(1 - \frac{i_2}{i_1}\right) \times \frac{h_{ie}i_3}{j\omega(L_2 + M)i_1\left(1 - \frac{i_2}{i_1}\right)} = 1$$

---

[4]Mutual inductance results when varying magnetic flux around the inductor cuts across the windings of a nearby inductor, thus inducing voltage in the second inductor. When L_1 and L_2 are close by, on increasing the current in L_1 leads to changing magnetic flux which cuts L_2, inducing a voltage in it. The coupling is given as

$$k = \left(\frac{lines\ of\ fluxes\ linking\ L_1\ and\ L_2}{lines\ of\ fluxes\ produced\ by\ L_1}\right)$$

The closer the coils are, more the lines of fluxes produced by L_1 are linked to L_2. Thus, inducing more voltage in L_2, thus showing good mutual inductance. The value of k can vary from zero to one. The mutual inductance is given as

$$M = k\sqrt{L_1 L_2}$$

which simplifies to

$$-g_m h_{ie} \frac{i_3}{i_1} = 1 \tag{15.37}$$

Applying KVL to the input loop, we have

$$h_{ie} i_3 + j\omega(L_1 + M)(i_3 - i_2) = 0$$

The ratio of the two currents involved in this equation can be written as

$$\frac{i_3}{i_2} = \frac{j\omega(L_1 + M)}{j\omega(L_1 + M) + h_{ie}}$$

Similarly, applying the KVL for the capacitor loop, we have

$$\frac{1}{j\omega C} i_2 + j\omega(L_2 + M)(i_2 - i_1) + j\omega(L_1 + M)(i_2 - i_3) \;=\; 0$$

$$i_2 \left[\frac{1}{j\omega C} + j\omega(L_1 + L_2 + 2M) \right] - j\omega(L_2 + M)i_1 - j\omega(L_1 + M)i_3 \;=\; 0$$

Substituting the value of current i_3 in this equation in terms of i_1, we have

$$i_2 \left[\frac{1}{j\omega C} + j\omega(L_1 + L_2 + 2M) + \frac{\omega^2(L_1 + M)^2}{j\omega(L_1 + M) + h_{ie}} \right] \;=\; j\omega(L_2 + M)i_1$$

$$i_2 \left[\frac{1 - \omega^2 C(L_1 + L_2 + 2M)}{j\omega C} + \frac{\omega^2(L_1 + M)^2}{j\omega(L_1 + M) + h_{ie}} \right] \;=\; j\omega(L_2 + M)i_1$$

This calls for some mathematics for simplifying. We have $\frac{i_1}{i_2}$ equal to

$$\left[\frac{h_{ie} - \omega^2 C(L_1 + L_2 + 2M)h_{ie} + j\omega(L_1 + M) - j\omega^3(L_1 + M)(L_1 + L_2 + 2M) + j\omega^3 C(L_1 + M)^2}{j\omega h_{ie} C - \omega^2(L_1 + M)C} \right]$$

$$= j\omega(L_2 + M)\frac{i_1}{i_2}$$

Giving, $\frac{i_2}{i_1}$ as

$$\frac{j\omega(L_2 + M)[h_{ie} + j\omega(L_1 + M)](j\omega C)}{h_{ie} - \omega^2 C(L_1 + L_2 + 2M)h_{ie} + j\omega(L_1 + M) - j\omega^3(L_1 + M)(L_1 + L_2 + 2M) + j\omega^3 C(L_1 + M)^2}$$

Since the ratio, $\frac{i_3}{i_2}$ and $\frac{i_2}{i_1}$ are known, we can compute the ratio $\frac{i_3}{i_1}$ using

$$\frac{i_3}{i_1} = \frac{i_3}{i_2} \times \frac{i_2}{i_1}$$

Thus, $\frac{i_3}{i_1}$ is equal to

$$\frac{-j\omega^3(L_1 + M)(L_2 + M)C}{h_{ie} - \omega^2 C(L_1 + L_2 + 2M)h_{ie} + j\omega(L_1 + M) - j\omega^3(L_1 + M)(L_1 + L_2 + 2M) + j\omega^3 C(L_1 + M)^2}$$

Use this equation in eqn(15.37) to obtain

$$jg_m\omega^3(L_1+M)(L_2+M)h_{ie}C =$$

$$h_{ie}-\omega^2C(L_1+L_2+2M)h_{ie}+j\omega(L_1+M)-j\omega^3(L_1+M)(L_1+L_2+2M)+j\omega^3C(L_1+M)^2$$

Collecting the real terms of the above equation we get the frequency of the output wave as

$$h_{ie}-\omega^2C(L_1+L_2+2M)h_{ie}=0$$

$$\omega=\frac{1}{\sqrt{(L_1+L_2+2M)C}} \tag{15.38}$$

Collection of the imaginary terms lead to

$$g_m\omega^3(L_1+M)(L_2+M)h_{ie}C-\omega(L_1+M)+\omega^3(L_1+M)(L_1+L_2+2M)-\omega^3C(L_1+M)^2=0$$

Simplified further

$$g_m\omega^2(L_2+M)h_{ie}C-1+\omega^2(L_1+L_2+2M)-\omega^2C(L_1+M)=0$$

Using eqn(15.38), which gives $\omega^2(L_1+L_2+2M)=1$, the above equation simplifies to

$$g_m\omega^2(L_2+M)h_{ie}C-1+1-\omega^2C(L_1+M) \;=\; 0$$
$$g_m(L_2+M)h_{ie}-(L_1+M) \;=\; 0$$

This gives

$$g_mh_{ie}=\frac{(L_1+M)}{(L_2+M)} \tag{15.39}$$

Also, since

$$g_mh_{ie} \;=\; \frac{i_c}{v_{be}}\times\frac{v_{be}}{i_b}$$
$$\;=\; h_{fe}$$

eqn(15.39) also implies

$$h_{fe}=\frac{(L_1+M)}{(L_2+M)} \tag{15.40}$$

Equation(15.40) gives the current gain that the transistor used to design the internal amplifier should have. The mathematics related to Hartley oscillaor though involved in fairly simple to design and tune.

Advantages of Hartley Oscillators

- The frequency is simply varied by the varying the value of C in the tank circuit.

- The output amplitude remains constant when tuned over the frequency range as long as the feedback ratio of L_1 to L_2 remains constant.

Disadvantages of Hartley Oscillators

- The output is rich in harmonic content and therefore not suitable where a pure sine wave is required.

Colpitt's Oscillator

Colpitts oscillators are similar to the shunt fed Hartley circuit except that instead of having a tapped inductor, Colpitts oscillator utilizes two series capacitors in its LC circuit. With the Colpitts oscillator the connection between these two capacitors is used as the center tap for the circuit. The tap between the two capacitors is grounded and the feedback is obtained from the coupling capacitor, C_1. The output from the Colpitts oscillator is collected across capacitor C_2.

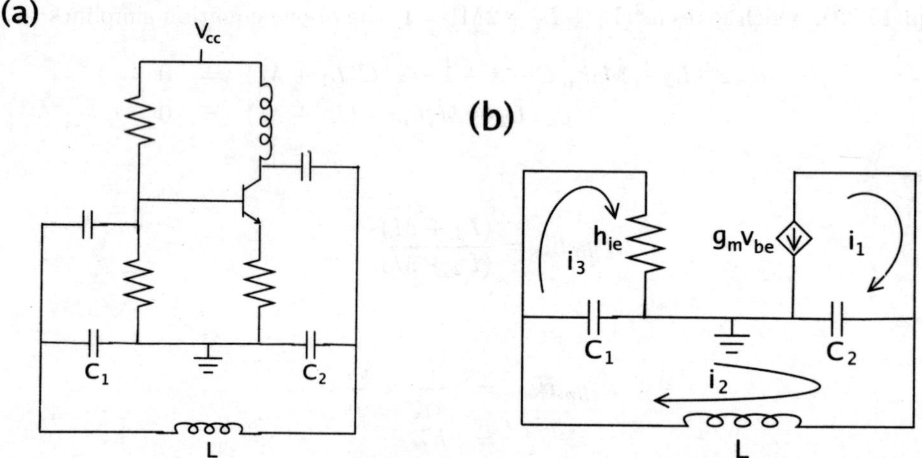

Figure 15.10: *The (a) Colpitts oscillator circuit with (b) the small signal equivalent circuit.*

The steps followed in the Hartley oscillator section will be repeated here also. The output voltage, as stated, is collected across C_2, hence

$$
\begin{aligned}
v_o &= X_{C_2}(i_1 - i_2) \\
&= X_{C_2} i_1 \left(1 - \frac{i_2}{i_1} \right)
\end{aligned}
$$

where the output current i_1 is provided by the current source (of the small signal transistor model)

$$
i_1 = -h_{fe} i_b = -g_m v_{be}
$$

Using this, the output voltage and voltage gain is given as

$$v_o = -g_m v_{be} X_{C_2}\left(1 - \frac{i_2}{i_1}\right)$$

$$A_v = \frac{v_o}{v_{be}} = -g_m X_{C_2}\left(1 - \frac{i_2}{i_1}\right)$$

The amount of feedback done is given as

$$\beta = \frac{v_{be}}{v_o} = \frac{h_{ie} i_3}{i_1 X_{C_2}\left(1 - \frac{i_2}{i_1}\right)}$$

Using the voltage gain and feedback ratio terms in Barkhausen's criteria

$$-g_m h_{ie}\frac{i_3}{i_1} = 1 \tag{15.41}$$

The voltage loop equation for the input loop is written as

$$h_{ie} i_3 + X_{C_1}(i_3 - i_2) = 0$$

which gives the ratio, i_3/i_2 as

$$\frac{i_3}{i_2} = \frac{X_{C_1}}{X_{C_1} + h_{ie}} \tag{15.42}$$

The loop equation for the tank circuit is given as

$$X_L i_2 + X_{C_2}(i_2 - i_1) + X_{C_1}(i_2 - i_3) = 0$$
$$i_2\left[X_{C_1} + X_{C_2} + X_L\right] - i_1 X_{C_2} - i_3 X_{C_1} = 0$$

Using eqn(15.42), the above equation modifies to

$$i_2\left[X_{C_1} + X_{C_2} + X_L\right] - i_1 X_{C_2} - \frac{i_2 X_{C_1}}{(1 + X_{C_1} h_{ie})} = 0$$

$$i_2\left[X_{C_1} + X_{C_2} + X_L - \frac{X_{C_1}}{(1 + X_{C_1} h_{ie})}\right] = i_1 X_{C_2}$$

This gives the ratio of the two currents i_2 and i_1 as

$$\frac{i_2}{i_1} = \frac{X_{C_2}(1 + X_{C_1} h_{ie})}{X_{C_2} + X_L + h_{ie} X_{C_1}(X_{C_1} + X_{C_2} + X_L)}$$

The ratio of currents i_3 and i_1 is determined using the obtained ratios

$$\frac{i_3}{i_1} = \frac{i_3}{i_2} \times \frac{i_2}{i_1}$$

giving

$$\frac{i_3}{i_1} = \frac{X_{C_2}}{X_{C_2} + X_L + h_{ie}X_{C_1}(X_{C_1} + X_{C_2} + X_L)}$$

This ratio is used in eqn(15.41)

$$\frac{-g_m h_{ie} X_{C_2}}{X_{C_2} + X_L + h_{ie}X_{C_1}(X_{C_1} + X_{C_2} + X_L)} = 1$$

which on re-writing gives an expression

$$\frac{-g_m h_{ie}}{1 + \frac{X_L}{X_{C_2}} + h_{ie}\frac{X_{C_1}}{X_{C_2}}(X_{C_1} + X_{C_2} + X_L)} = 1$$

Separating the imaginary and real terms we have

$$h_{ie}\frac{X_{C_1}}{X_{C_2}}(X_{C_1} + X_{C_2} + X_L) = 0 \tag{15.43}$$

$$\frac{-g_m h_{ie}}{1 + \frac{X_L}{X_{C_2}}} = 1 \tag{15.44}$$

The imaginary terms (eqn 15.43) lead to the expression for frequency at which the output oscillates

$$X_{C_1} + X_{C_2} + X_L = 0$$
$$-\frac{1}{\omega C_1} - \frac{1}{\omega C_2} + \omega L = 0$$
$$-\left(\frac{C_1 + C_2}{\omega C_1 C_2}\right) + \omega L = 0$$
$$(C_1 + C_2 - \omega^2 L C_1 C_2) = 0$$
$$\omega = \sqrt{\frac{1}{L}\frac{C_1 + C_2}{C_1 C_2}}$$

While the real terms of eqn(15.44) helps in identifying the type of transistor required for designing the oscillator's internal amplifier.

$$-g_m h_{ie} = 1 - \omega^2 L C_2$$
$$-g_m h_{ie} = 1 - \frac{C_1 + C_2}{C_1}$$

giving,

$$g_m h_{ie} = \frac{C_2}{C_1}$$

or

$$h_{fe} = \frac{C_2}{C_1}$$

This expression gives the minimum value of h_{fe} required for designing a Colpitts oscillator. However, note a typical transistor has $h_{fe} \approx 100$, hence a designer of Colpitts oscillator always selects C_2 hundred times C_1.

The Hartley and Colpitts oscillator are now being replaced by crystal oscillators that we study next.

15.4 Crystal Oscillators

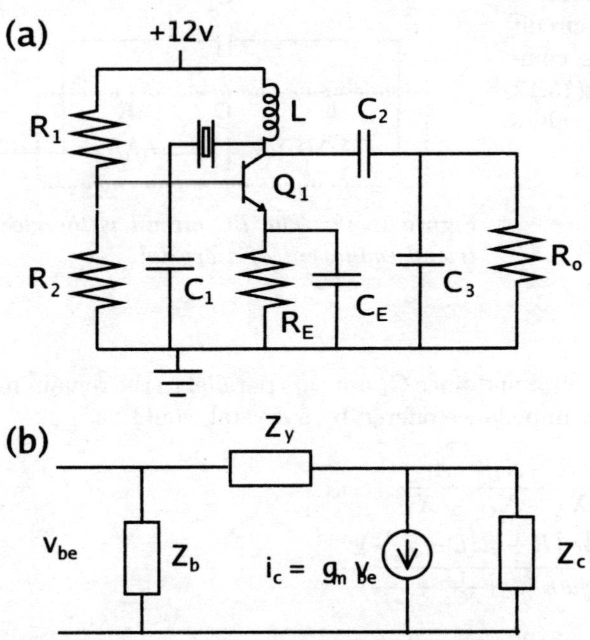

Figure 15.11: *The (a) Pierce crystal oscillator and (b) it's transconductance model.*

The frequency of oscillation is this type of oscillators is controlled by a crystal. The advantage of a crystal oscillator is it's good frequency stablility. Crystals used to fabricate these oscillators are **piezo-electric crystals**.[5] When a voltage is applied on the crystal, the crystal either elongates or is compressed, i.e. electrical variations induces a mechanical change in the crystal. When an AC signal is applied, the crystal alternatively compresses and stretches, in other words it vibrates. The natural frequency of the crystal's vibrations is found to be more constant than the oscillations in a tank/ LC circuit. The frequency of oscillation can be varied by the crystal's size. The thinner the crystal is, the faster it vibrates. Common crystals used are tourmaline, Rochelle salts,[6] and quartz.

A common and simple oscillator circuit which is designed using a quartz crystal is shown in fig(15.11a). This circuit is called *Pierce oscillator*. It is essentially a common emitter amplifier with a tuned (tank) circuit for a load and a quartz crystal as a feedback element. As stated earlier, the crystal vibrates at its own frequency, and the LC circuit is on the in the collector circuit to adjust the amplitude of oscillations. The resistances R_1, R_2 and R_3 are there to bias the transistor.

---

[5]The effect, discovered by Pierre Curie in 1883.

[6]Rochelle salt is a colorless to blue-white orthorhombic crystalline salt with a saline, cooling taste. It is also called Seignette salt after Pierre Seignette, an apothecary of La Rochelle, France, who was the first to make it. Chemically, it is potassium sodium tartrate. It is soluble in water and slightly soluble in alcohol, melts at about 75°C, has specific gravity 1.79, and exhibits double refraction. It is used in medicine as a mild purgative, often in the form of Seidlitz powders. It is an ingredient of Fehling's solution. It is used in silvering mirrors.

Usually, the collector current is kept approximately one milliamp, and an emitter voltage of 2 to 3 volts is maintained. The capacitors C_5 and C_6 provide a total capacitance which will resonate with 'L' (i.e. $C_5 + C_6$ is the C of the tank circuit acting as the load) at a frequency approximately 10% below the desired crystal frequency. The load resistance is kept across the divided capacitor in accordance to the required output voltage. The AC equivalent circuit of the Pierce oscillator as shown in fig(15.11b). The collector output circuit, Z_c, is a tank circuit made up of the tuning coil L and tank capacitors C_5 and C_6 which along with the load resistance provides some equivalent parallel collector resistance R_L. Note that near the tank circuit's resonant frequency, the inductive and capacitive reactances nearly cancel, leaving R_L as the dominant collector load. In practice, the resonant frequency of the tank circuit should be chosen to be approximately 10% lower than the operating frequency of the crystal in order to present a slightly capacitive susceptance.

The feedback circuit consists of the crystal impedance Z_y in series with the transistor base circuit impedance. The crystal impedance can be computed using the equivalent circuit shown in fig(15.12). For a typical 12Mhz Crystal, approximate values are:

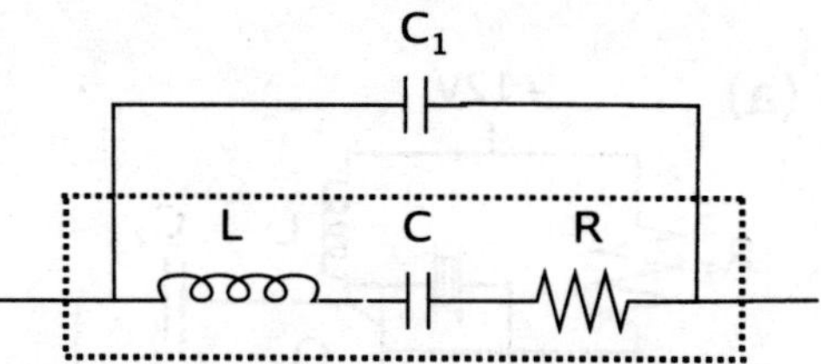

$$R = 100\Omega$$
$$L = 0.66H$$
$$C = 2.65fF$$
$$C_1 = 5pF$$

Figure 15.12: *The LC circuit is the electricial equilavent of a crystal.*

The capacitor 'C', is also called the series capacitor and since C_1 appears parallel to the remaining elements is called the parallel capacitor. The Impedance offered by a crystal would be

$$X = \frac{X_{C_1}(R + X_L + X_C)}{R + X_L + X_{C_1} + X_C}$$

$$= \frac{-j\frac{1}{\omega C_1}(R + j\omega L - j\frac{1}{\omega C})}{R + j\omega L - j(\frac{1}{\omega C_1} + \frac{1}{\omega C})}$$

Since the resistance is usually very small, we shall neglect it, hence

$$X = \frac{\frac{1}{\omega C_1}(\omega^2 - \frac{1}{LC})}{j\left[\omega^2 - \left(\frac{1}{LC_1} + \frac{1}{LC}\right)\right]}$$

Two reasonant frequencies can be defined in the above equation. Namely,

$$\omega_s^2 = \frac{1}{LC}$$

and

$$\omega_p^2 = \frac{1}{L}\left(\frac{1}{C_1} + \frac{1}{C}\right)$$

Hence, the impedance offered by the crystal can be summerised as

$$X = \frac{\frac{1}{\omega C_1}(\omega^2 - \omega_s^2)}{j(\omega^2 - \omega_p^2)} \quad or$$

$$= -\frac{j}{\omega C_1}\left(\frac{\omega^2 - \omega_s^2}{\omega^2 - \omega_p^2}\right) \tag{15.45}$$

where ω_s and ω_p are the series and parallel resonant frequencies respectively. Fig(15.13) shows the variation of the crystal's impedance with frequency. For the series resonant frequency, the impedance offered by the crystal is zero. For frequencies below ω_s, the crystal offers capacitive impedance, i.e.

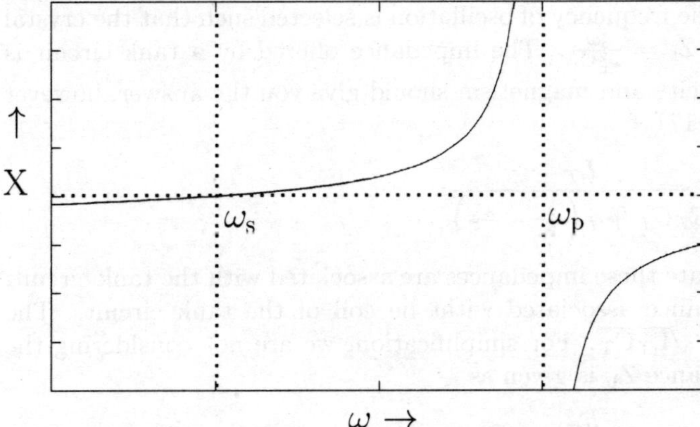

Figure 15.13: *Variation of the crystal's impedance with frequency (using eqn 15.45).*

$$X = \frac{1}{j\omega C_1}\left(\frac{\omega_s^2}{\omega_p^2}\right)$$

while for frequencies between the range $\omega_s < \omega < \omega_p$, the impedance offered is inductive:

$$X = -\frac{1}{j\omega C_1}\left(\frac{\omega^2}{\omega_p^2}\right)$$

$$= \frac{j\omega}{\omega_p^2 C_1}$$

Since the crystal will operate between it's series and parallel resonant frequencies, it presents an inductive reactance. The base circuit impedance Z_b is the parallel combination of the biasing resistors R_1 and R_2, the input impedance of the transistor (h_{ie}) and the phase shift capacitor C_2. It is this capacitance, that is varied to satisfy Barkhausen's criteria. The gain of the internal amplifier without feedback can be evaluated using the circuit shown in fig(15.11b). As stated earlier, it is the crystal offering impedance Z_y that acts as the feedback, hence for calculating the gain without feedback, open Z_y. The output voltage is given as

$$V_o = -g_m V_{be} Z_c$$

from which the voltage gain is found to be

$$A_v = -g_m Z_c$$

For sustained oscillations the Barkhausen's criteria ($A_v \beta = 1$) has to be met. Along with the internal amplifier's gain, we require an expression for the feedback fraction. The feedback fraction can be easily calculated using the simple current division equation on the AC model of the circuit (fig 15.11b). The current in impedances Z_b and Z_y is

$$I_{fb} = \left(\frac{Z_c}{Z_b + Z_y + Z_c}\right) g_m V_{be}$$

Assuming the feedback current to be very small, we approximate $g_m V_{be}$ to be the output current, hence

$$\beta = \left(\frac{Z_c}{Z_b + Z_y + Z_c}\right)$$

Barkhausen's criteria demands

$$A_v\beta = \frac{-g_m Z_c^2}{Z_b + Z_y + Z_c} = 1$$

or

$$Z_b + Z_y + Z_c + g_m Z_c^2 = 1$$

As stated in the course of discussion, the frequency of oscillation is selected such that the crystal imparts an inductive impedance, i.e. $Z_y = \frac{j\omega}{\omega_p^2 C_1}$. The impedance offered by a tank circuit is given as (previous knowledge of electricity and magnetism should give you the answer, however tank circuits are discussed in Chapter 17)

$$Z_c = \frac{L_T}{R_T C_T + j\left(\frac{\omega}{\omega_o} - \frac{\omega_o}{\omega}\right)}$$

where the subscript 'T' is used to indicate these impedances are associated with the tank circuit. Also, the resistance, R_T is the resistance associated witht he coil of the tank circuit. The resonant frequency (ω_o) is given as $1/\sqrt{L_T C_T}$. For simplification, we are not considering the load resistance R_L. Finally, the impedance Z_b is given as

$$Z_b = \frac{h_{ie}}{1 + j\omega h_{ie} C_2}$$

This is obtained from the assumption that $h_{ie} \ll R_1$ & R_2 and h_{ie} comes parallel to C_2. Combining these expressions in our condition for sustained oscillations

$$\frac{h_{ie}}{1 + j\omega h_{ie} C_2} + \frac{j\omega}{\omega_p^2 C_1} + \frac{L_T}{R_T C_T + j\left(\frac{\omega}{\omega_o} - \frac{\omega_o}{\omega}\right)} = 0 \qquad (15.46)$$

and

$$g_m Z_c^2 = 1 \qquad (15.47)$$

It should be evident that we have just separated the imaginary and real numbers. Equation(15.46) would relate to the frequency of oscillation and give the value of C_2 that should be placed in the circuit. Using appropriate assumptions (remember the tank circuit should have a resonant frequency ω_o 10% less than the crystal's frequency, also R_T the resistance of the inductor coil is negligible) eqn(15.46) reduces to

$$-\frac{1}{\omega_p h_{ie} C_2} + \frac{1}{\omega_p C_1} - \frac{L_T}{\left(\frac{\omega_p}{0.9\omega_p} - \frac{0.9\omega_p}{\omega_p}\right)} = 0$$

$$-\frac{1}{\omega_p h_{ie} C_2} + \frac{1}{\omega_p C_1} - 4.7 L_T = 0$$

giving

$$C_2 = \frac{C_1}{h_{ie}(1 - 4.7\omega_p L_T C_1)} \approx \frac{C_1}{h_{ie}}$$

Hence, to make this circuit work, a judicious choice of Z_b (inturn C_2) has to be made. A proper selection of C_2 introduces a phase difference of 2π between the output and the input signal at the crystal's natural frequency.

Precaution should be maintained while using the crystal. Crystals will overheat or crack when fed with too much voltage. The current flowing through a crystal generally should not be more than 100mA (0.1A).

Exercise

Q1. What is an oscillator?

Q2. What type of oscillator, LC or RC is used in the audio frequency range?

Q3. Which components provide the regenerative feedback signal in the phase-shift oscillator?

Q4. Draw a circuit diagram of a phase shift oscillator and calculate it's frequency of oscillation.

Q5. Find the expression for the frequency of oscillation of a Hartley oscillator. Also determine the minimum value of h_{fe} required for obtaining sustained oscillations.

Q6. Prove that the frequency with which a four section RC circuit's output would oscillate is given as

$$f = \frac{1}{2\pi\sqrt{10/7}RC}$$

Q7. Discuss the Barkhausen's criterion of oscillation.

Q8. Explain the working of, and obtain the expression for the frequency of oscillation of a Colpitts oscillator.

Q9. Why is a high-gain transistor used in the phase-shift oscillators designed with BJT?

Q10. Which RC network provides better frequency stability, three-section or four-section?

Q11. Derive the expression for the frequency and the condition of sustained oscillations using a generalized circuit.

Q12. What are piezo-electric crystals? What is the purpose of using them in oscillators?

Q13. Explain the working of a Wien bridge oscillator. Determine the frequency of oscillation.

Q14. Is it possible to have a two-section phase shift network instead of a three section network in an RC phase shift oscillator? Give reasons to support your answer.

Q15. In an RC phase shift network, what happens if R and C positions are interchanged.

Q16. How does an oscillator differ from an amplifier?

Q17. On what factors does the frequency depend for a piezo-electric crystal?

Q18. What is a piezo-electric crystal?

Q19. What is the advantage of a crystal oscillator over a LC oscillator?

Q20. Draw the equivalent circuit for the quartz crystal. Based on this circuit, discuss how the impedance of the crystal varies with frequency. Indicate the region on the graph that the crystal would be used as an oscillator.

Q21. Why is a Wien bridge oscillator preferred to a RC phase shift oscillator.

Q22. Explain the working of a Wein bridge oscillator. Derive the frequency of oscillation.

Chapter 16
Power Amplifiers

The common emitter transistor amplifier is the most used configuration by users because it has the best power amplification factor. That is, give a small amount of AC power at the input and collect a larger power at the output side of the amplifier. As simple as that. But is the conservation of energy being violated here?

Well we can rest assured that the law of conservation can not be violated. We do however appreciate that energy of one form can be converted to another (*a la* chemical energy to heat energy etc). The question is, at whose expense does the transistor circuit amplify the input AC power? The answer to this question is that the required additional power is collected from the DC power supply (V_{cc}) used to bias the transistor. Just as in the case of a factory worker, who is paid to do a job, the transistor is paid DC power to amplify the AC signal. Carrying the similarity of the factory worker further, we are interested in looking at the efficiency of the transistor in converting the DC power to AC power. We define efficiency as

$$\eta = \frac{P_{ac}}{P_{dc}} \times 100$$

where P_{ac} is the amount of AC power received at the output and P_{dc} is the DC power consumed by the transistor circuit. Before calculating the efficiency of the CE amplifier we have studied, trivially one can appreciate to achieve good efficiency our circuit should consume less DC power from the source. Prior to that one may also maximize the utility of the circuit by maximizing the input signal. We hence consider stepwise, what one can do to attain large power on the output side.

16.1　First Step to Increase Output Power

Consider, the requirement of large power at the output. This invariably demands

$$v_o = A_v v_i$$

(correspondingly for large power, large output current would also be demanded). One way would be to keep on increasing A_v, the transistor's gain. However, we have already established a limit on this. Then the only recourse would be to increase the input AC voltage, v_i. As explained in Chapter 8, input voltage changes the base current to the transistor amplifier. This can be pictured as the operating point ('Q' point) shifting along the length of the load line. The larger the input voltage, the more the 'Q' point moves towards the cut-off and saturation region of the transistor's output IV characteristics. Driven into these region, the output waveform is clipped. Even if one does not drive the transistor into it's cut-off or saturation region, if driven near to these regions the non-linearity of the IV characteristics in itself is going to introduce distortions in the output. Since, the input waveform is large and we are working in the transistor's non-linear region, analysis of the transistor circuits using small signal hybrid model would not be technically valid (while the model was developed assumption was that the transistor is a linear device where superposition principle could be used). The non-linearity as explained earlier, introduces harmonics in the output. The output voltage can be written in terms of Fourier series as

$$v_o = a_o + a_1 sin\omega t + a_2 sin2\omega t + a_3 sin3\omega t + ...$$

The DC component is ofcourse removed by the coupling capacitor (even if the coupling capacitor is not placed, the following mathematics is not going to change). Then the AC power developed at the output would be

$$P_{ac} = a_1^2 + a_2^2 + a_3^2 + ...$$

where components a_i, with $i \geq 2$ are the higher harmonics which are essentially giving distortion in the output. Thus, **distortion (in %)** can be defined as

$$D = \frac{\sqrt{a_2^2 + a_3^2 + a_4^2 + ...}}{a_1^2} \times 100$$

$$= \sqrt{\left(\frac{a_2}{a_1}\right)^2 + \left(\frac{a_3}{a_1}\right)^2 + \left(\frac{a_4}{a_1}\right)^2 + ...} \times 100$$

$$= \sqrt{D_2^2 + D_3^2 + D_4^2 + ...} \times 100(\%)$$

$$(16.1)$$

Note distortion is defined in terms of the ratio of the voltage/ current amplitudes of harmonics w.r.t. the fundamental frequency signal. Thus, the distortion in power is given as

$$P_{ac} = a_1^2 \left[1 + \left(\frac{a_2}{a_1}\right)^2 + \left(\frac{a_3}{a_1}\right)^2 + ...\right]$$

$$= a_1^2(1 + D^2)$$

How serious this distortion is can be understood by taking an example.

Example 16.1: Distortion contributions by the third and higher harmonics can be neglected. Consider 10% distortion being contributed by the second harmonic, what is the percentage of distortion introduced in the output.

Distortion (a_2/a_1) from the second harmonic is 10% or 0.1. Hence, the distortion contribution in output power is given as (neglecting distortions from higher harmonics)

$$P_{ac} = a_1^2(1 + 0.01) = 1.01a_1^2$$

Thus, 10% distortion only represents a 1% distortion in the output power.

Distortion in output power is 1%

Class A amplifiers

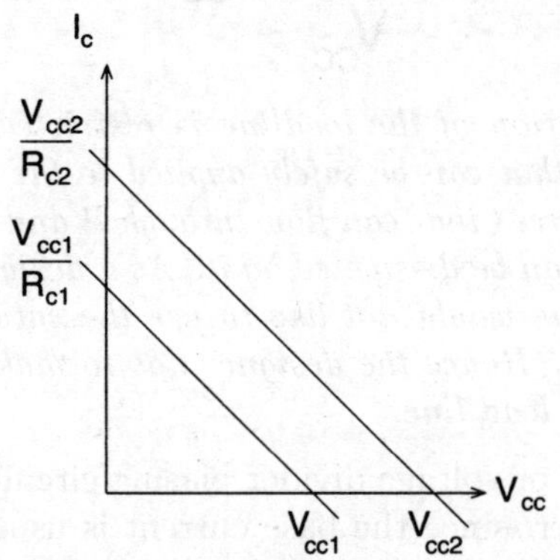

Figure 16.1: *The length of the loadline can be varied by varying* V_{cc}, R_L *or both.*

To prevent any asymmetrical clipping remember the operating or 'Q' point was selected on the load line at the center of the transistor's active region. Such transistor circuits are classified as **Class A** transistor amplifiers. Also, as explained earlier, while using a transistor as a power amplifier, we want the output signal to be large. For this we use the whole length of the load line for the 'Q' point to sweep as the AC input is fed. Thus, if larger output voltage is required, it is logical to ask for a greater length of load line. As shown in fig(16.1), this can be attained by increasing V_{cc} and I_{Csat}. If we keep the new load line parallel to the initial load line, we just have to vary V_{cc} keeping the load resistance the same. Remember, $I_{Csat} = V_{cc}/R_c$. For lines not to be parallel, we have to change R_c, (slope of the load line is given as $-1/R_c$). Thus by varying V_{cc}, R_c or both, the length of the load line increases, but can this be done without any limit?

One limiting fact would be, a very large V_{cc} can either lead to ***punch through*** or avalanche breakdown of the reverse biased collector-base junction. Also, the manufacturer specifies the maximum current that a transistor can handle, thus indirectly limiting the maximum supply voltage. Along with these limiting values, the manufacturer also specifies the maximum power that the transistor can handle. These limiting features are shown in fig(16.2). The load line has to lie within the shaded region of the IV characteristics of fig(16.2).

Assuming that the saturation and the cut-off region of the transistor are negligible, Then the output voltage can swing from zero (0) to $+V_{cc}$ (see fig 16.3), symmetrically about $V_{cc}/2$

(since the 'Q' point is half way). Thus, the output waveform can be described by equation

$$v_o(t) \;=\; \frac{V_{cc}}{2} sin\omega t$$
$$\;=\; V_{max} sin\omega t$$

(16.2)

The rms output voltage can be easily calculated, from which the output power is also trivially computable as

$$P_{ac} = \frac{V_{max}}{\sqrt{2}} \times \frac{I_{max}}{\sqrt{2}}$$

We have already shown above

$$V_{max} \;=\; \frac{V_{cc}}{2}$$
$$I_{max} \;=\; \frac{V_{max}}{R_c}$$

$$P_{ac} = \frac{V_{cc}^2}{8R_c}$$

The transistor amplifier works because of the DC supply (V_{cc}) fed to it. However, the power consumed by the transistor circuit would also depend on the current drawn by the circuit.

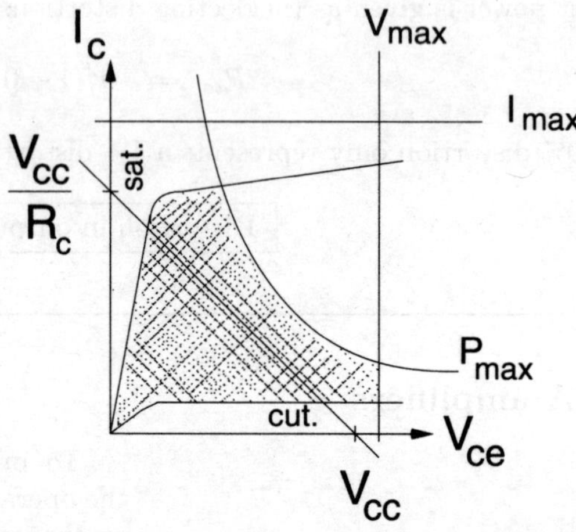

Figure 16.2: *The selection of the loadline is restricted by the maximum voltage that can be safely applied to the device, the maximum current that can flow through it and the maximum power that can be dissipated by it. As a designer of an analog circuit one would not like to use the saturation and cut-off region. Hence the designer has to make a judicious choice of the load line.*

The amplifier circuit, draws two currents (we are talking of voltage divider biasing circuits), I_B, the base current and I_{CQ}, the collector current. However, since the base current is usually very small (in μA), for calculations we neglect any contribution by the biasing circuit (R_1 and R_2). The DC power consumed by the circuit is

$$P_{dc} \;=\; V_{cc} \times I_{CQ}$$
$$\;=\; V_{cc}\frac{I_{Csat}}{2}$$
$$\;=\; \frac{V_{cc}^2}{2R_c}$$

The efficiency of our "Class A" transistor amplifier can now be worked out. Efficiency is given as

$$\eta \;=\; \left(\frac{V_{cc}^2}{8R_c} . \frac{2R_c}{V_{cc}^2} \right) \times 100$$
$$\;=\; 25\%$$

Thus, only 25% of the given (consumed) DC energy is converted to AC energy. A very poor efficiency. Where is the energy then being wasted? One obvious culprit could be the loss of energy interms of heat dissipation across the collector or the load resistance, R_c. Then the other contribution of wasteful consumption could be the transistor itself (V_{CE}). In the following sections, we investigate how to make the amplifier more efficient, by overcoming the stated problems.

16.2   Second Step to Increase Output Power

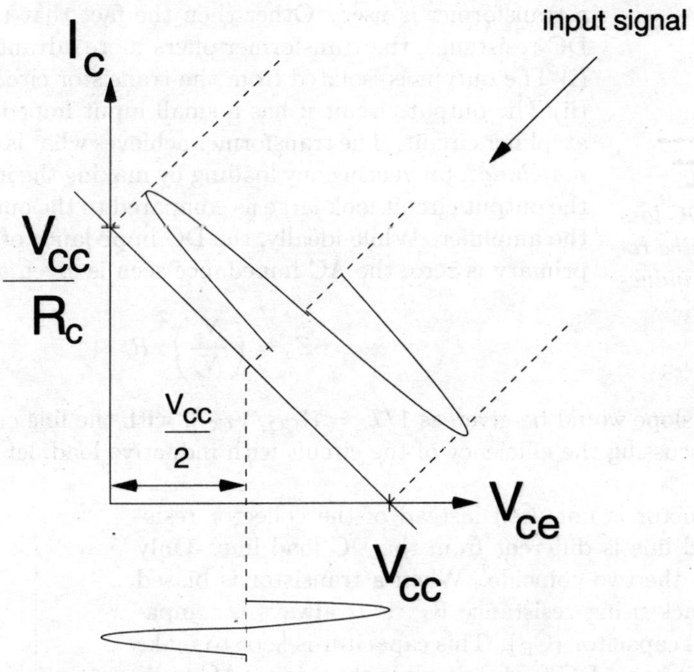

Figure 16.3: *The length of the loadline can be used by the varying input signal (results in base current variation). This results in output voltage varying from 0 to V_{cc}.*

As can be seen, there is some wastage of input DC power across the load resistance R_c. To overcome this, one might suggest making it zero. As far as DC power wastage is concerned, it's alright. However, where does the output voltage (that you take across the load resistance) develop? If R_c is made zero, the AC output voltage also falls zero. Thus, what we require is a DC impedance of the collector (load) should be zero, however, it's AC impedance should be non-zero. An inductor fits the bill. Figure(16.4) shows the required circuit. In the next section we investigate the nature of the load line with the replacement of the load resistance by an inductor.

AC load line

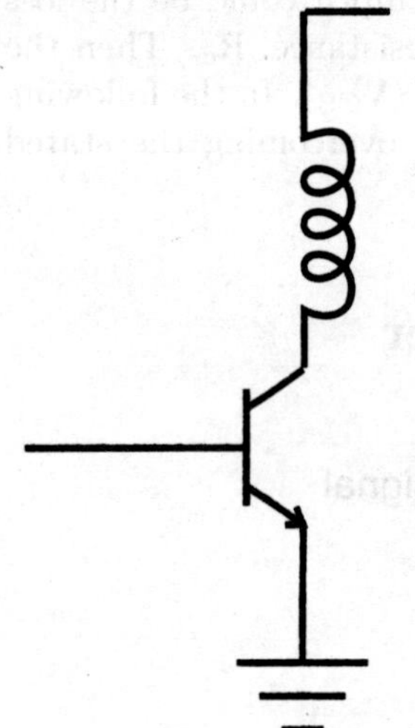

Figure 16.4: *The DC loss across the load can be reduced by using an inductive load.*

Consider the load line given by eqn(8.13). Since, the DC resistance in the circuit is zero, the slope of this line is infinite. Hence, the load line is parallel to the 'Y' axis and passes through the 'Q' point selected by the biasing circuit. Equation (8.12) gives the point of intersection of the load line on the 'X' axis as

$$V_{CE} = V_{CC}$$

Figure(16.5) shows the graphical representation of the DC loadline for an inductive load. For practical circuits, instead of an inductor, a transformer is used. Other then the fact that the coil offers zero DC resistance, the transformer offers more advantages, namely
(i) The output is isolated from the transistor circuit
(ii) The output circuit if has a small input impedance can load the amplifier circuit. The transformer achieves what is called *"impedance matching"*, preventing any loading by making the input impedance of the output circuit look large as compared to the output impedance of the amplifier. While ideally, the DC impedance of the transformer's primary is zero, the AC impedance seen is given as

$$Z_c = \left(\frac{N_p}{N_s}\right)^2 R_c$$

The AC loadline's slope would be given as $1/Z_c = (I_{CQ}/V_{CC})$, with the line cutting the 'X' axis at V_{CC}. Before discussing the efficiency of the circuit with inductive load, let us investigate the loadline further.

Even if an inductor is not used instead of the collector resistance, the AC load line is different from the DC load line. Only in special cases do the two coincide. When a transistor is biased for negative feedback using resistance R_E, it is always accompanied with a parallel capacitor (C_E). This capacitor is kept to make sure that the circuit has a DC feedback while there is no AC feedback. Writing the output side equation for the two cases would exhibit the difference in load line for the two cases. Even if no emitter resistance is being used, i.e. the transistor is biased using the fixed bias circuit, the DC loadline and AC loadline differ. The power amplifier circuit is designed to drive power into an external load (R_L), that is DC isolated from the circuit using a coupling capacitor. However, as far as the AC signal is concerned, it views this load resistance parallel to the collector resistance, R_c (revisit Chapter X to rebrush this idea). Thus, the load as seen would be

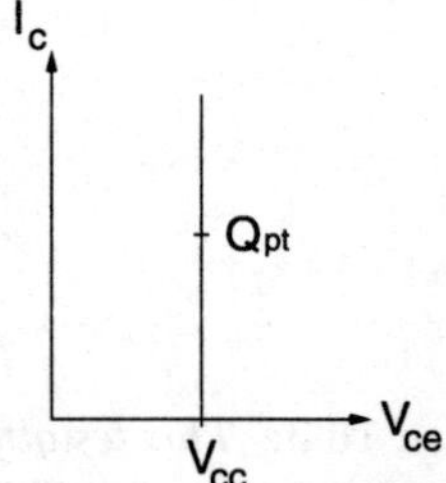

Figure 16.5: *The DC loadline for CE transistor circuit having a transformer instead of R_c as load in the collector arm.*

$$r_L = \frac{R_L R_C}{R_L + R_C}$$

The output voltage is than given as $v_o = -i_c r_L$, where the negative sign indicates phase reversal. The output is collected across the collector-emitter of the transistor, i.e. $v_{ce} = v_o$, shown as a swing along the 'X' axis of fig(16.3). Continuing with our analogy of the AC signal as a time varying DC signal, we may write

$$\Delta V_{CE} = -\Delta I_C r_L \tag{16.3}$$

The change in both current and voltage is with respect to the operating point ('Q' point). Hence, a positive input swing of AC would lead the 'Q' point to move from (I_{CQ}, V_{CEQ}) to $[i_{c(sat)}, 0]$ while the negative swing results it to move from (I_{CQ}, V_{CEQ}) to $[0, v_{ce(cut-off)}]$. Hence, using the co-ordinates of the signal in the positive swing, eqn(16.3) gives

$$(0 - V_{CEQ}) = -(i_{c(sat)} - I_{CQ})r_L$$

giving the new position, where the AC loadline cuts the 'Y' axis interms of known parameters as

$$i_{c(sat)} = I_{CQ} + \frac{V_{CEQ}}{r_L} \tag{16.4}$$

A similar exercise for the negative swing gives the position where the AC loadline cuts the 'X' axis

$$[v_{ce(cut-off)} - V_{CEQ}] = -(0 - I_{CQ})r_L$$

giving

$$v_{ce(cut-off)} = V_{CEQ} + I_{CQ}r_L \tag{16.5}$$

The operating point ('Q' point) in a Class A amplifier is kept in the middle of the active region of the transistor's output characteristics. A realistic assumption in the quantities defining the 'Q' point is given as

$$I_{CQ} = \frac{I_{sat}}{2} = \frac{V_{CC}}{2R_C} \tag{16.6}$$

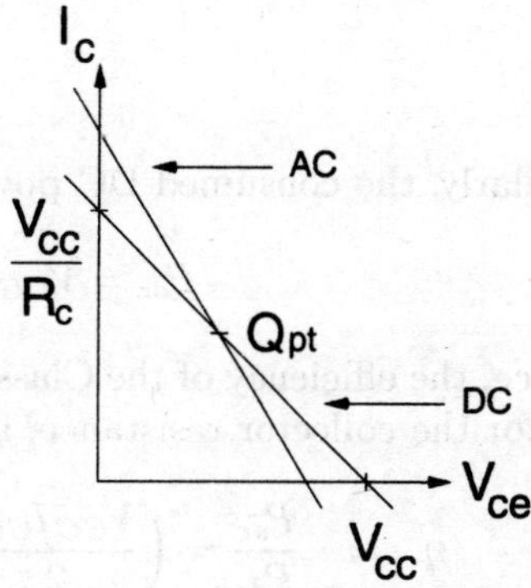

Figure 16.6: *The thick line shows the DC loadline of the transistor with fixed bias circuit while the thin line represents the AC loadline. The co-ordinates where the AC loadline cuts the axes is given by eqn(16.8) and eqn(16.9).*

and

$$V_{CE(cut-off)} = \frac{V_{CC}}{2} \tag{16.7}$$

Substituting eqn(16.6) and eqn(16.7) in eqns(16.4) and (16.5), the points on 'X' and 'Y' axis where the AC loadline cuts is given as

$$i_{c(sat)} = \frac{V_{CC}}{2R_C} + \frac{V_{CC}}{2r_L} \tag{16.8}$$

and

$$v_{ce(cut-off)} = \frac{V_{CC}}{2} + \frac{r_L}{2R_C}V_{CC} \qquad (16.9)$$

Since r_L is a parallel combination of resistances between R_C and R_L, the combination would be less than R_C, hence, the AC and DC loadlines would not coincide (see fig 16.6). However, if $R_L \gg R_C$, the two lines would coincide.

Figure(16.7) shows how the output voltage develops symmetrically about the DC loadline. From the graph it is evident that the peak to peak of the output wave is $2V_{CC}$. Since, we are still working with a Class A circuit, i.e. the 'Q' point is exactly half way of the transistor's active region (inturn AC loadline), the output current oscillates from 0 to $2I_{CQ}$. Hence, the AC power at the output is

$$P_{ac} = \left(\frac{1}{\sqrt{2}} \cdot \frac{2V_{CC}}{2}\right)\left(\frac{1}{\sqrt{2}} \cdot \frac{2I_{CQ}}{2}\right)$$

$$= \frac{V_{CC}I_{CQ}}{2}$$

Similarly, the consumed DC power by the circuit is

$$P_{dc} = V_{CC}I_{CQ}$$

Hence, the efficiency of the Class A power amplifier with inductor for the collector resistance, is given as

$$\eta = \frac{P_{ac}}{P_{dc}} = \left(\frac{V_{CC}I_{CQ}}{2}\right)\left(\frac{1}{V_{CC}I_{CQ}}\right) \times 100$$

$$= 50\%$$

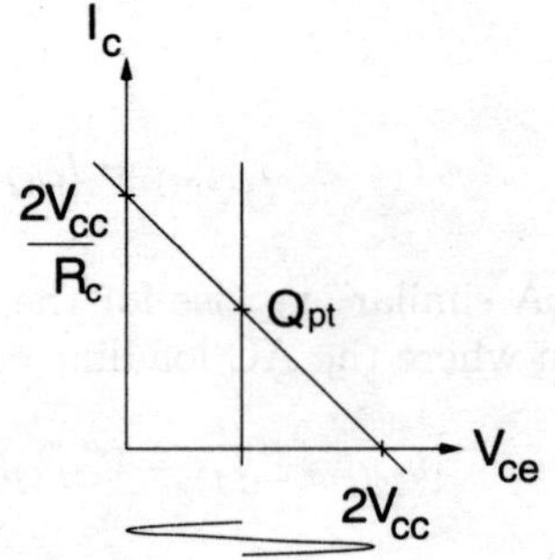

Figure 16.7: *The variation of output voltage with respect to the load line of amplifier circuit with inductive load (inductor instead of R_C).*

Though the efficiency has improved, it is still only 50%. That is of the power that the circuit absorbs from the DC power supply, it only converts 50% of that power into output AC power. We now investigate the steps that can be taken to further improve the efficiency of our amplifier circuit.

16.3 Third Step to Increase Output Power

From the above section, we understand that the load resistance dissipates 25% of the input DC power. Replacing R_c by an inductor/ transformer, we push the efficiency of the Class A transistor amplifier to 50%. Thus, even after the second step is taken to improve the efficiency of the transistor amplifier, the efficiency is not appreciable and 50% wastage is still present. But where? To understand this, we lookup the load line equation again

$$V_{cc} = V_{CEQ} + I_{CQ}R_C$$

$$(16.10)$$

The input DC power is given as

$$P_{dc} = V_{cc}I_{CQ}$$
$$= V_{CEQ}I_{CQ} + I_{CQ}^2 R_C$$

While the second term is the power dissipated across the load resistance, which we know how to minimize, it's the first term, $V_{CEQ}I_{CQ}$ which might be playing a culprit. $V_{CEQ}I_{CQ}$ is the DC power appearing across the transistor while it is biased. It is evident, since the current I_{CQ} is flowing all the time, the energy loss contributed by this would be substantial.

Class B Amplifiers

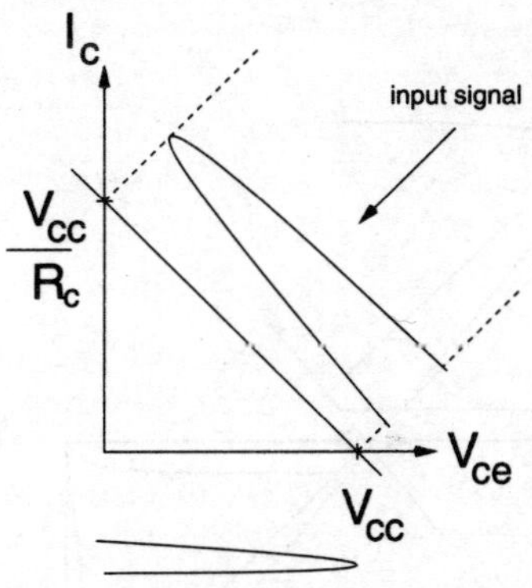

Figure 16.8: *The variation of output voltage with respect to the input voltage when the 'Q' point is selected in the transistor's cut-off region. Such amplifier's where the transistor is biased in it's cut-off region is an example of Class B amplifier.*

If the transistor base-emitter is not properly biased, i.e. I_B is kept zero, then I_C would also be zero. The base-emitter would only be biased when the input's positive cycle flows in, forward biasing the base-emitter. The input's negative cycle does not contribute any output. The circuit hence provides an output signal over one-half the input signal or only for 180°. The DC bias point is 0V or the 'Q' point lies in the cut-off region of the the transistor. Consider the power efficiency of the circuit

$$P_{ac} = V_{rms}I_{rms}$$

The output of a Class B power amplifier is similar to the output of the half wave rectifier. The expression for I_{dc} and I_{rms} of an half wave rectifier is listed in Table 5.5 (in terms of V_{dc} and V_{rms})

$$I_{dc} = \frac{I_{max}}{\pi}$$

Since, $I_{rms} = I_{max}/2$ (Table 5.2, data for half-wave rectifiers), we can write the above as

$$I_{dc} = \frac{2I_{rms}}{\pi}$$

The absorbed DC power is

$$P_{dc} = V_{cc}I_{dc}$$
$$= \frac{2I_{rms}V_{cc}}{\pi}$$

Using this expression we can evaluate the efficiency of the circuit. We obtain

$$\eta = V_{rms}I_{rms} \times \frac{\pi}{2I_{rms}V_{cc}} \times 100$$
$$= \frac{\pi V_{rms}}{2V_{cc}} \times 100$$

Again, the output AC swings from 0 to V_{cc}, hence, the AC output is given as

$$V_{rms} = \frac{V_{cc}}{2}$$

Hence, the power conversion efficiency is

$$\eta = \frac{\pi}{4} \times 100 = 78.5\%$$

This is appreciable, but then remember only 50% of the AC signal is present in the output, also making the output highly distorted. A way out would be to recreate the second half of the cycle (again using a Class B amplifier) and adding the two outputs. Exactly like the full wave rectifier circuit. Figure(16.9) shows the circuit of what is called a **Push-pull** amplifier with Class B transistor amplifiers achieving the idea suggested.

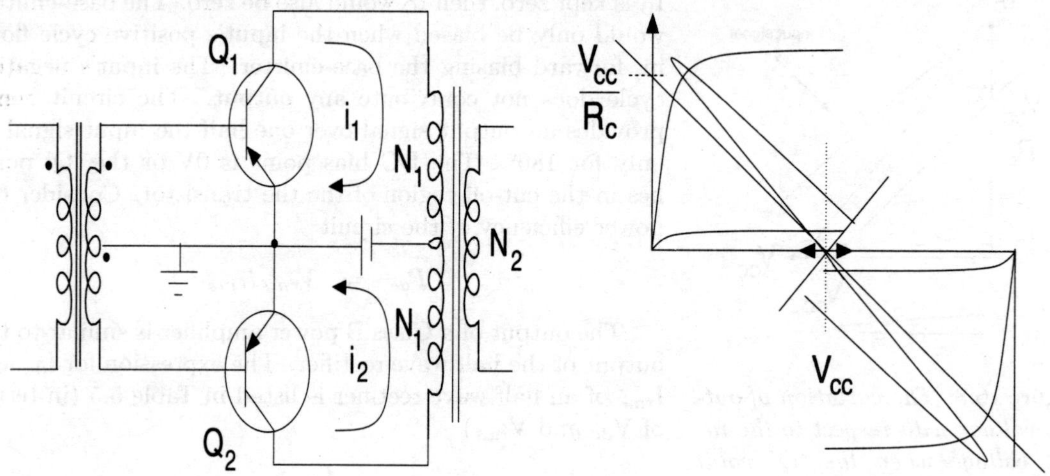

Figure 16.9: *Two Class B amplifier used in a push pull configuration. The effective use of the loadline by this circuit is also shown.*

Both the PNP transistors of fig(16.9) are biased in their cut-off region. Only one transistor would conduct at a given time. When the negative signal comes from the input, transistor Q_1's base-emitter (N-P in this case) is forward biased and the transistor conducts, contributing a current i_1 in the direction shown, giving a positive voltage in the output. Though the lower half of the transformer has sine signal in phase with the input wave and the sine wave as seen by Q_1 (note the polarity symbol/ dots), transistor Q_2 is reverse biased. The transistor does not conduct and contributes no current in the output. On the advent of the positive cycle at the input, the scenario changes and Q_1 switches off, while Q_2 is forward biased, since the center tap (connected to the emitter) is at a higher potential then the base. Transistor Q_2 contributes a current i_2 which sends the output negative. This one by one switching of the transistor is achieved by the apparent phase-shifting that the center-tapped transformer does between the two inputs. The output AC power output of this circuit is given as

$$P_{ac} = V_{rms} I_{rms}$$

Again using the information of the full wave rectifier from Table 5.5, we have

$$I_{dc} = \frac{2I_{max}}{\pi}$$

Since, $I_{rms} = I_{max}/\sqrt{2}$, we can write the above as

$$I_{dc} = \frac{2\sqrt{2}I_{rms}}{\pi}$$

The power consumed by the circuit from the DC source is

$$\begin{aligned}
P_{dc} &= V_{cc}I_{dc} \\
&= \frac{2\sqrt{2}I_{rms}V_{cc}}{\pi}
\end{aligned}$$

The efficiency is given as

$$\begin{aligned}
\eta &= V_{rms}I_{rms} \times \frac{\pi}{2\sqrt{2}I_{rms}V_{cc}} \times 100 \\
&= \frac{\pi V_{rms}}{2\sqrt{2}V_{cc}} \times 100
\end{aligned}$$

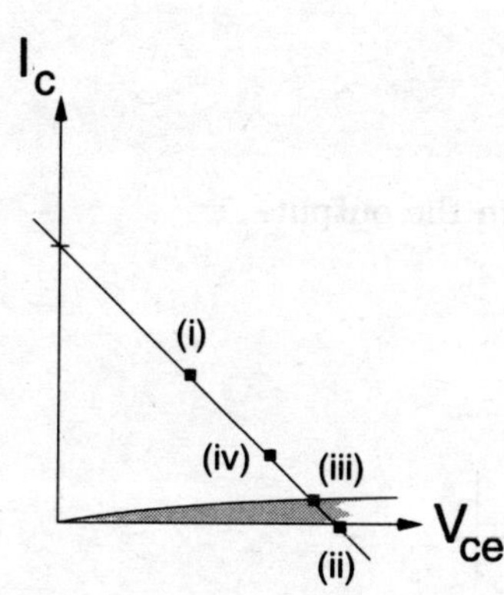

Figure 16.10: *The power amplifiers are basically classified on the basis of the position of the 'Q' point on the load line.*

Again, the output AC swings from 0 to $2V_{cc}$, hence, the AC output is given as

$$V_{rms} = \frac{V_{cc}}{\sqrt{2}}$$

Hence, the power conversion efficiency is

$$\eta = \frac{\pi}{4} \times 100 = 78.5\%$$

A substantial improvement in the efficiency of power conversion has been achieved. It should be noted that this improvement is not without merits and demerits.

Advantage of Push-pull Amplifiers

1) **Distortion is reduced.** The output of the push-pull amplifier is near sine like and hence is indicative of the fact that distortions are minimized. As explained above, the base currents of the the two transistors are out of phase by π, with respect to each other. Hence they may be represented as

$$\begin{aligned}
i_{b1} &= i_b sin(\omega t) \\
i_{b2} &= i_b sin(\omega t + \pi)
\end{aligned}$$

Since, each transistor only generates half of a sine wave, the output maybe represented as a Fourier series

$$\begin{aligned}
i_{c1} &= i_o + i_1 sin\omega t + i_2 sin2\omega t + i_3 sin3\omega t + \ldots\ldots \\
i_{c2} &= i_o + i_1 sin(\omega t + \pi) + i_2 sin(2\omega t + 2\pi) + i_3 sin(3\omega t + 3\pi) + \ldots\ldots
\end{aligned}$$

The output voltage depends on the current flowing on the primary of the second transformer. Since the two currents have opposite directions

$$v_o \quad \propto \quad i_{c1} - i_{c2}$$
$$= \quad k(i_{c1} - i_{c2})$$

Note, the DC term and the even harmonics are removed from the output.

$$v_o = 2k(i_1 sin\omega t + i_3 sin3\omega t + ...)$$

On re-writing we have

$$v_o = 2ki_1 \left[sin\omega t + \left(\frac{i_3}{i_1} \right) sin3\omega t + ... \right]$$

Consider that the distortions due to higher harmonics are negligible, the output voltage is purely sinusoidal.

$$v_o = 2ki_1 sin\omega t$$

Thus, the output of a push pull amplifier is quite free of distortions.

2) **Magnetization is removed**. Usually, transformers are not designed for carrying DC signals. Not only do they contribute to heating but also in transformers with core, they effect the value of the inductance related to the coil and inturn transformer performance. However, since the output currents of the two transistors flow in a opposite directions, the DC components cancel out (see mathematics above).

The Class B push-pull amplifier is not without disadvantage. The cut-off region is not a point region ($I_c = 0$) and occupies a definite area on the output IV characteristics of the transistor (see fig 16.10). An ideal Class B operation has 'Q' point represented by '(ii)' on the load line of the same figure. Infact the transistor only conducts for an input voltage greater than 0.7v (in case of silicon transistors) resulting in flow of current for only non zero base current. Hence for input voltages between 0v and 0.7v, the collector current would be zero. Hence when compared to the smoothly varying input voltage the push-pull amplifier's output would have a flat line (0v) between 0v and 0.7v (see fig 16.11). This is called the **cross-over distortion**. The distortion can be minimized by using transistor biased just above the cut-off region. Such biasing leads to Class AB amplifiers discussed in the next section. Also, a biasing circuit can be added to the push pull amplifier of fig(16.12) to bias the transistor's at point '(iii)' of the loadline. This is just at the edge of the cut-off region and just as the transistor gets a input voltage above zero volts, they conduct, thus minimizing the crossover distortion.

The use of the transformer to be discouraged in the designing of a push-pull amplifier because of the expense involved, the size and also the transformer might not be designed to carry any DC, direct current. Thus, a designer would avoid using a transformer and look for an alternative ways to input a sine wave. Below we discuss two such design variations.

Transistor Phase inverters

One can see from the explanation of the push pull amplifier circuit that the center-tapped transformer is essentially required to give two sine waves which (seemingly for the circuit on the secondary side) are out of phase. A simple CE mode amplifier operating in the Class A mode can be used to give two sine waves, one having a phase difference of 180° (or π) with respect to the other. Figure 16.12a shows that an input sine wave when collected at the collector of the transistor is out of phase with respect to the input. As studied in Chapter VIII the phase difference introduced is π. However, the same signal when collected at the emitter is in phase with the input. Thus, the output voltages, v_1 and v_2 (fig 16.12a) are out of phase with respect to each other and can form the inputs for the two transistors of a push pull amplifier.

Push-pull Amplifiers with Complementary Transistors

The push pull activity of the circuit is achieved by two transistors switching one by one. In the introductory push pull circuit and the circuit above, the switching was achieved using two input signals. Using an NPN and a PNP transistor, one can get the push pull activity with a single input. Figure(16.12b) shows the **complimentary push pull amplifier circuit** using NPN and PNP transistors. The input if positive, sends the NPN transistor into conduction while the PNP is in it's OFF state. The output follows the input. Similarly, on the coming of the negative cycle, the NPN switches OFF while the PNP conducts. Thus, instead of using two complimentary signals, two complimentary transistors/ switches are used to achieve the push pull activity.

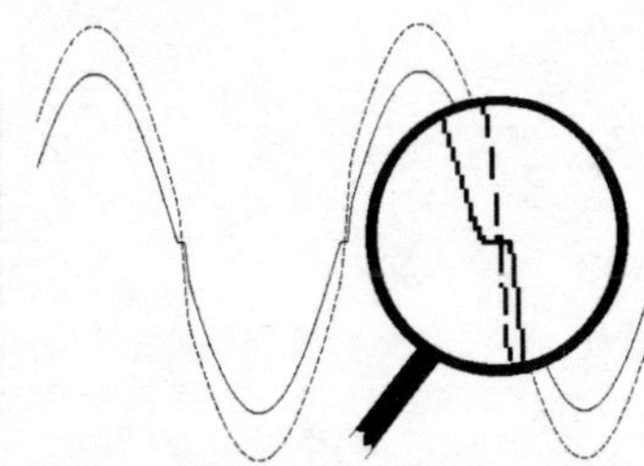

Figure 16.11: *The crossover distortion seen in a push-pull amplifier. The cross-over region (at 0v) has been magnified.*

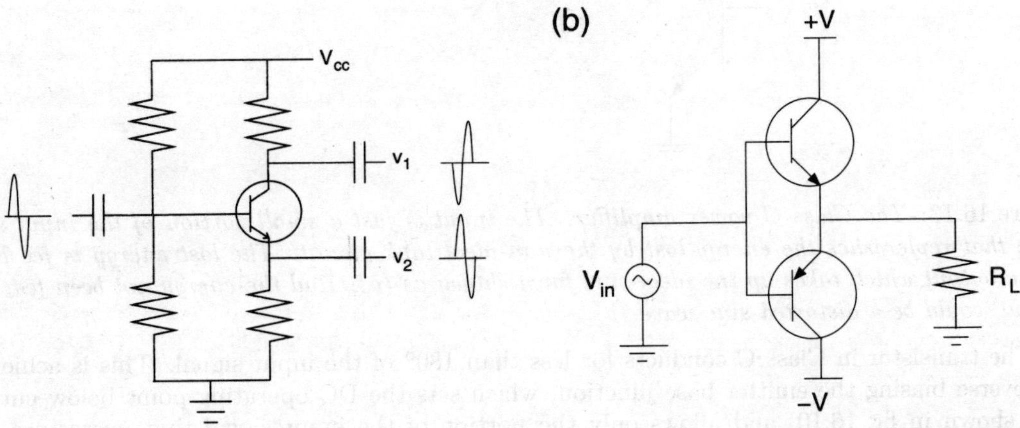

Figure 16.12: *The (a) transistor as a phase inverter and (b) the complimentary pair of transistor using a PNP and NPN transistor make up for a push pull amplifier that do not require a transformer.*

Class AB Amplifiers

Amplifiers designed for class AB operation are biased so that collector current flows for more than 180° but less than 360° of the input signal. This is achieved by biasing the transistor with the 'Q' point inbetween the half way mark (as in Class A) and the cut-off point (as in Class B). The 'Q' point for the Class AB amplifier is closer to cutoff (see point 'iv' of fig 16.10). Thus, fundamental difference between Class B and Class AB operation is biasing. In Class AB circuit, both transistors are "ON" for a brief moment in time around the zero-crossover point, while only one transistor is "ON" at any given time in a Class B circuit. The Class AB operated amplifier is commonly used as a push-pull amplifier to overcome the cross-over distortion of class B operation.

Class C Amplifiers

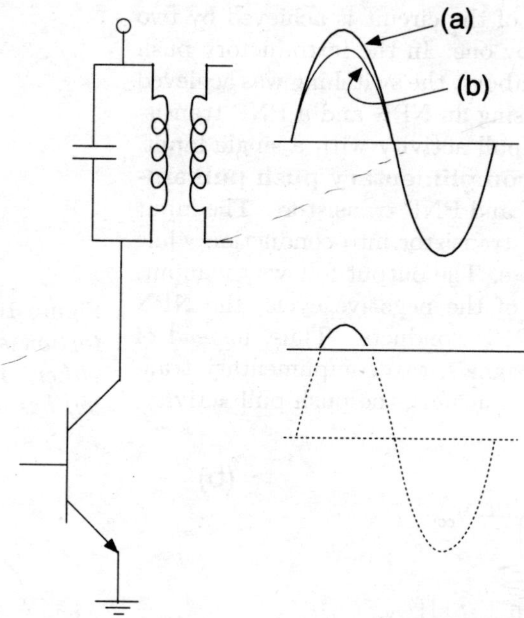

Figure 16.13: *The Class C power amplifier. The input is just a small portion of the input sine wave that replenishes the energy lost by the non-ideal tank circuit. The lost energy is fed back to the output which takes up the ideal sine form shown as (a). Had the energy not been fed, the output would be a distorted sine wave (b).*

The transistor in Class C conducts for less than 180° of the input signal. This is achieved by reverse biasing the emitter-base junction, which sets the DC operating point below cut-off (not shown in fig 16.10) and allows only the portion of the input signal that overcomes the reverse bias to cause collector current flow. The resulting output signal would be distorted with a biasing like this (only the tip of the sine wave would appear in the output). The missing part has to be recovered with a resonant circuit that converts the collector current pulses into a continuous sine wave.[1] Thus, with the tank circuit in the collector circuit or base circuit, Class C circuits are usually clubbed with oscillator or tuned amplifiers (see previous Chapter), which are commonly used in radio frequency (RF/ communication).

Exercise

Q1. Determine the maximum theoretical efficiency of a Class A power amplifier.

Q2. Calculate the effective impedance seen into the primary of a transformer with turn ratio of 20 when a speaker of 8Ω is connected to the circuit.

Q3. Name the three major classes of operation of an amplifier and give their characteristics. In which Class of operation does the crossover distortion take place?

Q4. Explain why Class B amplifiers are preferred over Class A for high-power applications such as audio power amplifiers.

Q5. A popular variation of the Class B amplifier is the Class AB amplifier, designed to eliminate any trace of crossover distortion. What is the difference between a Class B and a Class AB amplifier? Why do Class AB amplifiers have less crossover distortion than Class B amplifiers? Is there any disadvantage to changing from Class B to Class AB operation?

Q6. A simple yet impractical way to eliminate crossover distortion in a Class B amplifier is to add two small voltage sources to the circuit like that shown in fig(16.14). Explain why this solution works to eliminate crossover distortion.

Q7. Explain how even harmonics are eliminated in a push-pull configuration.

Q8. Derive the maximum theoretical efficiency of a Class B push-pull amplifier.

Q9. A push-pull class B amplifier uses dual power supplies of +30V and -30V. Calculate

   (i) the maximum power which may be delivered into an 8Ω load.
   (ii) the efficiency of the amplifier.

Q10. Explain why Class B and Class C operations are more efficient than that of Class A.

Q11. What is the advantage of complementary symmetry push pull amplifier? Explain it's operation.

---

[1]The tank circuit ideally should convert magnetic energy to electrical energy without any energy dissipation, giving a pure sinusoidal wave at the output. However, an ideal tank circuit only exists theoretically and energy dissipation takes place over the resistance inherently present with the inductor coil. If lost energy is supplied regularly to the system, oscillations would remain forever. This is similar to the child on a swing in the neighborhood play-ground. If left alone after a number of oscillations the swing comes to a rest. However, if on completion of one oscillation, the swing is given a small tap, it remains in oscillations with the same amplitude. This is subject to the condition the tap is just replacing the lost energy of the system.

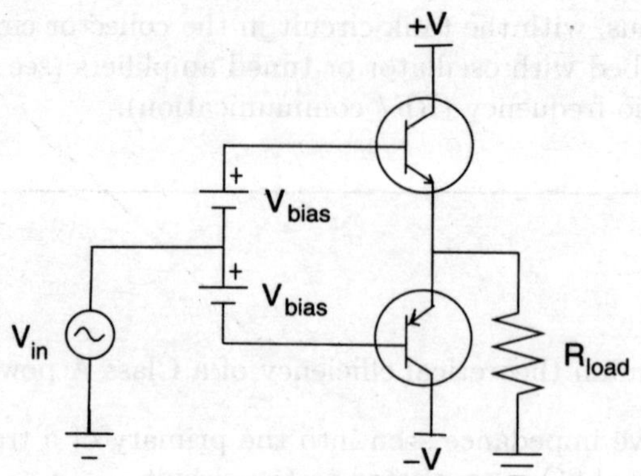

Figure 16.14: *Circuit for question 6.*

Q12. Indicate the biasing requirements for each of the following classes of power amplification: A, B, AB, C. Illustrate your answers by marking the position of the quiescent operating point for each case on a load line drawn on the BJT's IV characteristics.

Q13. What amplifier class of operation is the most inefficient but has the least distortion?

Q14. What two primary items determine the class of operation of an amplifier?

Q15. What is the name of the term used to describe the condition in a transistor when the emitter-base junction has zero bias or is reverse biased and there is no collector current?

Q16. What amplifier class of operation allows collector current to flow during the complete cycle of the input?

Chapter 17

Tuned Amplifiers

Amplifiers discussed until now are classified as untuned amplifiers. These amplifiers amplify all the signals having frequencies lying within the amplifier's large bandwidth. However, consider the situation where various frequencies are present but only one particular frequency is to be selected. This happens in radio-communication where so many channels (all at different frequencies) are present. Of these channels, a person might be interested in listening to a particular channel. He has to select the frequency of that channel to amplify. Since, there is a crowding of channels (for example popular FM channels available in India are 93.5, 98.0 and 102MHz) a large bandwidth amplifier would amplify all these channels, making confusion rather than music. Thus, to select a channel an amplifier is required whose bandwidth is small. As in the case of the stated channels, if one would like to hear the channel at 98MHz, the bandwidth of the amplifier should be around 6MHz. The narrow bandwidth provides the feature of selectivity or **"tunability"** to such an amplifier, which gives it the ability to select different frequencies.

The aim of this chapter is to study such tunable amplifiers (tuned amplifiers). Basic idea of these amplifiers are to use resonant circuits, which selects a frequency by providing a maximum impedance at the resonant frequency and this resonant circuits impedance acts as the load for the amplifier (See fig 16.13). Thus, good amplification is obtained only for the resonant frequency. Typically tank circuits (Capacitor in parallel to the inductor) are used. To understand how the bandwidth is reduced we investigate the features of the LC resonant circuits in some detail.

17.1   L and C in Series

The impedance offered by the circuit shown in fig(17.1) is given as

$$
\begin{aligned}
Z_T &= X_C + X_L + R \\
&= j\left(\omega L - \frac{1}{\omega C}\right) + R
\end{aligned}
$$

through which a current

$$
I = \frac{V}{Z_T} = \frac{V}{R + j\left(\omega L - \frac{1}{\omega C}\right)}
$$

flows. When the terms inside the bracket falls equal to zero, i.e. when $\omega L = 1/\omega C$, the current flowing in the circuit reaches a maximum since, at that corresponding frequency,[1] the total impedance offered is a minimum. Figure(17.1) also shows the variation of impedance offered and the current passing through the circuit with varying frequency. As can be seen the extrema in both conditions is at $f = 1/2\pi\sqrt{LC}$.

The series combination shows a frequency dependence, however as a load this would not serve our purpose since at resonant frequency for maximizing output we require the impedance offered to be a maxima.

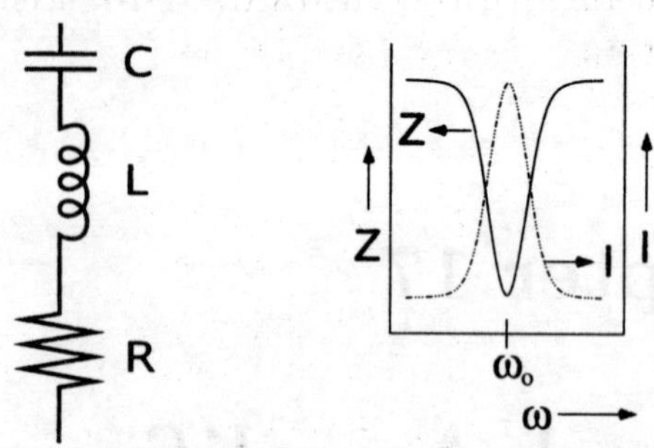

Figure 17.1: *A frequency dependent circuit, where the inductor and capacitor are kept in series. Note, even if no resistances are kept in the circuit, unless the inductor is an ideal inductor, the resistance of the wire which is coiled to give the inductance, would contribute in the circuit. Alongside, see the variation in impedance offered by the circuit along with the current flowing through it as a function of frequency.*

17.2   L and C in Parallel

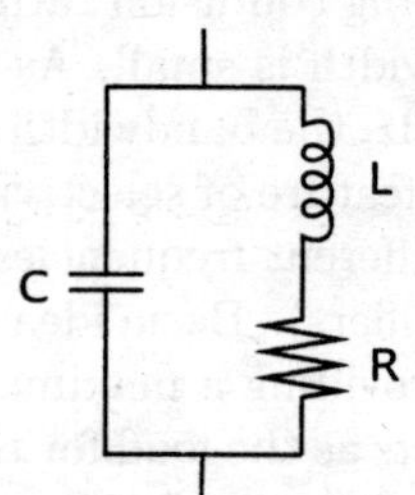

Figure 17.2: *A tank circuit, where the inductor and capacitor are kept parallel to each other.*

To obtain a maximum impedance, offered by a circuit as a function of frequency at resonant frequency, we use a load where the inductor is kept parallel to the capacitor (**tank circuit**). The physically inseparable resistance associated with the coil is considered in series with the inductance offered by the coil (see fig 17.2). The impedance offered by the circuit is given as

$$Z_T = \frac{(X_L + R)X_C}{X_L + X_C + R}$$
$$= \frac{-\frac{j}{\omega C}(R + j\omega L)}{R + j\left(\omega L - \frac{1}{\omega C}\right)}$$

On simplifying, we have

$$Z_T = \frac{\frac{L}{C} - \frac{jR}{\omega C}}{R + j\left(\omega L - \frac{1}{\omega C}\right)}$$

---

[1]The frequency is called the resonant frequency, given as

$$\omega L = \frac{1}{\omega C}$$
$$\omega^2 = \frac{1}{LC}$$

or

$$f = \frac{1}{2\pi\sqrt{LC}}$$

Considering the associated stray resistance is negligible as compared to ωC, the above equation reduces to

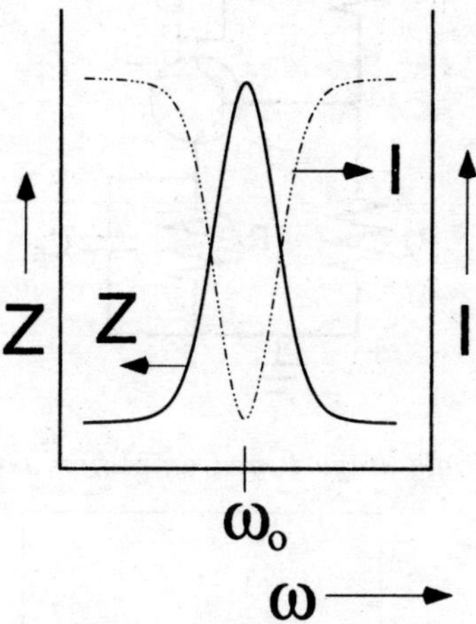

Figure 17.3: *The variation of impedance offered by a tank circuit with frequency along with how the current drawn by this circuit varies with frequency has been shown. The extrema occurs at the resonant frequency ω_o.*

$$
\begin{aligned}
Z_T &= \frac{\frac{L}{C}}{R + j\left(\omega L - \frac{1}{\omega C}\right)} \\
&= \frac{L}{RC + jC\left(\omega L - \frac{1}{\omega C}\right)}
\end{aligned}
\tag{17.1}
$$

At resonant frequency, the impedance offered by the circuit maximizes (corresponding the current drawn minimizes, see fig 17.3) and becomes

$$
Z_T = \frac{L}{RC}
\tag{17.2}
$$

The single-stage tuned amplifier circuit shown in fig(17.4) would have a voltage gain given as

$$
A_v = -\frac{h_{fe}}{h_{ie}} Z_T
\tag{17.3}
$$

which at resonant frequency would have a gain

$$
A_v = -\frac{h_{fe}}{h_{ie}}\left(\frac{L}{RC}\right)
$$

The frequency response of this amplifier circuit would be similar to the frequency response of the tank circuit shown in fig(17.5). For good selection (or tuning) of a certain frequency by the amplifier, the bandwidth of this circuit should be made narrow. We would has now have to develop a idea as to how to control the bandwidth of a tuned amplifier.

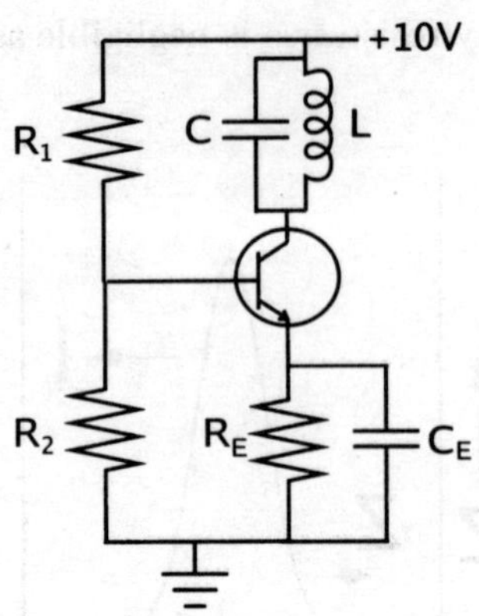

Figure 17.4: *A single stage tuned amplifier using a tank circuit.*

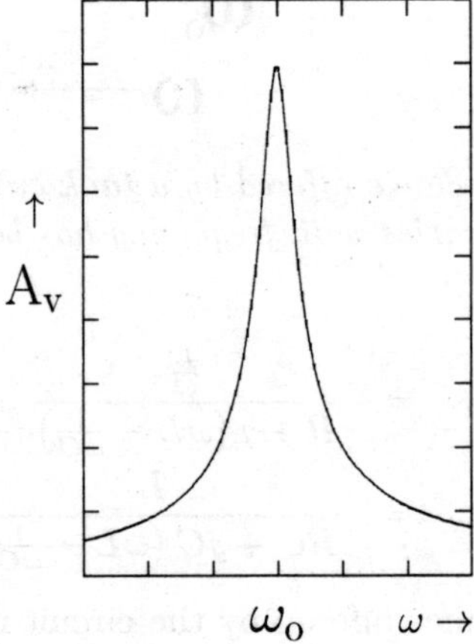

Figure 17.5: *The frequency response of a single stage tuned amplifier using a tank circuit.*

17.3   Stray Resistance Parallel or Series to the Inductor?

For calculating the bandwidth of the amplifier with a tank circuit load, it would be mathematically convenient to consider the stray resistance (R_s, the subscript 's' has now been added to show that initially the stray resistance was considered to be in series with the inductor) parallel to the inductor (it is placed as shown in fig 17.6, and now we call it R_p). But before proceeding to calculate the bandwidth let us understand the performance of this circuit and find the resonant frequency.

The impedance of the branch having a resistance in series with an inductor is given as

$$Z = R_s + j\omega L$$

Then the admittance would be given as

$$Y = \frac{1}{Z}$$

$$= \frac{1}{R_s + j\omega L} = \left(\frac{1}{R_s + j\omega L}\right)\left(\frac{R_s - j\omega L}{R_s - j\omega L}\right)$$

$$= \frac{R_s - j\omega L}{R_s^2 + \omega^2 L^2}$$

Since the stray resistance would be small, it would satisfy the condition $\omega^2 L^2 \gg R_s^2$, hence the admittance would reduce to

$$Y = \frac{R_s - j\omega L}{\omega^2 L^2}$$

$$= \frac{R_s}{\omega^2 L^2} + \frac{1}{j\omega L}$$

$$= \frac{1}{R_p} + \frac{1}{j\omega L}$$

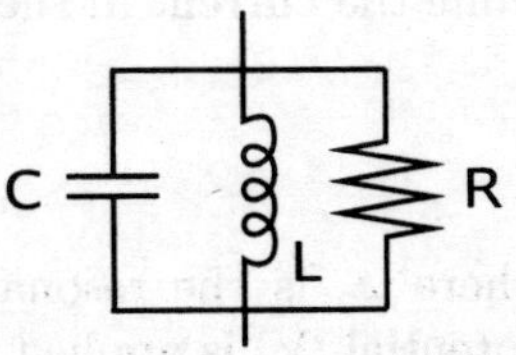

Figure 17.6: *A tank circuit, where the inductor and capacitor are kept parallel to each other. This this case the stray resistance is kept parallel to the inductor.*

That is, if you keep a resistance R_p parallel to the inductor (L), it would be equivalent to have a resistance R_s (where $R_s = \omega^2 L^2 R_p$) in series with the inductor. Since the resonant frequency of the resonant circuits were independent of the resistance, this circuit which behaves identically to the previous circuit would have a resonant frequency independent of the resistance. Even the response of the circuit's net impedance $|Z|$ to frequency would be identical (as shown in fig 17.3).

17.4 Quality Factor and Bandwidth

While the impedance of a parallel tank circuit and hence the amplification of an amplifier with the tank circuit as a load will be maximum at the resonant frequency, the use of this amplifier for selecting a channel would depend on the sharps of the peak. The maximum impedance offered by a tank circuit as stated (eqn 17.2), is at the resonant frequency and is given as

$$|Z| = \frac{L}{RC}$$

It would be of interest to study the variation of impedance (and hence gain) with frequency keeping L and C constant but with different resistances (R_s). While is is obvious that the maximum impedance offered at the resonant frequency would come down for increasing resistance, fig shows that the sharps of the peak also comes down. That is the ability of selecting a single frequency depends on the resistance and the ability decreases with increasing resistance. In other words the "*quality*" of selectivity or tunability comes down with resistance. A new figure of merit is introduced, called the quality factor and is defined (at resonant frequency) as

$$Q = \left(\frac{Amplitude\ of\ the\ oscillatory\ current\ in\ the\ capacitor}{Amplitude\ of\ the\ current\ supplied\ by\ the\ supply}\right)_{\omega=\omega_o} \tag{17.4}$$

In case of our L, C and R parallel circuit, the amplitude of current in the circuit at resonant frequency is given as

$$I = \frac{V}{Z_T} = \frac{V}{L/RC}$$

while the current in the capacitor would be

$$I_C = \frac{V}{1/\omega_o C}$$

where ω_o is the resonant frequency. The current division equation was not required as the potential 'V' is applied across C, L and R in parallel. Hence the quality factor of the circuit is given as

$$Q = \frac{\omega_o CV}{VRC/L}$$

giving

$$Q = \frac{\omega_o L}{R} \qquad (17.5)$$

Now we proceed to calculate the bandwidth of the tuned amplifier, using the example where R, L and C are parallel to each other. The admittance of the circuit is

$$\begin{aligned} A &= \frac{1}{R_p} + \frac{1}{X_C} + \frac{1}{X_L} \\ &= \frac{1}{R_p} + j\omega C - j\frac{1}{\omega L} \\ &= \frac{1}{R_p} + j\left(\omega C - \frac{1}{\omega L}\right) \end{aligned}$$

Hence, the impedance offered by the circuit combination is given as

$$\begin{aligned} Z_T &= \frac{1}{\frac{1}{R_p} + j\left(\omega C - \frac{1}{\omega L}\right)} \\ &= \frac{R_p}{1 + jR_p\left(\omega C - \frac{1}{\omega L}\right)} \\ &= \frac{R_p}{1 + j\omega R_p C\left(1 - \frac{\omega_o^2}{\omega^2}\right)} \end{aligned}$$

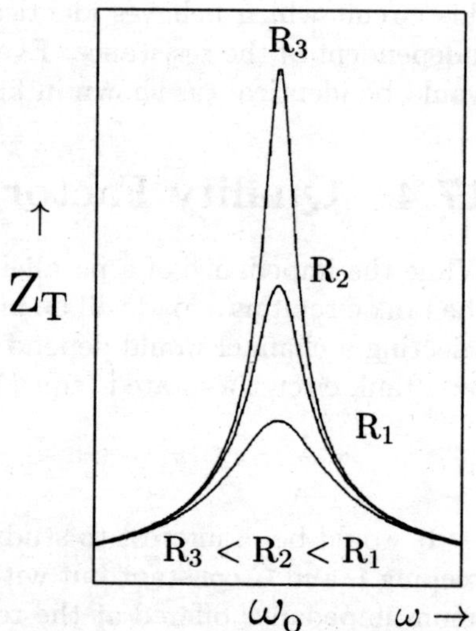

Figure 17.7: *The selectivity or sharpness of the tank circuit's frequency response depends on the resistance.*

Using eqn(17.2) and eqn(17.5), the above equation reduces to

$$\begin{aligned} Z_T &= \frac{R_p}{1 + jQ\frac{\omega}{\omega_o}\left(1 - \frac{\omega_o^2}{\omega^2}\right)} \\ &= \frac{R_p}{1 + jQ\left(\frac{\omega}{\omega_o} - \frac{\omega_o}{\omega}\right)} \end{aligned}$$

It has already been shown that the frequency response of the tuned amplifier depends on the frequency response of the tank circuit. Hence, the voltage gain's relation with frequency would be given as

$$A_v \propto \frac{R_p}{\sqrt{1 + Q^2 \left(\frac{\omega}{\omega_o} - \frac{\omega_o}{\omega} \right)^2}}$$

The bandwidth is defined in-between the two cut-off points where the voltage gain is only $1/\sqrt{2}$ of the maximum gain, hence the frequency of ω at which we have cut-off is given by

$$\sqrt{1 + Q^2 \left(\frac{\omega}{\omega_o} - \frac{\omega_o}{\omega} \right)^2} = \sqrt{2}$$

To find the two cut-off points we argue

$$Q \left(\frac{\omega}{\omega_o} - \frac{\omega_o}{\omega} \right) = \pm 1$$

giving

$$\left(\frac{\omega_1}{\omega_o} - \frac{\omega_o}{\omega_1} \right) = \frac{1}{Q} \tag{17.6}$$

and

$$\left(\frac{\omega_2}{\omega_o} - \frac{\omega_o}{\omega_2} \right) = -\frac{1}{Q}$$

equality the two equation we have

$$\left(\frac{\omega_1}{\omega_o} - \frac{\omega_o}{\omega_1} \right) = -\left(\frac{\omega_2}{\omega_o} - \frac{\omega_o}{\omega_2} \right)$$

Solving this we have

$$\omega_o^2 = \omega_1 \omega_2 \tag{17.7}$$

Substitute this result in eqn(17.6) after re-writing it as

$$\left(\frac{\omega_1^2 - \omega_o^2}{\omega_1 \omega_o} \right) = \frac{1}{Q}$$
$$\left(\frac{\omega_1^2 - \omega_1 \omega_2}{\omega_1 \omega_o} \right) = \frac{1}{Q}$$
$$\left(\frac{\omega_1 - \omega_2}{\omega_o} \right) = \frac{1}{Q}$$

Thus, the bandwidth of a single-stage tuned amplifier is given as

$$B.W = \frac{f_o}{Q} \tag{17.8}$$

This also affords a new definition of the circuit's/ amplifier's quality factor as

$$Q = \frac{f_o}{f_2 - f_1} \tag{17.9}$$

Example 17.1: A tuned-load common-emitter amplifier is to be used at a frequency of 10MHz. A 10μH inductor is to be used in the tuned circuit. The resistance of the inductor winding is 20Ω.

(i) What is the impedance of the tuned circuit at resonance?

(ii) If the collector-emitter capacitance of the transistor is 10pf, what parallel capacitance should be used for the tuned load?

(iii) There are 40 turns in the winding of the inductor. How many turns should there be in the winding of a secondary to do transformer matching to a 50Ω load? (Assume the impedance transformation through a transformer is given by the square of the turns ratio. Do you see why that is the case?)

The impedance at the resonant frequency is given in eqn(17.2). However, the value of C has not been given. This can be calculated from the equation for resonant frequency

$$f_o = \frac{1}{2\pi\sqrt{LC}}$$
$$C = \frac{1}{4\pi^2 f_o^2 L} = 25pF$$

Hence, the impedance offered at resonant frequency is $Z = 20K\Omega$.

The AC equivalent (small signal) circuit of fig(17.4) is shown in fig(17.8).

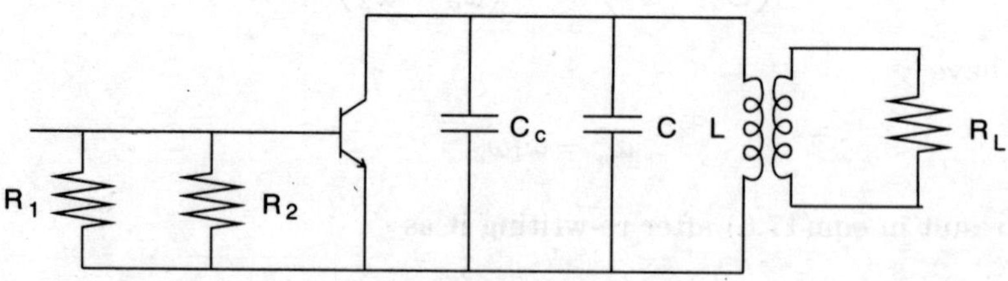

Figure 17.8: *The AC equivalent model of an amplifier with a tank circuit load.*

The collector-emitter capacitance (C_c) if present appears parallel to the tuned circuit's tank capacitor. The net capacitance should be 25pF, i.e.

$$C_c + C = 25pF$$

Since, C_c=10pF, the tank circuit should have a 15pF capacitor. For impedance matching

$$50 = n^2 \times 20000$$

where n(n_2/n_1) works out to be 0.05, thus if the primary has 40 winding, the secondary for proper impedance matching should have 2 loop winding.

Consider, a tuned amplifier (fig 17.4) with the resonant frequency as also the remaining specifications of Example 17.1 has to be designed to work with $I_c = 1mA$ and $V_{cc} = 10v$. For a transistor with $\beta = 100$, the base current would be $I_B = 10\mu A$ (10mA/β). Let us consider that a DC feedback of $V_{R_E} = 1v$ is sufficient, the

$$R_E = \frac{1}{1mA} = 1K\Omega$$

and

$$C_e = \frac{1}{\omega \frac{R_E}{10}} = 159pF$$

If the current in the biasing resistance R_1 is demanded as $100\mu A$, the current is R_2 would be $100 - I_B = 90\mu A$, therefore R_2 is

$$\frac{V_{BE} + V_{R_E}}{90\mu A} = \frac{1.6}{90}M\Omega = 17.8K\Omega$$

R_1 would be calculated as

$$\frac{V_{cc} - V_{R_2}}{100\mu A} = \frac{8.4}{100}M\Omega = 84K\Omega$$

The 'Q' factor (quality factor) of this tuned amplifier would be

$$Q = \frac{\omega_o L}{R_s} = 31$$

17.5   Double Tuned Amplifier

Two tank circuits can be coupled as shown in fig(17.9) to form what is called a double tuned amplifier. As can be seen, the first tank circuit is connected to the output of the first amplifier circuit, between the collector and grounded emitter (see the small signal AC model in fig 17.10). The second tank circuit is connected at the input (base-emitter port) of the second transistor amplifier. For analyzing this circuit, we follow steps similar to that used in Chapter 12 and

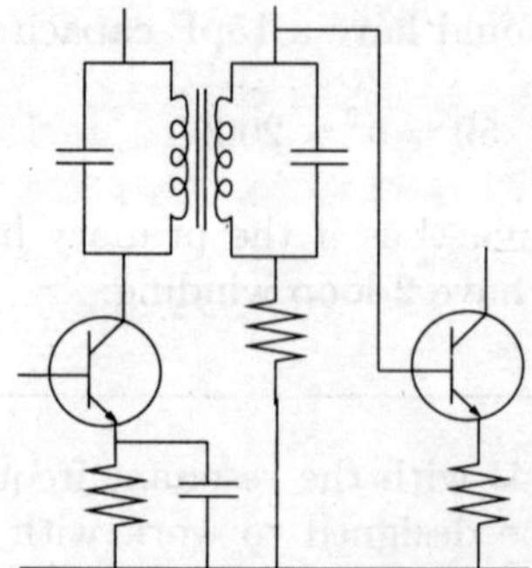

Figure 17.9: *The double tuned circuit uses two tank circuits with output of one stage being coupled to the input of the second stage using these two tank circuits. The shown circuit is not complete.*

convert the current source of stage one into a voltage source. The magnitude of the voltage source would be given as

$$v_1 = \frac{g_m V_{in}}{\omega C_p}$$

The voltage loop equations can be written as

$$-v_1 = \left[\frac{1}{j\omega C_p} + R_p + j\omega L_p\right] i_p + j\omega M i_s \tag{17.10}$$

and

$$v_o = [R_s + j\omega L_s]\, i_s + j\omega M i_p \tag{17.11}$$

The negative sign for the first loop voltage is as per the assumed direction of current i_p. The analysis of this circuit presents us with a wonderful situation, where we can put to test many things studied in the course of reading this book. We can now convert this equivalent circuit shown in fig(17.10b) to a 'T' equivalent circuit using ideas of section of Chapter 10. Add the terms $j\omega M i_p - j\omega M i_p$ to eqn(17.10), we have

$$-v_1 = \left[\frac{1}{j\omega C_p} + R_p + j\omega(L_p - M)\right] i_p + j\omega M(i_p + i_s) \tag{17.12}$$

Similarly, adding the terms $j\omega M i_s - j\omega M i_s$ to eqn(17.11) gives

$$v_o = [R_s + j\omega(L_s - M)]\, i_s + j\omega M(i_p + i_s) \tag{17.13}$$

Both equations (eqn 17.12 and 17.13) have term $j\omega M(i_p + i_s)$ common, implying that this branch is common to both loops. Using eqn (17.12 and 17.13) the equivalent circuit of fig(17.10b) can be redrawn as a 'T' equivalent circuit shown in fig(17.11). The purpose of reducing the equivalent circuit to the present form is to facilitate easy writing of the 'Y' parameters of this two port

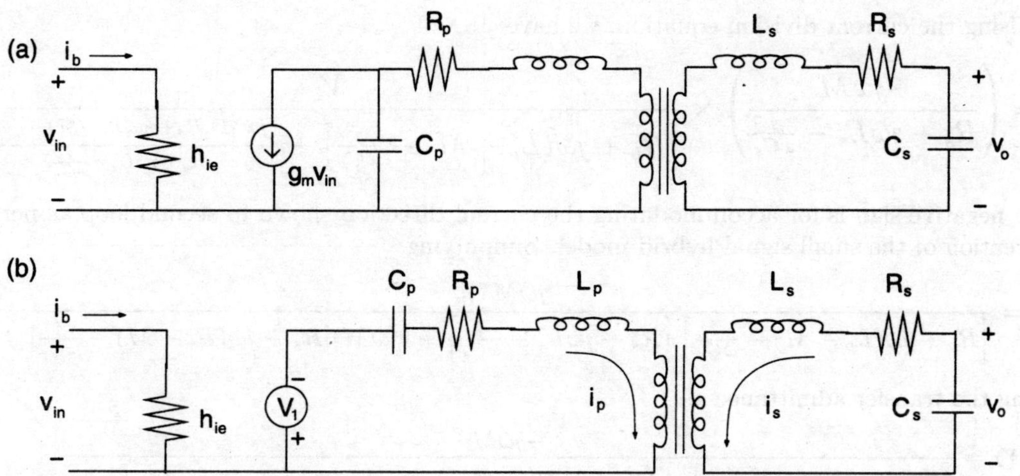

Figure 17.10: *The (a) hybrid equivalent circuit of the double tuned amplifier and in (b) the current source has been converted to a voltage source for purpose of calculation.*

network. The requirement of which would be appreciated as the current in the second loop is related to the voltage source present in the first loop by the transfer admittance (Y_T)

$$
\begin{aligned}
i_s &= Y_T v_1 \\
    &= \frac{g_m}{\omega C_p} Y_T v_{in}
\end{aligned}
$$

The transfer admittance (Y_T) can be calculated by first evaluating the current in the second loop when a voltage is applied in the first loop. The net current in the circuit, first and second loop is

$$
i_p = \frac{V_1}{\left[R_p + j\omega(L_p - M) - \frac{j}{\omega C_p}\right] + \dfrac{(j\omega M)\left[R_s + j\omega(L_s - M) - \frac{j}{\omega C_s}\right]}{R_s + j\omega L_s - \frac{j}{\omega C_s}}}
$$

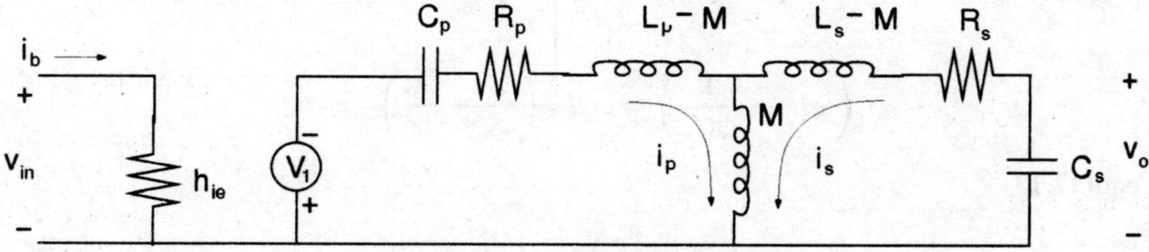

Figure 17.11: *The T equivalent circuit of a double tuned amplifier.*

Using the current division equation, we have

$$i_s = \left(\frac{-j\omega M}{R_s + j\omega L_s - \frac{j}{\omega C_s}}\right) \times \frac{V_1}{\left[R_p + j\omega(L_p - M) - \frac{j}{\omega C_p}\right] + \frac{(j\omega M)[R_s + j\omega(L_s - M) - \frac{j}{\omega C_s}]}{R_s + j\omega L_s - \frac{j}{\omega C_s}}}$$

The negative sign is for accommodating the current direction shown in second loop as per the convention of the small signal hybrid model. Simplifying,

$$i_s = \frac{-j\omega M V_{in}}{\left[R_p + j\omega(L_p - M) - \frac{j}{\omega C_p}\right]\left[R_s + j\omega L_s - \frac{j}{\omega C_s}\right] + (j\omega M)[R_s + j\omega(L_s - M) - \frac{j}{\omega C_s}]}$$

giving the transfer admittance as

$$Y_T = \frac{-j\omega M}{\left[R_p + j\omega(L_p - M) - \frac{j}{\omega C_p}\right]\left[R_s + j\omega L_s - \frac{j}{\omega C_s}\right] + (j\omega M)[R_s + j\omega(L_s - M) - \frac{j}{\omega C_s}]}$$

$$(17.14)$$

The expression is long and would require some simplification. Let us introduce a new variable, that gives the frequency deviation from the resonant frequency of the tank circuits, i.e. use

$$\frac{\omega}{\omega_o} = 1 + \delta \qquad (17.15)$$

or (assuming the deviation to be small)

$$\frac{\omega_o}{\omega} = (1 + \delta)^{-1}$$
$$= (1 - \delta) \qquad (17.16)$$

hence

$$\frac{\omega}{\omega_o} - \frac{\omega_o}{\omega} = 2\delta \qquad (17.17)$$

Now we write terms of eqn(17.14) in terms of the new variable. For which consider

$$\left(\omega L_s - \frac{1}{\omega C_s}\right) = \omega L_s \left(1 - \frac{1}{\omega^2 L_s C_s}\right)$$
$$= \omega L_s \left(1 - \frac{\omega_o^2}{\omega^2}\right)$$

or

$$\left(\omega L_s - \frac{1}{\omega C_s}\right) = \omega_o L_s \left(\frac{\omega}{\omega_o} - \frac{\omega_o}{\omega}\right)$$

using eqn(17.17)

$$\left(\omega L_s - \frac{1}{\omega C_s}\right) = 2\omega_o L_2 \delta$$

Substituting this expression in eqn(17.14) we have

$$Y_T = \frac{-j\omega M}{[R_p + j(2\omega_o L_p \delta - \omega M)]\,[R_s + j2\omega_o L_s \delta] + (j\omega M)[R_s + j(2\omega_o L_s \delta - \omega M)]}$$

Collecting the real and imaginary parts together

$$Y_T = \frac{-j\omega M}{[R_p R_s - 4\omega_o^2 L_p L_s \delta^2 + \omega^2 M^2] + j[2\omega_o L_s R_p \delta + 2\omega_o L_p R_s \delta]}$$

In double tuned amplifiers, even though the circuital elements are different, the combination is always selected such that

$$Q \;=\; \frac{\omega_o L_p}{R_p} = \frac{\omega_o L_s}{R_s}$$

$$\omega_o^2 \;=\; \frac{1}{L_p C_p} = \frac{1}{L_s C_s}$$

For removing the resistance terms from our transfer admittance expression, we write

$$Y_T = \frac{-j\omega k \sqrt{L_p L_s}}{\left[\frac{\omega_o^2 L_p L_s}{Q^2} - 4\omega_o^2 L_p L_s \delta^2 + \omega^2 k^2 L_p L_s\right] + j\left(4\frac{\omega_o^2 L_p L_s \delta}{Q}\right)}$$

Simplifying

$$Y_T = \frac{-j\omega k Q^2}{\sqrt{L_p L_s}[(\omega_o^2 - 4\omega_o^2 \delta^2 Q^2 + \omega^2 k^2 Q^2) + j4\omega_o^2 \delta Q]}$$

or

$$Y_T = \frac{\omega k Q^2}{\sqrt{L_p L_s}[4\omega_o^2 \delta Q - j(\omega_o^2 - 4\omega_o^2 \delta^2 Q^2 + \omega^2 k^2 Q^2)]}$$

Now the output voltage would be given as

$$\frac{j i_s}{\omega C_s} \;=\; \frac{j g_m}{\omega^2 C_p C_s} Y_T v_{in}$$

$$v_o \;=\; \left(\frac{j g_m}{\omega^2 C_p C_s} v_{in}\right) \times \frac{-j\omega k Q^2}{\sqrt{L_p L_s}[(\omega_o^2 - 4\omega_o^2 \delta^2 Q^2 + \omega^2 k^2 Q^2) + j4\omega_o^2 \delta Q]}$$

$$\;=\; \frac{g_m k Q^2 v_{in}}{\omega C_p C_s \sqrt{L_p L_s}[(\omega_o^2 - 4\omega_o^2 \delta^2 Q^2 + \omega^2 k^2 Q^2) + j4\omega_o^2 \delta Q]}$$

giving the voltage gain as

$$A_v = \frac{g_m k Q^2}{\omega C_p C_s \sqrt{L_p L_s}[(\omega_o^2 - 4\omega_o^2 \delta^2 Q^2 + \omega^2 k^2 Q^2) + j4\omega_o^2 \delta Q]}$$

Using the resonant frequency expression, the terms relating to the inductances and capacitances can be reduced to a neat expression

$$A_v = \frac{g_m \omega_o^4 k Q^2 \sqrt{L_p L_s}}{\omega[(\omega_o^2 - 4\omega_o^2 \delta^2 Q^2 + \omega^2 k^2 Q^2) + j4\omega_o^2 \delta Q]}$$

$$|A_v| = \frac{g_m \omega_o^4 k Q^2 \sqrt{L_p L_s}}{\omega \sqrt{16\omega_o^4 \delta^2 Q^2 + (\omega_o^2 - 4\omega_o^2 \delta^2 Q^2 + \omega^2 k^2 Q^2)^2}}$$

Remember, $\omega = \omega_o(1 + \delta)$, hence

$$|A_v| = \frac{g_m \omega_o k Q^2 \sqrt{L_p L_s}}{(1 + \delta)\sqrt{16\delta^2 Q^2 + [1 - 4\delta^2 Q^2 + (1 + \delta)^2 k^2 Q^2]^2}}$$

The approximation made in eqn(17.16) is only valid if $\delta \leq 0.015$, hence, terms appearing with one can be neglected. This gives

$$|A_v| = \frac{g_m \omega_o k Q^2 \sqrt{L_p L_s}}{\sqrt{16\delta^2 Q^2 + [1 - 4\delta^2 Q^2 + k^2 Q^2]^2}} \qquad (17.18)$$

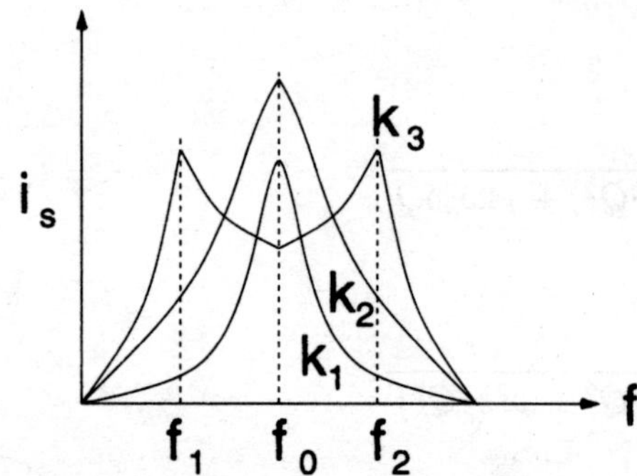

Figure 17.12: *The frequency response of a double tuned amplifier. The coupling coefficient $K_3 > K_2 > K_1$, with K_2 being the critical coupling.*

The square terms of δ have not been ignored since they appear along with the tank circuit's 'Q' factor (Quality factor) which is usually appreciably high (R_p and $R_s \to 0$). To identify at what frequencies would the gain be a maximum, we differentiate eqn(17.18) with respect to δ. We have

$$\delta(1 + 4\delta^2 Q^2 - K^2 Q^2)$$

giving maximizing conditions

$$\delta = 0 \qquad (17.19)$$

$$\delta = \pm\frac{1}{2Q}\sqrt{K^2 Q^2 - 1} \qquad (17.20)$$

Conditions (17.19) and (17.20) suggests that for $KQ \leq 1$ the amplifier has a single peak maximum at the resonant frequency ($\omega = \omega_o$). However, for $KQ > 1$, using eqn(17.20) we see that two maxima peaks occur at frequencies

$$f_1 = f_o\left(1 - \frac{1}{2Q}\sqrt{K^2 Q^2 - 1}\right)$$

$$f_2 = f_o\left(1 + \frac{1}{2Q}\sqrt{K^2 Q^2 - 1}\right)$$

Figure(17.12) graphically represents the mathematics developed above. Figure(17.12) plots the secondary current (i_s) instead of the gain, since the functional form of i_s (from Y_T) and gain are identical. The peak sharpens as 'K' increases, however after critical coupling, $K_c = 1/Q$, the peak splits into a doublet. The maximum voltage gain for $K \geq K_C$ is given as

$$A_{Vmax} = \frac{g_m \omega_o Q \sqrt{L_p L_s}}{2}$$

Critical coupling is said to occur when the resistance reflected back into the primary from the secondary is equal to the primary resistance R_p. For K's just greater than the crictical coupling condition (over coupled), the gain in between f_1 and f_2 is nearly equal to the maximum voltage gain (minimum occurring at $\delta = 0$, giving $A_{Vdip} = 2A_{Vmax}KQ/[1 + K^2Q^2]$), thus, for a large gain, the double tuned circuit effectively increases the bandwidth of the amplifier.

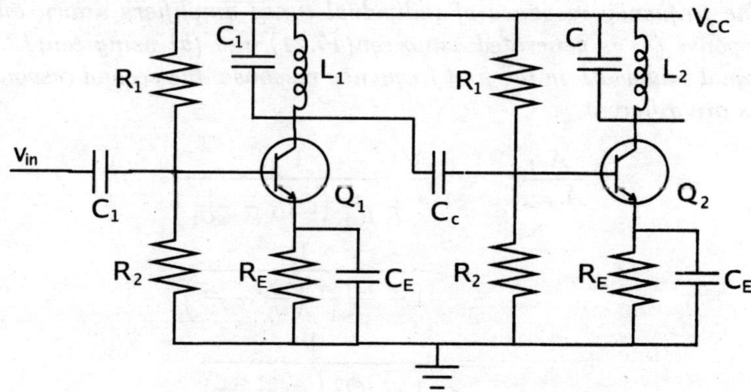

Figure 17.13: *The two stage staggered (tuned) amplifier.*

17.6 Staggered Amplifiers

The tuning of the input and the output circuits of a single stage tuned amplifier or the tuning of successive stages of cascaded tuned amplifiers to slightly different frequency gives a flat top frequency response. Such circuits are called staggered amplifier circuits, see fig(17.13). The gain of each stage can be written as (use eqn 17.1)

$$A_v = -\left(\frac{h_{fe}}{h_{re}}\right)\frac{L}{RC + jC\left(\omega L - \frac{1}{\omega C}\right)}$$

$$= \frac{A_{vres}}{1 + \frac{j}{R}\left(\omega L - \frac{1}{\omega C}\right)}$$

where A_{vres} (use eqn 17.2 and eqn 17.3) is the gain of the tuned amplifier at the resonant frequency. Until this stage we are not talking about L_1, L_2, C_1 and C_2 as we are developing general arguments. Now consider including the information that distinguishes the first tuned amplifier from the second, interms of the resonant frequency

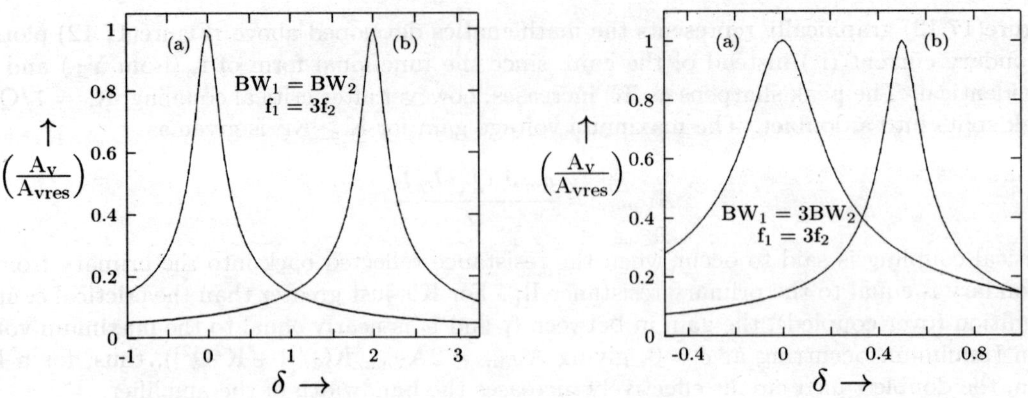

Figure 17.14: *The frequency response of individual tuned amplifiers drawn on the same scale (variable δ). Response (a) is generated using eqn(17.21) and (b) using eqn(17.23). While both amplifiers have equal bandwidth in the first frequency response, the second response is an example where bandwidths are different.*

$$\frac{A_{v1}}{A_{vres1}} = \frac{1}{1 + \frac{j}{R_{s1}}\left(\omega L_1 - \frac{1}{\omega C_1}\right)}$$

$$= \frac{1}{1 + \frac{j\omega_1 L_1}{R_{s1}}\left(\frac{\omega}{\omega_1} - \frac{\omega_1}{\omega}\right)}$$

$$= \frac{1}{1 + jQ_1\left(\frac{\omega}{\omega_1} - \frac{\omega_1}{\omega}\right)}$$

where ω_1 is the resonant frequency of the first stage tuned amplifier. Using eqn(17.15), the above equation can be written in terms of a deviation from the resonant frequency (see last section for detailed derivation)

$$\frac{A_{v1}}{A_{vres1}} = \frac{1}{1 + 2jQ_1\delta} \tag{17.21}$$

Similarly, for the second tuned circuit

$$\frac{A_{v2}}{A_{vres2}} = \frac{1}{1 + 2jQ_2\delta'} \tag{17.22}$$

δ' is also a deviation from the resonant frequency, f_2 (in this case the second resonant peak). To plot the two graphs (response) on the same scale, it would be necessary to relate δ' to δ. The relationship is trivially established as

$$\frac{\omega}{\omega_2} = 1 + \delta'$$

$$\frac{\omega}{\omega_1} \times \frac{\omega_1}{\omega_2} = 1 + \delta'$$

$$\delta' = \frac{\omega_1}{\omega_2}(1 + \delta) - 1$$

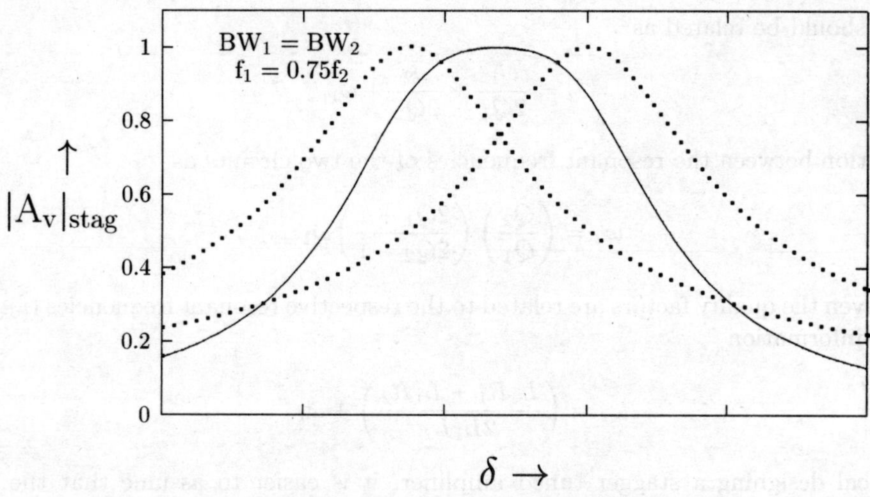

Figure 17.15: *The graph shows the frequency response of a two stage staggered (tuned) amplifier along with the frequency response of constituent tuned amplifiers.*

Hence eqn(17.22) can be written interms of variable δ as

$$\frac{A_{v2}}{A_{vres2}} = \frac{1}{1 + 2jQ_2\left[\frac{\omega_1}{\omega_2}\delta + \frac{\omega_1}{\omega_2} - 1\right]}$$

$$\frac{A_{v2}}{A_{vres2}} = \frac{1}{1 + 2j\frac{Q_2 Q_1}{Q_1}\left[\frac{\omega_1}{\omega_2}\delta + \frac{\omega_1}{\omega_2} - 1\right]}$$

$$\frac{A_{v2}}{A_{vres2}} = \frac{1}{1 + 2jQ_1\left[\frac{B.W_1}{B.W_2}\delta + \frac{B.W_1}{B.W_2} - \frac{Q_2}{Q_1}\right]} \tag{17.23}$$

Frequency response of two tuned amplifiers used in a stagger tuned amplifier shown in fig(17.14). The figures were generated using eqn(17.21) and eqn(17.23). On cascading the two tuned amplifier, the output gain would be given as

$$|A_v|_{stag} = \frac{A_{vres1}A_{vres2}}{\sqrt{1 + 4Q_1^2\delta^2}\sqrt{1 + 4Q_1^2\left[\frac{B.W_1}{B.W_2}\delta + \frac{B.W_1}{B.W_2} - \frac{Q_2}{Q_1}\right]^2}} \tag{17.24}$$

The frequency response as given by eqn(17.24) usually appears as an envolop on the two peaks as seen in fig(17.14). Hence, if the peaks are well resolved, that is are far apart, the stagger tuned amplifier would not have a flat frequency response. Infact there would be a pronounced dip between f$_2$ and f$_1$, the two resonant frequencies. For obtaining a flat frequency response, the separation between the two resonant peaks should be equal to the sum of half of their bandwidth. The bandwidth of a tuned circuit is given by eqn(17.8). The required separation is

$$B.W_1 + B.W_2 = \frac{f_1}{2Q_1} + \frac{f_2}{2Q_2}$$

Hence, ω_2 should be related as

$$\omega_2 = \frac{\omega_1}{2Q_1} + \frac{\omega_2}{2Q_2} + \omega_1 \tag{17.25}$$

giving relation between the resonant frequencies of the two circuits as

$$\omega_2 = \left(\frac{Q_2}{Q_1}\right)\left(\frac{2Q_1 + 1}{2Q_2 - 1}\right)\omega_1$$

However, even the quality factors are related to the respective resonant frequencies ($f_o = \omega_o L/R$). Using this information

$$\omega_2 = \left(\frac{L_2 R_1 + L_1 R_2}{2L_1 L_2}\right) + \omega_1 \tag{17.26}$$

For practical designing a stagger tuned amplifier, it is easier to assume that the two tuned amplifiers have equal bandwidth. Then

$$\frac{f_1}{Q_1} = \frac{f_2}{Q_2} \tag{17.27}$$

using the expression for quality factor, the above equation reduces to

$$\frac{R_1}{L_1} = \frac{R_2}{L_2} \tag{17.28}$$

Using this relation, eqn(17.26) reduces to

$$\omega_2 = \left(\frac{R_1}{L_1}\right) + \omega_1$$

$$\omega_2 = \omega_1\left[\left(\frac{R_1}{\omega_1 L_1}\right) + 1\right]$$

$$\omega_2 = \omega_1\left(\frac{1 + Q_1}{Q_1}\right) \tag{17.29}$$

This relation along with eqn(17.27), shows that for designing a stagger tuned amplifier with two tuned amplifiers with equal bandwidth, their quality factors should be related as

$$Q_2 = 1 + Q_1 \tag{17.30}$$

Equation(17.29) and eqn(17.30) were used to select the required (second) tuned amplifiers which on cascading gave a flat response in fig(17.15). Notice from eqn(17.24), the gain of the stagger tuned amplifier is very large. Figure 17.15 show the flatness of the response as also, the rapid increase from zero gain to maximum gain, as compared to that of the tuned amplifier. This skirt like response of tuned and double tuned amplifier implies a shallow selectivity. More tuned amplifiers can be staggered to make the frequency response nearly rectangular, giving sharp selectivity.

Exercise

Q1. What are tuned amplifiers?

Q2. Discuss the role of impedance variation with frequency in obtaining tuned amplifier's gain characteristics.

Q3. Find the bandwidth of a single-tuned amplifier.

Q4. What are stagger tuned amplifiers?

Q5. Discuss double-tuned amplifiers.

Q6. Why are double-tuned and stagger tuned amplifiers required?

Q7. What is the designing criteria of stagger tuned amplifiers? Explain what would happen if the design conditions are not met?

Q8. Draw the frequency response of a stagger tuned amplifier whose two individual tuned amplifiers have peaks far apart.

Q9. Enlist the advantages of a stagger tuned amplifier over a double tuned amplifier.

Chapter 18

The Thyristor

Until now the devices introduced in this text are classified as small signal devices due to their limitation in handling large currents and large voltages, or more appropriately large power. Their voltage, current and power handling capability being defined by V_{max}, I_{max} and P_{max} given in their data sheets provided by the device manufacturer. Typically, these devices were capable of handling milli-watts of power. In this chapter we discuss devices capable of handling power in watts. Such devices are called power devices. Study of these devices are done under the branch of electronics called power electronics. Power electronics, hence, is a branch of electronics that concerns itself with the conversion and control of electrical power.

Like other devices, here also we look for switching semiconductors, which has zero conduction drop across it (i.e. the voltage drop across the device when it conducts is zero) and zero leakage current at OFF condition (in other words when the switch is not conducting, no current flows through it). Such a device is losses and the efficiency is 100%. In 1956, Bell Telephone laboratory developed the PNPN device that had similar characteristics to the mercury arc tube thyristor that was in use till 1956 for controlling electric power. Since 1956, development in power devices lead to development in power electronics and via versa.

The semiconductor thyristor, also called Semiconductor controlled rectifier (SCR)[1] is a four layered and three pin devices, shown in fig(18.1). Because of the four layers, this devices has three junctions in it. When the anode is forward biased with respect to cathode, junction J_1 and J_3 are forward biased while J_2 is reverse biased (the gate is still grounded). This explains the I-V characteristics of the device, which in the first quadrant is identical to the 3rd quadrant of the normal PN diode. As can be seen from fig 18.2 (figure also shows the circuit required to experimentally determine the I-V characteristics of the SCR) even though the voltage supplied across the anode-cathode of the SCR is being increased, there is no flow of current (only the reverse leakage current through junction J_2) in the device. The resistance offered by the device

---

[1]Now a days its popularly called silicon controlled rectifiers, since SCR's are mostly fabricated with silicon. Si is preferred over Ge since the reverse saturation current for Si is far less, nearly zero even at high temperatures. This is a result of it's band gap being very large as compared to Ge. Since the band gap is large, charge carriers generated due to thermal excitation is also small. This gives the device good thermal stability. Also, the purification process of Si has improved immensely with improved technology. With low impurity levels in the Si crystal, the minority charge carriers which contribute to the reverse leakage current is reduced. This makes for a good switching action, with zero current when in 'OFF' state.

is quite high and the SCR is said to be in it's **forward blocking mode**. The resistance offered by the device, as stated, is large (ideally infinite). Hence it acts as a open switch, not allowing for any flow in current.

When the anode is made more and more positive with respect to the cathode, there reaches a stage that the SCR starts conducting. This results from the avalanche breakdown of the junction J_2. The voltage drop across the device drops to near zero with large current flowing through it. The device's impedance is near zero. The SCR is now said to have entered it's **forward conduction state** or is in it's 'ON' state. The forward biasing potential that send the SCR from it's forward blocking state to it's forward conducting state, with no role being played by the gate pin (i.e. I_g=0) is

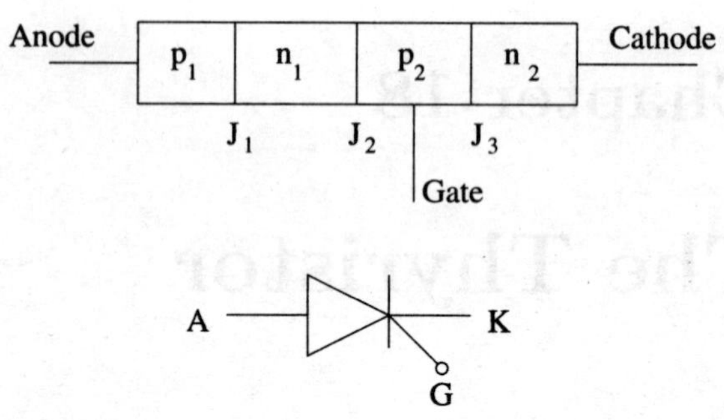

Figure 18.1: *A four layered, three pin SCR with its symbol*

called the **forward break-over voltage**, (V_{BO}). This switch like behavior enables it's use like a PN diode and hence name "rectifier". While the SCR goes from forward blocking state to it's forward conducting state, the SCR is only capable of maintaining itself in it's forward conducting state if the current passing through it has attained a minimum level called the **LATCHING CURRENT**, I_L. If the current through the SCR is less than I_L, when the device is being switched ON, the device immediately slips back into it's forward blocking state. This current level (i.e. I_L) has to be maintained for a minimum time, called the **turn on time**, t_{ON}. After the turn ON process is achieved ($t > t_{ON}$), a minimum level of current (I_H) has to be maintained in the SCR to keep the SCR in it's ON state. The SCR can be sent back into it's blocking state or OFF condition either by sending the SCR in reverse bias or by reducing the current below holding current.

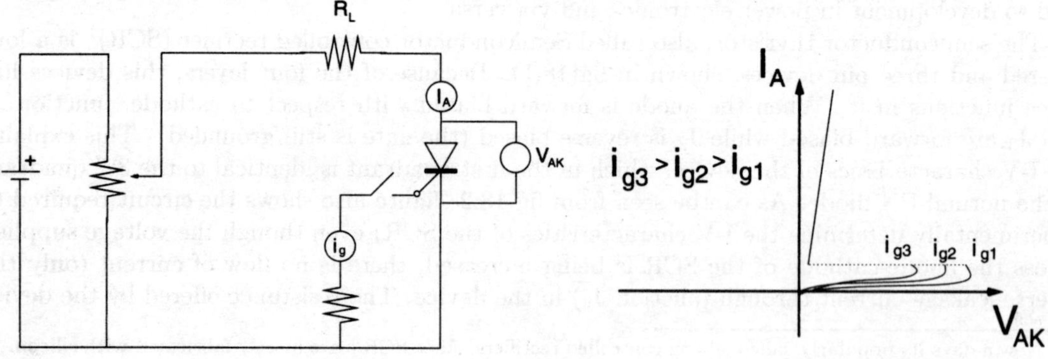

Figure 18.2: *Circuital diagram for measuring the I-V characteristics of an SCR along with the I-V characteristics of an SCR.*

When the voltage applied on the anode is reversed, i.e. the cathode is maintained at a higher potential then the anode, junctions J_1 and J_2 are reversed biased. With two junctions reverse biased, one can appreciate no or very little current would flow through the device. The SCR hence is said to be in the reverse blocking state. No reverse conducting state associated with an SCR.

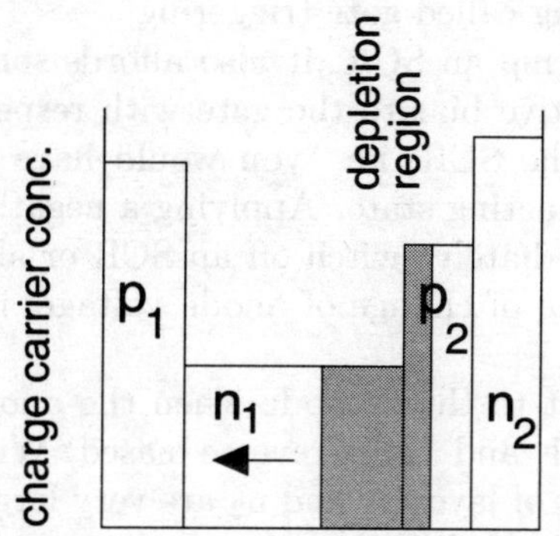

Figure 18.3: *Charge distribution among the four layers of the thyristor.*

At the breakover voltage, V_{BO}, the potential across the reverse biased junction, J_2 is large enough to initiate an avalanche breakdown in the reverse biased junction. Consider fig(18.3) which depicts the carrier concentration profile of a typical SCR. Since on increasing the potential (+ive) on the anode makes the junction J_2 more and more reverse biased, the depletion width of J_2 increases. The extent of the depletion width is different in the n_1 and p_2 layer (as is shown in fig 18.3). The extent to which the depletion width exists is given by the relationship

$$N_{A2}x_{PO2} = N_{D1}x_{NO1} \qquad (18.1)$$

The region with no free mobile charges (x_{NO1}) will be more in the n_1 region since it's doping concentration is far less than that of p_2 layer, i.e. $N_{A2} \gg N_{D1}$. Thus, an electron from n_1 layer has to travel less into the thickness of p_2 layer before it recombines unlike a hole of p_2 layer which has to travel more into the n_1 layer before it encounters an electron for recombination (it is for this reason the thickness of layer n1 is the largest). On further reverse biasing the depletion width grows. The growing of the depletion width increases the barrier potential, the potential appearing across the depletion width.

The increasing barrier potential enables the accelerating minority charge carriers to break covalent bondings present which leads to an avalanche. The barrier potential at which breakdown occurs is decided by the charge carrier concentration difference between the two layers (more the difference, larger the barrier potential required for breakdown). Thus, to increase power handling capacity, i.e. V_{BO} to be made large, the difference in carrier concentration between that of layer p_2 and n_1 has to be large. This in turn necessitates n_1 layer to be large (else method of breakdown would be of *punch through*, discussed in Chapter 5). Also, the large difference in charge carrier concentration resulting in a large depletion width would give rise to a large capacitance across the junction J_2, the switching speed of the SCR comes down. Thus, a trade off between speed and power handling would always be required. The voltage at which reverse breakdown will take place, thus dependent on the thickness of n_1 layer. Usually a thickness of 120mm gives V_{BO} of 2000V (for Si at T= 300K).

18.1 Gate Triggering

When a positive potential, V_{AK} is applied to the SCR, such that it is less then the break-over potential, one can still send the SCR into it's forward conduction state by applying a positive signal to the gate or the p_2 layer with respect to the cathode. This leads to an injection of electrons into the p_2 layer from the n_2 layer, which reduces the number of holes in the layer. As discussed in the previous section, this reduction in hole concentration in p_2 layer results in expansion of the depletion width of junction J_2. Thus for a lower voltage (less then V_{BO}), the depletion width has been increased. This increase leads to a increase in barrier potential which eventually leads to breakdown. A small signal as much as 70 to 150mA is enough to achieve this triggering into forward conduction state, the process being called gate triggering.

The gate thus provides for a controlled method of triggering an SCR. It also affords some protection to the circuit using an SCR. For example, a negative bias to the gate with respect to the cathode increases the forward blocking capacity of the SCR, i.e. you would have to apply $V_{AK} > V_{BO}$ for sending the SCR into it's forward conducting state. Applying a negative gate potential is sometimes used in practical circuits to immediately switch off an SCR or also to minimize the possibility of false triggering due to high rate of change of anode voltage, i.e. dv/dt.

Consider the situation gate is forward biased with respect to the cathode when the anode itself is reverse biased with respect to the cathode. Junctions J_1 and J_3 are reverse biased. With the doping profile being as it is, (shown in fig 18.3, i.e. doping of layer p_1 and n_2 are very large) the depletion region extends well into n_1 and p_2. When $+V_g$ is applied, junction J_3 disappears. The anode-cathode voltage, V_{AK} which initially was distributed across J_1 and J_3, now appears only across J_1. Though the voltage V_{AK} might not be enough for the breakdown of junction J_1, it would lead to increased reverse leakage current (I_{CBO}) which leads to heating and power loss. Usually $+0.25v$ (V_g) is a safe limit for a negative V_{AK}.

Another interesting aspect about gate triggering is that once the SCR is sent from it's forward blocking state to it's forward conducting state by the application of the gate trigger, the trigger signal can be removed and yet the SCR remains in it's conducting state. To appreciate this, however, one would have to go through the two transistor model of an SCR that we discuss below.

18.2   Two Transistor Model of Thyristor

The four layer SCR can be considered to be two transistors, an NPN and a PNP transistor wired back to back as shown in fig(18.4). The general equations relating the transistor's collector current to its emitter current are given as

$$I_E = I_C + I_B \tag{18.2}$$

and also by

$$I_C = \alpha I_E + I_{CBO} \tag{18.3}$$

where α is the DC common base current gain and I_{CBO} is the leakage current from the collector-base junction into collector (remember this leakage current appears due to the reverse biasing of the collector base junction, it is for this very reason emitter of PNP is shown as anode of SCR so that you have it forward biased with respect to the base). In our first transistor (PNP transistor) we re-write the two equation (eqn 18.2 and 18.3) as

$$I_{C1} = \alpha_1 I_{E1} + I_{CBO1} \tag{18.4}$$

Substituting this in eqn(18.2) we have

$$I_{B1} = (1 - \alpha_1)I_{E1} - I_{CBO1} \qquad (18.5)$$

As can be seen from fig(18.4) the base current of the first transistor is the same as the collector current of the second transistor, i.e.

$$I_{B1} = I_{C2}$$

Hence,

$$(1 - \alpha_1)I_{E1} - I_{CBO1} = \alpha_2 I_{E2} + I_{CBO2}$$

However, the emitter current of the PNP transistor and the emitter current of the NPN are the anode and cathode currents respectively of the SCR.

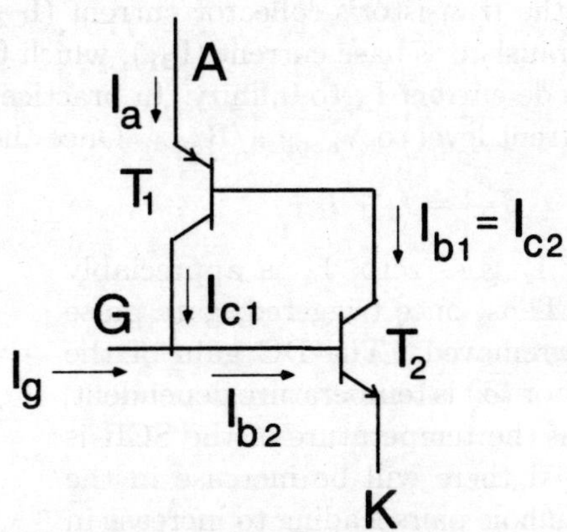

Figure 18.4: *Two transistor equivalent model of a SCR*

$$(1 - \alpha_1)I_A - I_{CBO1} = \alpha_2 I_K + I_{CBO2} \qquad (18.6)$$

But for the complete SCR, Kirchhoff's Current law gives

$$I_K = I_g + I_A$$

By substituting this expression of cathode current (I_K) in eqn(18.6), we have

$$(1 - \alpha_1)I_A - I_{CBO1} = \alpha_2(I_A + I_g) + I_{CBO2}$$

The net reverse leakage current in junction J_2 of the SCR being given as I_{CO},$(= I_{CBO1} + I_{CBO2})$. Substituting this and rearranging the above equation, we have

$$I_A = \frac{\alpha_2 I_g + I_{CO}}{[1 - (\alpha_1 + \alpha_2)]}$$

Initially consider the gate current to be zero, i.e.

$$I_A = \frac{I_{CO}}{[1 - (\alpha_1 + \alpha_2)]}$$

then with increasing I_A, α_1 and α_2 increases (The current gain α_1 and α_2 varies with the emitter current or anode current, the nature of variation is as shown in fig 18.5). This leads to an increase the emitter current. This leads to further increase in α_1 and α_2. This is an example of positive feedback or regenerative effect which ultimately leads to $\alpha_1 + \alpha_2 = 1$, resulting in a large anode current, I_A. Mathematically, though it should be infinite, the anode current is restricted by the external load applied to the anode circuit. The large anode current implies the turning ON of the SCR. If the gate current, I_g is applied, the regenerative condition is achieved even more easily.

The above paragraph completes a mathematical analysis of the positive feedback resulting from the back to back connection of the PNP and NPN transistors. On examining fig(18.4), it is clear, an increase in I_g leads to an increase in the base current (I_{B2}) of the NPN transistor, by which the transistor's collector current (I_{C2}) increases. This is nothing but an increase in the PNP transistor's base current (I_{B1}), which further pushes I_{B2}. This feedback goes on increasing the anode current I_A to infinity. In practice, this is avoided by an external load, which restricts the current level to $V_{applied}/R_{load}$. Once the feedback is established, in case of

$$I_{B2} = I_g + I_{C1}$$

even if I_g goes zero, I_A is appreciably large. Thus, once triggered, gate pulse maybe removed. The DC gain of the transistor (α) is temperature dependent, hence if the temperature of the SCR is increased there will be increase in the electron hole pairs leading to increase in the leakage current increases (i.e. I_{CO}), this leads to an increase in the anode current and condition for triggering SCR can be achieved. Thus, an SCR can be turned ON by thermal means also.

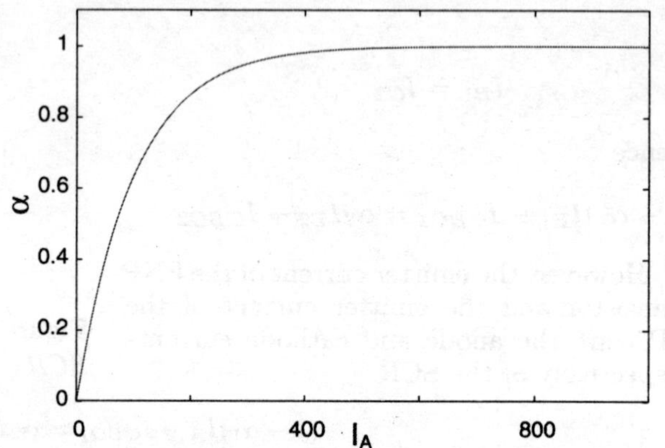

Figure 18.5: *Variation of DC gain with anode current.*

18.3   Transient Characteristics of SCR

In the above section we considered the steady state two transistor model, with DC currents flowing in the two transistors, however, when the DC voltage is just applied at t=0, the current changes with time. Till the circuit attains steady state, we say the circuit is in transit. The device behavior is different. The difference is due to the junctions present in the device. When an SCR anode is forward biased, while junctions J_1 and J_3 are forward biased, J_2 is reverse biased. A reversed biased junction has the characteristics of a capacitor. Infact, it is the depletion layer which gives the capacitance. Thus, the two transistor model is modified as shown in fig(18.6). If SCR is in blocking state, a rapidly rising voltage applied across the device would cause high current to flow through the junction capacitor

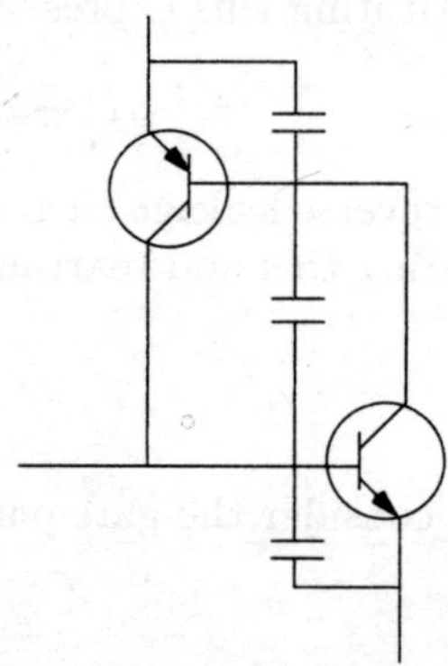

Figure 18.6: *Depletion regions capacitance gives rise to the transient behavior of the SCR.*

$$i = C dv/dt$$

i.e. large dv/dt leads to a large current. In the case of our SCR

$$i_{j2} = \frac{d(C_{j2}v_{j2})}{dt} = \frac{C_{j2}dv_{j2}}{dt} + \frac{v_{j2}dC_{j2}}{dt}$$

$$i_{j2} \sim \frac{C_{j2} dv_{j2}}{dt}$$

A large anode current is results is the SCR going from its forward blocking state to it's forward conducting state. Thus, this is another method of triggering the SCR, where current is made far greater then the latching current.

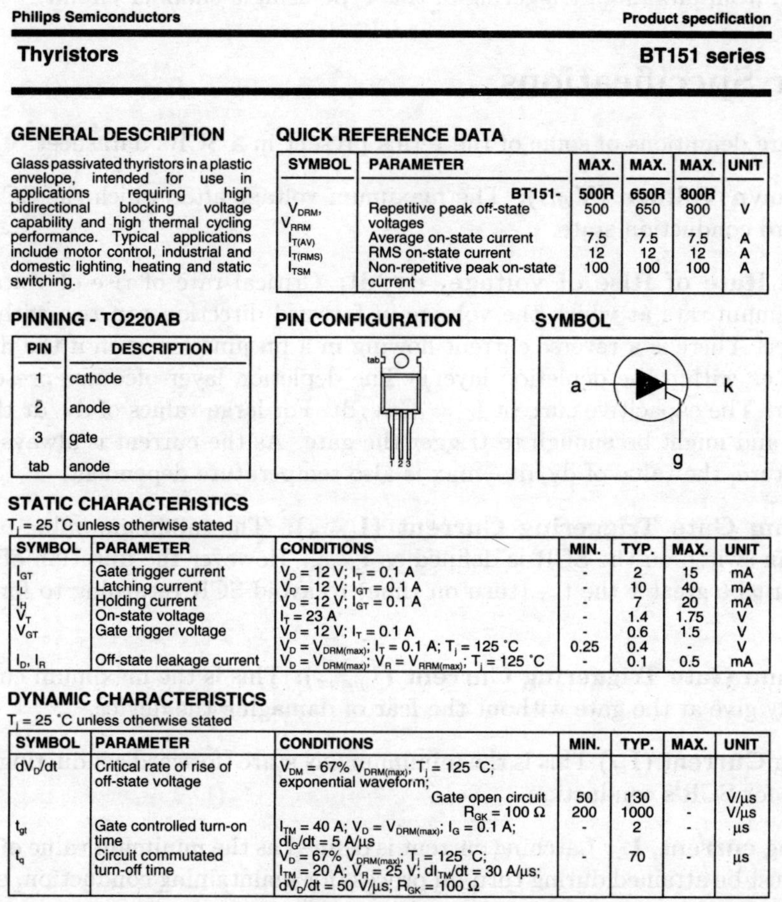

Philips Semiconductors   **Product specification**

Thyristors   **BT151 series**

GENERAL DESCRIPTION

Glass passivated thyristors in a plastic envelope, intended for use in applications requiring high bidirectional blocking voltage capability and high thermal cycling performance. Typical applications include motor control, industrial and domestic lighting, heating and static switching.

QUICK REFERENCE DATA

SYMBOL	PARAMETER	MAX.	MAX.	MAX.	UNIT
		BT151- 500R	650R	800R	
V_{DRM}, V_{RRM}	Repetitive peak off-state voltages	500	650	800	V
$I_{T(AV)}$	Average on-state current	7.5	7.5	7.5	A
$I_{T(RMS)}$	RMS on-state current	12	12	12	A
I_{TSM}	Non-repetitive peak on-state current	100	100	100	A

PINNING - TO220AB

PIN	DESCRIPTION
1	cathode
2	anode
3	gate
tab	anode

PIN CONFIGURATION

SYMBOL

STATIC CHARACTERISTICS

$T_j = 25\ ^\circ C$ unless otherwise stated

SYMBOL	PARAMETER	CONDITIONS	MIN.	TYP.	MAX.	UNIT
I_{GT}	Gate trigger current	$V_D = 12\ V; I_T = 0.1\ A$	-	2	15	mA
I_L	Latching current	$V_D = 12\ V; I_{GT} = 0.1\ A$	-	10	40	mA
I_H	Holding current	$V_D = 12\ V; I_{GT} = 0.1\ A$	-	7	20	mA
V_T	On-state voltage	$I_T = 23\ A$	-	1.4	1.75	V
V_{GT}	Gate trigger voltage	$V_D = 12\ V; I_T = 0.1\ A$	-	0.6	1.5	V
		$V_D = V_{DRM(max)}; I_T = 0.1\ A; T_j = 125\ ^\circ C$	0.25	0.4	-	V
I_D, I_R	Off-state leakage current	$V_D = V_{DRM(max)}; V_R = V_{RRM(max)}; T_j = 125\ ^\circ C$	-	0.1	0.5	mA

DYNAMIC CHARACTERISTICS

$T_j = 25\ ^\circ C$ unless otherwise stated

SYMBOL	PARAMETER	CONDITIONS	MIN.	TYP.	MAX.	UNIT
dV_D/dt	Critical rate of rise of off-state voltage	$V_{DM} = 67\%\ V_{DRM(max)}; T_j = 125\ ^\circ C$; exponential waveform;				
		Gate open circuit	50	130	-	V/µs
		$R_{GK} = 100\ \Omega$	200	1000	-	V/µs
t_{gt}	Gate controlled turn-on time	$I_{TM} = 40\ A; V_D = V_{DRM(max)}; I_G = 0.1\ A$; $dI_G/dt = 5\ A/µs$	-	2	-	µs
t_q	Circuit commutated turn-off time	$V_D = 67\%\ V_{DRM(max)}; T_j = 125\ ^\circ C$; $I_{TM} = 20\ A; V_R = 25\ V; dI_{TM}/dt = 30\ A/µs$; $dV_D/dt = 50\ V/µs; R_{GK} = 100\ \Omega$	-	70	-	µs

Figure 18.7: *A section of a SCR's datasheet issued by it's manufacturer.*

Methods of Turning ON an SCR

As discussed previously, the following methods are used to trigger on SCR in to forward conduction state:

1. Voltage Triggering.
2. Gate Triggering.

3. Thermal Triggering.

4. Radiation Triggering.

5. dv/dt Triggering: If the rate of rise of anode- cathode voltage is high, the charging current of the capacitive junction may be sufficient enough to turn on the SCR. The SCR is usually protected from unwanted triggering of this type using a snubber circuit.

Thyristor Specifications

Listed below are definitions of some of the terms present in a SCRs datasheet:

- **Breakdown Voltage (V_{BO}):** The maximum voltage after which the SCR is then sent to forward conduction state.

- **Critical Rate of Rise of voltage, dv/dt:** Critical rate of rise of voltage (dv/dt) is the maximum rate at which the voltage in forward direction can rise without triggering the device. There is a reverse current flowing in a pn junction even if the diode is reverse biased (i.e. within the depletion layer). The depletion layer ofcourse presents itself as a capacitor. The capacitive current $I_C = C dv/dt$. For large values of dv/dt the current will be large and might be enough to trigger the gate. As the current is always dependent on temperature, the value of dv/dt—max is also temperature dependent.

- **Minimum Gate Triggering Current (I_{gmin}):** The minimum value of gate current which can switch on the SCR is defined as I_{gmin}. However the duration of the triggering pulse must be greater the t_{on} (turn on time) to avoid SCR returning to forward blocking mode.

- **Maximum Gate Triggering Current (I_{gmax}):** This is the maximum current that you can safely give at the gate without the fear of damaging the device.

- **Holding Current (I_H)** This is the minimum ON state (forward conducting state) current required for SCR's conduction.

- **Latching current, I_L:** Latching current is defined as the minimum value of anode current which must be attained during turn-on process for maintaining conduction, even when gate signal is removed.

  The holding current is associated with the turn off process while the latching current is associated with the turn on process. Always, $I_L > I_H$.

- **di/dt Value and Hole storage Effect:** The maximum rate of change of current during the ON state which the SCR can handle is called the critical di/dt value of the device. The value is always specified at the maximum junction temperature. The spreading of charged carriers over the entire area of the SCR is relatively slow (**Explain Why?). If the current increases too fast, localized heating takes place. This is called 'HOLE STORAGE EFFECT'. Due to this localized heating device may be permanently damaged. Maximum available now-a-days is 200-250 A msec.

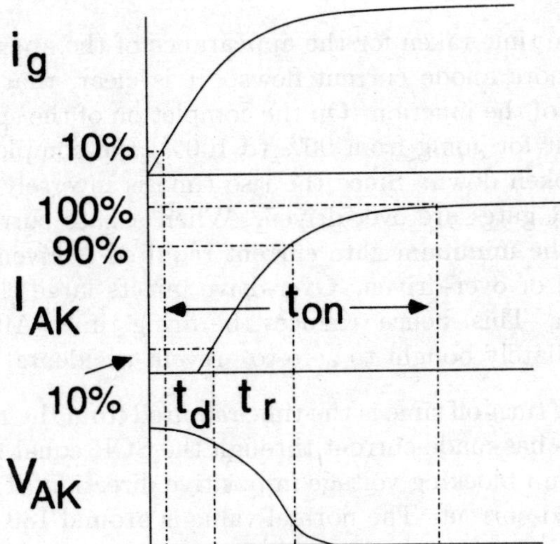

Figure 18.8: *Delay in SCR's conduction after applying gate trigger at t=0.*

- **Turn on Time:** After triggering the thyristor it does not instantaneously conduct but takes a finite time to reach full conduction. This time is called turn-on time. Figure(18.8) shows that the gate triggering pulse has been applied, the anode current starts rising from this instance and reaches 90% of V_{AK}/R (the maximum current in the circuit) only after a delay.

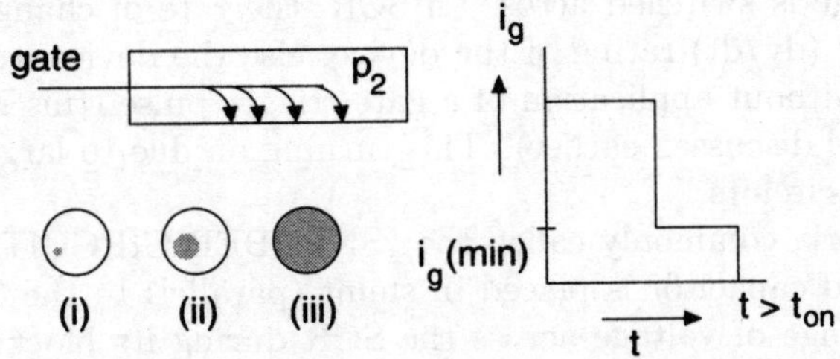

Figure 18.9: *Cross-section on SCR after (I) t_d, (II) t_r and (III) all the junction along cross-section has broken down with I_{AK} reaching 100%. An over-driving gate pulse is also shown.*

This time interval is further distinguished into three regions. The first, the time interval taken from when the gate current is 10% of its maximum and anode current is 10% of its maximum. This is called the delay time (t_d). A definite time (t_r), the rise time, is required for the anode current to rise from 10% to 90% of its maximum. The rise time is inversely proportional to the gate current. Thus, large pulse at the gate can reduce (t_r). Physically, the SCR has a circular cross-section with the gate electrode near one edge. When the gate current is provided, the junction breakdown (J_2) appears as a puncture near the gate electrode. It is through this small aperture that small anode current flows.

The delay time is the time taken for the appearance of the aperture. The aperture grows in size as a result more anode current flows. It is clear, that higher the gate current, easier is the opening of the junction. On the completion of the spread time (t_p, time taken for the anode current for going from 90% to 100%) the complete junction in the SCR's cross-section has broken down. Since the rise time is inversely proportional to the gate current, usually SCR gates are over-driven. When a gate current several times (2 to 3 times) higher than the minimum gate current required is given to the gate, the SCR is said to be hard fired or over-driven. Over-drive injects large electrons and hence makes for easier breakdown. This, hence, reduces the firing time. After hard driving the gate current is not immediately bought to zero to prevent accidental turn off.

- **Turn off Time:** The turn-off time is the time required from the zero current point (i.e. the commutation process has made current through the SCR equal to zero) to the time when the SCR regains its full blocking voltage in positive direction. This is only determined by the rate of rise of load-current. The normal value is around 180 msec.

- **Gate Power Loss:** This is the mean power loss due to gate current between the gate and the cathode.

- **Maximum On-state Voltage:** It is the voltage across the SCR once it is in forward conduction state. Usually, this is around 1.5v, but for all practical calculations can be neglected.

18.4   dv/dt Calculation Across a SCR

In practice a DC voltage is switched across an SCR, the rate of change of voltage across the SCR must be below the (dv/dt) rating of the device, else the device would be sent to forward conduction state even without application of a gate trigger pulse (this is ofcourse explained by the two transistor model discussed earlier). This turning on due to large dv/dt is also possible in rapidly changing AC circuits.

A suppression network, commonly called the "SNUBBER CIRCUIT", consisting of a series connected resistance and capacitor is placed in shunt (parallel) to the SCR. This RC network controls the rate of change of voltage across the SCR during its blocking state, and prevents accidental turning on. Consider the load in the circuit consists of inductance (L) and resistance (R). Also, initially there is no charge on the capacitor.

$$v_c = 0 \tag{18.7}$$

also, when the switch is closed, the SCR is in forward blocking state, hence the voltage equation becomes:

$$V_{dc} = (R_s + R)i + \frac{1}{C_s} \int i\, dt + L\frac{di}{dt} + v_c(t = 0)$$

differentiating with respect to time (t)

$$L\frac{d^2 i}{dt^2} + (R_s + R)\frac{di}{dt} + \frac{i}{C_s} = 0$$

or

$$\frac{d^2 i}{dt^2} + \frac{(R_s + R)}{L}\frac{di}{dt} + \frac{i}{LC_s} = 0$$

For simplifying the calculations let L = 0, it should be noted that L is a part of load that even if not present is kept by the designer to control the rising current in the SCR (next section).

$$\frac{di}{i} = -\frac{dt}{(R_s + R)C_s}$$

$$ln(i) = -t/(R_s + R)C_s + constt$$

$$i(t) = Aexp(-t/\tau)$$

where A is the integration constant, and $\tau = (R_s + R)C_s$ at t=0,

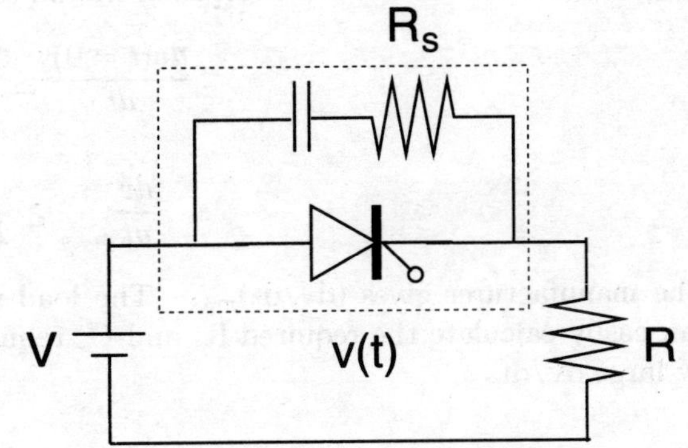

Figure 18.10: *A snubber circuit for dv/dt protection.*

$$A = i(t = 0) = \frac{V}{R_s + R}$$

since, the impedance offered by the rapidly charging capacitor is zero. Hence, the growing current is

$$i(t) = \frac{V}{(R_s + R)}exp(-t/\tau)$$

The KVL gives the voltage drop across the snubber circuit as

$$v(t) = V - R_L\frac{V}{(R_s + R)}exp(-t/\tau)$$

The rate of change of this voltage must be lower than the maximum dv/dt that is specified by the manufacturer to prevent accidental figuring. The rate of change of voltage appearing across the SCR is

$$\frac{dv(t)}{dt} = R\frac{V}{(R_s + R)\tau}exp(-t/\tau)$$

The rate of change will be maximum at t=0, therefore we are interested at the initial point (also, substituting the expression for τ),

$$\frac{dv(t = 0)}{dt} = \frac{VR}{(R_s + R)^2 C_s}$$

Consider R_s as negligible (usually R_s is kept just in case the user uses the circuit in no load condition, also to limit the discharge current of the capacitor). While the SCR is conducting, two currents would be present in the SCR

1. the load current and

2. the capacitor discharge current.

R_s restricts the discharging current so that the two currents together is lower than prescribed I_{max} of the SCR. The above expression should satisfy

$$\frac{dv(t=0)}{dt} \leq \frac{dv}{dt}_{max}$$

$$\frac{dv}{dt}_{max} \geq \frac{V}{RC_s} \qquad (18.8)$$

The manufacturer gives $(dv/dt)_{max}$. The load resistance is circuit designers choice, thus one can easily calculate the required R_s and C_s required to shunt the SCR and prevent turning on by large dv/dt.

18.5   di/dt Protection of SCR

A thyristor requires a minimum time to spread the current conduction uniformly throughout the junctions. If the rate of rise of anode current is very fast as compared to the spreading velocity of a turn-on process, a localized "hot-spot" heating will occur due to high current density, resulting in high temperature and causing device to fail.

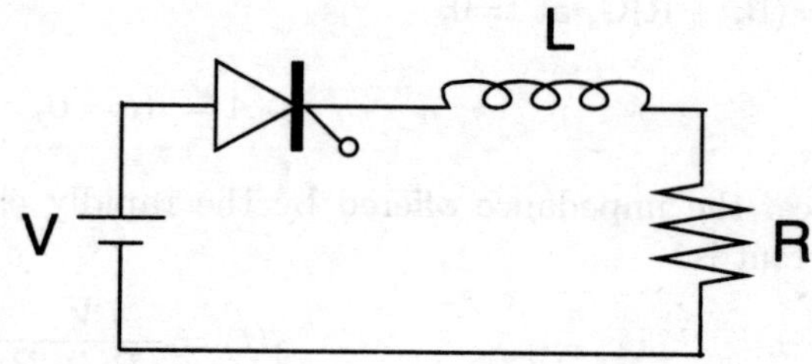

Figure 18.11: *An inductor is included in the load for di/dt protection.*

Large di/dt means charges 'q', are accelerating with large velocity and when they encounter J_2, which is reversed biased, they are rapidly slowed down. Excess energy (Kinetic Energy) is lost to lattice in terms of heat. This leads to a high temperature just outside junction J_2, which may even burn of the SCR (as also the possibility of turning on). Hence, in practice the SCR must be protected against high (di/dt). The loop equation for the circuit under consideration is given as

$$V = Ri + L\frac{di}{dt} \qquad (18.9)$$

solving which we have solution

$$i(t) = \frac{V}{R}[1 - exp(-t/\tau)]$$

where $\tau = L/R$. The integration constant is found by the initial condition, i(t=0)=0. The maximum change is ofcourse at t=0, therefore, from

$$\frac{di}{dt} = \frac{V}{R\tau}exp(-t/\tau)$$

$$\left(\frac{di}{dt}\right)_{max} = \frac{V}{L} \qquad (18.10)$$

For a mature reader this result comes directly from eqn(18.9). The rate of change of current i.e. di/dt is maximum at t=0, when all the potential drops across the inductor and not the resistance. Eqn(18.9) at t=0 reduces to

$$V = L\left(\frac{di}{dt}\right)_{t=0}$$

or

$$\left(\frac{di}{dt}\right)_{t=0} = \frac{V}{L}$$

The manufacturer data sheet specification gives the maximum di/dt capability, where the price of the SCR increases with increasing di/dt capability. By applying an appropriate inductor this growth of current can be controlled.

18.6 Gate-Cathode Characteristics

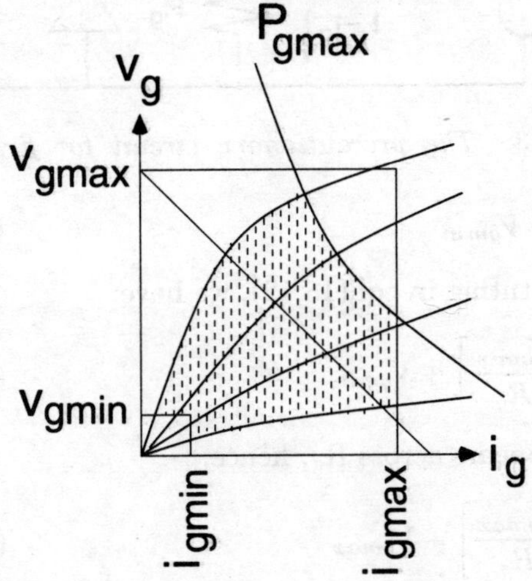

Figure 18.12: *The gate-cathode characteristics of an SCR.*

The gate and cathode of a SCR is just like a PN diode and hence it's IV characteristics would be similar to that of the diode (see fig 18.12). Only difference is in representation where the 'X'-axis represents current. This is just to highlight the fact that a SCR is a current controlled device, the ease with which triggering of the SCR occurs depends on the gate current. Also, flow of large current (I_{AK}) is controlled by the small gate current. The family of curves shown in fig(18.12) represents the considerable variation in gate-cathode IV characteristics within a single production batch. All SCR's of that production batch is assumed to lie within the shaded region shown in the figure.

18.7 Design of Firing Circuit Network

The SCR can be given a gate trigger by using a DC voltage source and a switch. All one has to do to fire the SCR is to make sure the SCR's anode is positive with respect to the cathode and then press the switch to fire the SCR. However, in this setup one has to keep in mind that the voltage appearing across the gate-cathode is less then v_{gmax}, the current going into the gate is less then i_{gmax} and the power dissipated across the gate-cathode is less then P_{gmax}. Also, to keep in mind is that for the SCR's gate to respond, the voltage applied across the gate-cathode is more then v_{gmin}, the current going into the gate is more then i_{gmin} and the power dissipated across the gate-cathode is more then P_{gmin}. Basically, one has to make sure that the applied gate signal should have value of voltage and current that lay somewhere in the shaded region of fig(18.12). This is similar to the process of biasing a transistor, which was done with proper selection of resistances in the circuit.

Thus on applying a DC voltage source to the gate with respect to the cathode, two resistances (R_s and R_g) are used for proper selection of voltage, current and power dissipation as suggested in the datasheet. The resistance R_s is kept in series with the voltage source and is connected to the gate to restrict the current flowing into the gate, while the gate-cathode voltage being applied appears across the resistance R_g (see fig 18.13).

For designing this firing circuit, one uses simple Kirchhoff's current law which suggests if the maximum possible gate current is entering the gate then the voltage appearing across R_g would be less. For proper triggering, it should be atleast V_{gmin}. Hence,

$$V_{gmin} = (I_1 - I_{gmax})R_g \quad (18.11)$$

And applying KVL in the first loop, we have

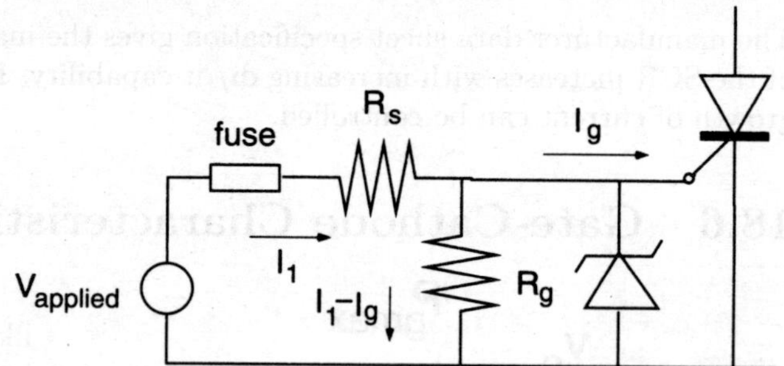

Figure 18.13: *The precautionary circuit for firing the SCR.*

$$V_{applied} = R_s I_1 + V_{gmin} \tag{18.12}$$

Finding the value of I_1 from eqn(18.11) and substituting in eqn(18.12), we have:

$$V_{applied} = R_s \left[I_{gmax} + \frac{V_{gmin}}{R_g} \right] + V_{gmin} \tag{18.13}$$

Similarly, when I_{gmin} flows into the gate, V_{gmax} appears across R_g, hence,

$$V_{applied} = R_s \left[I_{gmin} + \frac{V_{gmax}}{R_g} \right] + V_{gmax} \tag{18.14}$$

The two equations (eqn 18.13 and 18.14) form a set of simultaneous equations. Solving the simultaneous equation we have

$$R_s = \frac{V_{applied}(V_{gmax} - V_{gmin})}{(P_{gmax} - P_{gmin})}$$

and

$$R_g = \frac{V_{applied}(V_{gmax} - V_{gmin})}{V_{applied}(I_{gmax} - I_{gmin}) - (P_{gmax} - P_{gmin})}$$

Using the values of the manufacture sheet, R_s and R_g can be calculated. However, since usually the DC voltage used to provide the gate trigger is large, i.e. $V_{applied} > V_{gmax} > V_{gmin}$, the value of the resistances are easily computed as

$$R_s = \frac{V_{applied}}{I_{gmax}} \qquad (18.15)$$

and

$$R_g = \frac{V_{gmax}}{I_{gmax}} \qquad (18.16)$$

A fuse and a zener diode is used for added protection. The zener diode protects against crossing V_{gmax} by breaking down. Also, if accidental reverse biasing of gate-cathode is done (via applying the battery with opposite polarity), the zener diode provides a low resistance path, thus protecting the SCR's gate cathode. The fuse, obviously is kept to blow if the current is greater then I_{gmin}.

$\boxed{\text{Example 18.1:}}$ The SCR is to be gate triggered by a DC source and switch. The gate has to be protected since the maximum current it can handle is 100mA. The maximum gate voltage that can be given is 2V. If the DC source is 15v, find the value of the protective resistance so that the SCR will be fired.

Using eqn(18.15) and eqn(18.16), the required protective resistances work out to be

$$R_s = \frac{V_{applied}}{I_{gmax}} = 150\Omega$$

and

$$R_g = \frac{V_{gmax}}{I_{gmax}} = 20\Omega$$

$$\boxed{R_g = 150\Omega \text{ and } R_s = 20\Omega}$$

As discussed in the sections above, there are five methods of turning a SCR ON. Namely,

(i) Applying a forward potential across the anode of the SCR, greater than it's forward breakdown potential, (V_{BO}).

(ii) dv/dt triggering, i.e. apply a varying forward bias whose rate of change is greater than that specified by the manufacturer.

(iii) Heating the SCR to high temperature.

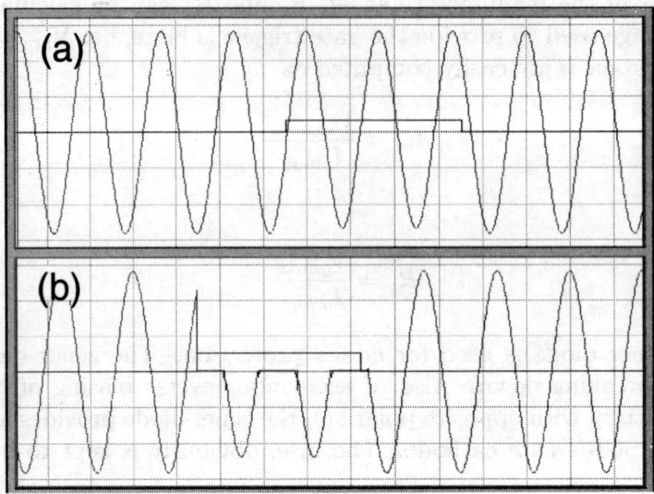

Figure 18.14: *The voltage across the SCR when triggered by a circuit shown in fig(18.13). In case (a) the values of the resistances used were $R_s = 200\Omega$, $R_g = 100\Omega$, while in case (b) $R_s = 20\Omega$, $R_g = 100K\Omega$. Note SCR in case (a) was not triggered.*

(iv) Triggering by incidenting light on the gate. Electron-hole pair in the semiconductor is generated by transfer of photon energy.

(v) Applying a small forward potential to the gate with respect to the cathode, called gate triggering.

Gate triggering is the most common method of turning ON the SCR, because this allows for good control over the instant of firing the SCR and is also the method that demands the least energy. While triggering the SCR with a DC source was discussed in the last example, we now investigate methods of triggering an SCR which is used in AC circuits.

18.7.1   Resistive Firing

Consider the circuit shown in fig(18.15). It is an example of resistive circuit used to fire the gate of the SCR.

The various resistances form a potential divider. When the appropriate gate voltage appears across resistance R_B, the SCR is triggered. The resistance, R_{min}, is kept to protect the gate. This minimum value of resistance should always be kept, so that even in the worst condition (R_v is removed or zero and no load is applied) the current to gate does not exceed rated value. Comparing with our previous analysis in the last section, R_{min} plays the role of R_s, hence (remember eqn 18.15)

$$R_{min} \geq \frac{V_{max}}{I_{gmax}}$$

The resistance R_B is to make sure that the voltage drop across the gate does not go beyond V_{gmax} under the worst conditions. The maximum drop occurs when $R_v=0$, thus

$$\frac{V_{max}}{R_{min} + R_B} R_B \leq V_{gmax}$$

Hence,

$$R_B \leq \frac{V_{gmax} R_{min}}{V_{max} - V_{gmax}}$$
$$\leq \frac{V_{gmax}}{I_{gmax}}$$

The result are same as that expressed in eqn(18.16) obtained for DC firing circuits.

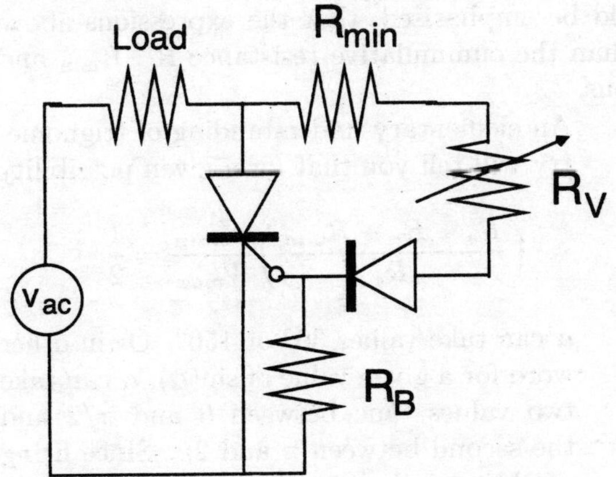

Figure 18.15: *A resistive network circuit for firing the SCR.*

When the SCR is ultimately triggered, the SCR enters it's forward conducting state with it's resistance reducing to zero. All the load current flows through the SCR, which shorts the resistive circuit kept for triggering the SCR. In the negative cycle. The SCR is reversed biased and it stops conducting. The current could flow from source through R_B, R_v etc, however, the reverse biased diode prevents this. Thus, the diode is kept to make sure the gate is not fed a negative cycle while the SCR itself is reverse biased by that very cycle. The waveform in fig(18.16) emphasises the working of this circuit.

Note from fig(18.16), the load current does not flow immediately after the SCR is forward biased by the AC's positive cycle. The instant it starts conducting is decided by the resistance, R_v. The delay from the instant the SCR is correctly biased to the instant the SCR starts conducting is called the firing angle (α) which as stated is decided by R_{min}. This resistance is kept as a variable resistance to vary the firing angle. The firing angle, or the instant the SCR conducts can be easily found by applying the potential divider equation for the resistive firing circuit as will be shown in the next section.

Mathematical Analysis

The various resistances combine to form a potential divider, and the voltage across the gate is given as

$$v_g = \left(\frac{R_B}{R_B + R_v + R_{min}} \right) V_{max} sin\Theta$$

The SCR is fired at the instant when the gate is fed v_{gmin}, hence the firing angle (α) is given as

$$sin\alpha = \left[\frac{R_B + R_v + R_{min}}{R_B}\right]\left(\frac{v_{gmin}}{V_{max}}\right)$$

or

$$\alpha = sin^{-1}\left(\frac{R_s + R_v + R_{min}v_{gmin}}{R_s V_{max}}\right)$$

where R_S is the value of R_B that is required to achieve V_{gmin}. Since v_{gmin}, R_{min}, R_s & V_{max} are fixed,

$$\alpha \propto sin^{-1}(R_v) \tag{18.17}$$

i.e. the firing angle is contolled by R_v. It should be emphasised, that the expressions above assume that the load impedance is far smaller than the cummulative resistance R_s, R_{min} and R_v and hence has been neglected in the derivations.

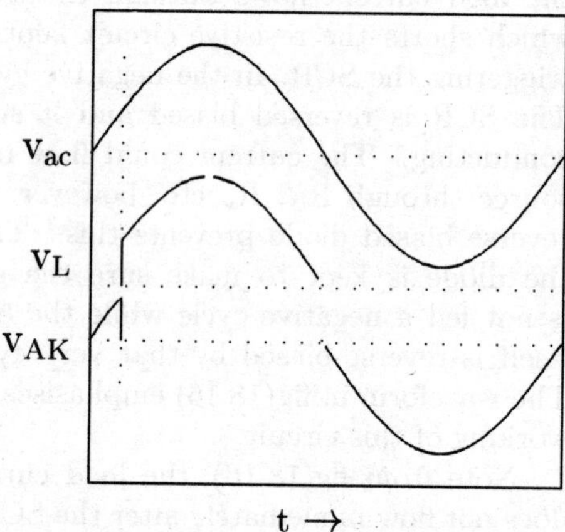

Figure 18.16: *Waveform of the load current and the voltage across the SCR is shown with respect to the input AC wave.*

An elementary understanding of trignometry will tell you that for a given possibility

$$\left(\frac{R_s + R_v + R_{min}}{R_s}\right)\frac{v_{gmin}}{V_{max}} = \frac{1}{2}$$

α can take values $30°$ or $150°$. Or in other word for a given value of $sin(\alpha)$, α can take two values, one between 0 and $\pi/2$ and the second between π and 2π. Since firing would take place in the first event/ instant itself, by resistive triggering, the SCR's firing can only take place between 0 and $\pi/2$. Thus, only 50% control is obtained. This is a major disadvantage of the method. The resistive firing though trivial, has the following disadvantages:

(i) The SCR can only be fired between 0 and $\pi/2$.

(ii) The firing angle 'α', becomes dependent on the temperature.

In the modified circuit shown in fig 18.17, one might say that the circuit should function similiarly to the last circuit. Well, yes, as far as the firing time/ angle is concerned there isn't any difference between the two circuits. However, note in the last case, when the SCR is fired, the firing circuit is shorted by the conducting SCR. This is not so in the new circuit, where, what ever drop is present in load is present across the firing network. Thus, some gate current will flow even when the SCR is conducting. This leads to power dissipation at the gate which is to be prevented.

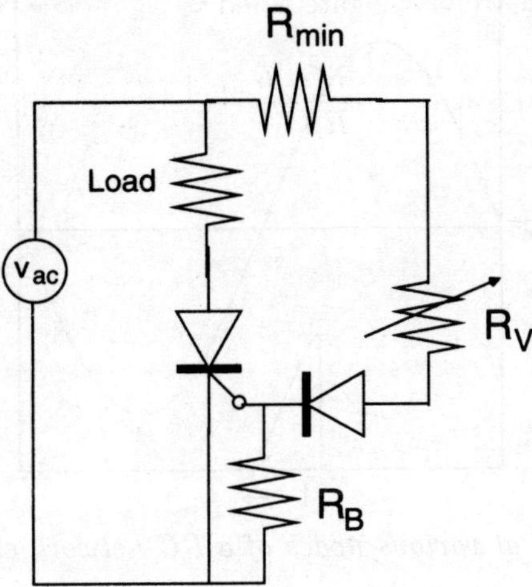

Figure 18.17: *Position of the load in a resistive network circuit for firing the SCR has been changed w.r.t. last circuit.*

18.7.2 RC Network

The limitations related with a resistive network, where the SCR can only be fired between 0 to $\pi/2$ can be removed by using a RC circuit, with the capacitor voltage appearing across the gate.

Consider the circuit shown. When the -ive cycle comes, diode D_1 is forward biased and the capacitor charges through the low resistance path provided by the diode (ideally the forward biased diode offers zero resistance), charging the capacitor to V_{max} at the instant the input attains the same (V_{max}). As the input voltage grows to 0, the capacitor retains it charge. On the arrival of the positive cycle, the capacitor charges to V_{max}, the in the opposite direction, i.e. from $-V_{max}$ to $+V_{max}$, linearly through the resistance. As soon as the capacitor reaches a potential equal to $v_g(min)$, the SCR is triggered. Note the wave diagram. As soon as the SCR is triggered, notice the change in the capacitor's voltage. Since the voltage drop across the conducting SCR goes zero, the capacitors volatge goes zero and remains zero till the end of the +ive cycle.

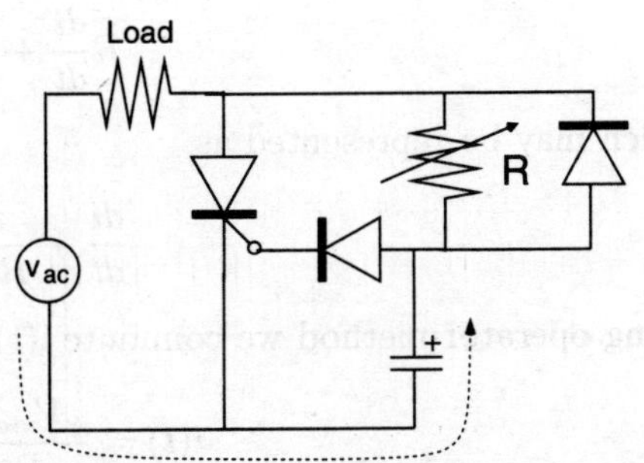

Figure 18.18: *A RC network circuit for firing the SCR.*

The firing angle depends on the charging of the capacitor which inturn depends on the value of R, thus allowing the possibility of triggering beyond $\pi/2$.

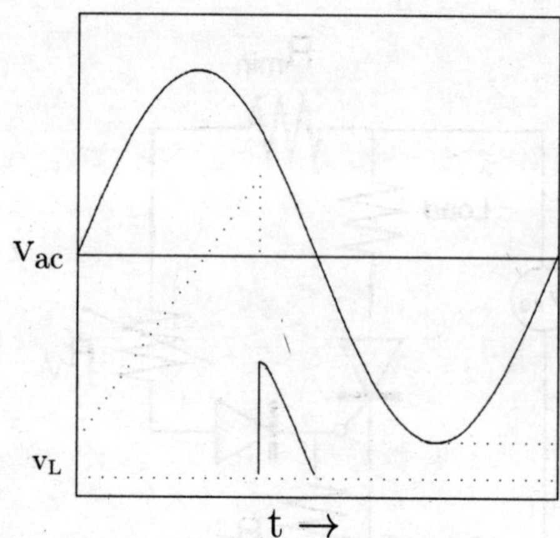

Figure 18.19: *The waveform at various nodes of a RC network circuit for firing the SCR.*

Mathematical Analysis

As explained in the last section the capacitor has been charged to $-V_{max}$ in the preceeding negative cycle. The loop equation in the +ive cycle is represented as

$$Ri + \frac{1}{C} \int idt + v_c(t = 0) = V_{max} sin\omega t$$

where, the initial condition has been incoperated in the loop equation, or

$$Ri + \frac{1}{C} \int idt - V_{max} = V_{max} sin\omega t$$

for solving this equation we differentiate this with respect to 't'

$$R\frac{di}{dt} + \frac{i}{C} = \omega V_{max} cos\omega t$$

which may be represented as

$$\frac{di}{dt} + \frac{i}{RC} = \frac{\omega V_{max}}{R} cos\omega t$$

Using operator method we commute i(t) by solving the above integral

$$i(t) = \frac{\omega V_{max}}{R} e^{-\frac{t}{RC}} \int e^{\frac{t}{RC}} cos\omega t dt \qquad (18.18)$$

The integral

$$\int e^{\frac{t}{RC}} cos\omega t dt$$

is a recurring integral if integrated by first function and second function, hence

$$P = \int e^{\frac{t}{RC}} cos\omega t dt$$

$$P = \frac{1}{\omega}e^{\frac{t}{RC}}sin\omega t + \frac{1}{\omega RC}\int e^{\frac{t}{RC}}sin\omega t\, dt$$

$$P = \frac{1}{\omega}e^{\frac{t}{RC}}sin\omega t + \frac{1}{\omega^2 RC}e^{\frac{t}{RC}}cos\omega t - \frac{1}{\omega^2 R^2 C^2}P$$

$$\left(\frac{1+\omega^2 R^2 C^2}{\omega^2 R^2 C^2}\right)P = \frac{1}{\omega^2 RC}e^{\frac{t}{RC}}(\omega RC sin\omega t + cos\omega t)$$

substituting

$$\omega RC = Asin\phi$$

and

$$Acos\phi = 1,$$

we have

$$A = \sqrt{1+\omega^2 R^2 C^2}$$

and

$$\phi = tan^{-1}\left(\frac{1}{\omega RC}\right)$$

Thus the above equation may be written as

$$P = \frac{RC}{\sqrt{1+\omega^2 R^2 C^2}}e^{\frac{t}{RC}}sin(\omega t + \phi)$$

Therefore, eqn(18.18) can be expressed

$$i(t) = \frac{\omega V_{max}C}{\sqrt{1+\omega^2 R^2 C^2}}sin(\omega t + \phi) + \frac{\omega V_{max}}{R}ke^{-\frac{t}{RC}}$$

where k is the integration constant. For evaluating this constant, we know that $i_c(t=0) = V_{max}/R$ (use the loop equation we started with). Hence,

$$\frac{V_{max}}{R} = \frac{\omega V_{max}C}{\sqrt{1+\omega^2 R^2 C^2}}sin\phi + \frac{\omega V_{max}}{R}k$$

$$\frac{\omega V_{max}}{R}k = \frac{V_{max}}{R} - \frac{\omega V_{max}C}{\sqrt{1+\omega^2 R^2 C^2}}sin\phi$$

Resulting in

$$i(t) = \frac{\omega V_{max}C}{\sqrt{1+\omega^2 R^2 C^2}}\left[sin(\omega t + \phi) - sin\phi e^{-\frac{t}{RC}}\right] + \frac{V_{max}}{R}e^{-\frac{t}{RC}} \qquad (18.19)$$

The variation in capacitor voltage would then be expressed as

$$v_c(t) = \frac{\omega V_{max}}{\sqrt{1+\omega^2 R^2 C^2}}\int\left[sin(\omega t + \phi) - sin\phi e^{-\frac{t}{RC}}\right]dt + \frac{V_{max}}{RC}\int e^{-\frac{t}{RC}}$$

or

$$v_c(t) = \frac{\omega V_{max}}{\sqrt{1+\omega^2 R^2 C^2}} \left[RCsin\phi e^{-\frac{t}{RC}} - \frac{cos(\omega t + \phi)}{\omega} \right] - V_{max}e^{-\frac{t}{RC}} + k$$

applying the initial condition discussed

$$k = -\frac{\omega V_{max}}{\sqrt{1+\omega^2 R^2 C^2}} \left(RCsin\phi - \frac{cos\phi}{\omega} \right)$$

On substituting our integration constant

$$v_c(t) = \frac{\omega V_{max}}{\sqrt{1+\omega^2 R^2 C^2}} RCsin\phi \left(e^{-\frac{t}{RC}} - 1 \right)$$

$$-\frac{V_{max}}{\sqrt{1+\omega^2 R^2 C^2}} [cos(\omega t + \phi) - cos\phi] - V_{max}e^{-\frac{t}{RC}}$$

Since the firing here is to be made more then $\pi/2$, R or C is obviously large hence the above equation reduces to

$$v_c(t) = V_{max}sin\phi \left(e^{-\frac{t}{RC}} - 1 \right) - V_{max}e^{-\frac{t}{RC}}$$

since RC is large ϕ tends to 0, hence using the series of expontential

$$v_c(t) = -V_{max} \left(1 - \frac{t}{RC} \right) = \frac{V_{max}}{RC}t - V_{max} \tag{18.20}$$

that is for an appropriate value of RC, the capacitor will charge linearly, the slope or how fast the linear varies being governed by R, C values. The SCR will only trigger after the gate is given v_{gmin}, thus the instant the SCR is triggered

$$t = RC \left(\frac{v_{gmin}}{V_{max}} + 1 \right) \tag{18.21}$$

can be varied to be greater than $\pi/2$.

18.8 Other Power Devices

18.8.1 Diacs

Apart from the SCR, in power electronics the other devices commonly used are the diacs and traics. Diacs are two pin, five layer power devices. They are essentially bi-directional avalanche diodes and hence derive their name from **Di**odes as **AC** switches, which can be switched from OFF state to ON state when either electrode is applied positive potential. This bi-directional character is very useful in AC circuits. Both the first and the third quadrant of the IV characteristics exists[2]. All this is because of the addition of the fifth layer. The forward blocking

---

[2]First and third quadrant only has meaning when we are talking of the characteristics with respect to one terminal. Note we shall not call the electrodes as anode or cathodes in this device since the device conducts in both directions. The terminals are named MT_1 and MT_2.

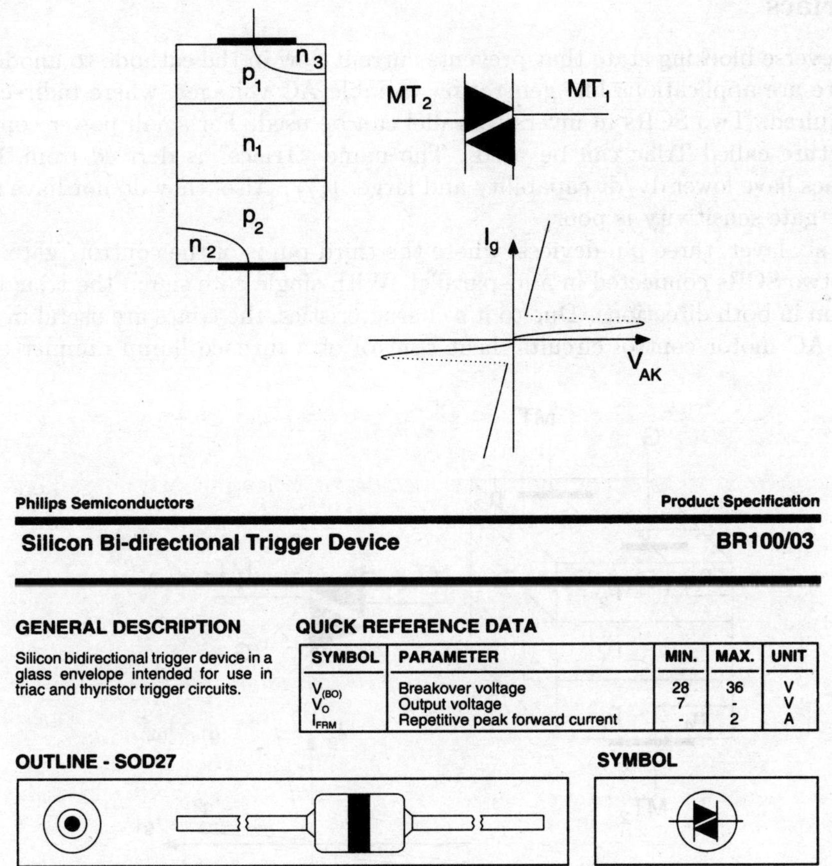

SYMBOL	PARAMETER	MIN.	MAX.	UNIT
$V_{(BO)}$	Breakover voltage	28	36	V
V_o	Output voltage	7	-	V
I_{FRM}	Repetitive peak forward current	-	2	A

Figure 18.20: *Block diagram of a diac along with it's symbol and I-V characteristics. Also, shown are some details listed in the diacs datasheet.*

state of the diac is explained as in the case of SCR. As can be seen from fig(18.20), at potentials below breakdown, junctions J_4 and J_2 are reverse biased while junctions J_3 and J_1 are forward biased. The reverse bias junction, J_4 however, does not contribute to/ or prevent flow of current, since the terminal metallization is such that the current takes the path of p_2 thus does not encounter the junction J_4. The second reversed biased junction prevents conduction. As the forward potential is increased, breakdown voltage approaches, giving forward conduction state in the same manner as in case of SCR. The breakdown mechanism is due to avalanche effect. When terminal T_1 is positive with respect to T_2 by a voltage greater than the breakdown voltage, the device will break-over into forward conduction state and the path of the current would be $p_2 - n_2 - p_1 - n_1$. Diacs as a device finds it's main use in firing/ triggering of triacs. Figure(18.20) also shows a section of the datasheet of the BR series diacs manufactured by Philips semiconductors.

18.8.2 Triacs

SCR have a reverse blocking state that prevents current flow in the cathode to anode direction. However, there are applications like generating variable AC voltages, where bidirectional conduction is required. Two SCRs in inverse parallel can be used. For small power considerations a single structure called Triac can be used. The name "Triacs" is derived from **Tri**ode **AC** switches. Triacs have lower dv/dt capability and larger t_{off}. Also, they do not have large V_{AK}, I_{AK} and their gate sensitivity is poor.

Triacs are six layer, three pin devices, where the third pin is of the control/ gate. Triacs are equivalent to two SCRs connected in anti-parallel. With single gate signal the triac is triggered into conduction in both directions. Due to it's characteristics, the triacs are useful in controlling AC power in AC motor control circuits, heat control of a furnace, lamp dimmer etc. Unlike

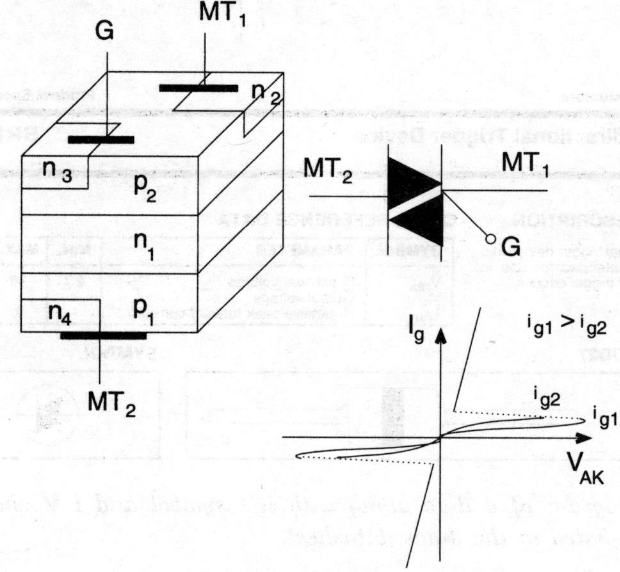

Figure 18.21: *Block diagram of a triac with it's symbol and I-V characteristics.*

the diac where the breakdown voltage is fixed (by the manufacturer), just as in the case of the thyrsistor the breakdown voltage can be reduced by increasing the gate triggering current. As is the case with SCR, this is the preferred method of using the Triac. Designing a circuit which uses triacs is far more involved than that with SCRs. This can be understood by the numerous modes in which a triac can be triggered. This is discussed in the following section. Figure(18.22) list some important information.

MODE I When electrode (again no notion of anode-cathode) T_2 is made +ive with respect to the electrode T_1, the triac is switched ON by applying +ive gate voltage (between T_1 and gate). This of course is same action as the SCR ($p_1 - n_1 - p_2 - n_2$). Only difference is the required gate current here is larger than that in SCR. The reason is due to the peculiar fashion in which the metallization is done two currents are present, (i) the $p_2 - n_2$ gate-cathode current and (ii) the ohmic current flowing through p_2 from the gate terminal to the terminal MT_1 which is smeared such that it is present on both p_2 and n_2 regions.

MODE II The device can also be switched on/ turned ON by applying a -ive signal to the gate with respect to terminal MT_1. In this case the device is switched on by an operation called **"junction gate operation"**.

Note here that the metallic contacts of the three terminals are such that the are in contact with two neighboring layers. The Gate electrode is in contact with layer p_2 and n_3, terminal T_1 is smeared over layers p_2 and n_2 and terminal T_2 over layer p_1 and n_4 (n_4 is not shown in fig 18.24 for clarity).

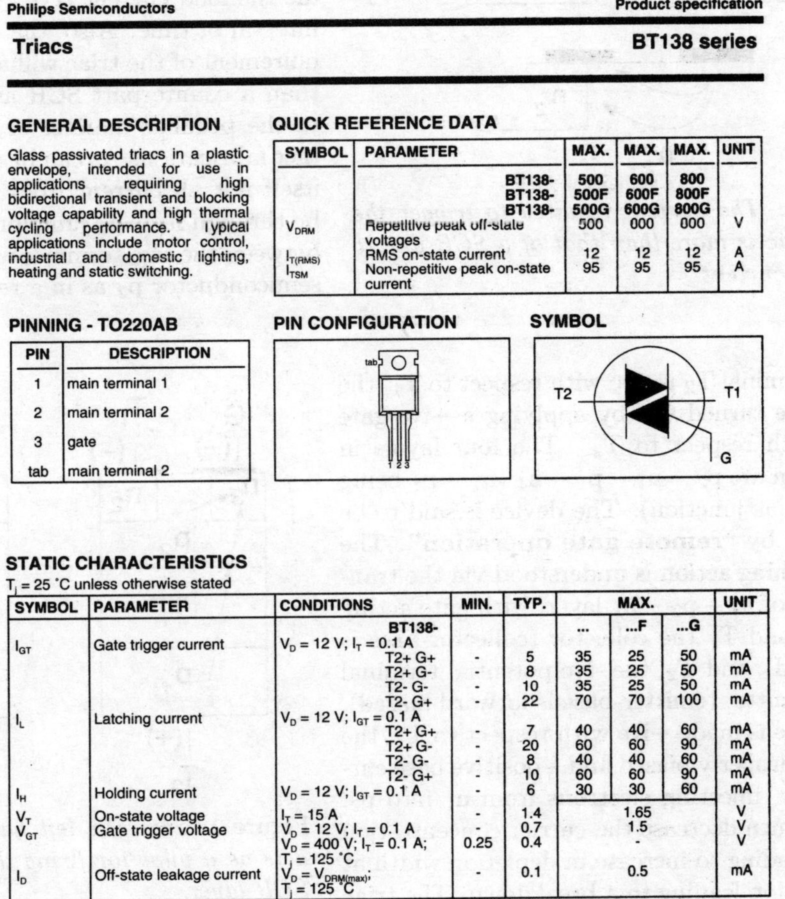

Philips Semiconductors   **Product specification**

Triacs   **BT138 series**

GENERAL DESCRIPTION

Glass passivated triacs in a plastic envelope, intended for use in applications requiring high bidirectional transient and blocking voltage capability and high thermal cycling performance. Typical applications include motor control, industrial and domestic lighting, heating and static switching.

QUICK REFERENCE DATA

SYMBOL	PARAMETER		MAX.	MAX.	MAX.	UNIT
		BT138- BT138- BT138-	500 500F 500G	600 600F 600G	800 800F 800G	
V_{DRM}	Repetitive peak off-state voltages		500	000	000	V
$I_{T(RMS)}$	RMS on-state current		12	12	12	A
I_{TSM}	Non-repetitive peak on-state current		95	95	95	A

PINNING - TO220AB

PIN	DESCRIPTION
1	main terminal 1
2	main terminal 2
3	gate
tab	main terminal 2

PIN CONFIGURATION

SYMBOL

STATIC CHARACTERISTICS

T_j = 25 °C unless otherwise stated

SYMBOL	PARAMETER	CONDITIONS	MIN.	TYP.	MAX. ...	MAX. ...F	MAX. ...G	UNIT
I_{GT}	Gate trigger current	V_D = 12 V; I_T = 0.1 A BT138-						
		T2+ G+	-	5	35	25	50	mA
		T2+ G-	-	8	35	25	50	mA
		T2- G-	-	10	35	25	50	mA
		T2- G+	-	22	70	70	100	mA
I_L	Latching current	V_D = 12 V; I_{GT} = 0.1 A						
		T2+ G+	-	7	40	40	60	mA
		T2+ G-	-	20	60	60	90	mA
		T2- G-	-	8	40	40	60	mA
		T2- G+	-	10	60	60	90	mA
I_H	Holding current	V_D = 12 V; I_{GT} = 0.1 A	-	6	30	30	60	mA
V_T	On-state voltage	I_T = 15 A	-	1.4	1.65			V
V_{GT}	Gate trigger voltage	V_D = 12 V; I_T = 0.1 A	-	0.7	1.5			V
		V_D = 400 V; I_T = 0.1 A; T_j = 125 °C	0.25	0.4	-			V
I_D	Off-state leakage current	V_D = $V_{DRM(max)}$; T_j = 125 °C	-	0.1	0.5			mA

Figure 18.22: *Some details listed in the triacs datasheet.*

Initially, since the gate and terminal T_1 act as a forward biased diode, a current flows from the layers $p_2 - n_3$. This lowers the hole concentration of layer p_2 leading to an increase in the depletion width of the reverse bias junction between n_1 and p_2. At appropriate depletion width, avalanche breakdown occurs and the four layers $p_1 - n_1 - p_2 - n_3$ act as a SCR. Only thing the triacs gate is acting as the SCR's cathode. It looks as if gate and terminal T_1 have interchanged their roles.

This results in the encircled region of fig(18.24), i.e. the left hand side of the triac structure to go in to zero resistance value. The gate hence is at the same potential as the terminal T_2. Thus, now gate becomes at a higher potential than terminal T_1. Thus, a current is set up in layer p_2 just as a normal SCR would be. The $p_1 - n_1 - p_2 - n_3$ SCR activity acts as a pilot to the $p_1 - n_1 - p_2 - n_2$ SCR, where the anode current of the pilot is the real gate current of the right hand SCR.

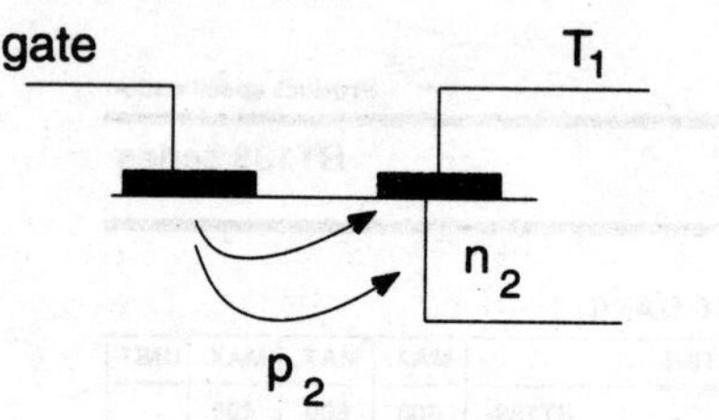

Figure 18.23: *The current required to trigger the gate of a triac is more than that of a SCR due to two possible current.*

This method requires gate circuit to handle the load current (I_{AK}) in it for a small interval of time. Also, the gate current requirement of the triac will always be higher than a counterpart SCR as explained due to the peculiar manner of electrode/ contact fabrication. As can be seen in this case itself not all current from T_1 to gate will be through $n_3 - p_2$ junction. Some current between the two terminal will flow through semiconductor p_2 as in a resistive path.

MODE III

When terminal T_2 is -ive with respect to T_1, the device can be turned ON by applying a +ive gate potential with respect to T_2. The four layers in this operation are $p_2 - n_1 - p_1 - n_4$ ($n_1 - p_1$ being the reverse bias junction). The device is said to be switched ON by **"remote gate operation"**. The remote switching action is understood via the transistor action of $n_2 - p_2 - n_1$ layer, with gate acting as the base and T_1 the collector (collector-base is reverse biased), and T_2, the -ive potential terminal acts as the emitter (emitter-base is forward biased). Since the gate is made +ive with respect to T_1, the transistor is properly biased and a positive base current will flow, injecting electrons from n_1 into p_2. This will in turn decrease the carrier concentration in layer n_1 leading to increase in depletion width at $p_1 - n_1$ junction leading to a breakdown. The triac is thus turned ON with current flowing in $p_2 - n_1 - p_1 - n_4$ layers.

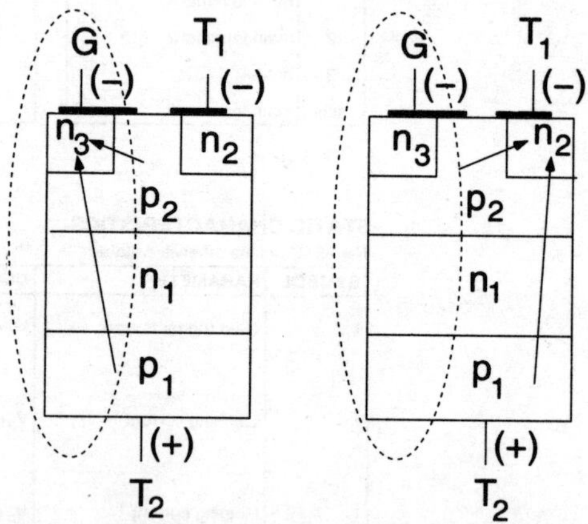

Figure 18.24: *The left hand of the triac acts as a pilot for firing the conventional SCR layer.*

MODE IV

If the gate is given negative potential, then transistor $n_3 - p_2 - n_1$ comes into action with terminal T_1 as the base. With the proper biasing of the transistor, current flows leading to

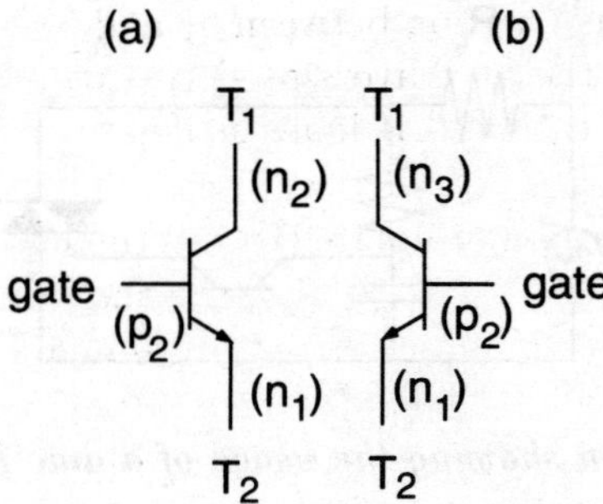

Figure 18.25: *In the third quadrant the firing of the triac takes place via remote gate operation where triggering in MODE III is explained by (a) $n_2 - p_2 - n_1$ transistor while (b) $n_3 - p_2 - n_1$ transistor explains triggering of triac in MODE IV.*

electrons entering into p_2 via n_1. This again leads to a reduction in charge carrier concentration in n_1 layer. Leading to an increase in p_1n_1 depletion width and an avalanche breakdown.

Thus there are four possible electrode potential combinations:

(1) T_2 positive and gate Positive with respect to T_1 results in first quadrant IV characteristics.

(2) T_2 positive and gate negative with respect to T_1 results in first quadrant IV characteristics.

(3) T_2 negative and gate positive with respect to T_1 results in third quadrant IV characteristics.

(4) T_2 negative and gate negative with respect to T_1 results in third quadrant IV characteristics.

The triac is rarely used in the first quadrant with negative gate trigger and the third quadrant with positive gate trigger.

Advantages of Triacs

1) They can be fired with +/-ive signals.

2) For AC circuits a single triac is enough instead of two SCRs.

3) in DC applications a diode is kept parallel to SCR to prevent accidental reverse breakdown (Since SCR is burnt on such break-over). PIV of such diodes are large as V_{DC} is usually large leading to increased cost. Since Triacs conduct both sides, it is preferred.

Disadvantages

1) Low dv/dt capability.

2) Rating is less compared to an SCR

3) Reliability is less.

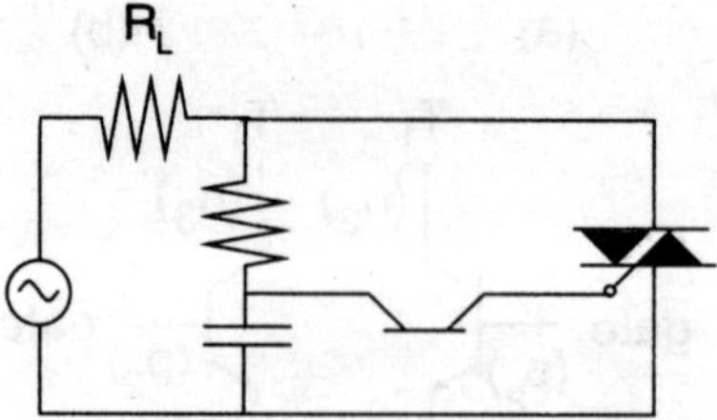

Figure 18.26: *Circuit diagram showing the usage of a diac in the firing circuit of a triac.*

4) Turn off is when V_{ac} across triac goes zero, making demands of turn-off process more
   precise.

The triac's firing in both directions is not symmetrical. Hence, the any ac circuit using triacs
would have large harmonics. The nature of functioning of switching devices results in deviation
from pure sine wave and hence harmonics. However, the amount of harmonics is significantly
reduced if the there is symmetry in the +ive and -ive cycle of the output. This is possible if the
triac is fired symmetrically. Since diacs breakdown at identical ($\pm$) potentials, if a diac is used
in the triacs firing circuit (fig 18.26), symmetrical output is obtainable. Diacs in such circuits
are similar to two zener diodes connected back to back.

18.8.3   Commutation

Commutation is the process of turning off a SCR, where additional components are used to cause
the transfer of current flowing through the SCR into other parts of the circuit, thus reducing
the current through the SCR to below it's holding limit. This forces the SCR into its forward
blocking state. Commutation based on this mechanism is called current commutation. A second
alternative would be to reverse bias the SCR and forcing it into its off state. This is called voltage
commutation. You will see common use of such commutations circuits in choppers and inverter
circuits, where an AC input is not being used.

Many commutation circuits have been developed and the objective of all the circuits is to
reduce the current in the SCR below holding current after a predetermined time.

Classification of the commutation circuits are done on the basis of arrangement of commutation circuit's components. Commutation circuits usually have a capacitor, an inductor, diodes
and sometimes even thyristors.

Commutation circuits are always used in DC circuits and not in AC circuits. This is because
in the negative cycle the anode is reversed biased and is case of forced commutation/ turn off.

Self Commutation (Class A)

The commutation of the SCR here takes place due to a current in the external circuitry forcing
the net current in SCR to be less than the holding current of the SCR. This type of commutation
is called current commutation. Here the commutating components are L and C which are
connected to the load as shown in fig(18.27).

Due to the LC circuit, an oscillatory current is set up. After a time interval π/ω the current falls to zero after which it is negative for time π/ω. If this interval is greater than t_{off} of the SCR, the SCR is successfully commuted off.

To analyze the circuit, we consider the voltage drop across the thyristor to be very small, since in forward conduction state the resistance of the SCR is near zero. Also, assume that the capacitor is not charged at time t=0.

$$V_s = V_L + V_C$$

$$V_s = L\frac{di}{dt} + \frac{1}{C}\int idt$$

differentiating this equation we have

$$L\frac{d^2i}{dt^2} + i/C = 0$$

$$\frac{d^2i}{dt^2} = -\frac{1}{LC}i$$

solving which we have

$$i = A sin\left(\frac{t}{\sqrt{LC}}\right) \quad : \quad \omega^2 = \frac{1}{LC}$$

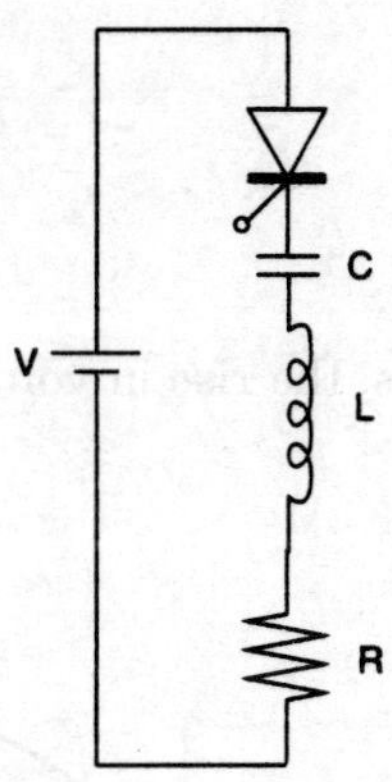

Figure 18.27: *Circuit having a Class A commutation process for switching of the SCR.*

Immediately, it follows that

$$\frac{T}{2} = \frac{\pi}{\omega} = \pi\sqrt{LC}$$

Thus, for successful turn-off of the SCR

$$\pi\sqrt{LC} \geq t_{off}$$

To complete the mathematical exercise, we calculate the constant A, using the knowledge $v_c(t = 0) = 0$ and hence at t=0 all the voltage drop is across the inductor, hence

$$v_L(t = 0) = L\frac{A}{\sqrt{LC}}cos\left(\frac{t}{\sqrt{LC}}\right)\Big|_{t=0} = V_s$$

Hence,

$$A = V_s\sqrt{\frac{C}{L}} : \quad Z_{LC} = \sqrt{\frac{L}{C}}$$

The voltage drop across the capacitor is given

$$v_c(t) = \frac{1}{C}\int idt = \frac{V_s}{Z_{LC}C}\int sin\frac{t}{\sqrt{LC}}dt$$

$$v_c(t) = -\frac{V_s}{Z_{LC}}\sqrt{\frac{L}{C}}\,cos\frac{t}{\sqrt{LC}} + k$$

or

$$v_c(t) = -V_s cos\frac{t}{\sqrt{LC}} + k$$

$$v_c(0) = 0 = -V_s + k: \quad k = V_s$$

Thus, the rise in voltage across the capacitor is given as

$$v_c(t) = V_s\left(1 - cos\frac{t}{\sqrt{LC}}\right)$$

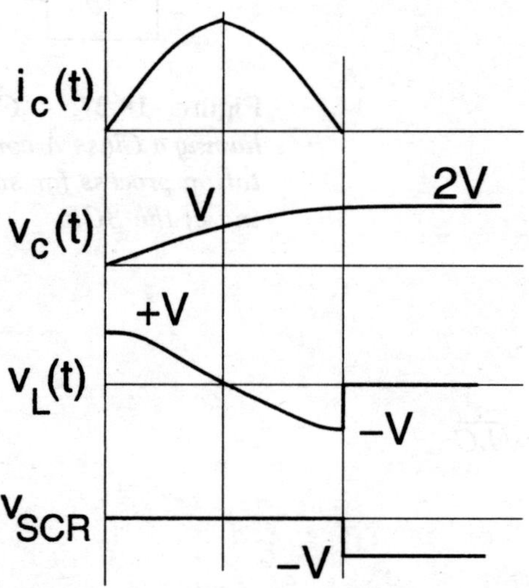

Figure 18.28: *Some important currents and voltages during Class A commutation process.*

i.e. when the SCR is triggered, a current flows through the LC circuit charging the capacitor C towards saturation voltage after which the current reverses and tries to flow through the SCR in opposite direction owing to the resonating effect of L and C. As a result of this the SCR is commutated off with current bring less then that of the holding value.

After $t = \pi\sqrt{LC}$, the voltage across the capacitor becomes $2V_s$. The variation is some important currents and voltages are given in fig 18.28. Note that on commutation $-V_s$ volts appear across the SCR. Also, at first trigger, $v_c(t = 0)$ which on successful commutation becomes $v_c(t = \pi\sqrt{LC}) = 2V_s$. Thus on next firing, since capacitor has not discharged, the voltage across the SCR is $-V_s$ volts, and hence it is incapable of conducting. The capacitor is discharged by placing a parallel R. This arrangement is called shunt capacitor. A disadvantage of this circuit is limited frequency to which circuit maybe used (capacitor has to be discharged before next trigger).

In case of R being present the circuit will be oscillatory but value of current will be dampened. The LCR circuit must be under-damped.

Class B (Another Current Commutation Circuit)

The circuit shown below (fig 18.29) is an example of a Class B commutation circuit. Initially the SCR is off. Hence the capacitor charges to a voltage E. When the SCR is triggered, current will flow through the load resistance R_L. Simultaneously, the capacitor discharges through SCR. After total discharge, the capacitor will charge to reverse polarity (path shown in fig 18.29). The

SCR, capacitor and inductor (L) form a resonant circuit. The oscillatory current will reverse direction of conduction and hence the current through the SCR would be $[I_L - i_c(t)]$. When this is equal or less than the holding current, the SCR will be turned off. Now to analyze the circuit we write the loop eqn

$$E = L\frac{di}{dt} + \frac{1}{C}\int i\,dt + Ri$$

let us consider that the load is very small, on differentiating we have

$$L\frac{d^2i}{dt^2} + \frac{i}{C} = 0$$

$$\frac{d^2i}{dt^2} = -\frac{1}{LC}i$$

which on solving gives

$$i = A\sin\left(\frac{t}{\sqrt{LC}}\right) \quad : \quad \omega^2 = \frac{1}{LC}$$

To compute the constant A, we find the voltage drop across the capacitor

$$v_c(t) = \frac{1}{C}\int i\,dt \quad = \quad -A\sqrt{\frac{L}{C}}\cos\frac{t}{\sqrt{LC}} + k$$

The boundary condition here is given as

$$v_c(t = 0) = -E$$

and

$$v_c(t = \pi\sqrt{LC}) = +E$$

Hence, we get the following simultaneous equation

$$-E = -A\sqrt{\frac{L}{C}} + k$$

and

$$+E = A\sqrt{\frac{L}{C}} + k$$

This gives k=0 and

$$A = \sqrt{\frac{C}{L}}E$$

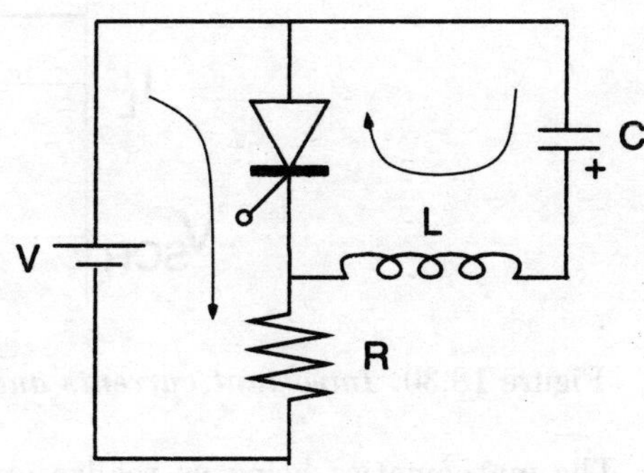

Figure 18.29: *A circuit which is an example for Class B commutation process.*

Hence

$$v_c(t) = -Ecos\frac{t}{\sqrt{LC}}$$

The oscillatory current is given as

$$i_c(t) = \sqrt{\frac{C}{L}}Esin\left(\frac{t}{\sqrt{LC}}\right)$$

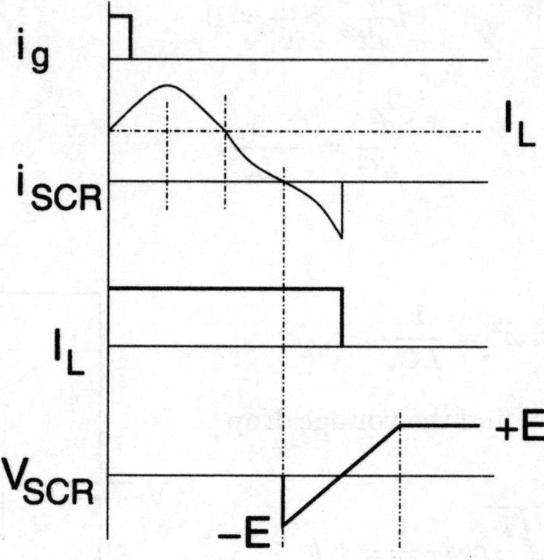

Figure 18.30: *Important currents and voltages during commutation process in Class B.*

The mathematics helps us realize and picturise the current and voltage levels at various instant of time during the process of commutation using Class B circuits (see fig 18.30). For worst design case, commutation should take place when the peak current appears, i.e.

$$\sqrt{\frac{C}{L}}E \geq \frac{E}{R}$$

An important aspect of this circuit is that in this circuit the load current does not flow through the commutating circuit and hence the rating and inturn cost of commutating circuit is low.

Class C Circuits

Class C & D are cases of voltage commutation and not current commutation as were the case in A & B, as can be seen in the following passages, the capacitor does not discharge through the SCR [i.e. current in SCR is not $I_L - i_c(t)$]. All it does is to reverse bias the SCR.

The circuit for a class C commutating circuit is shown in fig(18.31). Here an complimentary SCR is turned ON to commutate the first SCR (SCR$_1$), note both SCR's are load carrying and hence the termology main and auxiliary is not used.

When SCR_2 is conducting and SCR_1 is in its blocking state, the capacitor C is being charged by the current flowing through resistance R_1. Net current through SCR_2 is the sum of current through R_1 and R_2. After the capacitor is fully charged, the only current flowing is through R_2, and is given as V_{dc}/R_2. The polarity across the capacitor is as shown and is retained.

Now, for commutating, if SCR_1's gate is triggered, then the path ineffect looks as if the capacitor is in parallel connection to SCR_2. The current paths are shown in fig 18.32. The capacitor immediately reverse biases SCR_2 leading to its switching "OFF". The capacitor is charged to the reverse polarity through R_2 and SCR_2, after which the process repeats itself.

The total current in SCR_2 is $i_1 + i_2$. The current in capacitor is i_1 and can be computed from the solution of the equation below:

$$V_{dc} = i_1 R_1 + \frac{1}{C} \int i_1 dt$$

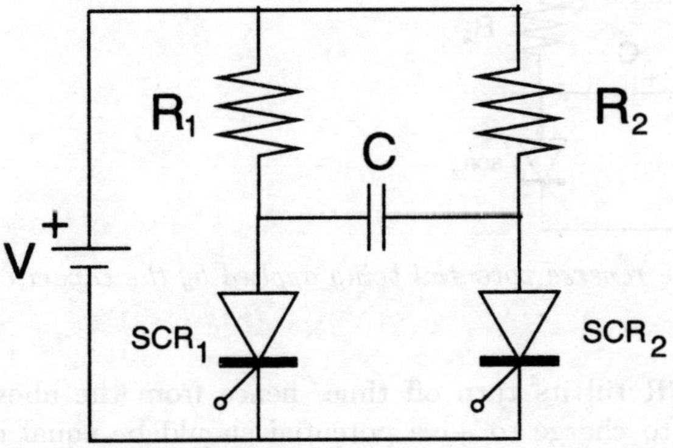

$$i_1 = A e^{\frac{-t}{R_1 C}}$$

The voltage across the capacitor is hence given as

$$v_c(t) = \frac{1}{C} \int i_1 dt = \frac{A}{C} \int e^{\frac{-t}{R_1 C}} dt$$

$$v_c(t) = -A R_1 e^{\frac{-t}{R_1 C}} + constt$$

considering the steady state where in the previous cycle the capacitor was already charged to the maximum reverse polarity,

Figure 18.31: *Example of an class C commutation circuit.*

$$v_c(t = 0) = -V_{dc} = -A R_1 + constt$$

Also, as 't' tends to infinity

$$v_c(t = \infty) = V_{dc} = constt$$

Thus,

$$A = \frac{2V_{dc}}{R_1}$$

giving the final expressions as

$$i_1(t) = \frac{2V_{dc}}{R_1} e^{\frac{-t}{R_1 C}} \tag{18.22}$$

$$v_c(t) = V_{dc} \left(1 - 2e^{\frac{-t}{R_1 C}} \right)$$

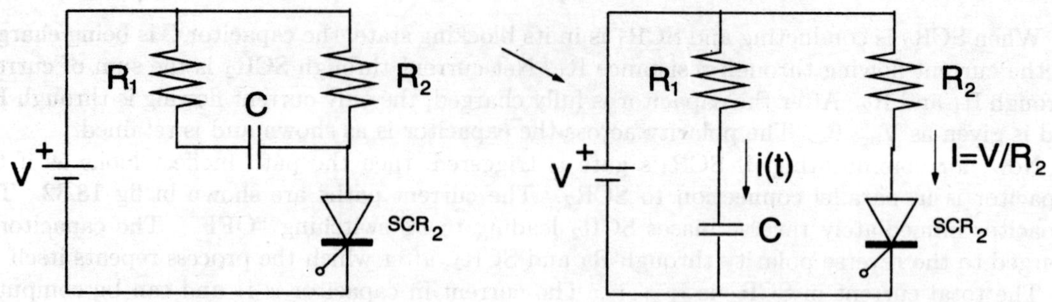

Figure 18.32: *Current path during the various modes in class C commutation.*

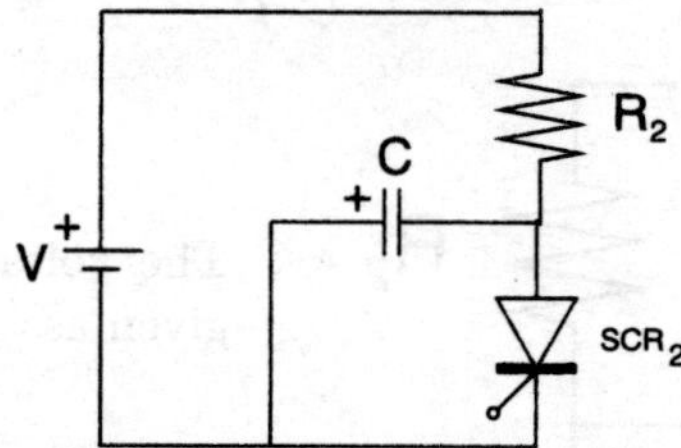

Figure 18.33: *Commutation of SCR_2 due to the reverse potential being applied by the capacitor.*

The capacitor should reverse bias the SCR till its turn off time, hence from the above expression, the time taken for the capacitor to charge to +ive potential should be equal or greater then t_{off}

$$0 = V_{dc}\left(1 - 2e^{\frac{-t_{off}}{R_1 C}}\right)$$

inturn

$$t_{off} = R_1 C Ln(2)$$

The current in SCR_2 as soon as SCR_1 switched OFF is given using eq(18.22)

$$i(t=0) = i_1(t=0) + i_2 = \frac{2V_{dc}}{R_1} + \frac{V_{dc}}{R_2}$$

The waveform for various circuit elements is as shown in fig(18.34). This circuit also finds use as static switches.

Class D (An Auxiliary SCR Switching a Charged Capacitor)

In such circuits (fig 18.35) there are two SCR's, of which only the main SCR carries the load current. The second SCR doesn't carry any load current and hence can be of lower rating. Initially, the capacitor is charged to voltage E with polarity shown in fig 18.35. Now, the main

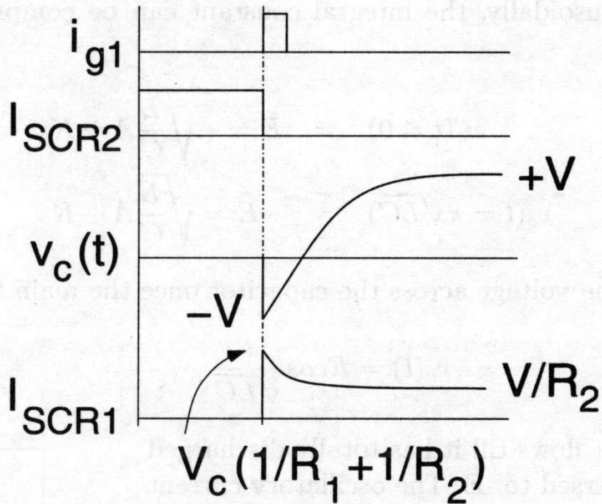

Figure 18.34: *The important currents and voltages during the commutation process of circuit shown in fig 18.31. This diagram is for the initial stage. Under steady state operation The current jumps to* $V_s \left(\frac{2}{R_1} + \frac{1}{R_2} \right)$ *and saturates at* $\left(\frac{V_s}{R_2} \right)$*.* **Observe difference between initial stage and steady stage.**

SCR (T_m) is turned on and hence voltage E is applied across load with a load current E/R_L flowing through it. As T_m is conducting, an oscillatory current is also set up because of the charged capacitor, which takes the path C, T_m, L, D and C. The auxiliary SCR is not in the picture at present. The current directions are shown in fig 18.36.

The net current through the main SCR is $I_L + i_c(t)$. The capacitor charges from +E to -E due to the oscillatory current. After completely attaining -E, the oscillatory current stops, due to the lack of path since the diode is now reverse biased. To commutate the main SCR, the auxiliary SCR is fired. This connects the capacitor in parallel to T_m and hence reverse biases it, thus, turning it off. Now the load current flows through the capacitor and T_a. The capacitor gets charged from -E to +E and since current goes zero, commutates T_a.

Mathematical Analysis: Class D

Assume that initially the capacitor is charged with +V (or +E) potential and as soon as the main SCR is fired, two currents flow in the main SCR, (i) the load current ($I_o = E/R$) and (ii) the capacitor current. The capacitor current is essentially a discharge current flowing in the tank circuit (LC circuit). The current obviously is oscillatory, i.e.

$$i_c(t) = A sin \frac{t}{\sqrt{LC}}$$

The voltage across the capacitor is given as

$$v_c(t) = \frac{1}{C} \int i_c(t) dt = -\sqrt{\frac{L}{C}} A cos \frac{t}{\sqrt{LC}} + k$$

Since, current varies sinusoidally, the integral constant can be computed from the boundary conditions

$$v_c(t=0) \; = \; E = -\sqrt{\frac{L}{C}}A + K$$

$$v_c(t=\pi\sqrt{LC}) \; = \; -E = \sqrt{\frac{L}{C}}A + K$$

giving the variation of the voltage across the capacitor once the main SCR is fired as

$$v_c(t) = E cos \frac{t}{\sqrt{LC}}$$

The capacitor current flows till it has totally discharged and it's polarity has reversed to -E. The oscillatory current is prevented to reverse direction because of the diode. The capacitor hence, holds -E, till the auxiliary SCR is fired. on firing the capacitor potential goes from -E to +E, with the charging taking place due to flow of current from source, through capacitor and resistance. That is from this charging action, the loop equation is

$$E = \frac{1}{C} \int i dt + Ri + v_c(t=0)(= -E)$$

The capacitor charges in the opposite direction on completion of charging, i.e.

$$v_c(t=\infty) = E$$

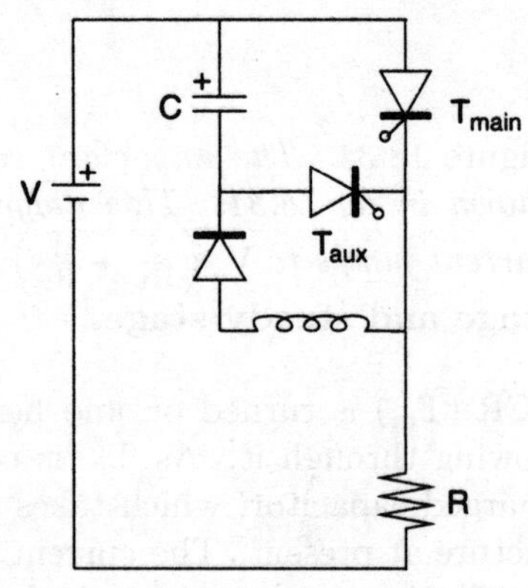

Figure 18.35: *Example of an class D commutation circuit.*

The charging capacitor current and voltage, hence is given as

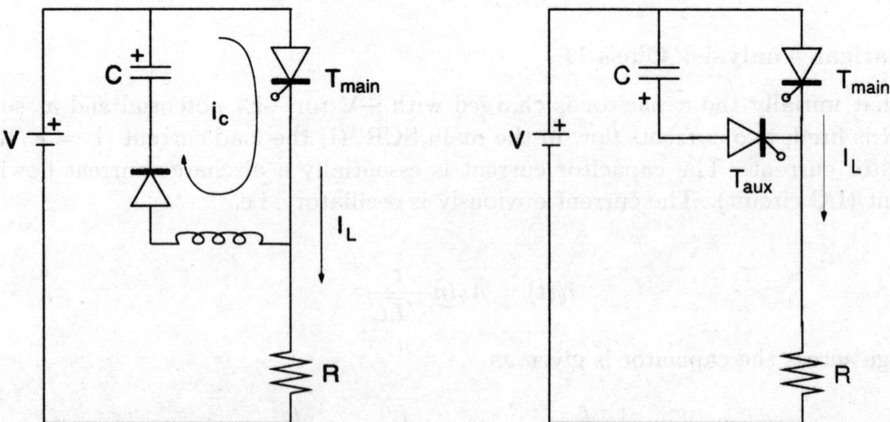

Figure 18.36: *Current flow during various modes of an class D commutation circuit.*

$$i_c(t) \;=\; \frac{2E}{R}e^{\frac{-t}{RC}} = 2I_o e^{\frac{-t}{RC}} \tag{18.23}$$

$$v_c(t) \;=\; 2E\left(\frac{1}{2} - e^{\frac{-t}{RC}}\right) \tag{18.24}$$

where I_o is the load current in the circuit.

Designing of Commutation elements

The value of 'C' to be selected for a given load (R) so that proper commutation takes place is obtained from eqn(18.24). For proper commutation, the capacitor should reverse bias, i.e. the capacitor's polarity should be negative for a time equal or greater then the turn-off time of the main SCR. Hence,

$$t_{off} = RCln(2)$$

For the case that a very large load is used in our Class D example, then eqn(18.23) reduces to $2I_o$. This possibility is shown as dotted line, instead of the exponential falling line, in the waveform for load current in fig(18.37). Since, the current is a constant with respect to time, from

$$v_c(t) = \frac{1}{C}\int i\, dt$$

the charging action of the capacitor gives a linear change of the capacitor voltage, i.e.

$$v_c(t) = \frac{2I_o t}{C}$$

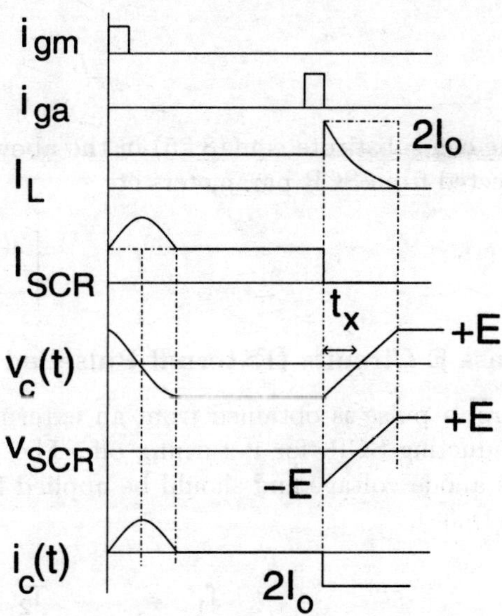

Figure 18.37: *Important currents and voltages during commutation process in Class D.*

In the time t_{off} (of the main SCR), the voltage across the capacitor varied from -E to zero. Also, $2I_o$ should be within the maximum permissible current limit of the auxiliary SCR. Hence, the proper commutation capacitor to be selected would be

$$C \geq \frac{2(t_{off})_{main}(I_{max})_{aux}}{E} \tag{18.25}$$

Similarly, the size of the inductor can be found from the maximum current limitation imposed by the main SCR. The largest current in the main SCR is given as

$$I_{main} = I_{osc} + I_o$$

where I_{osc} is the peak current due to the tank circuit ($A\sin\omega t$). Substituting this in the above equation, we have

$$I_{main} = \sqrt{\frac{L}{C}}E + I_o$$

Assuming the load current negligible as compared to the peak oscillatory current, then

$$(I_{max})_{main} \geq \sqrt{\frac{L}{C}}E$$

$$(I_{max})^2_{main} \geq \frac{L}{C}E^2$$

Thus, once the capacitor is selected, the size of the inductor would be given as

$$L \geq \left[\frac{E}{(I_{max})_{main}}\right]^2 C$$

One can substitute eqn(18.25) in the above equation to get an expression of the inductor to be selected from SCR parameters etc

$$L \geq \left[\frac{2(t_{off})_{main}(I_{max})_{aux}E}{(I_{max})^2_{main}}\right]$$

Class E Circuits (External Pulse led Commutation)

Here, a pulse is obtained from an external voltage source and applied to the cathode of the conducting SCR, for it turning off. The peak amplitude of the voltage must be greater than the anode voltage and should be applied for a period greater than t_{off}. Consider the circuit of fig(18.38),

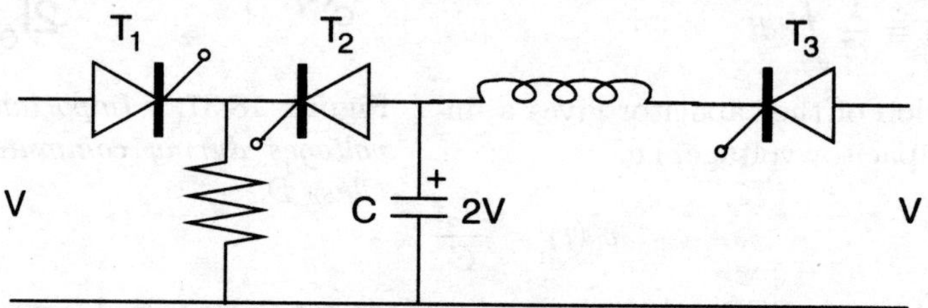

Figure 18.38: *An example of Class E commutation circuit.*

Initially T_1 and T_3 are conducting while T_2 is off. T_1 feeds the load resistance and T_3 has a Class A commutation circuit which leads to the charging of the capacitor to voltage 2E. Now, when commutation has to be done, T_2 is fired. This bring 2E at the cathode of T_1 while the anode is still at E. Thus, reverse biasing the SCR. If R is large, such that discharge is slow, then time for which SCR is reverse biased is greater than the SCR's turn-off time, hence successfully commutating the SCR.

Another method of commutating an SCR using external pulse would have a transformer's primary connected in series with the load. When switching off has to be done, a voltage pulse at the secondary is given such that the voltage across the primary (this would be the potential at the SCR's cathode) is greater than voltage appearing at the anode. This successfully commutates the SCR.

Class F (Line Commutation)

The automatic switching of of SCR's in AC circuits, on the appearance of the negative cycle, is called **line commutation** or **natural commutation**. The commutation is controlled by the line frequency. This method of commutation is important in phase controlled rectification circuits and is classified as Class F commutation.

18.9 Useful Circuits with SCR

The following section gives a very brief introduction to applications of SCR. These are by no means an exhaustive collection of SCR circuits but just being mentioned to give readers a favor of the maturing field called "Power Electronics".

18.9.1 Phase Control Rectifiers

Remember the expression for DC voltage obtained from an half wave rectifier circuit using a 'pn' diode (Chapter 5) which was obtained by solving the integral

$$V_{dc} = \frac{1}{2\pi} \int_0^\pi V_{max} sin\theta d\theta$$

which gave

$$V_{dc} = \frac{V_{max}}{\pi} \tag{18.26}$$

where V_{max} is the maximum/peak voltage. The DC output thus obtained is fixed. If you

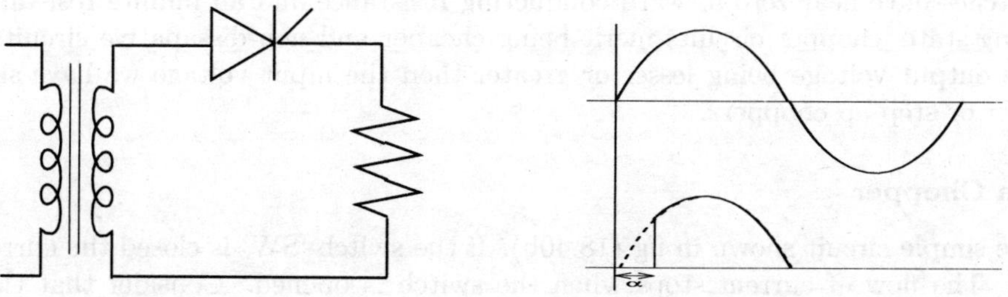

Figure 18.39: *Circuit for half-wave controlled rectifier.*

want to vary the DC output voltage as any practical circuit would be required to provide, you would have to vary the input source V_{max}. At home, on a 220v line this would prove difficult. However, by replacing the diode, which is simply a switch, with an SCR (a controllable switch), the output dc voltage can be varied. The variation can be from zero to a maximum that is same as that of a diode circuit.

Consider the circuit shown in fig (18.39). The diode of the half wave rectifier circuit has been replaced with an SCR. The variation in output voltage is shown in the figure. It can be noticed that the output is only present (i.e. at $t > 0$) when the SCR is fired. The firing ofcourse only results in conduction when SCR is in it's forward conducting state (that explains the $t > 0$ condition). As soon as the SCR is reverse biased, conduction stops.[3] Thus, the average area under the curve, inturn the DC volatge is now

$$V_{dc} = \frac{1}{2\pi} \int_{\alpha}^{\pi} V_{max} sin\theta d\theta \qquad (18.27)$$

α is called the firing angle.[4] The final expression is

$$V_{dc} = \frac{V_{max}(1 + cos\alpha)}{\pi} \qquad (18.28)$$

As can be seen the output DC voltage can be varied from V_{max}/π to 0 for $\alpha \to 0$ to π respectively.

18.9.2   Choppers

If you want a variable DC supply from a given DC source, you can do the following:

a) Use a potential divider. However, since current always flows in the potential divider, the method is not an economical method.

b) Use an inverter circuit to convert the DC to AC and then feed the AC output to a step up/ step down transformer. The new AC level then may be converted to required DC level using a phase controlled rectifier. As is understood from the description, large amount of circuital elements are required, hence making this option expensive.

c) Using choppers, which are nothing but switching circuits made of devices like SCR etc. The section below discusses how choppers achieve variable DC levels. Note since SCRs as switches have near zero forward conducting resistance and an infinite resistance in its blocking state, chopper circuits merit being cheaper and non-dissapative circuits. Based on the output voltage being lesser or greater then the input voltage we have step-down chopper or step up choppers.

Step Down Chopper

Consider the simple circuit shown in fig (18.40b). If the switch 'SW' is closed the current flows to the load. The flow of current stops when the switch is opened. Consider that the switch remains open for time t_2 (sec) and closed for t_1 (sec), then the total time of a single cycle is given as

$$T = t_1 + t_2$$

The average output voltage can then be calculated as

---

[3] Note we have only got a resistive load

[4] Let us say that the firing of the SCR takes place at the instance t_1. Then α is given as $\frac{2\pi}{T}t_1$. A firing angle may thus be defined as the angle between the instance SCR would conduct had it been a diode and the instant it is triggered.

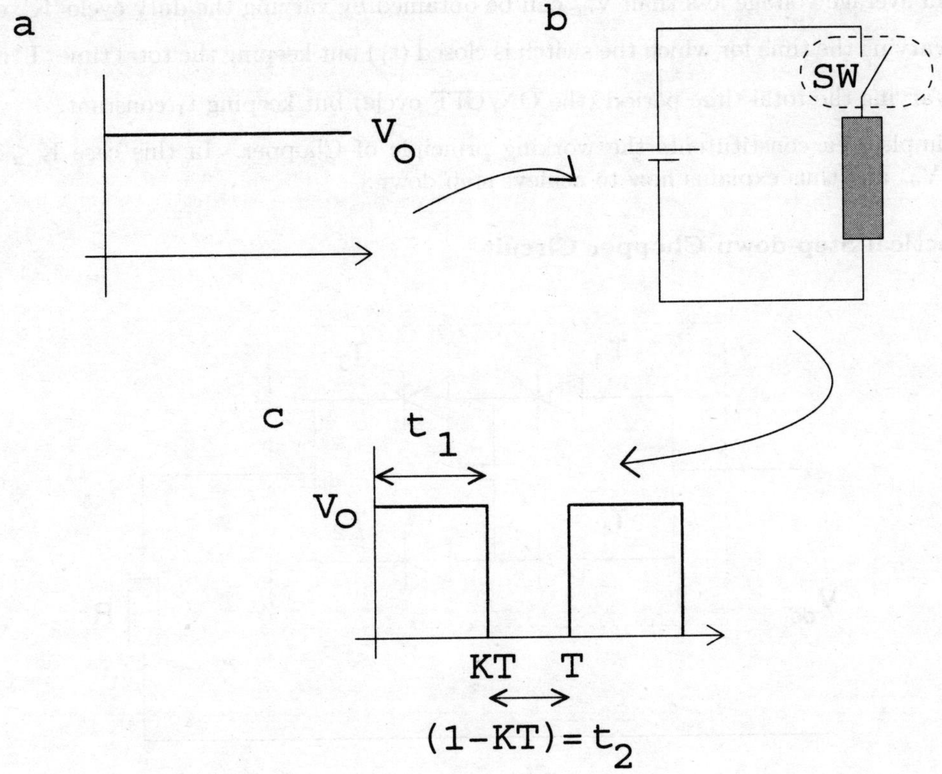

Figure 18.40: *The (a) DC input waveform, using a chopper in a simple circuit shown in (b) results in a modified output voltage at the load as shown in (c). The average output is a DC whose magnitude is less than the source voltage. Boxed item shows the chopper switch which is nothing but a high speed semiconductor switch.*

$$V_{av} = \frac{1}{T} \int_0^T V_{dc} dt$$

with two clearly demarkated regions of output (see fig 18.40c) the integral can be expanded to

$$V_{av} = \frac{1}{T} \left[\int_0^{t_1} V_{dc} dt + \int_{t_1}^{t_2} 0 dt \right]$$

which gives the average output voltage as

$$V_{av} = \frac{t_1}{T} V_{dc}$$

The ration t_1/T is called the the duty cycle (K). The above expression can be represented as

$$V_{av} = KV_{dc} \qquad (18.29)$$

Thus an average voltage less than V_{dc} can be obtained by varying the duty cycle 'K', either by

(i) Varying the time for which the switch is closed (t_1) but keeping the total time 'T' constant.

(ii) Varying the total time period (the ON/OFF cycle) but keeping t_1 constant.

This simple idea constitutents the working principle of Chopper. In this case $K \leq 1$, hence $V_{av} < V_{dc}$ and thus explains how to achieve step down.

A Practical Step-down Chopper Circuit

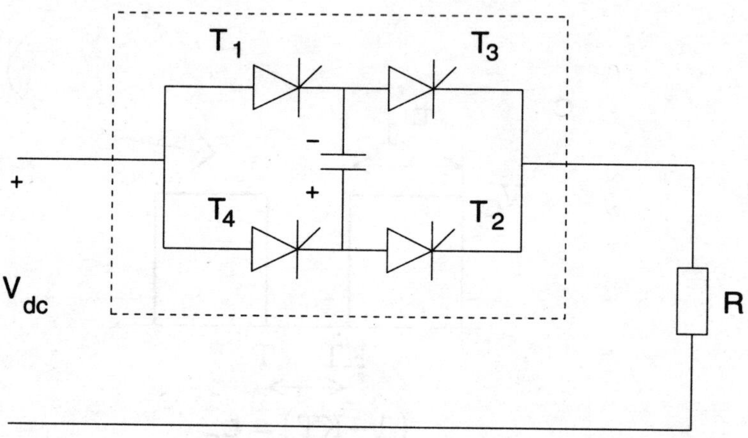

Figure 18.41: *The circuit of a load commutation chopper. Boxed item shows the chopper switch.*

This is one of the simplest chopper circuit using four SCRs and a capacitor. The circuital elements shown enclosed in the box (fig 18.41) is the chopper switch. Alternate SCRs are triggered simultaneously, i.e. either T_1 and T_2 or T_3 and T_4 are sent into conduction together. Assuming the load to be purely resistive (this assumption is made to simplify the mathematical computation), when any pair of SCRs are fired the current flows through the capacitor and resistance (see fig 18.42). As the current flows and the capacitor saturates, the current falls zero and naturally commutes (switches off) the two SCRs. Thus, the time taken for the current to fall zero and commutation takes place depends on the size of the load (R). Hence, the circuit is an example of **load commutation**. When SCRs T_1 and T_2 are fired, the current flows as shown in fig 18.42. Since in the last cycle, the capacitor would have charged to V_{DC} with the polarity shown, the load voltage at instant t=0 would be $2V_{DC}$. The variation in capacitor current is given by the solution of a first order differential equation and is of the form

$$i_c(t) = Ae^{-t/RC}$$

Since, at t=0, the capacitor would not offer any impedance, the current would be equal to $2V_{DC}/R$. Hence,

$$A = \frac{2V_{DC}}{R}$$

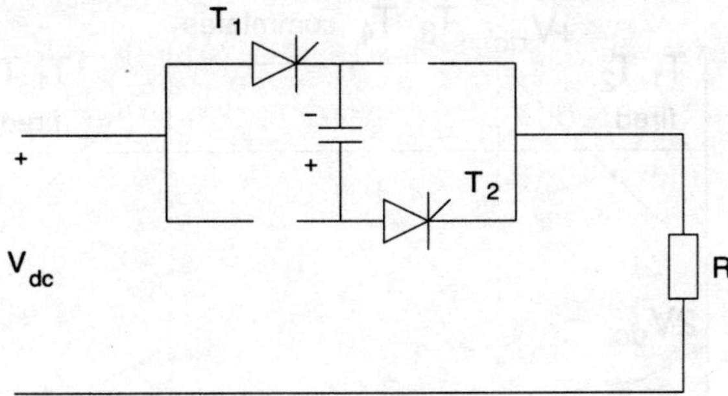

Figure 18.42: *The circuit as seen at the start of cycle, when T_1 and T_2 are fired.*

Thus, the variation of the capacitor current on firing SCRs T_1 and T_2 would be

$$i_c(t) = \frac{2V_{DC}}{R} \times e^{-t/RC}$$

We are now intersted in calculating the instant when the SCRs would commute. The above expression is of no use, since i_c goes to zero at infinity. Hence, we calculate the variation in capacitor voltage with time. The current in capacitor goes zero when it is completely charged, i.e in our case when the capcitor charges from $-V_{DC}$ to $+V_{DC}$. The capacitor voltage is given as

$$v_c(t) = \frac{1}{C}\int i\,dt \;=\; \frac{2V_{DC}}{RC} \times e^{-t/RC}$$
$$= \; -2V_{DC}e^{-t/RC} + K$$

The initial condition gives

$$-V_{DC} \;=\; -2V_{DC} + K$$
$$k \;=\; V_{DC}$$

Hence, the voltage across the capacitor varies as

$$v_c(t) = V_{DC}(1 - 2e^{-t/RC}) \tag{18.30}$$

The load current's variation would be given as

$$v_R = Ri_c(t) = 2V_{DC}e^{-t/RC} \tag{18.31}$$

If RC is large, eqn(18.30) and eqn(18.31) can be written as

$$v_c(t) \;=\; V_{DC}\left[1 - 2\left(1 - \frac{-t}{RC}\right)\right] = V_{DC}\left(\frac{t}{RC} - 1\right)$$
$$v_R \;=\; 2V_{DC}\left(1 - \frac{t}{RC}\right)$$

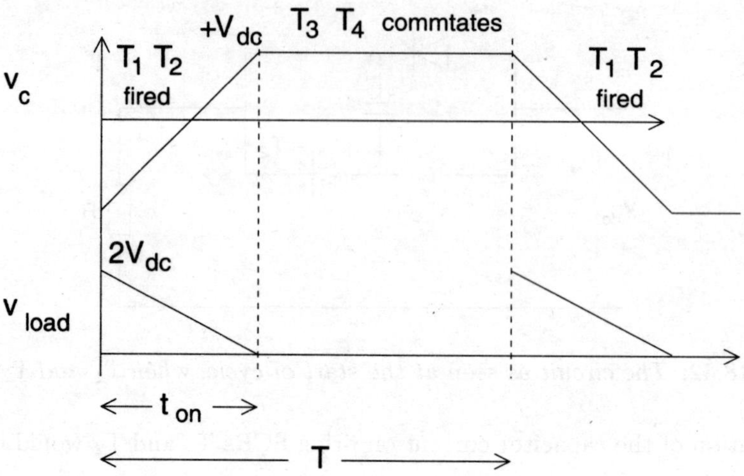

Figure 18.43: *The waveforms associated with a load commutation chopper.*

The linear variation in capacitor and load voltage is shown in fig(18.43). The average output voltage would be given as

$$
\begin{aligned}
V_{av} &= \frac{Area}{T} \\
&= \frac{1}{T}\left(\frac{1}{2} \times 2V_{DC} \times t_{on}\right) \\
&= \frac{V_{DC}t_{on}}{T}
\end{aligned}
\tag{18.32}
$$

Since, t_{on} depends on the RC combination, the variation of the average DC output is usually done by varying the total time period 'T'. The circuit is popular since no inductor is required and hence it can work at high frequencies with the second set of SCRs capable of being fired as soon as the capacitor has flipped its polarity.

For designing the commutation circuit, remember the maximum current in the circuit is just when the SCRs are switched ON and given as $2V_{DC}/R$. This should be less than the maximum current rating of the SCR, from which resistance in the circuit for a given SCR is decided as

$$
R_{min} = \frac{2V_{DC}}{I_{max}}
$$

The size of the capacitor to be selected is decided based on the maximum chopping frequency required

$$
f_{max} = \frac{1}{T_{min}}
$$

but $T \rightarrow T_{min}$ when $t_{on} + t_{off} \rightarrow t_{on}$. That is, to minimise the time period, T, is made equal to

t_{on}. Hence, the maximum chopping frequency (using fig 18.43) works out to

$$I_o = C\frac{dv}{dt} = \left| C\left(\frac{-V_{DC} - (-V_{DC})}{0 - t_{on}}\right) \right|$$

giving

$$\frac{1}{t_{on}} = \left(\frac{I_o}{2V_{DC}C}\right)$$

I_o is a DC current, based on the assumption of linear variation in capacitor voltage. Hence, if on designing, the maximum frequency required is f_{max}, then

$$f_{max} = \frac{1}{t_{on}} = \frac{I_o}{2V_{DC}C}$$

demanding

$$C = \frac{1}{t_{on}} = \frac{I_o}{2V_{DC}f_{max}}$$

18.9.3   Inverters

Inverters are circuits that convert DC power into AC power. For practical application the circuit is designed to deliver output AC either at fixed/constant voltage with variable frequency or variable voltage at fixed frequency. If required inverters can be designed to vary both. A variable output voltage can be obtained by

1. Varying the input voltage (DC level), and maintaining the gain of the voltage constant where the gain of the inverter is defined as

$$gain = \frac{(AC)\ output\ voltage}{(DC)\ input\ voltage}$$

2. Gain is varied, but the input is kept constant.

Below we study a simple circuit that would help us understand the basic principle behind inverter circuits.

The output across the load in a half-bridge inverter shown in fig18.44 is controlled by the firing of SCR_1 and SCR_2. When SCR_1 is fired, SCR_2 is in forward blocking state and hence doesn't conduct. The circuit hence effectively consists of the loop V/2, the conducting SCR_1 and the load. The drop across the SCR is negligible, hence all the source appears across the load with the polarity as shown in the fig(18.44). The second stage is the commutation of the conducting SCR and the firing of SCR_2. The result is that the polarity of the output changes. Thus, the output varies from +V/2 to -V/2. The output voltage is also shown in fig(18.44). The frequency of the AC output is controlled by the time period for which the SCR's are fired. It should be noted that in half-bridge inverters the two SCRs should not be conducting simultaneously. If so this would lead to shorting the power supply, with no current passing through the load (no output).

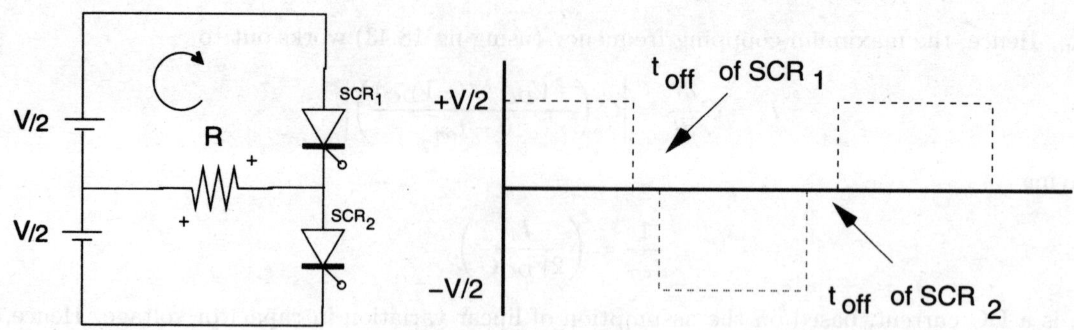

Figure 18.44: *A half bridge inverter with resistive load. Note the triggering circuits and turn-off circuits is not shown. Also shown is the output waveform as recieved at the resistance.*

Output voltage waveform of practical inverters are invariably non-sinusodial and hence contains harmonics. For low and medium power requirements this is acceptable. However, for high power applications only low distorted sinusodial waveform is accepted. High speed switching reduces the harmonics, thus calling for fast SCRs.

Practical Inverter Circuit: A Series Inverter

In this inverter circuit the commutating elements L and C are connected in series with the load, or in other words the load consists of RLC resonant circuit as in the Class A commutation circuit discussed earlier. SCR_1 and SCR_2 are alternatively turned "ON" by applying a triggering pulse to the gates depending on the frequency required by the output circuit. When SCR_1 is "ON", for t= T/2 (sec), current flows through the RLC circuit.

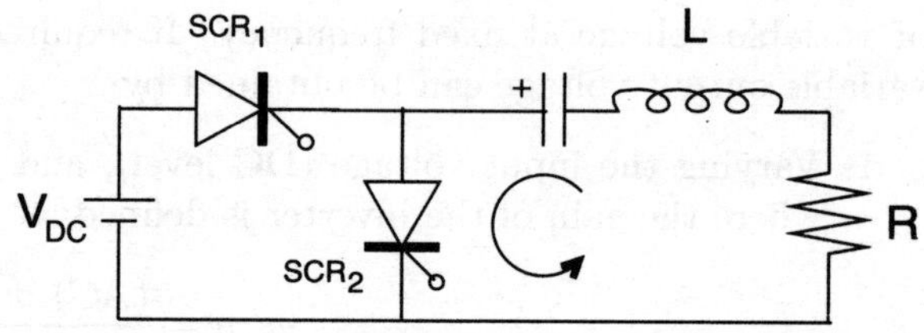

Figure 18.45: *A series inverter made using two SCR.*

cuit. The current varies sinusodially and falls to zero (time period decided by the RLC). Since the current through SCR_1 is below its holding current, the thyristor will commute off. At this point the capacitor has charged to its maximum potential with the shown polarity.

Now, after an interval of t_{off} of SCR_1, SCR_2 is fired. The RLC along with this SCR also forms an resonant circuit. The capacitor discharges in the path shown. As can be seen from fig 18.46, the load current reverses its direction and hence the output voltage is an AC. After the capacitor has fully discharged and charged with opposite polarity, the current falls to zero, commutating SCR_2. SCR_1 and SCR_2 should never be fired simultaneously of before the elapse of t_{off}, since this would only short the battery. The output waveform can be seen in fig(18.46a), from which the AC frequency is trivially understood to be

$$f = \frac{1}{2\left(\frac{T}{2} + t_{off}\right)}$$

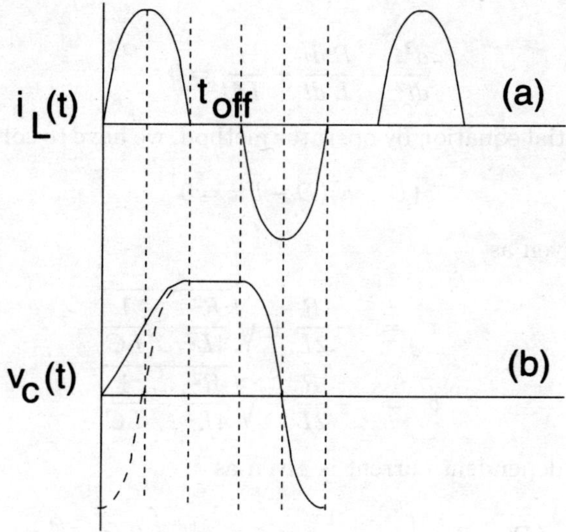

Figure 18.46: *Waveforms associated with the series inverter.*

$$= \frac{1}{T + 2t_{off}}$$

where 'T/2' is the time period of the half time period of the sine current and inturn the load (across R) voltage. Re-arranging the above equation to write interms of the resonant frequency (RLC), we have

$$f = \frac{1/T}{1 + 2\left(\frac{t_{off}}{T}\right)}$$

$$f = \frac{f_o}{1 + 2t_{off}f_o}$$

A limitation of this circuit is evident from this relationship. As can be seen the maximum operating frequency will always be less then the circuit's resonant frequency.

Capacitor Current and Voltage

A complete understanding of the circuit can be gained by analysing the circuit mathematically. The loop equation for this would lead to a second order differential equation

$$V_{DC} = Ri + L\frac{di}{dt} + \frac{1}{C}\int idt + V_c(t = 0) \tag{18.33}$$

which is evident on differentiating the above equation

$$L\frac{d^2i}{dt^2} + R\frac{di}{dt} + \frac{i}{C} = 0$$

On re-arranging

$$\frac{d^2i}{dt^2} + \frac{R}{L}\frac{di}{dt} + \frac{i}{LC} = 0$$

For solving this differential equation by operator method, we have to solve the quadratic equation

$$(D - a)(D - b)i = 0$$

where 'a' and 'b' are given as

$$a = -\frac{R}{2L} + \sqrt{\frac{R^2}{4L^2} - \frac{1}{LC}}$$

$$b = -\frac{R}{2L} - \sqrt{\frac{R^2}{4L^2} - \frac{1}{LC}} \tag{18.34}$$

respectively. The time dependent current is given as

$$(D - a)i = \frac{1}{D - b} \times 0 = e^{bt}\int 0 \times e^{-bt}dt$$

$$i(t) = \frac{A}{D - a} \times e^{bt} = Ae^{at}\int e^{(b-a)t}dt$$

$$i(t) = \frac{A}{b - a} \times e^{bt} + k$$

Subsituting

$$\omega = b - a = \sqrt{\frac{1}{LC} - \frac{R^2}{4L^2}}$$

we have

$$i(t) = \frac{-Ae^{\frac{-Rt}{2L}}}{2i\omega} \times e^{-i\omega t} + k$$

taking the real part

$$i(t) = \frac{Ae^{\frac{-Rt}{2L}}}{2\omega} \times sin(\omega t) + k \tag{18.35}$$

with the initial condition i(t=0)=0, k=0, the task remains to find the constant A. At t=0, the voltage drops only across the inductor and is given as

$$V_{DC} - v_c(t = 0) = L\left(\frac{di}{dt}\right)_{t=0}$$

i.e., Ri=0 as at t=0, i=0. Substituting eq(18.35) here we have

$$V_{DC} - v_c(t = 0) = L\frac{A}{2\omega}\left[-\left(\frac{R}{2L}\right)e^{-Rt/2L}sin(\omega t) + \omega cos(\omega t)e^{-Rt/2L}\right]_{t=0}$$

which reduces to

$$V_{DC} - v_c(t = 0) = \frac{AL}{2\omega}\omega$$

on re-arranging

$$A = \frac{2[V_{DC} - v_c(t = 0)]}{L}$$

thus eq(18.35) is

$$i(t) = \left[\frac{V_{DC} - v_c(t = 0)}{\omega L}\right] e^{-Rt/2L} sin(\omega t) \tag{18.36}$$

The exact solution of current hence would depend on the charge across the capacitor at the start of the cycle. While once the circuit is in perpetual oscillation (i.e. steady state) one expects same value of $v_c(t = 0)$ at the onset of each cycle. However, when the circuit is switched "ON" for the first time (initial cycle), capacitor is uncharged and $v_c(t = 0) = 0$.

(I) **Initial Cycle, When SCR$_1$ is Fired:**
  For the initial condition, i.e., when SCR$_1$ is fired for the first time and the capacitor was uncharged, $v_c(t = 0) = 0$, then eqn(18.36) reduces to

$$i(t) = \left[\frac{V_{DC}}{\omega L}\right] e^{-Rt/2L} sin(\omega t) \tag{18.37}$$

Voltage across capacitor after SCR$_1$ commutes off for initial condition

The voltage across the capacitor at any given instance may be represented as

$$v_c(t) = \frac{1}{C} \int i dt$$

initially $v_C(t = 0) = 0$, hence, we can find the variation of voltage across the capacitor with time using eqn(18.36). We have,

$$v_c(t) = \left[\frac{V_{DC} - v_c(t = 0)}{\omega LC}\right] \int e^{\frac{-Rt}{2L}} sin(\omega t) dt$$

The result of the integral is given as

$$
\begin{aligned}
P &= -\left(\frac{1}{\omega}\right) e^{\frac{-Rt}{2L}} cos(\omega t) - \left(\frac{R}{2\omega L}\right) \int e^{\frac{-Rt}{2L}} cos(\omega t) dt \\
&= -\left(\frac{1}{\omega}\right) e^{\frac{-Rt}{2L}} cos(\omega t) - \left(\frac{R}{2\omega L}\right) \left[\frac{1}{\omega} e^{\frac{-Rt}{2L}} sin(\omega t) + \frac{R}{2\omega L} \int e^{\frac{-Rt}{2L}} sin(\omega t) dt\right]
\end{aligned}
$$

or

$$\left(1 + \frac{R^2}{4\omega^2 L^2}\right) P = -\left(\frac{1}{\omega}\right) e^{\frac{-Rt}{2L}} \cos(\omega t) - \left(\frac{R}{2\omega^2 L}\right) e^{\frac{-Rt}{2L}} \sin(\omega t)$$

$$P = -\left(\frac{4\omega^2 L^2}{R^2 + 4\omega^2 L^2}\right) e^{\frac{-Rt}{2L}} \left(\frac{1}{2\omega^2 L}\right) [(2\omega L)\cos(\omega t) + (R)\sin(\omega t)]$$

Hence, $v_c(t)$ is

$$= -\left[\frac{V_{DC} - v_c(t=0)}{\omega LC}\right] \left(\frac{2L}{R^2 + 4\omega^2 L^2}\right) e^{\frac{-Rt}{2L}} [(2\omega L)\cos(\omega t) + (R)\sin(\omega t)] + K$$

$$= -\left[\frac{V_{DC} - v_c(t=0)}{\omega C}\right] \left(\frac{2}{R^2 + 4\omega^2 L^2}\right) e^{\frac{-Rt}{2L}} [(2\omega L)\cos(\omega t) + (R)\sin(\omega t)] + K$$

Consider the circuit is designed such that $4\omega^2 L^2 >> R^2$ (from $4L^2 >> R^2$), the above equation reduces to (also, the condition results in $\omega^2 LC = 1$)

$$= -\left[\frac{V_{DC} - v_c(t=0)}{2\omega L}\right] e^{\frac{-Rt}{2L}} [(2\omega L)\cos(\omega t) + (R)\sin(\omega t)] + K \tag{18.38}$$

As per initial condition

$$0 = -V_{DC} + K$$
$$K = V_{DC}$$

Therefore, the variation in the capacitor voltage, once the circuit is switched ON for the first time can be written as

$$= V_{DC} - \left[\frac{V_{DC}}{2\omega L}\right] e^{\frac{-Rt}{2L}} [(2\omega L)\cos(\omega t) + (R)\sin(\omega t)] \tag{18.39}$$

Eqn(18.39) describes the variation of capacitor voltage when SCR_1 is fired. After half the cycle, i.e. t=T/2 or $t = \frac{\pi}{\omega}$ (when SCR_1 commutes)

$$v_c\left(t = \frac{\pi}{\omega}\right) = V_{DC}\left[1 + e^{-R\pi/2\omega L}\right] \tag{18.40}$$

If the load resistance is zero, i.e. R=0, we have

$$v_c\left(t = \frac{\pi}{\omega}\right) = V_{DC}[1 + 1] = 2V_{DC}$$

This result is obviously similar since it is the result you obtained in Class A commutation circuit. Due to the resistance in the commutation circuit of SCR_1 (during the initial cycle), the capacitor is charged to voltage $V_1 = V_{DC}[1 + e^{-R\pi/2\omega L}]$.

(II) SCR_2 is conducting:
The DC supply is disconnected from the load when the second SCR (SCR_2) is fired. The

loop equation for this cycle is given as

$$0 = Ri + L\frac{di}{dt} + \frac{1}{C}\int idt + v_c(t=0)$$

$$0 = Ri + L\frac{di}{dt} + \frac{1}{C}\int idt + V_{DC}\left[1 + e^{-R\pi/2\omega L}\right]$$

On differentiating the above equation

$$L\frac{d^2i}{dt^2} + R\frac{di}{dt} + \frac{i}{C} = 0$$

Rather then repeat the steps to evaluate the integration constant etc, all we have to do is make appropriate substitutions in the general solution for current given as eqn(18.36). When SCR_2 is fired, the DC source is disconnected and hence in eqn(18.36) substitute $V_{DC} = 0$. The capacitor is aleady charged to $V_{DC}[1 + e^{-R\pi/2\omega L}]$ at the start of this cycle, hence

$$i(t) = -\left[\frac{V_{DC}(1 + e^{-R\pi/2\omega L})}{\omega L}\right]e^{-Rt/2L}sin(\omega t) \tag{18.41}$$

The capacitor voltage was evaluated in eqn(18.38) and in the second cycle, all we have to do is account for the voltage across the capacitor at the start of the second cycle. Hence, we have

$$v_c(t) = \left[\frac{V_{DC}(1 + e^{-R\pi/2\omega L})}{2\omega L}\right]e^{\frac{-Rt}{2L}}[(2\omega L)cos(\omega t) + (R)sin(\omega t)] + K$$

The constant 'K' can be evaluated using the initial condition. For the second cycle, at t=0, the capacitor voltage is given as $V_{DC}(1 + e^{-R\pi/2\omega L})$. Hence,

$$V_{DC}(1 + e^{-R\pi/2\omega L}) = V_{DC}(1 + e^{-R\pi/2\omega L}) + K$$

$$K = 0$$

The voltage across the capacitor when SCR_2 is fired is given as

$$v_c(t) = \left[\frac{V_{DC}(1 + e^{-R\pi/2\omega L})}{2\omega L}\right]e^{\frac{-Rt}{2L}}[(2\omega L)cos(\omega t) + (R)sin(\omega t)]$$

At the end of half cycle (π/ω), when the current goes zero and the SCR_2 switches off, the voltage across the capacitor would be given as

$$v_c(t) = -V_{DC}(1 + e^{-R\pi/2\omega L})e^{-R\pi/2\omega L}$$

$$= -V_{DC}(e^{-R\pi/2\omega L} + e^{-R\pi/\omega L})$$

For the given condition, the second term would fall off rapidly and would tend to zero. Hence, the capacitor would charge to

$$V_2 = -V_{DC}e^{-R\pi/2\omega L}$$

(III) SCR_1 is fired and steady state starts:

Since, the capacitor is already charged (with voltage across it being V_2), we have reached steady state condition on now firing SCR_1. With $v_c(t = 0) = V_2$, then eqn(18.36) reduces to

$$i(t) = \left[\frac{V_{DC}(1 + e^{-R\pi/2\omega L})}{\omega L}\right] e^{-Rt/2L} sin(\omega t) \tag{18.42}$$

As can be seen the current given by eqn(18.42) is not the same as that in eqn(18.37). However, it is equal and opposite to eqn 18.41, indicative of attainment of steady state.

The voltage across the capacitor can be calculated using eqn(18.38) and is given as

$$= -\left[\frac{V_{DC}(1 + e^{-R\pi/2\omega L})}{2\omega L}\right] e^{\frac{-Rt}{2L}}[(2\omega L)cos(\omega t) + (R)sin(\omega t)] + K$$

As per initial condition

$$-V_{DC}e^{-R\pi/2\omega L} = -V_{DC}(1 + e^{-R\pi/2\omega L}) + K$$
$$K = V_{DC}$$

Therefore, the variation in the capacitor voltage, on subsequent firings of SCR_1 can be written as

$$= V_{DC} - \left[\frac{V_{DC}(1 + e^{-R\pi/2\omega L})}{2\omega L}\right] e^{\frac{-Rt}{2L}}[(2\omega L)cos(\omega t) + (R)sin(\omega t)] \tag{18.43}$$

Eqn(18.43) describes the variation of capacitor voltage shown in fig(18.46b) and marked as (iii). After half the cycle, i.e. t=T/2 or $t = \frac{\pi}{\omega}$ (when SCR_1 commutes)

$$v_c\left(t = \frac{\pi}{\omega}\right) = V_{DC} + V_{DC}\left[1 + e^{-R\pi/2\omega L}\right] e^{\frac{-R\pi}{2\omega L}}$$
$$= V_{DC} + V_{DC}\left[e^{-R\pi/2\omega L} + e^{-R\pi/\omega L}\right]$$

As argued earlier, the term $e^{-R\pi/\omega L}$ as per conditions of circuit design falls off rapidly to zero. Hence, the maximum voltage across the capacitor is given as

$$V_3 = V_{DC} + V_{DC}e^{-R\pi/2\omega L} \tag{18.44}$$

Eqn 18.44 is same as eqn 18.41, indicating achievement of steady state. The capacitor charges to $V_{DC}(1 + e^{-R\pi/2\omega L})$ when SCR_1 is fired and $-V_{DC}e^{-R\pi/2\omega L}$ when SCR_2 is fired.

Exercise

Q1. Explain how junction J_2 of an SCR breaks down on the injection of electrons in the P_2 layer.

Q2. Give an experimental set-up for the determination of the I-V characteristics of an SCR. Define and display on the I-V characteristics, the SCR's forward break-over voltage, latching and holding current.

Q3. Using the two transistor model derive an expression for I_A. Explain how the SCR remains in it's forward conducting state even after the gate trigger is removed.

Q4. Is an SCR a voltage or current triggered device. Discuss.

Q5. What is a snubber circuit.

Q6. Explain how an SCR is triggered ON using a RC network. What is the advantage of this method over triggering with a resistive network? Draw the necessary waveform.

Q7. Why is germanium not used for the fabrication of an SCR?

Q8. What is a commutation circuit?

Q9. What would happen if the gate is given a negative bias with respect to the cathode when the anode is having a positive potential with respect to the cathode?

Q10. Discuss the various methods by which an SCR can be sent into it's forward conducting state.

Q11. Explain the class D commutation circuit of an SCR. Draw the various current and voltage waveform to explain the process.

Q12. Explain the operation of the auxiliary SCR switching a charged capacitor in Class D commutation circuits. Also, obtain an expression for L and C required for successful commutation.

Q13. Explain why the inner two layers (N_1 and P_2) of an SCR are lightly doped and made wide.

Q14. What is regenerative effect?

Q15. Give the transient analysis of a SCR using the two transistor model.

Q16. Find the values of protective circuital elements, if the SCR has the following ratings:

    (i) $V_s = 200V$.
    (ii) $(dv/dt)_{max} = 200V/\mu s$.
    (iii) $(di/dt)_{max} = 100A/\mu s$.
    (iv) $R_L = 10\Omega$

Q17. What are the requirements of firing an SCR?

Chapter 19

Integrated Circuits

After the invention of the transistor, new dreams and hopes where made. The idea of smaller computers and appliances was pushed forward. However, to achieve useful applications the number of transistors required varied from couple to hundreds (this was in 1960's, today a powerful microprocessor on an average requires millions of transistors). Wiring together these circuital devices was tedious. More importantly, as far as industry was concerned this required larger time for assembling circuits by skilled workers, thus not only increasing the cost of manufacturing but also increased cost of maintenance. Thus, the dream of miniaturization as also of large scale manufacturing was still far off.

In the summer of 1958 Jack Kilby at Texas Instruments found a solution to this problem. Texas Instruments (US) in those days were working on a project to build smaller electrical circuits. Mostly working alone, Kilby suggested instead of making discrete devices of silicon, make all the components on the same block (monolith) of silicon or any other preferred semiconductor material. In September 1958, Kilby had his first integrated circuit (IC) ready, which was essentially a phase shift oscillator. A transistor, about seven resistances and three capacitors were made on a signal block of silicon. By making all the parts out of the same block and adding the metal contacts required to connect the various devices, there was no more need for individual discrete components. No more wires and components had to be assembled manually. The circuits could be made smaller and the manufacturing process could be automated. For his contribution to science and technology, Jack Kilby was rightly awarded the Nobel Prize in Physics in the year 2000. After his contribution in the invention of the integrated circuit manufacturing process, Kilby led his team at Texas Instruments to develop the first hand-held calculator.

The miniaturization process now picked up with full earnestness. With new breakthroughs in physics, more and more devices were packed in into smaller and smaller area. So much so that in todays standards, a dust particle on top of the silicon block can spoil many devices which were supposed to be made in the dust particles shadow region.

Since tiny particles like a hair, a speck of dust, a dead skin cell, bacteria or even the single particles in tobacco smoke are huge objects that are big enough to ruin a chip, chip production takes place in a rooms specially maintained such that they are devoid of these particles. Such rooms are called **Clean rooms**. This is a specially designed room, where furniture are built from materials that don't give off particles, and where extremely effective air filters and air circulation systems change the air completely while also making sure large particles do not enter the room. To prevent contamination, workers wear special suits which are made of ultra clean material. Free movement in and out of the room is minimized. Such complexities are involved in making an IC that it would be interesting to get an overview of how industry makes an IC from sand.

Since the devices are restricted to the surface of the silicon block, the block needs large area but need not have much thickness. That is, integrated circuits are essentially two dimensional structures with area $\gg$ thickness. Hence, to increase profits, the silicon block is cut to have very small thickness. So much small that they are popularly called Si wafers.

Transferring the circuit designed on a paper onto the Si wafer surface is very much similar to the process of making positive prints in photography. The image is present in form of a negative, whose counterpart in IC fabrication technology is called mask, contains the information of the circuit. Information is transfered from the negative to the positive in photography using visible light, similarly information is transferred from the mask to the Si wafer using UV light[1] in IC fabrication. The reason why UV light is used is that, to pack more circuits in a given area, the amount of diffraction is to be minimized. Smaller is the wavelength of light under consideration, smaller is the amount of diffraction taking place. A point to appreciate is that silicon does not react to light, i.e. Si is not photo-sensitive and no changes are induced in it on exposing it to light. Thus, for image transfer of information from the mask to the wafer, is broadly a three step process. Firstly an oxide layer is grown on the silicon wafer which then is coated with a photo-sensitive material. Then information transfer is done from the mask on to the light sensitive material coated on the Si wafer. After this transfer, the same information recorded in the photo-sensitive material has to be transferred from the photo-sensitive material to the Si surface. This is usually done by removing the region of the photo-sensitive material which was exposed to light while the photo-sensitive material that laid in the shadow region of the mask remain the the Si surface. This exposes the underlying oxide covered wafer where light traveled through the mask. Through thus created windows in the photo-sensitive material, the oxide can be removed by chemical reaction. Since, the reactant can not react with the oxide covered by the photo-sensitive material they remain on the wafer. Thus, where ever a window existed in the photo-sensitive material, a window in the oxide layer also appears exposing the Si surface below it. The oxide layers left on the Si wafer are areas of the shadow region and are regions where circuital elements are not required. This oxide layers give mechanical protection to the Si below it when violet process of plasma irradiation takes place to push III or V group elements in the the exposed Si, thus creating circuital features.

The above is a very brief description of the process undertaken in fabricating an IC. However, while discussing the above in detail it is best to start the story from scratch, i.e. first how is a Si wafer made?

---

[1]light with wavelength in Ultra-Violet region

19.1 Making Silicon Wafer

19.1.1 Electronic Grade Silicon

Silicon is readily available in sand (quartzite). The chemical composition of quartzite is SiO_2. The extraction of Si from quartzite is quite elaborate and involves heating it with carbon products like wood in a furnace. A number of reactions takes place, but the overall reaction is

$$SiC + SiO_2 \rightarrow Si + SiO \ (gaseous) + CO \ (gaseous)$$

The Si formed is 98% pure and is said to be metallurgical grade, however, is still not good enough as far as electronics industry standards are concerned. Hence, after cooling, the large pieces of Si are broken down and treated with hydro-chloric acid

$$Si + 3HCl \rightarrow SiHCl_3 \ (gaseous) + H_2 \ (gaseous)$$

a reaction that takes place at 300°C. The products are both in gaseous state. On cooling to room temperature while hydrogen remains in it's gaseous state, $SiHCl_3$ or tri-chloro-silane liquidises. Thus by distillation pure tri-chloro-silane is collected. Highly pure form of Si, called **Electronic Grade Silicon (EGS)**[2] is produced by a reduction reaction of tri-chloro-silane in it's vapor state. In a vacuum chamber gaseous tri-chloro-silane and hydrogen are introduced. A chemical reaction results.

$$SiHCl_3 + H_2 \rightarrow Si \ (solid) + 3HCl \ (gaseous)$$

The two gases react resulting in Si, whose atoms condensates on a heated Si seed (in the shape of a rod of diameter 4mm, giving the process the name "slim rod" method) kept inside the vacuum chamber. While the HCl, still in it's gaseous state, is pumped out while the vacuum is being maintained. As the reaction proceeds, the diameter of the Si rod increases, resulting in a very pure sample of Si.

When solid matter forms, the atoms come and join together by forming chemical bonds. These bonding are highly directional, i.e. they point in a specific direction and hence material properties like resistivity are also different in different directions. When the atoms arrange themselves totally randomly within the solid, the are said to be in amorphous state, with average property the same in all the directions, however, with large local variations. This is unacceptable in IC industry where circuital features are so minute that local variation would dictate individual IC. Hence, it is better to use single crystals (atoms arranged periodically to large extents with highly directional properties) in IC industry where no local variations exist.

As stated the Silicon obtained by the above method is pure, however, it exists in it's polycrystalline state (atoms arranged periodically to small extents in dimensions). Thus, maintaining the purity, the EGS has to be converted to single crystal Si. We now discuss two of the most popular methods used by industry to prepare single crystal silicon.

---

[2]EGS is a polycrystalline state of Si with high purity. The impurity concentration of EGS is in **parts per billion or ppb.**

19.1.2   Czochralski Method

A small piece of single crystal Silicon is used as a seed and is held by the seed holder over the crucible such that the seed just touches the surface of the molten Si. At the point of contact, the seed becomes molten and hence captures atoms from the molten material in the crucible.

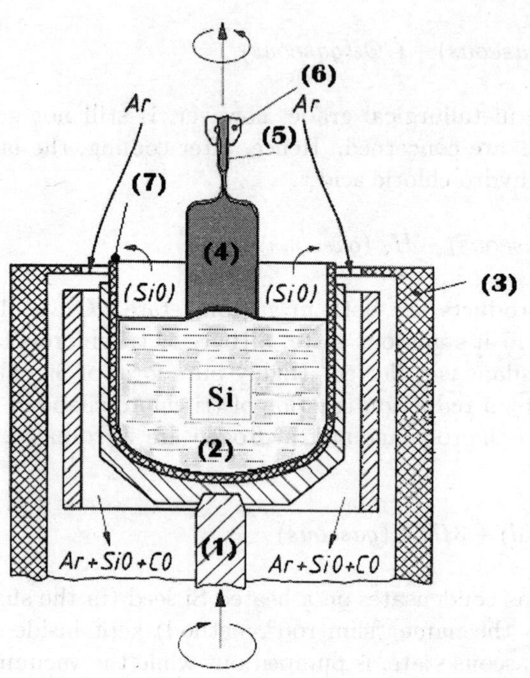

Figure 19.1: *The Czochralski apparatus for manufacturing single crystals of silicon. The important features of the apparatus are (1) The crucible shaft, (2) molten Si, (3) thermal shield, (4) single crystal being pulled out, (5) the seed, (6) it's holder and (7) the crucible.*

The seed holder is slowly pulled up. Along with it follows the seed and the viscous molten material from the surface. The typical rate of pulling is of few millimeters per minute. This allows for the slow cooling of the molten material as it is coming away from the surface of molten material. The slow cooling process is called annealing and this allows for the atoms of the newly solidifying material to arrange itself in an order identical to the seed. Hence, the newly formed material has the same crystal structure as that of the seed. The pulling action of the seed holder is accompanied with a rotation given to the seed holder (say anti-clockwise) and the crucible containing molten Si (this would be in the opposite direction to that of the seed holder, hence in our case clockwise rotation). This leads to an thinning of the molten Silicon that is being pulled out. The process is called **"necking"**. The seed is a single crystal Silicon. This may be obtained from crystals made in previous runs of Czochralski manufacture or by some other method. It is not necessarily to be a perfect single crystal and might have regions where the atomic arrangement is far from the ordering one expects in single crystals, i.e. there might be defects or dislocations present in the crystal structure of the seed. These defects and dislocations would propagate into the single crystal being formed by the pulling action. Experience shows that a small diameter seed reduces propagation of defects. Also, propagation of dislocations are minimized by the necking formed by the rotating of the seed and crucible. The Czochralski apparatus is usually 6.5m tall, with crucible large enough to accommodate 60Kg of Si charge (molten material), which in one pull can give an single crystal Si of diameter 10cm and upto 3m long. The whole crystal growth in this method is done in an inert gas (Argon) environment to prevent the charge and the graphite susceptor from reacting with atmospheric oxygen. Thus, the initial investment and the running (operating) cost makes the Czochralski Method expensive, ibid suitable for industrial purposes.

19.1.3   Float Zone Method

Float zone method is used to grow single crystals of silicon with far greater purity as compared to Czochralski method. The geometry of the apparatus is such that larger scale manufacturing of single crystals can not be done. Thus, this method is only used either for research or where more than manufacturing yield high purity is of more importance.

The rod shaped EGS obtained from slim rod technique is kept vertically on top of a single crystal silicon seed. The rod and seed is enclosed in an quartz tube which after evaluation is filled with Argon gas. An RF coil is slowly moved from the bottom of the tube to the top (see fig 19.2). During the process, a small region or small zone of the EGS is kept in molten state. While the RF coil is made to move in upward direction, the molten zone moves along with the RF coil in the upward direction.

The solid above the molten region is prevented from collapsing downwards by the surface tension of the molten material. As the zone moves upwards, the polycrystalline EGS enters a molten state at the zones advancing edge while a single crystal of silicon is left in the retreating end and grows with the same crystal structure as the seed. The very fact that a crucible as is not being used as in Czochralski

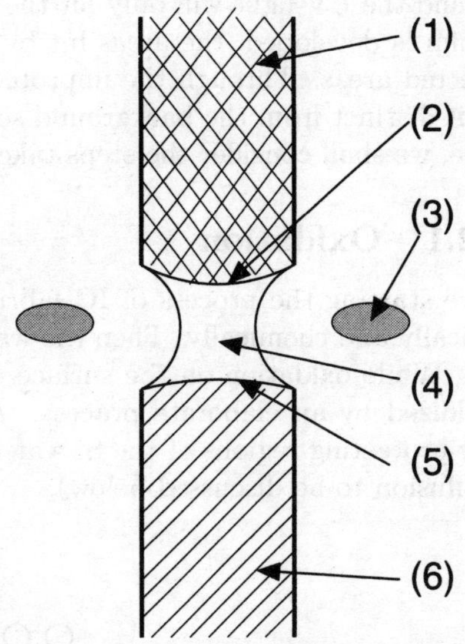

Figure 19.2: *The setup for the float zone method of crystal purification. The (1) original slim rod forms a leading melt edge region (2) as the RF heating coil (3) moves upward. As the RF coil moves upwards, the receding molten edge (5) solifies giving a pure crystal (6).*

method, increases the purity of the single crystal being formed in this method. An improved purity means higher resistivity, lower minority carriers, making the semiconductor useful in power devices (see Chapter 18). The solid cylinder of silicon is now cut into slices whose dimensions are in micro-meter. The thinness of such cut silicon sheets gives its popular name "silicon wafer". Circuits are fabricated on this wafer by a sequence of elaborate methods that we discuss below.

19.2   Making a Diode on the Si Wafer/ Lithography

Once you have a Si wafer (usually, since wafers are used to make semiconductor devices, during the manufacture of single crystals itself, dopant are added into molten Si to get either 'n' or 'p' type semiconductors from which wafers are made), you have to transfer your circuit onto this wafer. This is done by a process called **photolithography**. Lithography essentially means writing on stone. But what is the "pen" one uses to write on the silicon wafer? Today lithography is done by a high energy UV-light (ultra-violet "light", thus the term photo in the name) shone through a mask onto the silicon wafer that is covered with a photosensitive film. UV-light is used since it has a very low wavelength. Smaller is the wavelength, smaller is the markings that you can make on the wafer (Its like a thin tipped pen). The mask describes the circuit of the chip and the UV-light will only hit the areas of Si wafer that is not covered by the mask. When the film is developed, the areas hit by light are removed. Now the wafer has unprotected and protected areas. Through the unprotected openings new impurities can be added to make the circuit distinct from the background semiconducting material. All this must be very confusing. Hence, we shall consider the steps taken in making a simple diode in an IC.

19.2.1   Oxidation

Before starting the process of IC fabrication, the Si wafer's surface is throughly cleaned both physically and chemically. Then the wafer is given a coat of SiO_2. The process is called **Oxidation**. While oxidation on the surface of iron (rusting) is annoying, the surface of the Si wafer is oxidized by an elaborate process. An oxide layer serves as a stencil on the surface of the wafer protecting regions of the Si wafer that are not supposed to be doped (during the process of Diffusion to be discussed below).

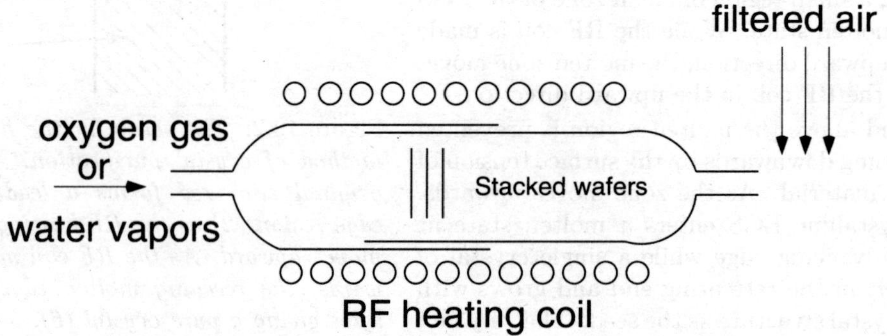

Figure 19.3: *The industrial oxidation process of silicon wafer surfaces.*

There are various methods present for wafer oxidation, however, the most popular and simple method is of **Thermal Oxidation**. As shown in fig(19.3), the wafers of Si are kept vertically in a quartz tube. While the wafers are being pushed into the tube, filter air is used to blow off any dust particle on the wafer surface. While the whole system is maintained at 900   1200°C, oxygen or water vapors are passed into the quartz tube which is casing the wafers. Oxidation takes place on the wafer surface, where the under going chemical reaction is given as

$$Si \ (soild) + O_2 \ (gas) \ \longrightarrow \ SiO_2 \ (solid)$$
$$Si \ (soild) + 2H_2O \ (gas) \ \longrightarrow \ SiO_2 \ (solid) + 2H_2 \ (gas)$$

19.2.2   Photo-resist

The top of the oxide layer is now coated with a light sensitive material called photo-resist. Consider a class of photo-resist called negative photo-resists. These are essentially made by short chained polymers combined by photo-sensitive compounds. When light falls on the negative photo-resist (-ive photoresist), a chemical reaction takes place by which the length of the polymer chains increases. This increase in polymer length depletes it's ability to dissolve. Thus, the polymers in the unexposed region are readily dissolved, exposing the underlying silicon-dioxide layer.

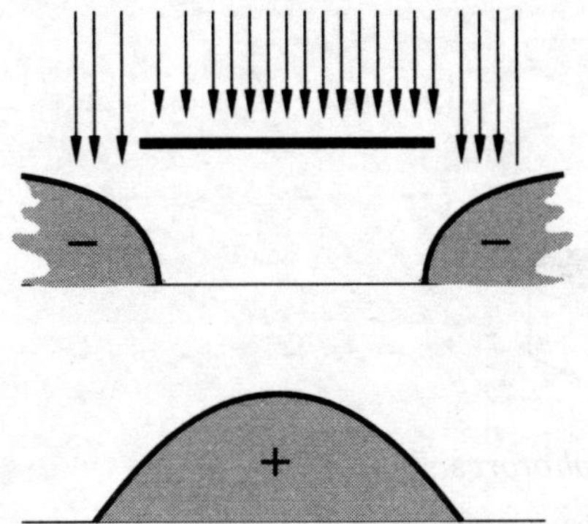

Figure 19.4: *The profile of the undisolved (a) negative and (b) positive photoresist obtained after the photolithographic process.*

While in the case of -ive resist, the unexposed resist gets dissolved and is washed away by the developer solution, in the case of +ive resist the reaction is such that light exposed region becomes more soluble to the developer. Figure(19.4) compares the profile of the exposed-unexposed regions of a -ive and +ive photoresist. Thus either photo-resist maybe used depending on the personnel preference of the manufacturer or the comparative advantage or disadvantage that the manufacturer might decide as per his requirement. For example +ive resist require more exposure energy and exposure time for proper performance. While the -ive resist on exposure becomes less soluble and more absorbent. Thus, swelling up as it absorbs water from the developer solution. The result of oxidation and coating with photo-resist is shown in fig(19.5).

19.2.3   Masking and Transfer

Figure(19.5) shows the profile of the silicon wafer after the process of growing an oxide layer and then a photo-resist layer at the top. The wafer is now ready for processing. Consider our objective to be fabricating a resistance 'R' on this wafer. This would involve creating a region ("bar") of 'n' type semiconductor within the 'p' type wafer. For trivial explanation of the fabricating process of a resistance on the wafer, remember the resistance is related to it's dimensions by

$$R = \rho \frac{l}{A} = \rho \frac{l}{w \times d}$$

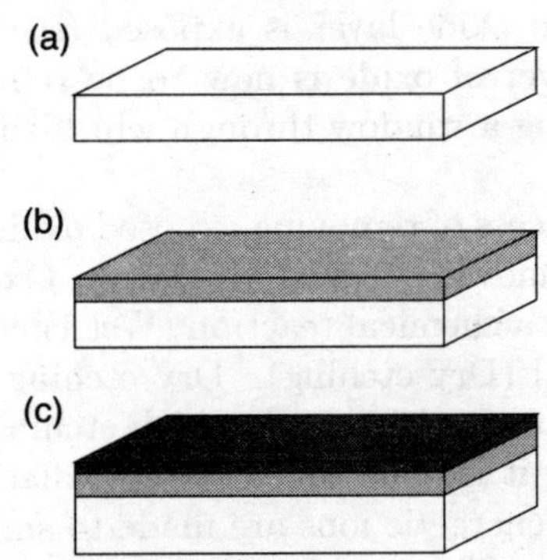

Figure 19.5: *The wafer (a) after oxidation (b) followed by coating with photoresist (c).*

The dimensions of the Mask, decides the length (l) and width (w) of the bar. The duration for which diffusion is done indicates the thickness (d) to which the 'n' type dopants have traveled.

Along with this, the resistivity (called sheet resistance in IC industry) of the doped region depends on the doping concentration. Thus, controlling these parameters, we can fabricate the required resistance. In the example of fig (19.6), the shape of the resistance is visible on the mask and it is transparent to UV light. On irradiation, light passes through the mask and falls on the photo-resist. We want the exposed region to be removed, creating a window through

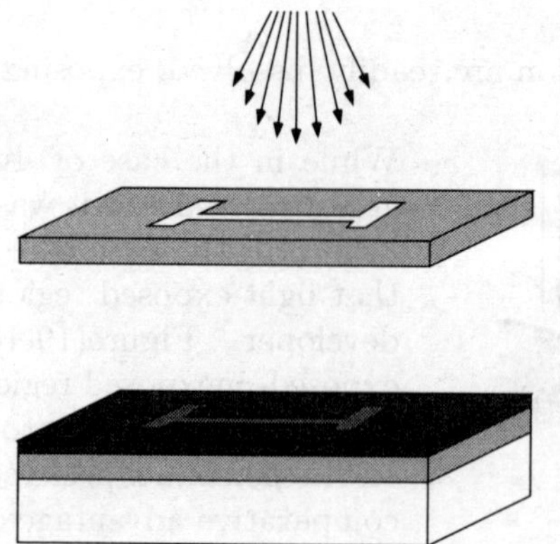

Figure 19.6: *The exposure of +ive photoresist.*

which 'n' type dopants can be bombarded to force them into the 'p' type wafer. Thus, the +ive photo-resist is required. After exposure, regions of the exposed +ive photo-resist is dissolved by the developer, exposing the oxide layer immediately below it (fig 19.7).

19.2.4   Etching

After the process of transferring the mask pattern on the photo-resist and then washing it with a suitable developer, the oxide layer is exposed (see fig 19.7a). The exposed layer of oxide is now to be removed, fig(19.7b), thus creating a window through which the silicon wafer is exposed.

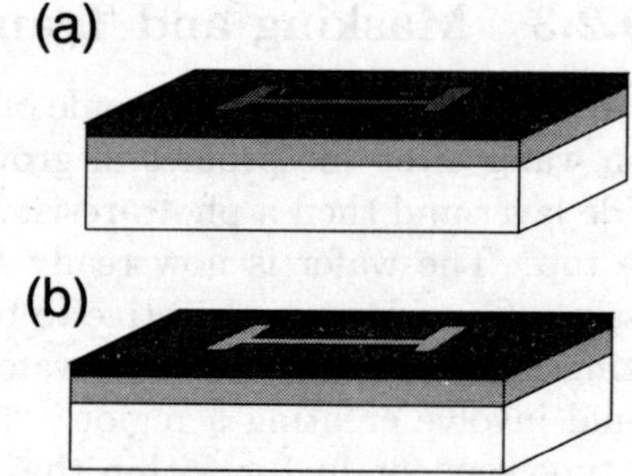

The process of removing exposed oxide region via layer by layer removal is called **Etching**. Oxide layer can be removed by a chemical reaction (Wet Etching) or by physical removal (Dry etching). Dry etching though is superior as to the precision with which etching can be done, is costly since it is done by plasma irradiation. In this process highly energetic ions are made to smash on the oxide layer surface. The transfer of momentum from the ions to

Figure 19.7: *The oxide layer is exposed through the window of dissolved photo-resist.*

surface atoms ensues the etching process. However, the cost and also slow speed of the process makes wet etching more popular. Wet etching essentially involves transferring the reactant to the reacting surface, as in this case, getting HF acid to the silicon dioxide layer. Followed by removing the product of the chemical reaction as it progresses. The chemical reaction that takes place during wet etching is given as

$$SiO_2 + 6HF \longrightarrow H_2SiF_6 + H_2O$$

Figure(19.8) shows our wafer after wet etching. Notice, etching has not only taken place in the 'z' direction but also in the 'x' direction, cutting into the oxide layer below the photo-resist. This is due to the isotropic reacting behavior of the ectant (HF), cutting equally in all directions. The area of Si exposed thus effectively increases. This contributes in the decreasing

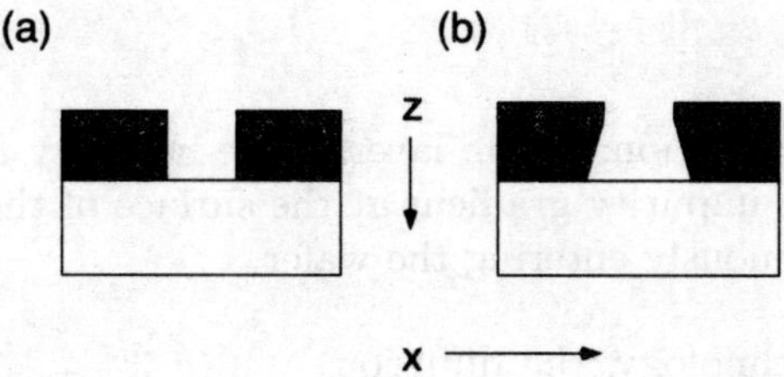

Figure 19.8: *The difference in oxide layer after (a) wet etching and (b) dry etching.*

of resolution. This becomes a serious problem if the concentration of the etchant is high or if the reaction is allowed to go on for a long time. This obviously is a disadvantage associated with wet etching and where dry etching is better then it's counterpart. After etching the photo-resist layer is completely removed and the resulting silicon wafer with it's oxidized layer is used for the next step of diffusion.

19.2.5   Diffusion

Diffusion is the processes by which chemical species or dopant are introduced into the silicon wafer to form the electronic structures such as resistance, diode, capacitor etc. Diffusion is the movement of a chemical species from an area of high concentration to an area of lower concentration. The controlled diffusion of dopant into silicon is the foundation of fabrication of devices during IC manufacture. In our example of fabricating a resistance on the given 'p' type wafer, dopant of the V group ('n' type) have to be diffused into the wafer. The fabrication process demands the concentration and depth to which diffusion takes place would be controllable. Industry uses two major methods to deposit impurities into a substance by thermal diffusion.

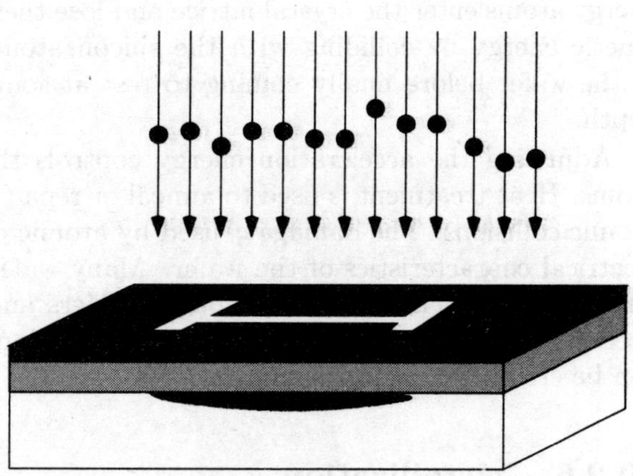

Figure 19.9: *Doping impurities through the window of etched* SiO_2.

- Predeposition Method

  A flux of impurities continuously arrives at the surface of the substrate such that the concentration gradient of the impurity remains constant at the surface of the wafer, i.e. more and more impurity is introduced at the surface such that even after impurity has entered the wafer, the concentration of impurity at the surface is the same.

- Redistribution Method

  Also called as drive-in diffusion, a thin layer of the impurity material is deposited on the wafer. In this case, the impurity gradient at the surface of the wafer decreases with time since impurity is continuously entering the wafer.

With improvement in technology, the diffusion process is being replaced by the ion-implantation technique. Ion implantation is the process of depositing a chemical species into a substrate by direct bombardment (see fig 19.9) of the substrate with high-energy ions of the chemical for deposition. The greatest advantage of ion implant over diffusion is it's more precise control for depositing dopant atoms into the substrate. During ion implantation, impurity atoms are vaporized and accelerated toward the silicon substrate. These high-energy atoms enter the crystal lattice and lose their kinetic energy by colliding with the silicon atoms of the wafer before finally coming to rest at some depth.

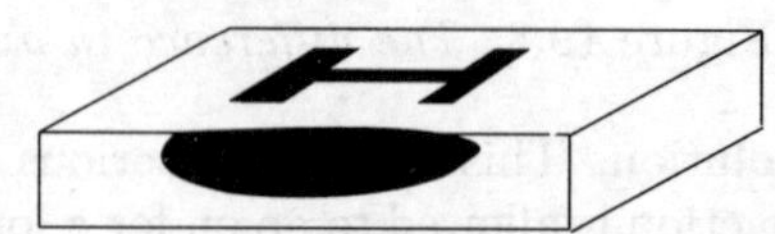

Figure 19.10: *The resistance is visible after the oxide layer has been completely etched. If more structures is to be fabricated on this wafer, it has to undergo many more cycles of oxidation, coating with photo-resist, exposure through mask, developing, etching and diffusion etc.*

Adjusting the acceleration energy controls the average depth of depositing the impurity atoms. Heat treatment is used to anneal or repair the crystal lattice disturbances caused by the atomic collisions. The damage caused by atomic collisions during ion implantation changes the electrical characteristics of the wafer. Many wafer atoms are displaced, creating deep electron and hole traps which capture mobile carriers and increase resistivity. Annealing is therefore needed to repair the lattice damage and put dopant atoms in substitutional sites where they can be electrically active again.

19.2.6   Metallization

Figure 19.11: *Electron microscope image of metal contacts made on surface of an IC.*

Once the resistance (or for that matter any feature) is fabricated, metal contacts have to be made so that they maybe connected to the circuit. The fabrication step where proper interconnection of circuit elements is made is called **Metallization**. Aluminum is a popular metal used for interconnection. The popularity of Aluminum for metallization is due to it's good adhesion to both silicon and silicon dioxide, it can be easily be vacuum deposited (since it has a low boiling point), and has high conductivity. In addition to pure aluminum, alloys of aluminum like Al-Cu or small amount of silicon in Aluminum are also used to form IC interconnections. Different alloys are used for different performance- related reasons. For example, small amounts of copper are added to reduce the potential for electro-migration effects (in which current applied to the device induces mass transport of the metal). Figure(19.11) is an electron microscope image of metal contacts made on the silicon wafer.

In the last paragraph one of the reasons for using Al was stated to be that it can be easily vacuum deposited. Vacuum deposition is a method in which materials (like in this case Al) are melted in vacuum. The Aluminum evaporates and moves in an upward direction. Since the enclosure is evacuated, the atoms of Aluminum do not react with oxygen and are deposited on any surface kept in it's path. The Aluminum condensates on the surface forming the required metal contacts. Metal layers are vacuum-deposited onto wafers by one of the following methods:

- Filament Evaporation

  Filament evaporation is accomplished by gradually heating a filament of the metal (Al) to be evaporated. The metal is placed in a basket. Electrodes are connected to either side of the basket and a high current passed through it, causing the basket to heat. As power (and therefore heat) is increased, the metallic filament partially melts and is eventually vaporized. In this way, atoms of aluminum break free from the filament and deposit onto the wafers. While filament evaporation is the simplest of all metallization approaches, problems of contamination during evaporation is a point of concern in IC fabrication.

- Electron-beam Evaporation

  In Electron-beam evaporation, an intense beam of electrons are focussed into a crucible that contains the aluminum. As the beam is directed into the source area, the aluminum is heated to its melting point, and eventually, evaporation temperature. The benefits or this technique are speed and low contamination, since only the electron beam touches the aluminum source material.

- Sputtering

  Aluminum sputtering is used commonly in IC metallization processes and is popular because the adhesion of deposited metals is excellent. RF sputtering is done by ionizing inert gas particles in an electric field (producing a gas plasma) and then directing them toward the Aluminum source or target (in form of a pellet), where the energy of these gas particles physically dislodges, or "sputters off", atoms of the source material.

Note the degree of difficulty in IC manufacture is not just the fabrication process but also designing. The features of the circuit should be such that all "nodes" of the circuit should be available at the surface for metallization and the interconnecting metal wires should not short while they criss-cross the wafer surface. Fabricating more features on the wafer involves repeated cycles of oxidation, coating photo-resists, exposing the mask (photo-lithography), developing the photo-resist, etching, diffusion etc. The more the number of cycles, the higher is the cost of production. The number of cycles hence have to be minimized. This is done by proper designing of the mask. Requiring for very skilled and highly trained personnel.

Exercise

Q1. How is Electronic Grade Silicon (EGS) obtained from metallurgical grade silicon?

Q2. Why is silicon preferred for IC fabrication?

Q3. Would an amorphous or polycrystalline silicon with isotropic behavior be better or single crystal with anisotropic characters be better for IC fabrication. Justify.

Q4. Define and explain "necking" in crystal growth.

Q5. Discuss the positive and negative resist.

Q6. Would X-ray be better than UV radiation, as a wavelength for exposing mask image on silicon surface. Explain why or why not.

Q7. Describe briefly the main steps in creating a structure on a single level of a semiconductor chip?

Q8. Describe how a capacitor could be formed in a semiconductor chip manufacturing process.

Q9. Give a short note on various metallization processes.

Q10. What is etching? Discuss which process of etching, *d*ry or wet is better. Justify.

Q11. Explain which method, *C*zochralski or Float zone is better for the manufacturing of very pure silicon wafers?

Further Reading

The following are list of books that are highly recommended for students who have enjoyed this book and want to learn more. Even if the reader feels confident with his understanding of the subject, it is always best to get a different prespective of the subject one has learnt.

1. Allen Mottershead, "Electronic Devices and Circuits, An introduction", Prentice-Hall India (1993).

2. Dennis Roddy and John Coolen, "Electronics: Theory, Circuits and Devices", Reston Publishing Company (Prentice-Hall, 1982).

3. John D. Ryder, "Electronic fundamentals and applications, Integrated and Discrete Systems", Prentice-Hall India (1987).

4. Jacob Millman and Christos Halkias, "Integrated Electronics: Analog and Digital Circuits and Systems", McGraw-Hill International Book Company (1972).

5. S. M. Sze, "Physics of Semiconductor Devices" (2nd ed), Wiley Eastern Ltd (1983).

6. Joseph Edminister and Mahmood Nahvi, "Electric Circuits", Tata McGraw-Hill (2002).

7. M. E. Van Valkenburg, "Network Analysis", Prentice-Hill India (2000).

8. Vasudev K. Aatre, "Network Theory and Filter Design", Wiley Eastern Ltd, (1990).

9. Ken Sander, "Electric Circuit Analysis, Principles and Applications", Addison-Wesley (1992).

10. Ben Streetman, "Solid State Electronic Devices", Prentice-Hill India (1995).

11. D. C. Sarkar, "Transisitor Physics and Circuit Design", S. Chand and Company (1985).

12. Ralph J. Smith, "Electronics: Circuits and Devices", Wiley International (1973).

13. L. Barnes, "Transistors for Technical Colleges", Iliffe Books, London (1965).

14. Robert L. Boylestad and Louis Nashelsky, "Electronic Devices and Circuit Theory", Prentice Hall India (2000).

15. Paul Horowitz and Winfield Hill, "The Art of Electronics", Cambridge University Press, Great Britain (1995).

16. Henry Jacobowitz, "Electronics, Made Simple", Rupa & Co, India (1994, reprint).

Index